Dynamics and Control of Electric Transmission and Microgrids

Dynamics and Control of Electric Transmission and Microgrids

K. R. Padiyar
Professor Emeritus
Indian Institute of Science
Bangalore, India

Anil M. Kulkarni
Professor
Electrical Engineering Department
Indian Institute of Technology Bombay
Mumbai, India

WILEY

IEEE PRESS

This edition first published 2019

Registered Offices
John Wiley & Sons, Inc., 111 River Street, Hoboken, NJ 07030, USA
John Wiley & Sons Ltd, The Atrium, Southern Gate, Chichester, West Sussex, PO19 8SQ, UK

Editorial Office
The Atrium, Southern Gate, Chichester, West Sussex, PO19 8SQ, UK

For details of our global editorial offices, customer services, and more information about Wiley products visit us at www.wiley.com.

Wiley also publishes its books in a variety of electronic formats and by print-on-demand. Some content that appears in standard print versions of this book may not be available in other formats.

Library of Congress Cataloging-in-Publication Data

Names: Padiyar, K. R., author. | Kulkarni, Anil M., author.
Title: Dynamics and control of electric transmission and microgrids /
K. R. Padiyar, Anil M. Kulkarni.
Description: First edition. | Hoboken, NJ : John Wiley & Sons, Inc., [2019] |
Includes bibliographical references and index. |
Identifiers: LCCN 2018033206 (print) | LCCN 2018037544 (ebook) | ISBN 9781119173397 (Adobe PDF) |
ISBN 9781119173403 (ePub) | ISBN 9781119173380 (hardcover)
Subjects: LCSH: Electric power systems–Control. | Electric power transmission. |
Microgrids (Smart power grids)
Classification: LCC TK1007 (ebook) | LCC TK1007 .P329 2018 (print) | DDC 621.319–dc23
LC record available at https://lccn.loc.gov/2018033206

Cover Design: Wiley
Cover Image: © hrui/Shutterstock

Set in 10/12pt WarnockPro by SPi Global, Chennai, India
Printed in Singapore by C.O.S. Printers Pte Ltd

10 9 8 7 6 5 4 3 2 1

Contents

Preface

Although power systems have been in operation for more than a century, they continue to grow in size and extent, not only by expansion in demand, but also through system interconnections which bring about reduction in costs and improvement in the reliability of supply. For example, in India, five regional electricity networks have been interconnected to form an all-India grid. However, interconnections can also result in security issues when the system is subjected to large disturbances. To mitigate these problems and improve security, major qualitative changes have occurred in the operation and control of grids over the past couple of decades. The key technologies that have been developed and deployed include (i) renewable energy systems, (ii) voltage-source converters for high power applications, and (iii) wide-area measurement systems (WAMS). This book has been motivated by the need to integrate these components in the conventional power system analysis framework in order to understand their impact on the stability of power systems.

A significant proportion of power in many large power grids is now from renewable energy sources (solar, wind, and hydro) and this is likely to grow further. In contrast to the conventional large and centralized power stations, renewable energy generation is mostly in the form of a large number of relatively smaller and distributed energy sources. Since these sources are environment friendly and come in smaller units, they can often be deployed near the consumers (loads). The placement of smaller energy resources (including storage devices) near the loads has spurred research into the autonomous operation of small grids (microgrids), which can be connected to the bulk power grid or can operate as independent islands.

The optimal utilization of the transmission system using flexible AC transmission systems (FACTS) controllers and DC transmission systems is attracting renewed interest due to the need for flexible power exchanges between regions. The availability of modular multilevel converter topology promises to extend the benefits of voltage source converters (VSC) technology to high power transmission. VSCs have also found application in the grid interfaces of renewable energy systems and DC transmission from off-shore wind-energy farms to the mainland. Monitoring and control of both microgrids and bulk transmission grids is now feasible through time-synchronized local and WAMS.

The consequences of these changes on the dynamics of the grid are yet to be completely understood. What we know is that (i) a significant proportion of power exchanges in the grid are likely to be mediated through power electronic interfaces, (ii) some of the energy sources may not be "dispatchable" (power extraction from the energy source is not scheduled, but is extracted as per the availability), and (iii) it is now feasible to have centralized monitoring and supervision of the local controllers and protection systems associated with various power system components. The blackouts in the Indian grid in 2012 have highlighted the continuing vulnerability of large grids to angular stability problems. At the same time, the proliferation of energy systems that are interfaced to the power grid through power electronic converters has increased focus on the stability of a wider spectrum of transients, especially those associated with converter–controller–network–mechanical system interactions. Higher bandwidth models of many power system components and their controllers are required for the analysis of these interactions.

With this background in mind, the coverage in this book includes the modeling and control of wind and solar energy systems, VSC-based controllers, WAMS, energy storage systems, and microgrids (in Chapters 4, 5, 9, 10, 11, and 12), besides the conventional components and their controllers like synchronous machines, AC transmission lines, excitation, and prime-mover systems (Chapters 2 and 3). Simplifications in the models based on the dynamics of interest are indicated in the discussions. The treatment of conventional stability issues like power swing damping, loss of synchronism, voltage stability, and subsynchronous resonance (Chapters 6, 7, 8, and 10) is done with a view of the recent developments.

The target audience for this book is senior undergraduate and graduate students of electrical engineering, researchers, and practicing engineers involved in design and operation. Besides a good grounding in classical power system subjects, the repertoire of a power systems analyst has to now include a good knowledge of power electronics, analytical techniques like matrix transformations, eigenvalue-analysis, control system design, and signal processing. In addition, basic knowledge of the attributes, capabilities, and limitations of WAMS and renewable/energy storage systems is essential. The coverage of this book reflects this thinking.

The title of the book essentially implies the major role played by network controllers (as opposed to the generator controls) in the optimal and secure system operation. The technological and analytical developments that are yet to come, such as reliable and cost-effective wide band-gap power semiconductors, sensor networks, and cyber-physical systems, are expected to bring about self-healing smart grids.

Although we have tried to cover most of the relevant topics, it is difficult to do complete justice to them all in a single book. Inevitably, we have had to select only a few topics for detailed treatment. However, we hope that in situations where our treatment has been brief, the reader will be able to explore the topic further through the references provided at the end of the chapters.

Acknowledgements

We would like to thank John Wiley & Sons for their support throughout the process of writing this book. In particular, we would like to thank Mr Peter Mitchell and the team of project editors who coordinated our project schedule and manuscript preparation.

Several past and present students of IIT Bombay have assisted us in the preparation of the manuscript. Ajinkya Sinkar diligently carried out many of the computer simulations reported in this book. Mukesh Das, Vedanta Pradhan, Kunal Salunkhe, Adarsh Rajagopal, Kevin Gajjar, Pragati Gupta, Huma Khan, and Vinay Chindu also helped in some of the examples presented in the book and reviewing the initial drafts. Santosh Singh helped in surveying various electrical storage technologies. K.N. Shubhanga and Kalyan Dasgupta reviewed portions of the initial drafts of Chapters 4, 7, and 9, while Mahipalsinh Chudasama checked some of the results presented in Chapter 10. We thank them for their enthusiastic support, which made our task much lighter.

Harshala Ramane prepared most of the figures in the book, while Kaustav Dey, Vinay Chindu, and Joel Jose helped in the arduous task of compiling and formatting the final draft of the manuscript. We express our gratitude to them for their painstaking efforts.

K. R. Padiyar
I wish to thank Ms. Ella Mitchell from John Wiley, Chichester who requested me to write a new book on power system dynamics. Since the first edition of my earlier book on the subject was co-published by John Wiley Singapore, I readily agreed. I thank my former student Anil Kulkarni for agreeing to join this book writing project and contributing Chapters 4, 6, 7, 9, 10, and 11. He also undertook the responsibility of the final proof-reading and preparation of the final manuscript for submission. My former student K. Saichand made available the results from a case study on the VSC-HVDC link reported in the appendix of Chapter 5. My thanks are also due to Mrs. Sudha Aithal for typing my part of the manuscript with painstaking care.

Finally, I wish to acknowledge the support, patience, and understanding of my wife Usha.

Anil M. Kulkarni
My colleagues Professors B.G. Fernandes and Himanshu Bahirat reviewed the initial draft of the chapters on wind and solar energy, and provided suggestions and clarifications on some technical points. Professor Vivek Agarwal suggested some useful references on energy storage devices. I thank my colleagues for these inputs, and IIT Bombay for providing a conducive environment for my academic pursuits.

Finally, I wish to thank my family: my wife, Rama Kulkarni, and my daughters Shachi and Ruchi for their patience and support during my preoccupation with the book, and my mother for her encouragement for all my academic endeavors.

K. R. Padiyar
Anil M. Kulkarni

1

Introduction

1.1 Present Status of Grid Operation

1.1.1 General

An electric power grid is a network of synchronized power providers and consumers that are connected by transmission and distribution lines and operated by one or more control centres. In the past, electric grids meant mainly high-voltage transmission grids. However, more recently microgrids at power distribution level have been researched and developed.

Electrical energy emerged about 125 years ago for street lighting. However, due to its tremendous versatility and the ability of power engineers to deliver it at affordable cost and high reliability, electrical energy is now regarded as a necessity (along with food, clothing, shelter, education, and health care) by citizens all over the world. Electrification was considered the greatest engineering achievement of the twentieth century (ahead of the internet). Electricity provided a clean alternative to steam engines in factories for providing motive power. It is now used in homes and offices for cooking, heating, and cooling, and to power appliances and gadgets that are used for cutting down manual labour and improving the quality of service in many spheres of life, including health care, education, and entertainment. In the modern world electricity drives communication equipment and information technology. The application of electrical drives in rail and road transportation can reduce environmental pollution and improve efficiency of energy usage. About 40% of the total energy used in developed countries is electrical energy [1].

The supply of reliable power at affordable cost has been the guiding principle in planning power (supply) systems all over the world. Until the 1990s, the electricity supply industry (ESI) was a regulated monopoly in which each utility took care of the planning, design, and operation of the generation, transmission, and distribution facilities under its control in the designated area. The tariff was decided on the basis of an equitable return on investment, taking into account the capital, operation, and maintenance costs. Such utilities are essentially vertically integrated utilities (VIU). The exchange of energy among VIUs was based on mutual needs with the help of tie lines (interconnections). While costs can come down due to sharing of generation reserves and load diversity, the reliability of the overall power system can be improved due to interconnections.

Dynamics and Control of Electric Transmission and Microgrids, First Edition.
K. R. Padiyar and Anil M. Kulkarni.

While most of the utilities in the USA were privately owned, in the UK, India, and elsewhere they were publicly owned. The desire to introduce competition (in comparison with communication systems) in power generation with the expectation of providing a choice for the customers and reducing costs led to the unbundling of generation, transmission, and distribution. Also, deregulation was introduced in which a large customer could buy power from any competing supplier who had access to transmission facilities irrespective of ownership. In India, power trading was introduced in 2003 to encourage generation by private parties who can sell the power to the highest bidder in the energy market through power exchanges.

A major motivation for the introduction of competition was necessitated by the search for unconventional solutions in tackling the problems of growth in power demand as fossil fuels become scarce and environmental and climate change issues restrict the choices. Nuclear power does not have the drawback of increasing the carbon foot print; however, safety issues limit its application. Large hydroelectric plants have the drawbacks of environmental issues and displacement of populations. Wind and solar energy are renewable; however, they are difficult to dispatch due to variability of generation. While energy storage technologies (such as batteries) can help to stabilize the output of renewable generation, they are not fully mature and need further development.

Long distance transmission is not feasible unless voltage levels can be raised. This led to the development of AC generation and transmission, as transformers can be built to operate at very high efficiencies and thus stepping up or down of voltages becomes feasible. The power losses in a conductor are proportional to the square of the current, which is inversely proportional to the voltage level for a specified power level. The combined losses in the transmission and distribution networks are generally kept below 10% of the power generated. AC generators can also be built for high power ratings, unlike DC generators where commutator and brushes limit the power level.

AC transmission operates at high or extra-high voltage (EHV) from 132 kV to 765 kV (phase to phase). Recently ultra-high-voltage (UHV) lines have also been under consideration. Typically, all long distance AC transmission lines use overhead conductors which are strung from towers by insulator strings (to avoid contact with the metallic towers, which are grounded). In urban areas, underground cables are used for short distances. Although cables (laid in trenches) are immune to weather fluctuations, the faults in cable transmission tend to be permanent (not self-clearing) and take more time to repair.

If a symmetrical three-phase AC line is injected with a positive sequence voltage, the currents flowing through the conductors also belong to the positive sequence. Thus, normally it is adequate to analyze the single-phase (positive sequence) network consisting of the transmission lines in a power system. The series impedance of a line is due to the resistance and (positive sequence) inductance of the line. The series impedance of the line is $(R + jX)$, where R is the resistance, $X = 2\pi fL$ is the reactance, and L is the inductance. Typically $R \leq \frac{X}{10}$. The positive sequence shunt capacitance of a line is about 11 nanofarad/km (11×10^{-9} F/km). It should be noted that the three parameters R, L, and C are distributed in nature. The analysis of a line with distributed parameters can be quite complex. However, in steady state with sinusoidal voltages and currents they can be represented by phasors which vary continuously with distance from the sending end. AC transmission over 1000 km is not feasible unless special reactive power compensation methods are employed. One method is to apply the Static VAr Compensator (SVC) at regular intervals, which reduces the effective line length.

For a lossless line, there is a specific loading on the line, called surge impedance loading (SIL), in which the voltage magnitude remains identical along the line. The power factor is also unity at all points along the line. This implies that reactive powers supplied at both ends of the line are zero. This loading is an economic loading as not only is the voltage profile is flat along the line, but the reactive power requirements are also zero. The SIL is given by

$$SIL = \frac{V^2}{Z_n}, \quad Z_n = \sqrt{\frac{L}{C}} \tag{1.1}$$

where V is the rated (line to line) voltage and Z_n is the surge impedance, which varies between 200 and 400 Ω. The value decreases as the voltage level increases. For a 400 kV line, $Z_n \approx 300$ Ω and SIL is about 500 MW. It should be noted that SIL is not a limit on the power flow in the line. The maximum limit of power flow is determined by the peak temperature permissible for the aluminium conductor (75° C), ambient temperature, and the power loss in the conductor. This is called as the thermal limit, which is not fixed, but varies with the season. In winter, with lower ambient temperatures, higher loadings are feasible. The dynamic rating of a line based on temperature sensors is a better option to make the best use of available transmission capacity. The increase in the peak temperature of the conductor results in increasing the sag of the conductor. This can result in short circuits if there is vegetation below the line. This was an initiating cause for the major power blackout in the USA and Canada in August 2003.

For line lengths above 300 km, stability becomes the major criterion in determining the power transfer capability of an AC line. For line lengths in the range of 100–300 km, the voltage variation can determine the power transfer limit. The thermal limit determines the power transfer below 100 km. The loadability of an AC line as a function of length is shown in Figure 1.1. The power flow limit decreases as the line length increases due to the stability criterion. It should be noted that power flow in an AC line cannot be controlled or regulated unless special controllers using high power electronics are employed. These are known by the generic name FACTS (flexible AC transmission system) controllers [2, 3].

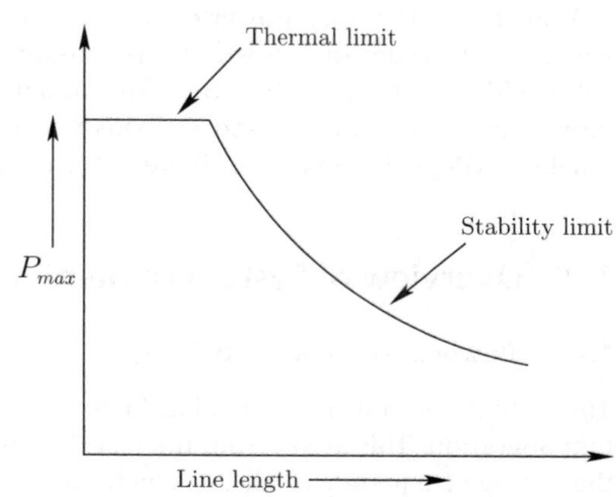

Figure 1.1 The loadability of an AC line as a function of length.

1.1.2 HVDC Transmission [4]

Unlike AC transmission, the power transfer capability in DC transmission is independent of the line length. Thus high-voltage direct current (HVDC) transmission is invariably used in transmitting power from remote hydro plants to load centres which may be 2000 km away. Already ultra-high-voltage direct current (UHVDC) transmission at ±800 kV is employed in China and India, transmitting about 6000 MW.

HVDC transmission also has the following advantages:

1) A two-conductor, bipolar (with positive and negative polarities with respect to ground) line can carry approximately the same amount of power as a three-conductor AC line (for similar current carrying capacity of the conductors and insulation levels). This implies the HVDC line requires smaller right of way (RoW) compared to AC transmission.
2) A bipolar DC line can be operated in the monopolar (with single conductor) mode with ground return.
3) A DC line can interconnect two AC systems operating at two different frequencies.
4) DC cables (unlike AC cables) do not require reactive power compensation, which makes them uniquely suited for underwater transmission.
5) The power electronic converters at the two ends of the line are fast acting and permit rapid control of power. This capability can be used to improve system stability.

A major drawback of HVDC transmission is the limitation of its application at present for point-to-point transmission. The introduction of voltage source converter HVDC (VSC-HVDC) is expected to overcome some of the drawbacks of line commutated converter HVDC (LCC-HVDC).

1.1.3 Reliability of Electricity Supply

For customers connected to low or medium voltage feeders, the reliability indices are the frequency and duration of the power outage. We can also define the Average Service Availability Index (ASAI), which considers the total duration of power interruptions in a year. In the USA, prior to restructuring, the best index was 1 hour in a year, which translates to 99.99% availability.

Apart from reliability, power quality (PQ) is a major issue in power distribution. PQ encompasses issues such as variations in frequency, voltage magnitude (sags and swells), voltage flicker, and departure from sinusoidal waveform (which results in current harmonics), which can affect motor loads and increase losses. Thus, power quality and reliability (PQR) is a major attribute of the energy supply that can impact costs.

1.2 Overview of System Dynamics and Control

1.2.1 Power System Stability [5–7]

The stability of power systems has been and continues to be of major concern in system operation. This arises from the fact that in steady state (under normal conditions), the average frequency of all the synchronous generators must remain identical in the system. This is termed the *synchronous operation* of a system. Any disturbance, small or large, can affect the synchronous operation. The system load is always undergoing changes. Generally, the load variations tend to be small or slow (when they are large). A large disturbance can be a single- or three-phase fault in a transmission line followed by switching out the faulted line. A line may also trip (due to overloading) to prevent damage to the conductors.

Although the stability of a system is an integral property of the system, for the purposes of system analysis it is divided into two broad classes [8]:

1) Steady-state or small signal stability
   A power system is steady-state stable for a particular steady-state operating condition if following any small disturbance, it reaches a steady-state operating condition which is identical or close to the pre-disturbance operating condition.
2) Transient stability
   A power system is transiently stable for a particular steady-state operating condition and for a particular (large) disturbance or sequence of disturbances if following that (or a sequence of) disturbance(s), it reaches an acceptable steady-state operating condition.

It is important to note that while the steady-state (small signal) stability is a function only of the operating condition, transient stability is a function of both the operating state and the disturbance(s). It should also be noted that while the system can be operated even if it is transiently unstable, small signal stability is necessary at all times. In general, this depends on the system loading. An increase in the load can bring about the onset of instability.

Transient instability caused by a major disturbance, such as a line fault followed by clearing of the fault (by tripping the line, which increases the postfault impedance viewed from the generator terminals), was a major problem in maintaining system stability. The loss of synchronous operation adversely affected the reliability of remote generating stations connected to load centres. The problem was tackled by:

1) reducing the clearing time of the circuit breakers and high speed reclosing
2) introducing switching stations in long lines to limit the increase in the reactance of the postfault system
3) increasing the speed of the response of the excitation system
4) application of braking resistors in hydro stations and fast valving in thermal stations.

Fast-acting static excitation systems using high-gain automatic voltage regulator (AVR) were introduced in the 1960s. Although this helps in improving transient stability, the small signal stability is adversely affected. There were spontaneous oscillations in the power flows in lines that increased in magnitude for no apparent reason. This problem is serious, as undamped oscillations can result in loss of synchronous operation. The solution was to introduce power system stabilizers (PSS) that apply limited modulation of the terminal voltage using a control signal based on the rotor speed (or frequency) and power output.

In the late 1970s it was observed that under certain conditions a disturbance can cause uncontrolled decrease in voltage magnitude, leading to voltage collapse. This phenomenon also leads to power blackouts, although the generators remain in synchronism. This has been termed a voltage stability problem in contrast to the angle stability problem that was prevalent. The voltage stability can also be divided into small signal and transient stability. The primary factor that is responsible for voltage collapse is the mechanism for load restoration (inherent or controlled). Thus, induction machines, HVDC converters and on-load tap changers (OLTC) contribute to the voltage instability problem [9].

The system frequency is regulated under normal conditions by load frequency control (LFC) or automatic generation control (AGC). However, under disturbed conditions that lead to uncontrolled tripping of tie lines, islands can be formed with excess load. The reduction of frequency below 5% of the rated frequency can result in damage to steam turbines apart from affecting the customer equipment. Uncontrolled variation of frequency is termed frequency instability.

1.2.2 Mathematical Preliminaries [9, 10]

A nonlinear, continuous time dynamic system is described by the differential equation

$$\dot{x} = f(x, u) \tag{1.2}$$

where f and x are column vectors of dimension n. u is a vector of dimension r and can be viewed as the input vector. If u is a constant vector, the system is said to be autonomous. If the elements of u are explicit functions of time t, then the system is said to be non-autonomous. If the initial conditions are specified, that is,

$$x(t_o) = x_o \tag{1.3}$$

then the solution of (1.2) is expressed as $\phi_t(x_o)$ to show explicitly the dependence on initial conditions. (Note that if u is assumed to be constant, it can be viewed as a vector of parameters and the dependence of the solution on u need not be shown explicitly).

f is called the vector field and $\phi_t(x_o)$ is called the trajectory through x_o. $\phi_t(x)$, where $x \in \mathbb{R}^n$, is called the flow.

For non-autonomous systems, the trajectory is also a function of time and is expressed as $\phi_t(x_o, t_o)$, which indicates that the solution passes through x_o at $t = t_o$.

With some restrictions on f, the solution of (1.2) has the following properties:

1) The solution exists for all t.
2) At any time t, $\phi_t(x) = \phi_t(y)$ if and only if $x = y$. Also, as $\phi_{(t_1+t_2)} = \phi_{t_1} \cdot \phi_{t_2}$, it follows that a trajectory of an autonomous system is uniquely specified by its initial condition and that distinct trajectories do not intersect.
3) The derivative of a trajectory with respect to the initial condition exists and is non-singular.

For t and t_o fixed, $\phi_t(x_o)$ is continuous with respect to the initial state x_o. This implies that equilibrium point satisfies

$$0 = f(x_e, u) \tag{1.4}$$

Equation (1.4) shows that x_e is a function of u. In general, there are several equilibrium points which are obtained as real solutions of (1.4).

Stability of Equilibrium Point

An equilibrium point x_e is said to be asymptotically stable if all nearby trajectories approach x_e as $t \to \infty$. It is unstable if no nearby trajectories remain nearby. An unstable equilibrium point is asymptotically stable in reverse time (as $t \to -\infty$). An equilibrium point is non-stable (also called the saddle point) if at least one of the nearby trajectories approaches x_e in forward time (as $t \to \infty$) and if at least one trajectory approaches x_e in reverse time (as $t \to -\infty$).

The stability of an equilibrium point can be judged by the solution of the linearized system at x_e. Letting

$$x = x_e + \Delta x \tag{1.5}$$

and substituting in (1.2) gives

$$\dot{x}_e + \Delta\dot{x} = f(x_e, u) + \left[\frac{\partial f(x, u)}{\partial x}\right]_{x=x_e} \Delta x \tag{1.6}$$

From (1.4) and (1.6) we get

$$\Delta\dot{x} = [A(x_e, u)]\Delta x \tag{1.7}$$

where A is a $n \times n$ matrix whose elements are functions of x_e and u. The (i, j)th element of $[A]$ is given by

$$A_{ij}(x_e, u) = \frac{\partial f_i}{\partial x_j}(x_e, u) \tag{1.8}$$

For a given x_e and u, the matrix A is constant. The solution of the linearized state equation (1.7) is given by

$$\begin{aligned}\Delta x(t) &= e^{A(t-t_0)}\Delta x(t_0)\\ &= c_1 e^{\lambda_1 t} v_1 + c_2 e^{\lambda_2 t} v_2 + \ldots + c_n e^{\lambda_n t} v_n\end{aligned} \tag{1.9}$$

where $c_1, c_2, \ldots, c_n$ are constants depending on the initial conditions. λ_i and v_i are the ith eigenvalue and the corresponding eigenvector of matrix $[A]$. It is assumed that all eigenvalues are distinct.

From (1.9), it can be seen that if $\Re[\lambda_i] < 0 \; \forall \; \lambda_i$, then for all sufficiently small perturbations from the equilibrium point x_e, the trajectories tend to x_e as $t \to \infty$. Hence, x_e is asymptotically stable.

If $\Re[\lambda_i] > 0$ for all λ_i then any perturbation leads to the trajectory leaving the neighborhood of x_e. Hence x_e is unstable. If there exists i and j such that $\Re[\lambda_i] < 0$ and $\Re[\lambda_j] > 0$ then x_e is a saddle point. If $\Re[\lambda_i] \neq 0 \; \forall \; \lambda_i$ then the equilibrium point is said to be hyperbolic.

No conclusion can be drawn regarding the stability of an equilibrium point if it is not hyperbolic and has no λ_i with real part greater than zero.

A stable or unstable equilibrium point with no complex eigenvalues is called a node.

Remarks

1) For all practical purposes, an equilibrium point (EP) that is not stable can be termed as unstable. For a hyperbolic equilibrium point, the number of eigenvalues with positive real parts determines its type. A Type 1 unstable EP (UEP) has one eigenvalue in the right half of the s plane (RHP). An EP with all eigenvalues in the RHP is called a source.
2) Equilibrium points are also termed fixed points. A stable EP (SEP) is also called a sink.

Steady-State Behavior

The steady-state behavior of an autonomous system is obtained from the asymptotic behavior of the system trajectories assuming that the difference between the trajectory and its steady state is called *transient*.

It is obvious that stable equilibrium points are subsets of steady-state behavior. In addition, a system may also exhibit limit cycles. A limit cycle is an isolated periodic solution (with the trajectory forming a closed curve in state space).

There can be more complex behavior, such as chaos, which does not have any fixed pattern in the steady-state solution. In general, chaotic systems exhibit sensitive dependence on initial conditions and the spectrum of the steady-state solution has a broad-band noise-like component [10].

1.2.3 Power System Security

Although power system planners try to design the system to handle various contingencies that may occur during system operation, there could be several unforeseen contingencies that can lead to catastrophic failures resulting in major blackouts spread over large areas and affecting millions of customers. It may take hours or days to restore full power to customers. A major blackout in the north-east USA occurred in November 1965. About 30 million people were affected in eight states and parts of Canada. The initiating event was the faulty operation of a relay in a transmission line carrying 300 MW from a power plant on the Niagara river. This tripped the line, transferring the load carried by the line to the parallel lines, which all tripped out. This resulted in a surge of power flow into the east–west transmission lines in New York state that resulted in tripping of the interconnecting line and seven generating units feeding the north-east grid. Loss of these plants placed a heavy drain on the generators in New York city. The result was a complete collapse of the system. It took over 12 hours to restore power to New York city [11]. This major blackout showed that reliable design is not adequate in maintaining system security during system operation. System security implies taking steps to monitor security and prevent catastrophic failures when a major contingency occurs. The system operating states can be divided into five broad categories, as shown in Figure 1.2 [12]. The definitions of these categories are based on answering the following questions:

1) Can the system meet the load demand? Mathematically, this is equivalent to verifying whether the following equality constraint is satisfied:

$$S_{Li} = S_{Li}^d, \quad i = 1, 2, \dots, N \tag{1.10}$$

   where S_{Li} is the complex (load) power supplied at bus i and S_{Li}^d is the complex demand at bus i of the N bus system. Note that the two quantities need not be equal as shortage of generation and/or transmission congestion can restrict the power supplied and limit it to a value less than the unrestricted demand.
2) Is all equipment (generators, transformers, and transmission lines) operating within its feasible region (when it can operate continuously without any damage caused by heating or overvoltages)? For example, the operating region of a generator in the P–Q plane is determined by constraints of limits on the armature current, field current, prime mover output, and stability considerations. Similarly, the insulation and power transfer characteristics force the transmission line voltages to remain in a feasible

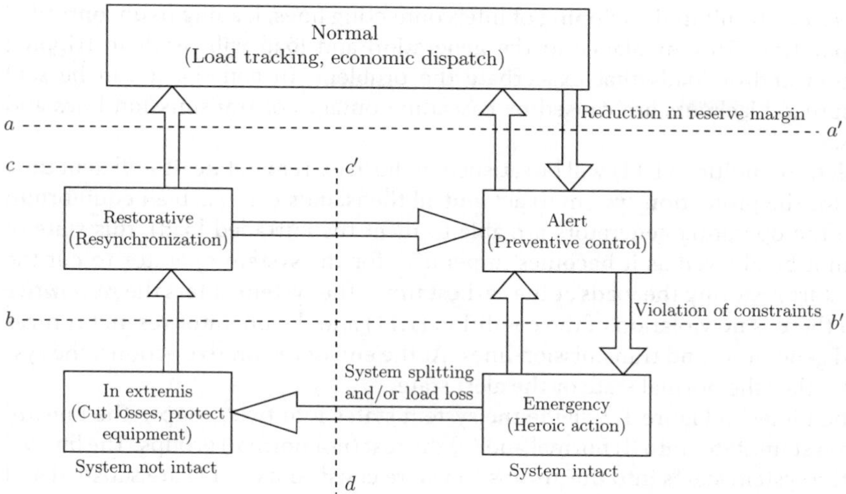

Figure 1.2 The system operating states. Source: [12] Fig. 1, p. 49, reproduced with permission ©IEEE 1978.

range. The power flow in an AC line is limited by the temperature rise of conductors, voltage regulation, and stability constraints depending on the line length. Mathematically, the following is the constraint equation of an equipment (or component) k operating in the feasible region:

$$X_k \in A_k, \quad k = 1, 2, \ldots, N_c \tag{1.11}$$

where X_k is the state of the component k and A_k is its feasible region. N_c is the total number of components.

3) Assuming the answers to the above two questions are affirmative, does the system (at the current state of operation) have sufficient margin such that any major credible contingency (loss of a heavily loaded line or a generator) will not result in overloading of any of the equipment? If the answer is affirmative, the system is said to satisfy security constraints and the operating state is classified as *normal*. If the answer is negative, the operating state is said to be in the *alert* state.

When the system is in the alert state and a credible contingency occurs, the system can enter into an *emergency* state that is characterized by one or more piece of equipment getting overloaded and operating outside its feasible region. This state of affairs cannot continue as the protection system will act automatically (with or without some time delay). If there is a fault (short circuit) in a transmission line, the protection system acts within three to five cycles. However, overloading of a line may be permitted for a few minutes. In either case, the protection system acts decisively to trip the component.

When the constraint (1.11) is violated, even if primary protection fails to act, back-up protective relays operate to disconnect the component. As the protection starts to operate, the system enters the *in extremis* state and this results in the equality constraints (1.10) being violated in addition to the constraint (1.11). Depending on the severity of the disturbance, load shedding will be limited to a localized area or widespread. The disturbances can result in the tripping of interconnecting lines, leading to uncontrolled system separation. The imbalance in the generation and load will result in tripping of generation and/or loads that exacerbate the problem. In general, it can be said that widespread blackouts are caused by cascading outages of transmission lines and generators.

Although the condition (1.11) will be satisfied in the in extremis state (if sufficient time is allowed for the protection system to act and all the islands can reach an equilibrium state when the operating generators are able to meet the curtailed load), this state of affairs cannot be allowed as it becomes imperative for the system operator to cut the losses and start restoring the loads at the earliest time. The system enters the *restorative* state when there is no violation of the condition (1.11) and the operator resynchronizes the tripped generators and transmission lines. At the end of restorative actions, the system enters either the normal state or the alert state.

The dashed lines in Figure 1.2 divide the system states into two groups. The line aa' divides the system states into (i) normal and (ii) the rest (not normal) groups. The line bb' separates the system states into the groups (i) where constraints (1.11) are satisfied and (ii) where constraints (1.11) are not satisfied. The line $cc'd$ separates the system states into groups (i) where the system is intact and (ii) where it is not intact.

It is obvious that apart from the normal and alert states, the remaining three states are temporary (or transient). It should be noted that the loss of transient stability following a major disturbance can also result in the transition from the alert state into the emergency state. When there is a danger of loss of stability, there is hardly any time for the system operator to act to avert the situation. In general, the operator has little control over the system when it enters the emergency state. That is why the improvement of system security has been predicated on the preventive control by redeploying generation and transmission resources such that the system does not remain in the alert state. The preventive control requires checking for security constraints by conducting on-line security analysis by considering a list of credible contingencies. Static security analysis involves performing fast on-line power flow computations based on the operating data and knowledge of short-term load forecasts.

1.3 Monitoring and Enhancing System Security

Modern energy control (or management) centres have computer-based supervisory control and data acquisition systems (SCADA) for monitoring security. The measurements of power and reactive power flows, voltage magnitudes, and breaker status are gathered at remote terminal units (RTU) and the data are telemetered to control centres. Although the raw data can be used directly for preliminary checking of security by observing violations of the inequality constraints (which define the feasible operating regions of various equipment) if any, it is necessary to filter out bad data and also compute power flows in lines where the measurements are not available. This is possible by state estimation (SE), where states are defined by voltage magnitudes and phase angles at all the buses. For an N bus system, there are $(2N - 1)$ states as one phase angle can be assumed arbitrarily (as zero). To ensure complete and accurate SE, there is a need for redundancy in the measurements, and the placements of meters should be such that the network is observable. On the basis of SE, it is possible to compute power and reactive power at all the load buses and in all lines. Unfortunately, SE does not consider the system dynamics and is not updated sufficiently fast to accommodate sudden and major disturbances.

The security analysis function in energy control centres refers to static security analysis where the major assumption is that any contingency under consideration does not result in loss of stability. Thus, dynamic security analysis is not performed. This is a seriously flawed approach and is justified only on the grounds that an accurate dynamic security analysis is not feasible unless an on-line transient stability type of simulation can be performed. While attempts have been made to apply parallel processing techniques to speed up computations, this approach has not been entirely successful. Transient energy function (TEF) methods have been tried out in Canada, the USA, and Japan for both dynamic security assessment (DSA) and preventive control [13].

1.4 Emergency Control and System Protection

Instead of relying purely on preventive control, it makes lot of sense to detect emergency states fast enough and deploy emergency control or remedial action schemes to steer the system to an alert or normal state. This approach is based on the utilization of technological advances in communications, computers, and control. A major development is the technology of synchronized phasor measurements that enable direct and fast measurement of voltage phase angles (in addition to their magnitudes) that were earlier estimated by SE. This can help in real-time transient stability prediction [13]. The essential feature of the phasor measurement unit is the measurement of positive sequence voltages and currents in a power system in real time with precise time synchronization. This is achieved with a global positioning system (GPS). GPS is a US government-sponsored program that provides worldwide position and time broadcasts free of charge. The current wide area communication technologies (such as SONET) are capable of delivering messages from one area of a power system to multiple nodes of the system in less than 6 ms [14, 15].

The emergency state is associated with the operation of protection systems based on the relays that monitor the voltages and currents at different locations in the system. Misoperation of protection systems is often a contributing factor for the cascading outages that lead to system collapse. Among possible failures in the protection systems, the most troublesome are the "hidden" failures (HF), which are defined as "a permanent defect that will cause a relay system to incorrectly and inappropriately remove circuit element(s) as a direct consequence of another switching event" [16].

It should be possible to eliminate hidden failures by modifying the protection system and introducing new concepts and technologies. It must be realized that equipment protection must be distinguished from system protection. While equipment protection must be designed to protect an equipment such as a transmission line from high fault currents that can damage the conductors, out-of-step protection is essentially a system protection scheme against transient (synchronous) instability. The complexities of a large power system imply the necessity for adaptive protection [17].

The other technological innovations that can be applied for emergency control are based on power electronic converter controls used in FACTS and HVDC controllers.

The speed of control actions enables their application not only for the improvement of first swing stability but also for damping of initial large oscillations [18, 19].

1.5 Recent Developments

Synchronized phasor measurement units (PMUs) were introduced in the early 1980s and have become a mature technology with several applications. These include (i) power system monitoring, (ii) power system protection, and (iii) power system control [20]. In the 2003 north-east USA blackout and 1996 US West Coast blackouts, the PMU monitoring capabilities were essential for the quick and accurate postmortem analysis of events. The US–Canada Task Force on 14 August 2003 blackout made a recommendation to "require use of time-synchronized data recorders" [21]. With PMU installation cost ranging from 10,000 to 70,000 dollars (depending on the utility, location, and availability of communication channels) placements of PMUs in the optimum locations is one of the first steps of a wide-area measurement system (WAMS). Prior to synchronized phasor measurements, the system state (voltage phasors at all buses) could not be measured directly, but inferred from the unsynchronized power flow measurements. The assumption was that the system did not change during two consecutive measurements. The direct integration of a few synchronized phasor measurements into an existing nonlinear (state) estimator is straightforward, but does not improve the quality of the estimator. If PMU data is time tagged and used directly for state estimation, the drawbacks of the earlier static state estimator can be overcome. Since the voltage and current phasors of the network are linearly related, the estimation problem is a linear weighed least squares problem which requires no iterations. The major issues are the number of PMUs required and their placement. Measuring line currents can extend the voltage measurements to buses where no PMU is installed. Using a model of the transmission line, line current can be used to compute the voltage at the other end of the line.

1.5.1 Power System Protection

Traditional out-of-step relays are found to be unsatisfactory in highly interconnected networks [21]. The traditional out-of-step relays often misoperate: they fail to determine correctly whether or not an evolving electro-mechanical swing is stable or not. Wide-area measurements of positive sequence voltages (and hence swing angles) provide a direct way to determine stability using real-time data instead of precalculated relay settings. The loss of synchronism in a power systems is governed by the following propositions.

Proposition 1.1 [13] However complex, the mechanism of loss of synchronism in a power system originates from the irrevocable separation of its machines into two groups.

Proposition 1.2 [22,23] There is a unique cutset (called the critical cutset) consisting of transmission lines and transformers (or series elements) connecting the two groups of machines that separate. The cutset angles (the difference between the angles of the terminal buses of each member of the cutset) become unbounded.

Proposition 1.1 has been utilized extensively to determine the mode of instability (MOI) that characterizes the loss of synchronism when a disturbance occurs. However, the determination of MOI is not straightforward and several empirical approaches have been suggested and tried.

Unlike Proposition 1.1, which is justified based on observations from a large number of stability studies, it is possible to prove Proposition 1.2 [22, 23].

Detection of Instability by Monitoring Critical Cutset [20]

If we know the critical cutset, it is quite straightforward to monitor the angles across each member of the cutset using PMUs and wide-area measurements. Strictly speaking, there is no need to use GPS technology to measure the angle across a transmission line. The phasor measurements of line current and the bus voltage at either end of the line coupled with the knowledge of the series impedance of the line is adequate to compute on-line the angle across a line. As a matter of fact, the criterion for instability requires measurements of bus frequencies and line flows. The frequency monitoring network (FNET) can be employed for the purpose of wide-area frequency monitoring [24].

1.5.2 Development of Smart Grids

On 14 August 2003 there was a major blackout that affected the USA and Canada. In the same year, there were blackouts in the UK, Sweden, and Italy. These incidents have led to a rethinking on the design of energy management systems (EMSs). In the late 1990s, the concept of distributed generation (DG), based on combined heat and power (CHP) and renewable energy (solar and wind), had taken shape to empower customers to tackle the problems of power interruptions and improve resilience to overcome blackouts caused by bad weather conditions. In this context, the concept of a smart grid was promoted by the US Department of Energy (DoE) to modernize the electric grid [24]. According to the DOE, "Smart Grid generally refers to a class of technology people are using to bring utility electric delivery systems into the 21st century, using computer-based remote control and automation. These systems are made possible by two-way communication technology and computer processing that has been used in other industries". According to the International Electrotechnical Commission (IEC), "The smart grid comprises everything related to an electrical system in between any point of generation and any point of consumption. Through the addition of Smart Grid technologies, the grid becomes more flexible, interactive and is able to provide real time feedback. It is an electricity network that can intelligently integrate the actions of all users controlled to it—generators, consumers and those that do both—in order to efficiently deliver sustainable, economic and secure electricity supplies. Smart technologies improve the observability and/or controllability of the power system. Thereby, smart grid technologies help to convert the power grid from a static infrastructure to be operated as designed, to a flexible, 'living' infrastructure operated proactively" [25]. The components of smart grid technology are [26]:

1) advanced metering infrastructure (AMI) systems that measure, collect, and analyse energy usage
2) distribution automation (DA), which provides the extension of an intelligent control over electrical power grid functions to distribution level and beyond

3) distributed generation (DG) and energy storage (DS), which combined are known as distributed energy resource (DER) and provide the energy in microgrids
4) substation automation, which involves the automation of electric power distribution and transmission substations
5) FACTS, which is a power electronics-based system to enhance controllability and increase the power transfer capability of network
6) demand response (DR) systems that manage customer consumption of electricity in response to supply conditions, based on the price signals.

1.5.3 Microgrids [27]

The distribution system provides major opportunities for smart grid concepts. This includes the integration of high levels of DERs using microgrids. Basic objectives are improved reliability, high penetration of renewable energy sources, dynamic islanding, and improved generation efficiencies through the use of waste heat. DERs encompass a wide range of prime mover technologies such as internal combustion (IC) engines, gas turbines, microturbines, photovoltaic systems, fuel cells, wind power, and DC storage. Most emerging technologies have an inverter to interface with the distribution system. These technologies have lower emissions and the potential to have lower cost, negating traditional economies of scale. The applications include power support at substations, deferral of transmission and distribution (T&D) upgrades, high fuel efficiency (through capturing waste heat), use of renewable energy, higher power quality and a smarter distribution system, which enables the application of multiple, clustered microgrids.

1.5.4 Role of System Dynamics and Control

The power system dynamics and control is undergoing revolutionary changes with the introduction of WAMS based on PMU and sensor networks, application information and communication technologies, expanding contribution of renewable energy, and availability of network controllers such as FACTS and HVDC converters. In the past, the study of dynamics and control was restricted to synchronous generators. The grid was often assumed to be passive (in particular the distribution network). The application of several prime mover technologies in DER has introduced new complexities. The electric grid is formed by interconnecting a large number of DERs through the installation of power delivery networks, which require long-term planning. It would be more appropriate to route power flow using network controllers (such as FACTS) to ensure reliable and efficient system operation. Since the trend is towards the application of multiple DERs of modest ratings, it may be valid to compare the operation of communication and electricity networks.

1.6 Outline of Chapters

Chapter 2 describes grid characteristics and operation, particularly for overhead lines. The circuit models of transmission/distribution networks are presented, with the derivation of the network equations in the Kron (Park) reference frame and also in the stationary reference frame (α, β, o components). Grid operation with frequency and power controls is described. The dynamic characteristics of the grid that determine the response to the injection of voltages at frequencies above or below the nominal value, injection of reactive currents, are analyzed in detail as these affect the subsynchronous resonance (SSR) and reactive power compensation based on SVC or STATic (Synchronous) COMpensator (STATCOM). The response of the grid under switching (electromagnetic) transients is also analyzed. The chapter concludes with discussion of the reliability and planning of transmission and distribution networks.

Chapter 3 introduces modelling and simulation of synchronous generator dynamics, which is affected by excitation and prime mover controllers. Park's transformation is applied for the derivation of the machine model, which is used for stability analysis as well as the study of electromagnetic transients, including the simulation of SSR phenomena. The dynamics of the synchronous generator is illustrated using the example of a single-machine finite bus (SMFB) system when the finite bus voltage magnitude and frequency are variable. The analysis of the SMFB system can be easily extended for the study of the dynamics of multi-machine systems.

Chapter 4 presents the modeling of wind-turbine generating systems. The types of generator and power electronic configurations for grid interfacing are described. The modeling of a doubly-fed induction generator-based system in the D-Q frame and its control using rotor-side and grid-side power electronic converters is discussed in detail.

FACTS and HVDC transmission controllers can be applied to improve the performance of power grids. In Chapter 5, the various FACTS controllers are briefly described based on a unified power flow controller (UPFC) model. The thyristor-controlled series capacitor (TCSC) and STATCOM, which are used extensively, are taken up for detailed analysis of their performance. The development of HVDC transmission over the years is described in some detail, leading to VSC-HVDC with modular multilevel converters (MMC). The modelling of HVDC transmission for the enhancement of system stability is described.

Chapters 6 and 7 present the analysis of the angular stability problem in power systems. Chapter 6 focuses on relative angular oscillations, which are also known as power swings. Power swings can be negatively damped under certain operating conditions, resulting in sustained or growing oscillations of the relative angles and power flows. This problem is analyzed using a small-signal model of a power system, which is obtained by linearizing the nonlinear differential equations of the system at the equilibrium point. The problem of poor damping or negative damping of power swings can be alleviated by the use of power swing damping controllers (PSDCs), which modulate the set-point of the power electronic actuators in the power systems. Examples of such actuators are the generator excitation system, FACTS, and HVDC systems. Chapter 6 also presents the typical structure of a PSDC and the general principles of design using a multi-modal decomposition approach. The choice of feedback signals for PSDCs and the effect of PSDCs on the damping of swing modes are illustrated using examples.

Chapter 7 presents the analysis of the phenomena of loss of synchronism (LoS) in a synchronous grid. This problem is manifested as a monotonic increase in the angular separation between the rotors of synchronous machines. In multi-machine systems, groups of machines separate from each other when LoS occurs. The grouping is decided by the nature and location of the disturbance, system parameters, and operating conditions. These aspects are illustrated using the numerical simulation of the dynamic response of a single-machine infinite bus system and a ten-machine system. The chapter also describes the energy function method. This is an approximate method for stability assessment which does not require the simulation of the entire time response of the system.

Voltage stability and control issues are presented in Chapter 8. The definitions of voltage stability, comparison of voltage and angle stability, and the factors affecting instability and voltage collapse are discussed. It is pointed out that a major factor is the dynamics of load restoration. An integrated analysis of voltage and angle stability is presented based on a three-bus system example. In general, the swing mode affects the angle stability while the exciter mode affects the voltage stability. An analysis of the small signal voltage instability decoupled from angle instability is presented at the end of the chapter with a discussion on the control of voltage instability.

Chapter 9 presents the technology and potential applications of WAMS. This technology makes available time-synchronized measurements across the system for various monitoring and control applications, especially those involving the slower electro-mechanical phenomena. In this chapter we present the various methods for swing-mode estimation, including ringdown waveform analysis, ambient data-based estimation, and probing signal analysis, and illustrate these using examples. The potential application of WAMS for damping control, emergency control, supervision of protective relays, and parameter estimation are also described.

While Chapters 6 to 9 focus on the slower electro-mechanical and voltage dynamics of the system, Chapter 10 analyzes another class of dynamic phenomena involving faster interactions between the rotating turbine-generator mechanical system (torsional dynamics), the electrical network, and the controllers of power electronic systems. A striking example of this is the phenomenon of SSR. The key modeling issues when analyzing these faster interactions are highlighted in this chapter. Analytical/numerical methods of modeling power electronic systems for subsynchronous interactions are also described. SSR mitigation techniques, such as subsynchronous damping controllers (SSDCs), and the modeling of grid-connected power electronic systems are also presented in this chapter.

Chapter 11 presents the technologies associated with solar-photovoltaic (PV) energy conversion and energy storage systems. The characteristics of solar-PV systems, the power electronic configurations for the interfacing of these systems to a power system, and their control strategies are presented in this chapter. The characteristics of various energy storage technologies and their suitability for different grid applications are also discussed in the chapter.

Microgrids essentially consist of distributed generation and storage which are provided in the low or medium voltage distribution systems. The concept, architecture, operation, and control of microgrids are discussed in detail in Chapter 12. The chapter also includes a description of the energy management system that must consider the operating mode of a microgrid in a grid-connected or islanded mode. The dynamic modelling of DERs is also presented for a class of major energy resources. Some operating problems and possible solutions are discussed in addition to the adaptive network protection in microgrids. The chapter concludes with a discussion of future trends.

Appendices A to C present important and specific issues that affect problems of transient stability (LoS), grid synchronization, and current regulation, followed by an algorithm for load compensation.

Appendix D presents the symmetrical component transformation of variables, which is commonly used to study three-phase systems under balanced and unbalanced operating scenarios. The per-unit representation of power system data and its use in network calculations is also presented.

References

1 Padiyar, K.R. (2014) *Understanding the Structure of Electricity Supply*, BS Publications, Hyderabad.
2 Hingorani, N.H. and Gyugyi, L. (2000) *Understanding FACTS – Concepts and Technology of Flexible AC Transmission Systems*, IEEE Press, New York.
3 Padiyar, K.R. (2007) *FACTS Controllers in Power Transmission and Distribution*, New Age International, New Delhi.
4 Padiyar, K.R. (2015) *HVDC Power Transmission Systems*, New Age, New Delhi, 3rd edn.
5 Byerly, R.T. and Kimbark E.W. (Editors) (1974) *Stability of Large Electric Power Systems*, IEEE Press, New York.
6 Kundur, P. (1994) *Power System Stability and Control*, McGraw-Hill, New York.
7 Padiyar, K.R. (2002) *Power System Dynamics: Stability and Control*, BS Publications, Hyderabad, 2nd edn.
8 Bose, A., Concordia, C., Dunlop, R. *et al.* (1982) Proposed terms & definitions for power system stability. *IEEE Transactions on Power Apparatus and Systems*, **PAS-101** (7), 1894–1898.
9 Kundur, P., Paserba, J., Ajjarapu, V. *et al.* (2004) Definition and classification of power system stability. IEEE/CIGRE joint task force on stability terms and definitions. *IEEE Transactions on Power Systems*, **19** (3), 1387–1401.
10 Guckenheimer, J. and Holmes, P. (1983) *Nonlinear Oscillations, Dynamical Systems, and Bifurcations of Vector Fields*, Springer-Verlag, New York, 3rd edn.
11 Glover, J.D. and Sarma, M.S. (2002) *Power System Analysis & Design, SI Version*, Thomson Brooks/Cole, New Delhi, 3rd edn.
12 Fink, L.H. and Carlsen, K. (1978) Operating under stress and strain. *IEEE Spectrum*, **15** (3), 48–53.
13 Pavella, M., Ernst, D., and Ruiz-Vega, D. (2000) *Transient Stability of Power Systems – A Unified Approach to Assessment and Control*, Kluwer Academic, Boston.
14 Begovic, M., Novosel, D., Karlsson, D. *et al.* (2005) Wide-area protection and emergency control. *Proceedings of the IEEE*, **93** (5), 876–891.
15 Adamiak, M.G., Apostolov, A.P., Begovic, M.M. *et al.* (2006) Wide area protection – technology and infrastructures. *IEEE Transactions on Power Delivery*, **21** (2), 601–609.
16 De La Ree, J., Liu, Y., Mili, L. *et al.* (2005) Catastrophic failures in power systems: causes, analyses, and countermeasures. *Proceedings of the IEEE*, **93** (5), 956–964.
17 Centeno, V., Phadke, A.G., Edris, A. *et al.* (1997) An adaptive out-of-step relay. *IEEE Transactions on Power Delivery*, **12** (1), 61–71.
18 Padiyar, K.R. and Rao, K.U. (1997) Discrete control of series compensation for stability improvement in power systems. *International Journal of Electrical Power & Energy Systems*, **19** (5), 311–319.

19 Krishna, S. and Padiyar, K.R. (2005) Discrete control of unified power flow controller for stability improvement. *Electric Power Systems Research,* **75** (2), 178–189.

20 De La Ree, J., Centeno, V., Thorp, J.S., and Phadke, A.G. (2010) Synchronized phasor measurement applications in power systems. *IEEE Transactions on Smart Grid,* **1** (1), 20–27.

21 US–Canada Power System Outage Task Force (2004), Final report on August 14, 2003 blackout in United States and Canada: Causes and recommendations.

22 Padiyar, K.R. (2013) *Structure Preserving Energy Functions in Power Systems: Theory and Applications,* CRC Press, Boca Raton.

23 Krishna, S. and Padiyar, K.R. (2009) On-line dynamic security assessment: Determination of critical transmission lines. *Electric Power Components and Systems,* **38** (2), 152–165.

24 US Department of Energy (2008), *The smart grid: An introduction.*

25 International Electrotechnical Commission (2010) *Smart grid standardization road map.*

26 Bush, S.F. (2014) *Smart grid: Communication-enabled Intelligence for the Electric Power Grid,* John Wiley & Sons, New York.

27 Lasseter, R.H. (2011) Smart distribution: Coupled microgrids. *Proceedings of the IEEE,* **99** (6), 1074–1082.

2

Grid Characteristics and Operation

In this chapter, a three-phase AC transmission grid is considered for study. The modeling, analysis and investigation of the dynamic characteristics is undertaken. It should be noted that the transmission grid (and even the distribution grid in rural areas) is made up of overhead transmission lines.

2.1 Description of Electric Grids

An electric power grid is a network of synchronized power providers and consumers that are connected by transmission and distribution lines and operated by one or more control centres. In the past, electric grids meant mainly transmission grids at high or extra-high voltage (up to 800 kV, line to line). In recent times, microgrids at the power distribution level (below 40 kV, line to line) have been researched and developed. While distribution networks are typically made up of radial lines operating at medium voltages, transmission networks are made of a number of lines that will provide more than one path for power flow between any two buses in the network. This arrangement provides redundancy, such that removal of one line does not affect the system reliability. As a matter of fact, contingency analysis is carried out during system planning and operation to determine the total (maximum) transfer capability (TTC) during a specific contingency condition.

The operation of the electric grid is subject to the following fundamental laws of electricity [1, 2]:

1) Faraday's law
This states that the voltage measured around any closed path is equal to the time rate of change of magnetic flux normal to the surface bounded by the path. The voltage around a loop is the line integral of the electric field intensity around the closed path, and the flux is the surface integral of the flux density over the area enclosed by the path. Thus

$$\int_L \overline{E}.dl = -\frac{\partial}{\partial t}\int_S \mu\overline{H}.dS \tag{2.1}$$

2) Ampere's law
This states that the work done in taking a magnetic pole around a closed loop in a magnetic field is equal to the current passing through the surface enclosed by the

Dynamics and Control of Electric Transmission and Microgrids, First Edition.
K. R. Padiyar and Anil M. Kulkarni.

closed loop. Since current may be either conduction current or displacement current (as that flowing through a capacitor) or both, the equation describing Ampere's law is

$$\int_L \overline{H}.dl = \int_S \sigma \overline{E}.dS + \frac{\partial}{\partial t}\int_S \epsilon \overline{E}.dS \tag{2.2}$$

The first term on the right-hand side of (2.2) represents the conduction current. The second term is the displacement current.

3) Gauss's law
This states that the net outward electric flux through any closed surface is equal to the net charge enclosed within the surface. The equation describing Gauss's law is

$$\int_S \epsilon \overline{E}.dS = q = \psi \tag{2.3}$$

The symbols used in the above equations are:
$\overline{E}$: electric field intensity in V/m
$\overline{H}$: magnetic field intensity in A/m
$\overline{L}$: length in meters
$\overline{S}$: area in square meters
σ: conductivity in mho per meter
q: charge in coulombs
ψ: lines of electric flux
μ: permeability in H/m
The permeability of free space is

$$\mu_0 = 4\pi \times 10^{-7} \text{ H/m}$$

ϵ = permittivity in F/m. The value for free space is approximately

$$\epsilon_0 = \frac{1}{36\pi} \times 10^{-9} \text{ F/m}$$

It should be noted that the velocity of propagation of electromagnetic waves is

$$v = \frac{1}{\sqrt{\mu_0 \epsilon_0}} = 3 \times 10^8 \text{ m/s}$$

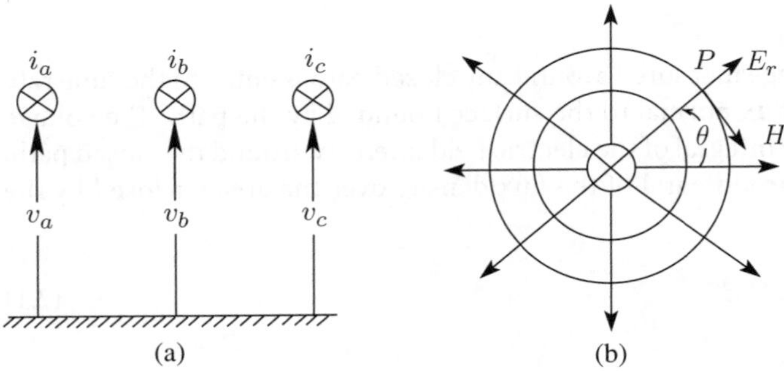

Figure 2.1 A three-phase overhead line. (a) Typical configuration; (b) Electric and magnetic fields around a conductor.

Figure 2.1a shows a typical configuration of conductors of a three-phase, high-voltage transmission line. The voltages (with respect to the ground) are v_a, v_b, and v_c for the three phases (a, b, and c). The currents flowing in the conductors are i_a, i_b, and i_c. If the three conductors in Figure 2.1a are sufficiently far apart, the electromagnetic fields near any one conductor would be unaffected by the presence of the others and would be symmetrical, as shown in Figure 2.1b. The charges and currents of the conductor would be symmetrically distributed and if the conductors were perfect (zero resistance), then there would be no electric field along the wire in the direction of the current. The electric and magnetic fields are mutually perpendicular everywhere in the space outside the conductor. The field configuration shown in Figure 2.1b is called a transverse electromagnetic field (TEM).

The assumption about the symmetrical distribution of charge and current on a conductor is reasonably accurate for overhead lines with relatively large ratios of spacing to diameter. For multiconductor cables, this assumption is not applicable and the effect of the nonsymmetrical distribution of charge and current must be considered in calculating the impedances. The changes in impedances when the symmetry does not hold are due to the proximity effect. Neglecting this effect, the method of computing the circuit parameters for transmission lines is given in [2].

2.2 Detailed Modeling of Three-Phase AC Lines [3]

The voltages and currents (flowing in a transmission line) are functions of both time and the position (determined by the distances from the sending end) along the line. Thus, we can describe the voltage and current at a distance x from the sending end by the following partial differential equations [3]:

$$-\frac{\partial v(x,t)}{\partial x} = [R]i(x,t) + [L]\frac{\partial i(x,t)}{\partial t} \tag{2.4}$$

$$-\frac{\partial i(x,t)}{\partial x} = [C]\frac{\partial v(x,t)}{\partial t} \tag{2.5}$$

where v and i are three-dimensional vectors, and $[R]$, $[L]$, and $[C]$ are 3×3 matrices defining transmission line resistances, inductances, and capacitances per unit length of the lines, respectively. These matrices are symmetric. These equations can be solved either in the frequency domain (using Fourier transform) or in the complex frequency domain using Laplace transformation.

The time-varying electric field produces a time-varying magnetic field, which in turn generates an electric field, and so on, resulting in the propagation of energy. This propagation phenomenon for traveling waves can occur in both space and on the transmission line.

While field equations are applicable in general, the circuit equations are more convenient when the voltage v and current i are well defined (as in the case of a three-phase transmission line). Here, we can define resistances, inductances, and capacitances and apply circuit equations. The resistances include the earth conduction effect. The solution of the partial differential equations (2.4) and (2.5) are required when we want to compute electromagnetic transients following a switching operation (closing of a breaker with or without trapped electric charges on the conductors).

The computation of overvoltages caused by lightning strikes and switching actions is required for satisfactory design of the transmission line and connected equipment such as switchgear (circuit breakers, isolators).

In steady state, the currents and voltages in a line are sinusoidal of rated frequency and can be represented by phasors. During disturbances, the magnitudes and phase (angles) of these phasors vary and can lead to low-frequency oscillations in power flow and voltages in the lines.

2.3 Circuit Models of Symmetric Networks

The self and mutual inductances of a transmission line can be calculated based on the configuration of the line. Similar comments apply to the capacitances of the line. Although the matrices $[L]$ and $[C]$ in (2.4) and (2.5) are symmetric, it is possible that the self inductances are unequal in the three phases (a, b, and c) and mutual inductances L_{ab}, L_{bc} and L_{ca} are also not equal. For a fully transposed line which results in a symmetric three-phase network, we can expect $L_{ab} = L_{bc} = L_{ca}$ and $L_a = L_b = L_c$. With these relations, the matrix $[L]$ becomes a cyclic symmetric matrix. Similar comments also apply to the capacitance matrix $[C]$. The resistance matrix can be assumed to be diagonal, neglecting the effect of earth conduction, and with symmetry, $R_a = R_b = R_c$.

It can be shown that for a symmetric three-phase network, the positive sequence (single-phase) network is identical to the negative sequence (single-phase) network and they are uncoupled. The zero-sequence network is also uncoupled with the positive and negative sequence network. If there are no zero-sequence voltages in the network, then the zero-sequence network is unexcited and thus can be ignored.

The single-phase positive (or negative) sequence network that can exhibit the response of a three-phase, symmetric transmission line network is shown in Figure 2.2.

It should be noted that a transmission line is represented by a linear network with constant values of circuit parameters. Typically, a single-phase distributed parameter network representing a line is represented by a cascade of π networks. The number of π sections depends on the nature of the response under study. If we are interested in only low-frequency response (which is adequate in dynamic studies), a single π section is adequate to represent a line.

A single-phase π equivalent of a transmission line is shown in Figure 2.3. However, this can also represent a three-phase line when the coefficient matrices, inductance $[L]$, resistance $[R]$, and capacitance $[C]$, are all 3×3 matrices. These are defined as

$$[L] = \begin{bmatrix} L_s & L_m & L_m \\ L_m & L_s & L_m \\ L_m & L_m & L_s \end{bmatrix}, \; [R] = \begin{bmatrix} R_s & R_m & R_m \\ R_m & R_s & R_m \\ R_m & R_m & R_s \end{bmatrix}, \; [C] = \begin{bmatrix} C_s & C_m & C_m \\ C_m & C_s & C_m \\ C_m & C_m & C_s \end{bmatrix}$$

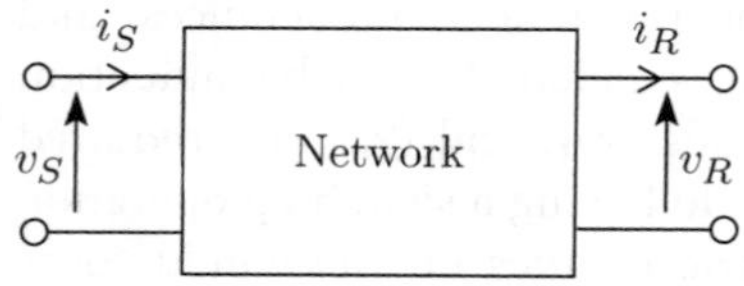

Figure 2.2 A two-port network.

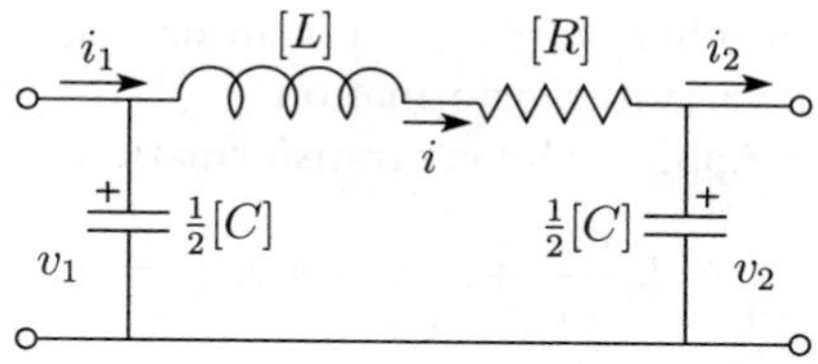

Figure 2.3 π equivalent of a transmission line.

The network equations are

$$[L]\frac{di}{dt} + [R]i = v_1 - v_2 \tag{2.6}$$

$$\frac{1}{2}[C]\frac{dv_1}{dt} = i_1 - i \tag{2.7}$$

$$\frac{1}{2}[C]\frac{dv_2}{dt} = i - i_2 \tag{2.8}$$

where v_1, v_2, i_1, i_2, and i are three-dimensional vectors, with phase variables as elements. For example,

$$i^t = \begin{bmatrix} i_a & i_b & i_c \end{bmatrix}, \quad v_1^t = \begin{bmatrix} v_{1a} & v_{1b} & v_{1c} \end{bmatrix}, \quad v_2^t = \begin{bmatrix} v_{2a} & v_{2b} & v_{2c} \end{bmatrix}$$

2.4 Network Equations in *DQo* and $\alpha\beta o$ Components [4–7]

1) For a symmetric network excited at the generator nodes by a positive sequence of sinusoidal voltages, the currents and voltages in the network in steady state are also sinusoidal (positive sequence). It can be shown that the frequencies of the voltages at all nodes in the network, in steady state, are identical and equal to the operating frequency (say ω_o).

For a connected network, it is obvious that the entire network is to be transformed using a single transformation with reference to a common, synchronously rotating reference frame. Such a transformation is termed Kron's transformation [4] and defined as:

$$\begin{bmatrix} f_a \\ f_b \\ f_c \end{bmatrix} = \sqrt{\frac{2}{3}} \begin{bmatrix} \cos\theta_o & \sin\theta_o & \frac{1}{\sqrt{2}} \\ \cos\left(\theta_o - \frac{2\pi}{3}\right) & \sin\left(\theta_o - \frac{2\pi}{3}\right) & \frac{1}{\sqrt{2}} \\ \cos\left(\theta_o + \frac{2\pi}{3}\right) & \sin\left(\theta_o - \frac{2\pi}{3}\right) & \frac{1}{\sqrt{2}} \end{bmatrix} \begin{bmatrix} f_D \\ f_Q \\ f_o \end{bmatrix} \tag{2.9}$$

$$= [C_K] f_{DQo}$$

where $f_{DQo}^t = \begin{bmatrix} f_D & f_Q & f_o \end{bmatrix}$. It should be noted that f can be any variable – voltage or current. θ_o is defined as

$$\theta_o = \omega_o t + \delta_o \tag{2.10}$$

where ω_o is the average (synchronous) frequency in the network in steady state and δ_o is a constant. There is no loss of generality if we assume $\delta_o = 0$. $[C_K]$ is defined such that

$$[C_K]^{-1} = [C_K]^t \tag{2.11}$$

In other words, $[C_K]$ is an orthogonal matrix and satisfies the condition for a power invariant transformation.

Applying Kron's transformation to (2.6) results in

$$L_1 \frac{di_D}{dt} + \omega_o L_1 i_Q + R_1 i_D = v_{1D} - v_{2D} \tag{2.12}$$

$$L_1 \frac{di_Q}{dt} - \omega_o L_1 i_D + R_1 i_Q = v_{1Q} - v_{2Q} \tag{2.13}$$

$$L_o \frac{di_o}{dt} + R_o i_o = v_{1o} - v_{2o} \tag{2.14}$$

The last equation can be neglected if no zero-sequence voltages or currents are present. L_1, R_1, L_o, and R_o are defined by

$$L_1 = L_s - L_m, \quad L_o = L_s + 2L_m$$
$$R_1 = R_s - R_m, \quad R_o = R_s + 2R_m$$

It should be noted that L_1 and R_1 are positive sequence (or negative sequence) quantities. Applying Kron's transformation to (2.7) and (2.8) gives

$$\frac{1}{2} C_1 \frac{dv_{1D}}{dt} + \frac{\omega_o}{2} C_1 v_{1Q} = i_{1D} - i_D \tag{2.15}$$

$$\frac{1}{2} C_1 \frac{dv_{1Q}}{dt} - \frac{\omega_o}{2} C_1 v_{1D} = i_{1Q} - i_Q \tag{2.16}$$

$$\frac{1}{2} C_1 \frac{dv_{2D}}{dt} + \frac{\omega_o}{2} C_1 v_{2Q} = i_D - i_{2D} \tag{2.17}$$

$$\frac{1}{2} C_1 \frac{dv_{2Q}}{dt} - \frac{\omega_o}{2} C_1 v_{2D} = i_Q - i_{2Q} \tag{2.18}$$

The zero-sequence variables are normally neglected and their equations can be omitted. C_1 is positive sequence capacitance given by

$$C_1 = C_s - C_m$$

Denoting

$$\hat{i} = i_Q + j i_D, \quad \hat{v}_1 = v_{1Q} + j v_{1D}, \quad \hat{v}_2 = v_{2Q} + j v_{2D}$$

then equations (2.12) and (2.13) can be expressed as

$$L_1 \frac{d\hat{i}}{dt} + (R_1 + j\omega_o L_1)\hat{i} = \hat{v}_1 - \hat{v}_2 \tag{2.19}$$

2.4.1 Transformation to Park (*dqo*) Components [6]

During a transient, we can replace (2.10) by

$$\theta = \omega_o t + \delta$$

where δ is variable. Thus, the radian frequency is variable and is given by

$$\omega = \frac{d\theta}{dt} = \omega_o + \frac{d\delta}{dt}$$

The transformation $[C_P]$ obtained by replacing θ_o by θ in $[C_K]$ is called Park's transformation. The relationship between Kron and Park transformation is given by

$$[C_K] = [C_P][T_1] \tag{2.20}$$

where

$$[T_1] = \begin{bmatrix} \cos\delta & \sin\delta & 0 \\ -\sin\delta & \cos\delta & 0 \\ 0 & 0 & 1 \end{bmatrix}$$

From (2.20), we have

$$\begin{aligned} f_D &= f_d \cos\delta + f_q \sin\delta \\ f_Q &= -f_d \sin\delta + f_q \cos\delta \end{aligned} \tag{2.21}$$

From (2.21), we can obtain

$$f_Q + jf_D = e^{j\delta}(f_q + jf_d) \tag{2.22}$$

From (2.22), we can express (2.19) as

$$e^{j\delta}\left[L_1\frac{d\hat{i}'}{dt} + j\frac{d\delta}{dt}L_1 i' + (R_1 + j\omega_o L_1)\hat{i}'\right] = (\hat{v}_1' - \hat{v}_2')e^{j\delta} \tag{2.23}$$

where

$$\hat{i}' = i_q + ji_d, \quad \hat{v}_1' = v_{1q} + jv_{1d}, \quad \hat{v}_2' = v_{2q} + jv_{2d}$$

Simplifying (2.23) we get

$$L_1\frac{d\hat{i}'}{dt} + (R_1 + j\omega L_1)\hat{i}' = \hat{v}_1' - \hat{v}_2' \tag{2.24}$$

Equation (2.24) can also be derived directly by applying Park's transformation. Equation (2.22) is a very useful relation and can be represented by the phasor diagram shown in Figure 2.4.

2.4.2 Steady-State Equations

Neglecting transients, the equations that are applicable in steady state are obtained by neglecting variations in *D-Q* components in (2.12) to (2.18) and neglecting zero-sequence components. These complex equations are

$$(R_1 + j\omega_o L_1)\hat{I} = \hat{V}_1 - \hat{V}_2 \tag{2.25}$$

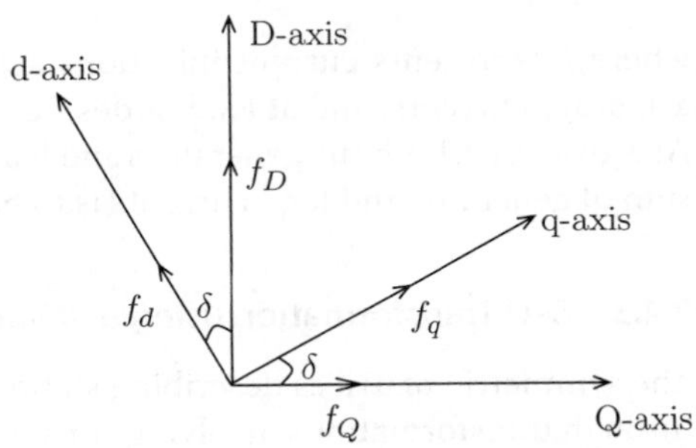

Figure 2.4 Phasor diagram.

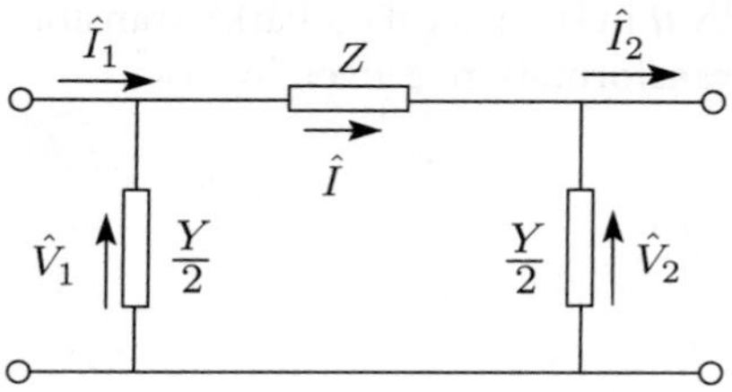

Figure 2.5 Steady-state equivalent circuit (positive sequence).

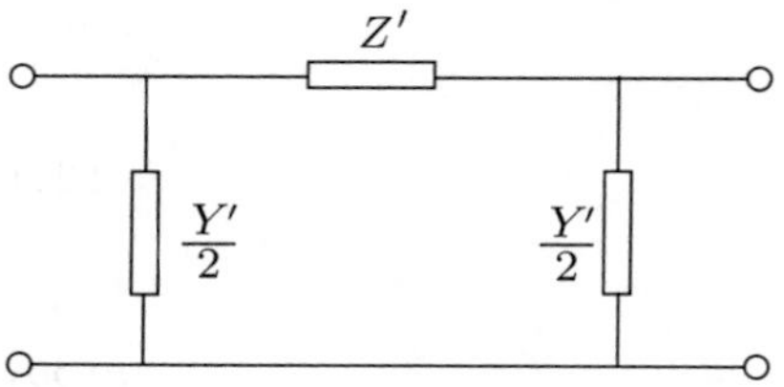

Figure 2.6 Exact π equivalent circuit.

$$j\omega_o \frac{C_1}{2}\hat{V}_1 = \hat{I}_1 - \hat{I} \tag{2.26}$$

$$j\omega_o \frac{C_1}{2}\hat{V}_2 = \hat{I} - \hat{I}_2 \tag{2.27}$$

These equations describe the single-phase circuit shown in Figure 2.5 with sinusoidal excitation. It should be noted that the constancy of i_D and i_Q implies sinusoidal currents in the line. In Figure 2.5, Z and Y are defined by

$$Z = R_1 + j\omega_o L_1, \quad Y = j\omega_o C_1$$

For long transmission lines, the exact π equivalent shown in Figure 2.6 is applicable in steady state, where

$$Z' = \frac{Z \sinh\sqrt{ZY}}{\sqrt{ZY}}, \quad Y' = \frac{Y \tanh(\sqrt{ZY}/2)}{(\sqrt{ZY}/2)}$$

R_1, L_1, and C_1 are calculated by multiplying the positive sequence R, L, and C parameters per unit length by the length of the line.

The AC system consisting of transmission lines, transformers, and other impedance elements can be represented by a single-phase equivalent network (in positive sequence parameters). The equations for such a network are expressed conveniently using a bus admittance matrix Y as

$$[Y]V = I \tag{2.28}$$

where I represents current injections at the nodes. At generator nodes I consists of armature currents, and at load nodes I consists of load currents (treated as injections). At a given bus, both the generators and loads may be present, in which case the algebraic sum of generator and load currents is to be considered.

2.4.3 *D-Q* Transformation using α-β Variables

The symmetric matrices describing stationary three-phase networks can be decoupled through transformations involving constant real matrices. The most well known among these is Clarke's transformation using α-β variables [5]. This uses a power invariant transformation given by

$$\begin{bmatrix} f_a \\ f_b \\ f_c \end{bmatrix} = [C_C] \begin{bmatrix} f_\alpha \\ f_\beta \\ f_o \end{bmatrix} \tag{2.29}$$

where

$$[C_C] = \begin{bmatrix} \sqrt{\frac{2}{3}} & 0 & \frac{1}{\sqrt{3}} \\ -\frac{1}{\sqrt{6}} & -\frac{1}{\sqrt{2}} & \frac{1}{\sqrt{3}} \\ -\frac{1}{\sqrt{6}} & \frac{1}{\sqrt{2}} & \frac{1}{\sqrt{3}} \end{bmatrix}$$

Equation (2.6) is transformed to the three decoupled equations given below:

$$L_1 \frac{di_\alpha}{dt} + R_1 i_\alpha = v_{1\alpha} - v_{2\alpha} \tag{2.30}$$

$$L_1 \frac{di_\beta}{dt} + R_1 i_\beta = v_{1\beta} - v_{2\beta} \tag{2.31}$$

$$L_o \frac{di_o}{dt} + R_o i_o = v_{1o} - v_{2o} \tag{2.32}$$

The advantage of using Clarke's $\alpha\beta o$ components is that a three-phase network is transformed to three decoupled networks: α, β, and zero sequence. Out of these, the α and β networks are identical and the zero-sequence network can be generally neglected (in the absence of zero-sequence currents). Thus, the circuit shown in Figure 2.3 is transformed to the two decoupled circuits shown in Figure 2.7. Although the two circuits are identical, it should be noted that the currents and voltages are different in the α and β circuits.

The transformation from $\alpha\beta o$ to DQo components is given by

$$\begin{bmatrix} f_\alpha \\ f_\beta \\ f_o \end{bmatrix} = [T_2] \begin{bmatrix} f_D \\ f_Q \\ f_o \end{bmatrix} \tag{2.33}$$

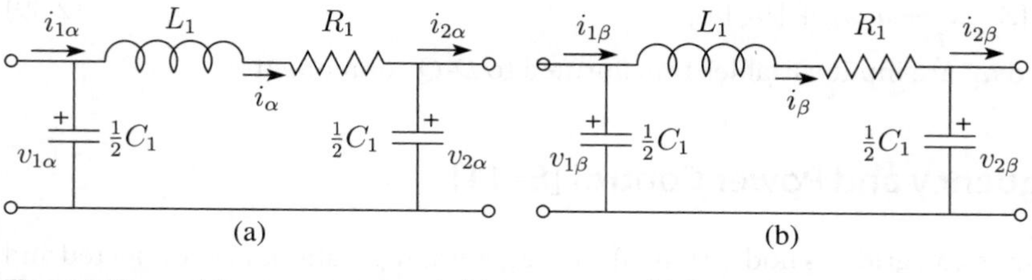

Figure 2.7 α and β sequence network. (a) α sequence; (b) β sequence

where

$$[T_2] = \begin{bmatrix} \cos\theta_o & \sin\theta_o & 0 \\ -\sin\theta_o & \cos\theta_o & 0 \\ 0 & 0 & 1 \end{bmatrix}$$

It should be noted that $[T_2]$ is similar to $[T_1]$ in that δ is replaced by θ_o. It is not difficult to see that the transformation from $\alpha\beta o$ to dqo components is given by

$$\begin{bmatrix} f_\alpha \\ f_\beta \\ f_o \end{bmatrix} = [T_3] \begin{bmatrix} f_d \\ f_q \\ f_o \end{bmatrix} \tag{2.34}$$

where

$$[T_3] = \begin{bmatrix} \cos\theta & \sin\theta & 0 \\ -\sin\theta & \cos\theta & 0 \\ 0 & 0 & 1 \end{bmatrix}$$

The advantage of using α-β variables for a stationary network is that the state (differential) equations for the network can be obtained on a single-phase basis. For the single-phase α network, the general equations are

$$\dot{x}_\alpha = [A_N]x_\alpha + [B_N]u_\alpha \tag{2.35}$$

where x_α are the state variables, which consist of inductor currents and capacitor voltages (note that only those inductors which form part of cotree (links) and capacitors which form part of tree are considered). The equations for the β network can be expressed as

$$\dot{x}_\beta = [A_N]x_\beta + [B_N]u_\beta \tag{2.36}$$

The structure of (2.36) follows from the fact that the β network is identical to the α network. u_α and u_β are input variables (in α-β components) which may include voltage and current sources in the network. It is convenient to apply D-Q transformation to (2.35) and (2.36). Expressing x_α and x_β in terms of x_D and x_Q as

$$\begin{aligned} x_\alpha &= \cos\theta_o x_D + \sin\theta_o x_Q \\ x_\beta &= -\sin\theta_o x_D + \cos\theta_o x_D \end{aligned} \tag{2.37}$$

the transformed network equations are

$$\dot{x}_D = [A_N]x_D - \omega_o x_Q + [B_N]u_D \tag{2.38}$$

$$\dot{x}_Q = [A_N]x_Q + \omega_o x_D + [B_N]u_Q \tag{2.39}$$

where u_D and u_Q are input variables transformed to D-Q components.

2.5 Frequency and Power Control [8–11]

An AC transmission grid has nodes (buses) where generating stations are connected and other nodes where loads are connected. The load buses could be away from generating stations by distances of the order of hundreds of kilometres. As power (and reactive power) at load buses vary with time, the generation should also vary to meet the load requirements. This is achieved by automatic generation control (AGC), whose objectives are [8]:

1) to regulate the system frequency at nominal value (50 or 60 Hz)
2) to maintain the schedules of interchange of power between control areas in an interconnected power system
3) to ensure economic dispatch that will minimise production costs.

It should be noted that in steady state, the bus voltages and currents in an AC grid are all sinusoidal. A sinusoidal quantity is characterized by three parameters: magnitude, phase, and frequency. As the load varies, these parameters also vary. However, it can be shown that the frequencies at all connected buses are identical. This follows from the fact that in a line (say connected between buses 1 and 2), the power flow (neglecting the line losses) is given by

$$P_{12} = \frac{V_1 V_2}{X_{12}} \sin(\delta_1 - \delta_2) \tag{2.40}$$

The power is assumed to flow from bus 1 to bus 2. The subscripts 1 and 2 refer to the values at the respective buses and X_{12} is the reactance of the line. If P_{12} remains constant in steady state, this implies $(\delta_1 - \delta_2)$ is also constant and hence

$$\frac{d\delta_1}{dt} = \frac{d\delta_2}{dt} = 2\pi\Delta f \tag{2.41}$$

when Δf is the variation from the nominal or operating frequency. It is not difficult to observe that the frequencies at all connected buses are identical and equal to $f_o + \Delta f$, where f_o is the (nominal) operating frequency.

Consider a network having the same topology as the transmission network, but with voltages replaced by frequencies at the buses and the currents replaced by power flows in individual branches (elements). If the bus frequencies are all identical, then this analogous network can be reduced to a single bus, as shown in Figure 2.8. Here, only the variations in power and frequency (which are expressed per unit) are shown.

In Figure 2.8, $\Delta P_{a\Sigma}$ is the sum of the accelerating powers acting on the synchronous generator rotors. $\Delta P_{m\Sigma}$ and $\Delta P_{L\Sigma}$ are the sums of variations in the mechanical and load powers, respectively. Figure 2.8 shows the effect of total accelerating power on the change in the frequency ($\Delta\overline{\omega}$). H_T and D_T are the sums of inertia constants and per-unit damping of individual generators (expressed on a common base) in the

Figure 2.8 A single-bus equivalent of power system.

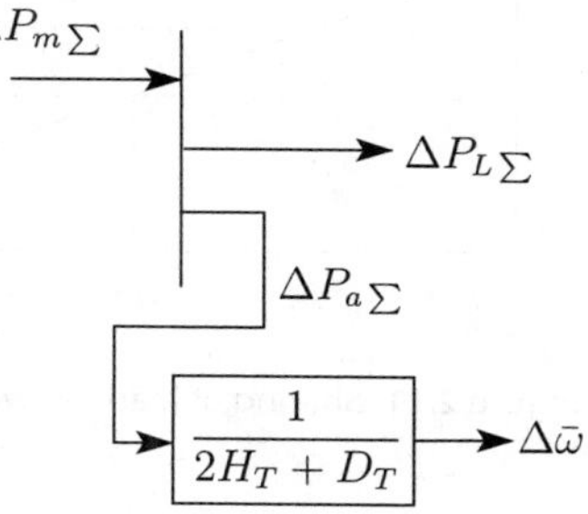

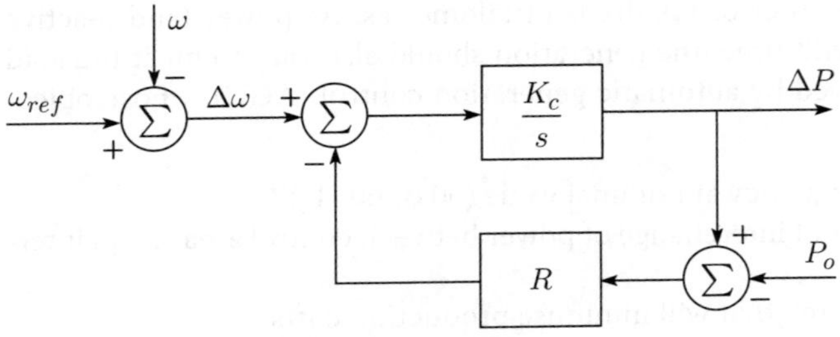

Figure 2.9 Block diagram of governor.

system. It should be noted that the load powers are also functions of the frequency. The mechanical power output of a prime mover (steam, gas or hydro turbine) is affected by the governor, as shown in Figure 2.9.

The pure integral control is not feasible for stable operation of more than one unit connected in parallel. The frequency as a function of the power output of a generator has a drooping characteristic, as shown in Figure 2.10. The value of R in Figure 2.9 is the negative slope of the governor characteristic shown in Figure 2.10. If two generators are connected in parallel, there will always be a unique frequency at which they share the load change depending on their speed droop characteristics (see Figure 2.11). In Figure 2.11, the two units are initially operating at the nominal frequency (f_o). When there is a load increase of ΔP_L, both units slow down, resulting in increased outputs and reach a final steady-state frequency (f_1), such that $\Delta P_1 + \Delta P_2 = \Delta P_L$.

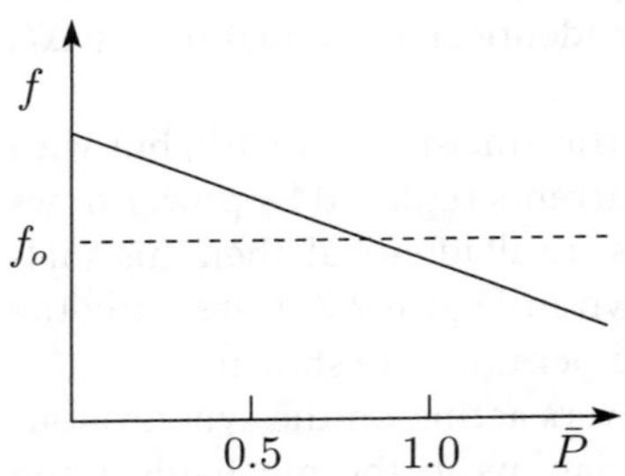

Figure 2.10 Speed droop characteristic.

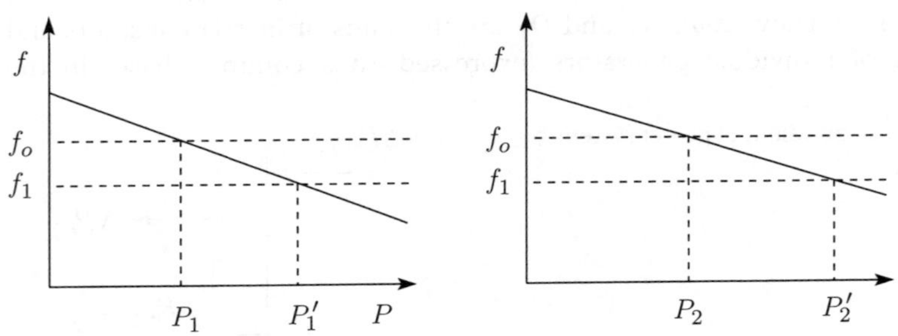

Figure 2.11 Sharing of load by two generating units.

2.5.1 Tie-Line Bias Frequency Control [10, 11]

The AGC is applied to a part of the overall system, designated the control area. The selection of the control area may be based on the ownership or other considerations when regional transmission operator (RTO) or independent system operator (ISO) function in a restructured power system. A control area is normally connected to the adjacent control areas by interconnecting (tie) lines. A control area may be importing or exporting power to the other control areas as firm (on contractual basis) power or wheeling power from one area to another (again based on prior agreements). In both cases, the net interchange of power (the sum of powers exported over the tie lines) must be kept constant under normal conditions (Note: this assumes that the power losses in the system are approximately constant).

Analysis of a Two-Area System

Consider a two (control) area system with a single tie line, as shown in Figure 2.12. The loads in each area have a constant component and a frequency dependent component. Thus,

$$\begin{aligned} \Delta P_{L1} &= \Delta P_{L10} + D_1 \Delta\overline{\omega}_1 \\ \Delta P_{L2} &= \Delta P_{L20} + D_2 \Delta\overline{\omega}_2 \end{aligned} \tag{2.42}$$

In steady state, $\Delta\overline{\omega}_1 = \Delta\overline{\omega}_2 = \Delta\overline{\omega}$. Hence,

$$\begin{aligned} \Delta P_{m1} - \Delta P_{L1} &= \Delta P_t \\ \Delta P_{m2} - \Delta P_{L2} &= -\Delta P_t \end{aligned} \tag{2.43}$$

From governor characteristics,

$$\Delta P_{m1} = -\frac{\Delta\overline{\omega}}{R_1}, \quad \Delta P_{m2} = -\frac{\Delta\overline{\omega}}{R_2} \tag{2.44}$$

Substituting equations (2.42) and (2.44) in (2.43) we get

$$-\Delta P_t - \Delta P_{L10} = \left(\frac{1}{R_1} + D_1\right)\Delta\overline{\omega} \tag{2.45}$$

$$\Delta P_t - \Delta P_{L20} = \left(\frac{1}{R_2} + D_2\right)\Delta\overline{\omega} \tag{2.46}$$

If $\Delta P_{L20} = 0$, we get

$$\Delta\overline{\omega} = \frac{-\Delta P_{L10}}{\left(\frac{1}{R_1} + \frac{1}{R_2} + D_1 + D_2\right)} \tag{2.47}$$

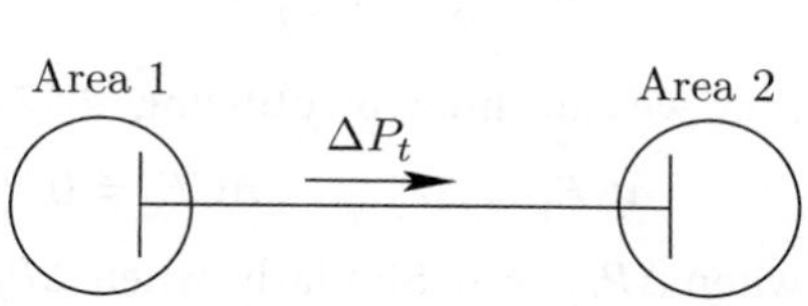

Figure 2.12 A two-area system.

Table 2.1 Required control action in two-area system

$\Delta\bar{\omega}$	ΔP_t	ΔP_{L1}	ΔP_{L2}	Control action
−	−	+	0	Increase generation in area 1
+	+	−	0	Decrease generation in area 1
−	+	0	+	Increase generation in area 2
+	−	0	−	Decrease generation in area 2

and

$$\Delta P_t = \frac{-\Delta P_{L10}\left(\frac{1}{R_2} + D_2\right)}{\left(\frac{1}{R_1} + \frac{1}{R_2} + D_1 + D_2\right)} \tag{2.48}$$

The tie-line bias frequency control has the following objectives:

1) to hold the system frequency at the specified nominal value
2) to maintain the correct value of interchange power between control areas.

To implement the control, we note the following facts:

1) If frequency has decreased and net interchange of power (leaving the system) has increased, a load increase has occurred outside the system.
2) If frequency has decreased and the net interchange power (leaving the system) has decreased, a load increase has occurred in the system.
3) If frequency has increased and the net interchange has increased, a load decrease has occurred in the system.
4) If frequency has increased and the net interchange has decreased, a load decrease has occurred outside the system.

The above four facts are summarized in Table 2.1 along with the required control actions for a two-area system.

The required change in the generation is called the area control error (ACE). For the two-area system, the equations for ACE are

$$\begin{aligned} ACE_1 &= -\Delta P_t - B_1\Delta\bar{\omega} \\ ACE_2 &= \Delta P_t - B_2\Delta\bar{\omega} \end{aligned} \tag{2.49}$$

where B_1 and B_2 are called frequency bias factors. If we select B_1 and B_2 as

$$\begin{aligned} B_1 &= \left(\frac{1}{R_1} + D_1\right) \\ B_2 &= \left(\frac{1}{R_2} + D_2\right) \end{aligned} \tag{2.50}$$

then we can show (by utilizing (2.47) and (2.48)) that

$$ACE_1 = \Delta P_{L10}, \quad ACE_2 = 0 \tag{2.51}$$

when $\Delta P_{L20} = 0$. Similarly, when $\Delta P_{L10} = 0$, we can show that

$$ACE_2 = \Delta P_{L20}, \quad ACE_1 = 0 \tag{2.52}$$

2.6 Dynamic Characteristics of AC Grids

The electric grid has generators connected at specific nodes (buses) and also loads connected at different buses. For stability analysis, the loads are modeled by algebraic equations representing the relationships of load active and reactive powers as functions of bus voltage magnitudes and frequency. The synchronous generator dynamics is represented by the rotor swing equations and the effects of excitation and prime-mover controls. Typically, the dynamics involve frequencies not exceeding 5 Hz. In such cases, the grid can be assumed to be in quasi-steady state and is represented by bus admittance matrix. Although switching transients in transmission lines have frequencies of the order of several kilohertz, they can be ignored in the stability analysis.

However, when the grid contains HVDC converters or FACTS controllers which have fast response, the grid dynamics represented by three-phase circuit models are necessary. If zero-sequence components can be neglected, either α-β or d-q components can be applied. Since d and q components represent a phasor, we can state that *dynamic* phasor representation is required (involving differential equations in d-q components).

In the analysis of subsynchronous resonance, dynamic phasor representation is required to analyse the negative damping torque due to subsynchronous frequency currents.

2.6.1 Grid Response to Frequency Modulation [12]

Consider a synchronous generator connected to a large system through a long transmission line which may be compensated by series compensation. Neglecting transient or subtransient saliency, it is possible to represent the generator stator by a voltage source behind a reactance. The voltage at the internal bus of the generator has constant frequency in steady state (assuming the rotor speed is constant). However, if the rotor has subsynchronous frequency oscillations superimposed on the constant speed of rotation, the internal bus voltage experiences frequency modulation. Here we consider a single-machine system connected to an infinite bus through a series (capacitor) compensated AC line. It can be shown that the subsynchronous frequency oscillations of the generator can be sustained or even negatively damped when the frequency of rotor oscillation (f_m) is the complement of the electrical resonant frequency (f_{er}), that is

$$f_m \approx f_o - f_{er}$$

Assuming that the generator rotor oscillates (about a constant speed of ω_o) sinusoidally, the per-unit speed ($\overline{\omega}$) is given by

$$\overline{\omega} = \overline{\omega}_o + A\sin(\omega_m t) \tag{2.53}$$

where ω_m is the oscillation frequency of the rotor about a synchronously rotating axis in rad/s.

The single-phase equivalent circuit of the generator connected to an infinite bus through a series compensated line is shown in Figure 2.13. This consists of a voltage source (e_g) behind a transient inductance. $e_{g\alpha}$ and $e_{g\beta}$ are defined by

$$\begin{aligned} e_{g\alpha} &= \overline{\omega}E'\sin(\omega_o t + \delta) \\ e_{g\beta} &= \overline{\omega}E'\cos(\omega_o t + \delta) \end{aligned} \tag{2.54}$$

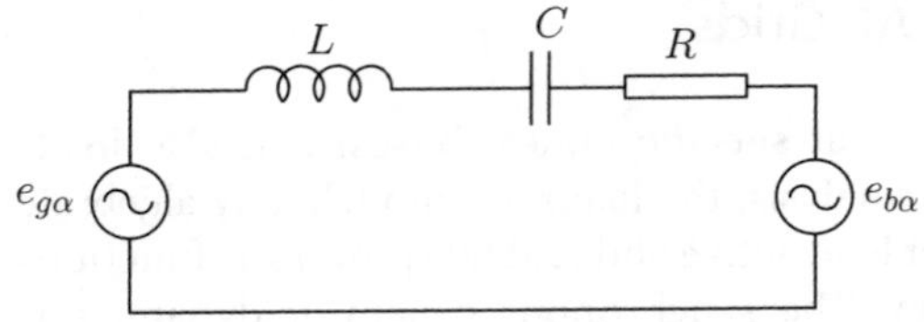

Figure 2.13 α-sequence stator equivalent circuit.

where E' is the voltage behind the transient inductance (L'). The infinite bus voltage (e_b) is defined by

$$\begin{aligned} e_{b\alpha} &= E_b \sin(\omega_o t) \\ e_{b\beta} &= E_b \cos(\omega_o t) \end{aligned} \tag{2.55}$$

L is the total inductance in the circuit shown, including the transient inductance (L') of the generator, the (leakage) inductance of the transformer, and transmission line inductance. R is the total resistance and C is the series capacitance. The β sequence equivalent circuit is similar to the α sequence except that the voltage sources are replaced by $e_{g\beta}$ (for the generator) and $e_{b\beta}$. Since the α and β sequence networks are linear, we can apply superposition theorem. It can be shown that $e_{g\alpha}$ and $e_{g\beta}$ contain subsynchronous frequency ($\omega_o - \omega_m$) and supersynchronous frequency ($\omega_o + \omega_m$) components which give rise to armature currents having similar frequency components. However, the electrical torques have the same frequency (ω_m) of the oscillation. It can be shown (see Appendix 2.B) that the electrical torque due to subsynchronous frequency currents is given by

$$T_e^{sub} = E' i_q^{sub} = -\frac{A(E')^2}{2\omega_m Z_{sub}}(\omega_o - \omega_m)\sin(\omega_m t + \phi_{sub}) \tag{2.56}$$

and the electrical torque due to supersynchronous frequency currents is given by

$$T_e^{sup} = E' i_q^{sup} = \frac{A(E')^2}{2\omega_m Z_{sup}}(\omega_o + \omega_m)\sin(\omega_m t - \phi_{sup}) \tag{2.57}$$

It should be noted that both components of the torque (T_e^{sub} and T_e^{sup}) have the same frequency of oscillation as that of the rotor. The damping torque coefficient (T_D) is the net component of the torque in phase with the oscillation, divided by A. It can be shown that

$$T_D = -\frac{(E')^2}{2\omega_m}\left[\frac{(\omega_o - \omega_m)\cos\phi_{sub}}{Z_{sub}} - \frac{(\omega_o + \omega_m)\cos\phi_{sup}}{Z_{sup}}\right] \tag{2.58}$$

$Z_{sub}\angle\phi_{sub}$ and $Z_{sup}\angle\phi_{sup}$ are the impedances of the network (viewed from the generator internal bus) at subsynchronous and supersynchronous frequencies, respectively.

The net damping tends to be negative due to the fact that

$$\begin{aligned} &Z_{sub} \ll Z_{sup} \quad \text{and} \\ &\phi_{sub} \approx 0 \quad \text{while} \quad \phi_{sup} \approx 90^\circ \end{aligned}$$

in a series compensated system when a torsional oscillation frequency is approximately equal to the complement of the electrical resonance frequency, that is

$$f_o - f_m \approx f_{er}$$

Figure 2.14 shows the phasor diagram giving the position of the torque components in relation to the rotor velocity. What is interesting is that the supersynchronous frequency

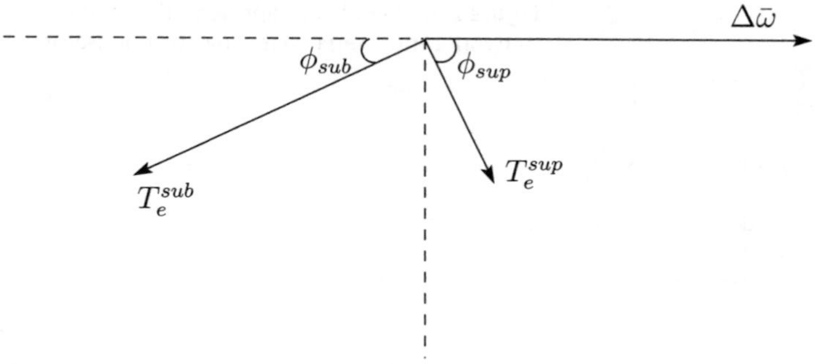

Figure 2.14 Phasor diagram showing torque components.

currents in the network give rise to positive damping torque (although of small amplitude). It is the subsynchronous frequency component of network currents that cause negative damping. The smaller the oscillation frequency, higher the negative damping. Thus, the first torsional mode (with the smallest frequency) can cause the most severe problem if network impedance is a minimum at the frequency corresponding to the complement ($\omega_o - \omega_m$) of the torsional mode frequency ω_m.

2.6.2 Grid Response to Injection of Reactive Current [13, 14]

Fast regulation of voltage at suitable locations in the grid can enhance the power transfer capability in the AC network. The shunt FACTS controllers, Static VAr Compensator (SVC) and static compensator (STATCOM), can achieve this. SVC is based on variable reactive impedance (thyristor-controlled reactor (TCR) in shunt with a fixed capacitor (FC) or thyristor switched capacitor (TSC)). The STATCOM is based on a voltage source converter (VSC). In both cases, the controller enables the injection of reactive current at the bus where the SVC or STATCOM is connected.

The objective of shunt FACTS controller is to regulate the bus voltage. It has been observed in the James Bay system in Quebec province of Canada, involving shunt compensation of the 735 kV line, that a 90 Hz frequency component resulting from parallel resonance between the line and shunt capacitor can result in adverse interaction with the voltage regulator associated with the SVC. To overcome this problem, a notch filter was provided in the voltage measuring circuit [15]. This phenomenon due to the network characteristics can be explained as follows.

Consider a simplified system as seen from the terminals of the static shunt compensator (shown in Figure 2.15).

The positive sequence driving point impedance $Z_{Th}(s)$ in the Laplace domain at the SVC bus is given by

$$Z_{Th}(s) = \frac{(R + sL)}{(s^2LC + sRC + 1)} \tag{2.59}$$

The equations in D-Q (synchronously rotating reference frame) variables are given by

$$\begin{bmatrix} V_{SD}(s) \\ V_{SQ}(s) \end{bmatrix} = \begin{bmatrix} Z_{DD}(s) & Z_{DQ}(s) \\ Z_{QD}(s) & Z_{QQ}(s) \end{bmatrix} \begin{bmatrix} I_{RD}(s) \\ I_{RQ}(s) \end{bmatrix} \tag{2.60}$$

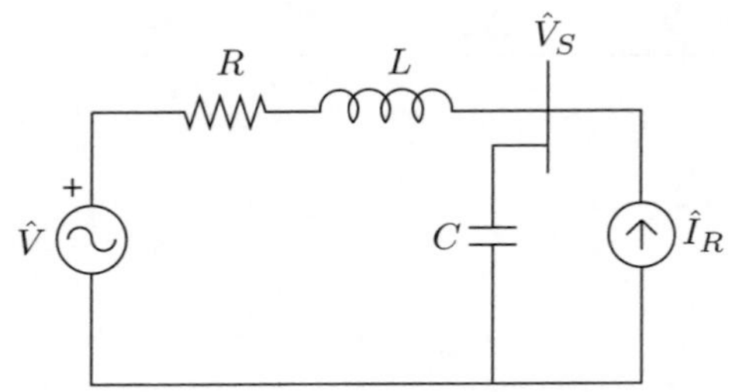

Figure 2.15 Simplified representation of the network connected to a reactive compensator.

where

$$Z_{DD}(s) = Z_{QQ}(s) = \frac{Z_{Th}(s + j\omega_o) + Z_{Th}(s - j\omega_o)}{2} \tag{2.61}$$

$$Z_{QD}(s) = -Z_{DQ}(s) = j\frac{Z_{Th}(s + j\omega_o) - Z_{Th}(s - j\omega_o)}{2} = -Im[Z_{Th}(s + j\omega_o)] \tag{2.62}$$

The D-Q components of the SVC bus voltage (V_S) and current (I_R) are shown in Figure 2.16. From Figure 2.16, we have

$$V_{SD} = V_S \sin\theta, \quad V_{SQ} = V_S \cos\theta \tag{2.63}$$

From (2.63), we get

$$\Delta V_S = \sin\theta_o \Delta V_{SD} + \cos\theta_o \Delta V_{SQ} \tag{2.64}$$

If we assume that the current injected by the SVC is purely reactive, we have

$$I_{RQ} = -I_R \sin\theta, \quad I_{RD} = I_R \cos\theta \tag{2.65}$$

From (2.65), we get

$$\begin{bmatrix} \Delta I_{RD}(s) \\ \Delta I_{RQ}(s) \end{bmatrix} = \begin{bmatrix} \cos\theta_o \\ -\sin\theta_o \end{bmatrix} \Delta I_R(s) + I_{Ro} \begin{bmatrix} -\sin\theta_o \\ -\cos\theta_o \end{bmatrix} \Delta\theta(s)$$

$$= \begin{bmatrix} \cos\theta_o \\ -\sin\theta_o \end{bmatrix} \Delta I_R(s) \tag{2.66}$$

We have assumed $I_{Ro} = 0$. Linearizing (2.60), and substituting from (2.64) and (2.66), we get

$$\frac{\Delta V_S(s)}{\Delta I_R(s)} = Z_{QD}(s) = \frac{KN(s)}{D(s)} \tag{2.67}$$

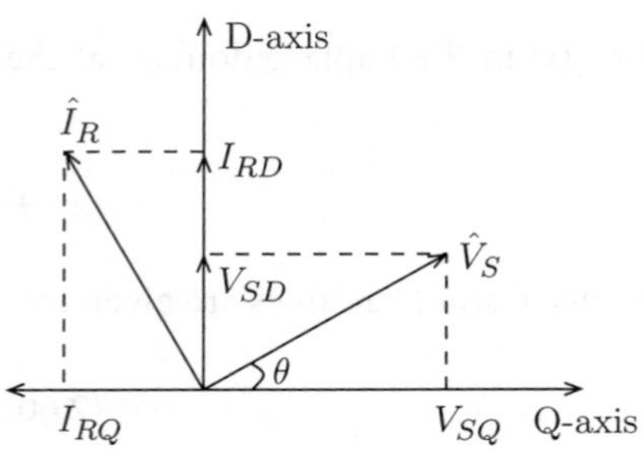

Figure 2.16 Phasor diagram showing voltage and current components.

where $N(s)$ and $D(s)$ are the numerator and the denominator polynomials, and K is a constant. These are defined below:

$$K = (\omega_o L)\omega_n^2$$
$$D(s) = [s + \sigma + j(\omega - \omega_o)][s + \sigma - j(\omega - \omega_o)] \times [s + \sigma + j(\omega + \omega_o)][s + \sigma - j(\omega + \omega_o)]$$
$$N(s) = s^2 + 2\frac{R}{L}s + \frac{R^2}{L^2} + \omega_o^2 - \frac{1}{LC}$$
$$\sigma = \frac{R}{2L}, \quad \omega = \sqrt{\omega_n^2 - \sigma^2}, \quad \omega_n = \frac{1}{\sqrt{LC}}$$

When $R \approx 0$, then

$$N(s) \approx s^2 + \omega_o^2 - \frac{1}{LC} = s^2 + \omega_0^2 - \omega_n^2 \tag{2.68}$$

Since $\omega_n > \omega_o$, the numerator polynomial $N(s)$ has a pair of real zeros given by

$$s = \pm\sqrt{\omega_n^2 - \omega_o^2} \tag{2.69}$$

Thus, the transfer function $\frac{\Delta V_S(s)}{\Delta I_R(s)}$ is a non-minimum phase type with a zero in the right half of the (complex) plane (RHP) (on the positive real axis). The control of plants with non-minimum phase type is always problematic and requires compensation to enable the use of P-I control with high gains.

Remarks

1) The four poles of the transfer function $\frac{\Delta V_S(s)}{\Delta I_R(s)}$ all lie in the left half of the (complex) plane (LHP) and are given by

$$p_1, p_2 = -\sigma \pm j(\omega - \omega_o), \quad p_3, p_4 = -\sigma \pm j(\omega + \omega_o)$$

where $-\sigma \pm j\omega$ are the poles of the impedance function $Z_{Th}(s)$.

2) For $R \approx 0$, it can be shown that the transfer function reduces to

$$\frac{\Delta V_S(s)}{\Delta I_R(s)} \approx \frac{(\omega_o L)\omega_n^2(s^2 - \omega_n^2 + \omega_o^2)}{[s^2 + (\omega_n - \omega_o)^2][s^2 + (\omega_n + \omega_o)^2]} \tag{2.70}$$

Substituting $s = 0$, we get

$$\frac{\Delta V_S(s)}{\Delta I_R(s)} = -X_{Th} \tag{2.71}$$

where

$$X_{Th} = \frac{\omega_o L}{1 - \omega_o^2 LC} = \frac{(\omega_o L)}{\omega_n^2 - \omega_o^2}\omega_n^2$$

is the reactance of the parallel combination of L and C, evaluated at the operating frequency ω_o.

The adverse control interactions with network resonance are functions of the effective short-circuit ratio (ESCR) and the operating point. The lower value of ESCR and the SVC operation at capacitive region worsen the control stability problem. The controller gain needs to be reduced to overcome the problem. This may not be practical as the response of the SVC needs to be fast at lower values of ESCR. One of the solutions suggested in [13] is to provide a compensator in cascade with the voltage regulator, which rejects the complement of the network resonance frequency component. The transfer function of the compensator is assumed to be

$$G_{comp}(s) = \frac{(s+\sigma_1)^2 + \omega_r^2}{(s+\sigma_2)^2 + \omega_r^2} \tag{2.72}$$

where $\sigma_1 < \sigma_2$. This function represents a notch filter that rejects the component of frequency of ω_r. By arranging $\omega_r = \omega_n - \omega_o$, the interaction between the network resonance (of frequency ω_n) and the controller can be avoided. It should be noted that a component of frequency of ω_n in the AC voltage is transformed to a frequency of ω_r after demodulation in the voltage measurement circuit.

Example 2.1 A voltage regulator with the transfer function $G_R(s) = -\frac{K_i}{s}$ is used to control the reactive current injected into the static compensator bus. The plant (system) transfer function is given by (2.67) (note the gain is assumed to be negative as the reactive current has the opposite sign to that of the SVC susceptance).

(a) Plot the root loci as K_i varies from zero to a high value.
(b) Repeat (a) if a compensator with a transfer function given in (2.72) is connected in cascade with the voltage regulator.

Data: $\omega_o = 314.16$ rad/s, $\omega_o L = 0.45$, $R = 0.05$, $\omega_n = 1.5\omega_o$, $\sigma_1 = 30$, $\sigma_2 = 300$.

Solution
The root loci for case (a) is shown in Figure 2.17a. The root loci for case (b) is shown in Figure 2.17b.

It is interesting to observe that without the compensator, the system becomes unstable when $K_i > 36$. The provision of the compensator increases the value of instability gain to 1250. In this case, two complex roots cross the imaginary axis at $\omega = 292$ rad/s.

The provision of the compensator in the voltage control system is an alternative to providing a notch filter (tuned to ω_n) on the AC side before the voltage input signal is demodulated using a rectifier.

2.7 Control of Power Flow in AC Grids [14, 16]

2.7.1 Power Transfer Capability of a Line

The positive sequence equivalent circuit of an AC transmission line can be represented by the two-port network shown in Figure 2.2. The sending end voltage ($\hat{V}_S$) and current ($\hat{I}_S$) are expressed in terms of the receiving end voltage and current, as

$$\hat{V}_S = A\hat{V}_R + B\hat{I}_R \tag{2.73}$$

$$\hat{I}_S = C\hat{V}_R + D\hat{I}_R \tag{2.74}$$

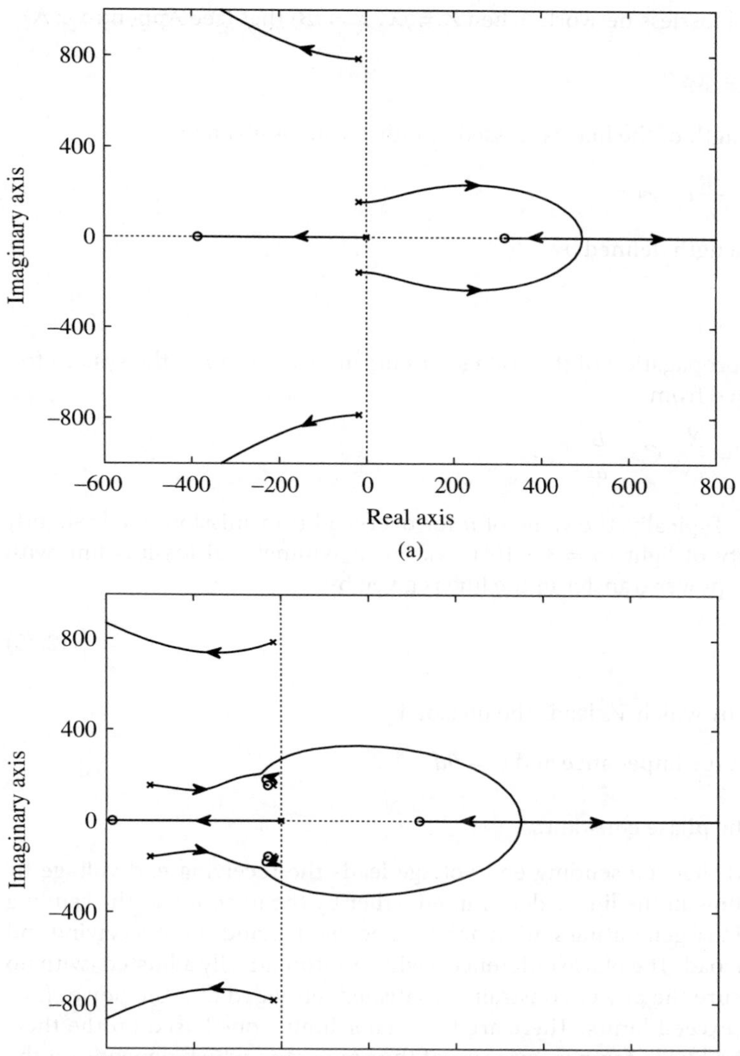

Figure 2.17 Root loci of a static compensator with a voltage regulator. (a) Case (a); (b) Case (b)

The two-port network can also be represented by the π equivalent shown in Figure 2.6. The ABCD constants satisfy the following relations:

$$A = D = 1 + \frac{Y'Z'}{2}, \quad B = Z'$$

$$AD - BC = 1$$

It can be shown for a lossless network (when $Z = jX, Y = jB$) that (see Appendix 2.A)

$$A = 1 + \frac{YZ}{2} = \cos\theta$$

θ is the electrical length of the line expressed in radians and is given by

$$\theta = \omega\sqrt{LC} = \frac{2\pi}{\lambda}d$$

where λ is the wavelength defined by

$$\lambda = \frac{u}{f}$$

u is the velocity of propagation of the voltage or current wave and f is the system frequency. u is calculated from

$$u = \frac{d}{\sqrt{LC}}, \; L = \frac{X}{\omega}, \; C = \frac{B}{\omega}$$

d is the line length. Typically, the value of u for overhead transmission line is slightly less than the velocity of light ($u = 3 \times 10^8$m/s). For a symmetrical lossless line with $|\hat{V}_S| = |\hat{V}_R| = V$, the power transfer in the line is given by

$$P = \frac{V^2 \sin\delta}{Z_n \sin\theta} \tag{2.75}$$

where δ is the angle by which $\hat{V}_S$ leads the phasor $\hat{V}_R$.

$Z_n = \sqrt{\frac{L}{C}}$ is the surge impedance and $\theta = \beta d$.

$\beta = \frac{2\pi}{\lambda}$ is called the phase constant.

It should be noted that the sending end voltage leads the receiving end voltage by angle δ. The power flow in the line is determined either by the injection at the sending end if it is connected to a generating station or by the power demand at the receiving end if it is connected to a load. The phase difference δ will be automatically adjusted (with no control effort) to ensure the power constraint is satisfied, provided that the power flow in the line does not exceed limits. There are two major limits, one based on the thermal limit (determined by the temperature rise of the conductor, which depends on the current magnitude and ambient temperature) and the other based on the stability limit (based on the maximum value of δ that can be tolerated without loss of steady-state stability). Theoretically $\delta_{\max} = 90°$ for a lossless line, but in practice the voltage regulation at the sending and receiving ends is not ideal due to the system impedances viewed from the two ends of the line. Thus, the power transfer capability of a transmission line as a function of the line length is shown in Figure 2.18.

It is interesting to note that for a lossless line (assumption valid for EHV and UHV transmission lines), the receiving end complex power, expressed in terms of the surge impedance loading (SIL) is independent of the line (electrical) parameters [16]. Thus,

$$\frac{P_R + jQ_R}{SIL} = \frac{j\left[\left(\frac{V_S}{V_R}\right)^* - \cos\beta d\right]}{\sin\beta d}|V_R|^2 \tag{2.76}$$

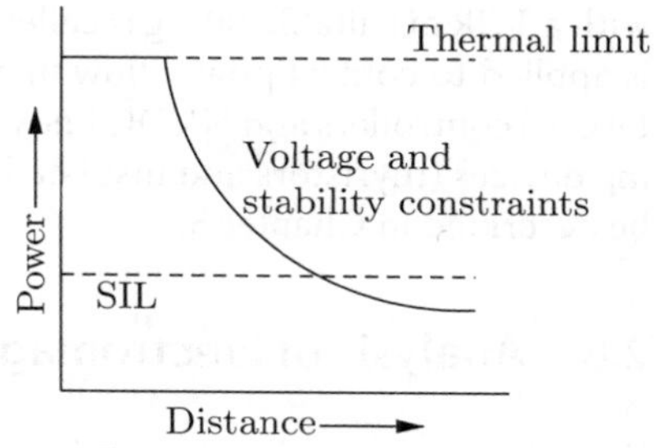

Figure 2.18 Power transfer capability of a line.

The above expression depends only on the line length and the terminal voltages. The $SIL = \frac{V^2}{Z_n}$ where V is the rated voltage. If $|\hat{V}_S| = |\hat{V}_R| = V$, then

$$\frac{P_R}{SIL} = \frac{\sin\delta}{\sin\theta}$$

2.7.2 Power Flow in a Line connected to an AC Transmission Grid

A transmission line in an AC network is a series element connected between two nodes of the network (note that we are considering a single-phase positive sequence network, although the extension to a three-phase network can be easily done if we wish to study unbalanced power flows in a distribution network). The power flow in a line is dependent on the power (and reactive power) injections in the nodes of the network and Kirchoff's laws. In a radial network, only Kirchoff's current law (KCL) applies, whereas in a mesh type of network both KCL and Kirchoff's voltage law (KVL) apply to govern the power flow. For reasons of reliability, the transmission network has meshes so that the path of power flow to a load bus is not interrupted by the failure of any line.

Tripping of a line due to a fault can result in overloading of some lines, which can lead to a catastrophic event of blackouts, which must be avoided. During system operation, contingency analysis is carried out to evaluate system security. If system dynamics can be ignored, this is termed static security analysis based on power flow analysis. To speed up contingency analysis, simplified power flow (such as DC load flow) is often used. If a contingency leads to overloading of line(s), it is relieved by modifying the power injections of generators and/or loads at appropriate buses. If there are some HVDC lines in the system, they inject power into the AC network and rescheduling of power flows in the HVDC lines can be used to relieve overloading of AC lines. This approach is faster and does not involve economic penalties when compared to rescheduling of generators.

To strengthen the transmission network, series (capacitor) compensation is used to increase power flow in AC lines. Shunt compensation (inductive or capacitive) is applied to regulate voltages at load buses. Fast regulation of bus voltages at appropriate locations in a long line can help to increase power transfer capability. For example, voltage regulation at the midpoint of a long line is ideally equivalent to the reduction of line length by a factor of two. The voltage regulation at n locations (which are equidistant) in a line is equivalent to reducing the line length by a factor of $\frac{1}{n+1}$ and thus improving the power transfer capacity (see Figure 2.18).

SVCs are made of TCRs in parallel with TSCs. SVCs are variable impedance devices that belong to the first generation of FACTS controllers. A series capacitor in parallel with a TCR of suitable rating is called a thyristor-controlled series capacitor (TCSC) and is applied to control power flow in a line by varying the effective series compensation. FACTS controllers and HVDC lines have the capability of fast control due to fast switching devices (thyristors and insulated-gate bipolar transistor (IGBT) devices). These will be described in Chapter 5.

2.8 Analysis of Electromagnetic Transients

Disturbances occur in a transmission network due to faults, switching operations, and lightning surges that can result in transient overvoltages and currents in the lines. The system has to be properly designed and also protected against these disturbances. This is achieved by proper coordination of the insulation and design of protection schemes. The simulation of electromagnetic transients becomes essential for such studies in addition to the investigation of interference in neighboring communication lines, and hazards to personnel livestock and equipment.

There are transients related to the electro-mechanical oscillations in the generators (induced by faults and switching). Their frequencies do not normally exceed 5 Hz and they are investigated in transient stability studies. On the other hand, electromagnetic transients cover a range from microseconds (study of transient recovery voltages), milliseconds (switching surges) to cycles (study of ferroresonace). Physical simulators using miniaturized circuit components were built for the study of the electromagnetic transients. For AC systems, these were called transient network analyzer (TNA) and DC-TNA (systems with HVDC links). With the introduction of fast digital computers, programs have been developed with the common acronym EMTP (Electromagnetic Transients Program). The earliest version is the BPA-EMTP developed by Dommel [17].

2.8.1 Modeling of Lumped Parameter Components

The transmission network can be represented by lumped parameter (R-L-C) elements. The storage elements, inductance (L) and capacitance (C), are modeled by first-order differential equations given by

$$\begin{aligned} L\frac{di_L(t)}{dt} &= v_L(t) \\ C\frac{dv_C(t)}{dt} &= i_C(t) \end{aligned} \tag{2.77}$$

By integration of the above equations we get

$$\begin{aligned} i_L(t) &= i_L(t-h) + \frac{1}{L}\int_{t-h}^{t} v_L(\tau)\, d\tau \\ v_C(t) &= v_C(t-h) + \frac{1}{C}\int_{t-h}^{t} i_C(\tau)\, d\tau \end{aligned} \tag{2.78}$$

By applying the trapezoidal rule of integration to the integrals in (2.78), we get

$$\begin{aligned} i_L(t) &= i_L(t-h) + \frac{h}{2L}[v_L(t) + v_L(t-h)] \\ v_C(t) &= v_C(t-h) + \frac{h}{2C}[i_C(t) + i_C(t-h)] \end{aligned} \tag{2.79}$$

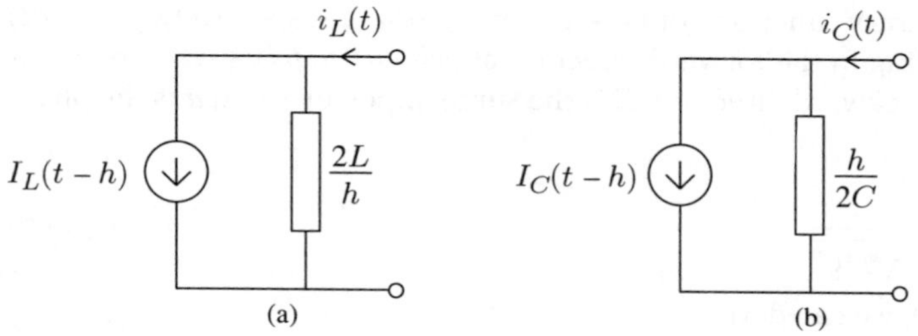

Figure 2.19 Equivalent circuit. (a) Inductor; (b) Capacitor

Equation (2.79) represents the equivalent circuits for the inductor and a capacitor shown in Figure 2.19.

Here, both the inductor and capacitor are represented by a resistor in parallel with a current source that depends on the past history.

The current sources $I_L(t-h)$ and $I_C(t-h)$ are defined by

$$\begin{aligned} I_L(t-h) &= i_L(t-h) + \frac{h}{2L} v_L(t-h) \\ I_C(t-h) &= i_C(t-h) + \frac{2C}{h} v_C(t-h) \end{aligned} \tag{2.80}$$

Thus, a linear *R-L-C* lumped parameter network can be modeled by a resistive network excited by current sources that depend on the past history of currents and voltages of energy storage elements *L* and *C*. It is possible to derive the nodal equations in the matrix form as

$$[G]\ \underline{v}(t) = \underline{i}(t) - \underline{I} \tag{2.81}$$

where $[G]$ is the conductance matrix, and $\underline{v}(t)$ and $\underline{i}(t)$ are column vectors of node voltages and currents injected, respectively. $\underline{I}$ is the equivalent source current vector which also depends on past history of currents and voltages in the inductors and capacitors in the original *R-L-C* network.

2.8.2 Modeling of a Single-Phase Line

Consider a lossless single-phase line with inductance L' and capacitance C' per-unit length. The line is assumed to be connected across nodes k and m with ground return. The voltage v and current i in the line are functions of the time and distance x (measured from one end, say node k). The line is a distributed parameter component and is described by the following partial differential equations

$$\frac{\partial v}{\partial x} = -L' \frac{\partial i}{\partial t} \tag{2.82}$$

$$\frac{\partial i}{\partial x} = -C' \frac{\partial v}{\partial t} \tag{2.83}$$

The general solution of the above equations is

$$i(x,t) = f_1(x - at) + f_2(x + at) \tag{2.84}$$

$$v(x,t) = Zf_1(x - at) - Zf_2(x + at) \tag{2.85}$$

where f_1 and f_2 are arbitrary functions of $(x - at)$ and $(x + at)$, respectively. $f_1(x - at)$ represents a wave traveling in the forward direction at velocity a. $f_2(x + at)$ represents a wave traveling in the backward direction. Z is the surge impedance and a is the phase velocity defined by

$$Z = \sqrt{\frac{L'}{C'}}, \quad a = \frac{1}{\sqrt{L'C'}} \tag{2.86}$$

From (2.84) and (2.85), we can derive

$$v(x,t) + Zi(x,t) = 2Zf_1(x - at) \tag{2.87}$$

$$v(x,t) - Zi(x,t) = -2Zf_2(x + at) \tag{2.88}$$

It is interesting to note that $(v + Zi)$ is constant when $(x - at)$ is constant and, similarly, $(v - Zi)$ constant when $(x + at)$ is constant. This implies that

$$\begin{aligned} v_m(t) - Zi_m(t) &= v_k(t-\tau) + Zi_k(t-\tau) \\ v_k(t) - Zi_k(t) &= v_m(t-\tau) + Zi_m(t-\tau) \end{aligned} \tag{2.89}$$

where $\tau = \frac{d}{a} = d\sqrt{L'C'}$ is the travel time for the wave from one end to the other end. From the above equations, we obtain

$$\begin{aligned} i_k(t) &= \frac{1}{Z}v_k(t) + I_k(t-\tau) \\ i_m(t) &= \frac{1}{Z}v_m(t) + I_m(t-\tau) \end{aligned} \tag{2.90}$$

where

$$\begin{aligned} I_k(t-\tau) &= -\left(\frac{1}{Z}\right) v_m(t-\tau) - i_m(t-\tau) \\ I_m(t-\tau) &= -\left(\frac{1}{Z}\right) v_k(t-\tau) - i_k(t-\tau) \end{aligned} \tag{2.91}$$

Equations (2.90) and (2.91) represent the equivalent impedance network which describes the lossless line at its terminals (see Figure 2.20).

2.8.3 Approximation of Series Resistance of Line

If the distributed resistance is to be taken into account, this can be done by considering the line as lossless and adding lumped resistances at both ends of the line. To improve the

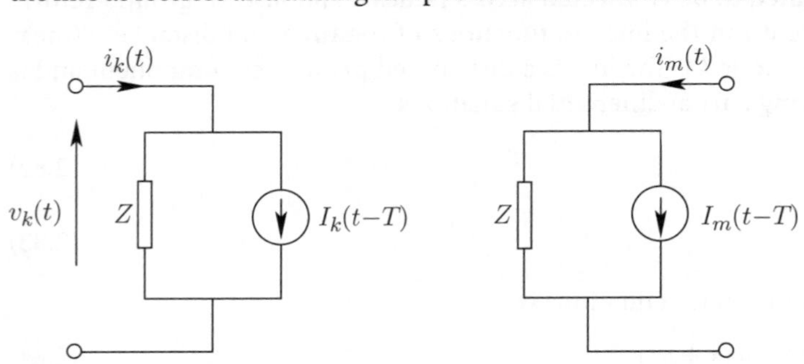

Figure 2.20 Equivalent impedance network for a lossless line.

approximation, the line can be divided into several sections and the resistances inserted at the end of each section.

Remarks

1. If we assume a distortionless line where $R'/L' = G'/C'$ (R' and G' are distributed series resistance and distributed shunt conductance, respectively), then the modified expressions for the current sources are obtained as

$$I'_k(t-\tau) = \exp\left(-\frac{R'}{L'}\tau\right) I_k(t-\tau)$$

$$I'_m(t-\tau) = \exp\left(-\frac{R'}{L'}\tau\right) I_m(t-\tau) \tag{2.92}$$

   where $I_k(t-\tau)$ and $I_m(t-\tau)$ are as defined in (2.91). However, power transmission lines are not distortionless, since $G' \approx 0$ in the absence of corona losses. Even if corona losses (which are weather dependent) are to be considered, they are nonlinear functions of the voltage.
2. Representation of distributed series resistance is relatively simpler compared to the problem of frequency dependence of R' and L' due to earth conduction.
3. Any branch with a nonlinear element can be handled by applying the compensation theorem where solution of the network is divided into two parts, one with a linear element and the other with a nonlinear element. When the network contains more than one nonlinear element, a similar approach can be used if the network can be torn into subnetworks, each containing only one nonlinear element. If the network contains lossless lines, this is feasible.

2.8.4 Modeling of Lossless Multiphase Line

Equations (2.82) and (2.83) are applicable if the scalars v, i are replaced by vectors, and scalars L' and C' are replaced by matrices. By differentiating (2.82) and (2.83), we can obtain

$$\frac{\partial^2 v(x,t)}{\partial x^2} = [L'][C']\frac{\partial^2 v(x,t)}{\partial t^2} \tag{2.93}$$

$$\frac{\partial^2 i(x,t)}{\partial x^2} = [C'][L']\frac{\partial^2 i(x,t)}{\partial t^2} \tag{2.94}$$

The complication introduced by mutual couplings between the phases can be overcome if the phase variables are transformed to modal variables by a similarity transformation and the resulting matrices become diagonal. For example, we have shown that for a symmetric three-phase network $\alpha\beta o$ transformation results in two uncoupled α and β sequence networks (which are identical) and a zero-sequence network. For a general case, when $L_{ab} \neq L_{ac} \neq L_{bc}$, we can still obtain modal transformation that diagonalizes the relevant matrices and decouples the equations. Each of the decoupled equations (in modal variables) can be solved using the algorithm for the single-phase line. The parameters to be used are the travel time and surge impedance for the specific mode. The transformation matrices relating phase variables to modal variables are defined below:

$$v_{ph} = [T_v]v_{mode}, \quad i_{ph} = [T_i]i_{mode} \tag{2.95}$$

The columns of the matrices $[T_v]$ and $[T_i]$ are the eigenvectors, which are not unique unless they are normalized. It can be shown that

$$[T_i] = [C'][T_v] \tag{2.96}$$

2.8.5 Modeling of Multiphase Networks with Lumped Parameters

If we consider a three-phase line connected across nodes k and m, represented by series resistance and inductance (with mutual couplings), then the equations are

$$[R]i_k + [L]\frac{di_k}{dt} = v_k - v_m \tag{2.97}$$

where i_k is the vector of currents in the line flowing from node k to m. By applying the trapezoidal rule of integration to (2.97), we obtain the following equations:

$$i_k(t) = [S]^{-1}(v_k(t) - v_m(t)) + I_k(t - \Delta t) \tag{2.98}$$

$$I_k(t - \Delta t) = [H]([v_k(t - \Delta t) - v_m(t - \Delta t)] + [S]I_k(t - 2\Delta t)) - I_k(t - 2\Delta t) \tag{2.99}$$

where

$$[S] = [R] + (2/\Delta t)[L], \quad [P] = [R] - (2/\Delta t)[L] \; [H] = ([S]^{-1} - [S]^{-1}[P][S]^{-1})$$

Discussion

In network theory, there are standard techniques for the formulation of state space equations based on the network topology and application of graph theory. This involves the identification of a maximal number of capacitors as tree branches and the inductors in the link or cotree branches. The state vector consists of voltages across the capacitors in the tree branches and currents in the inductors in the link or cotree branches.

In contrast, the approach used in EMTP does not require the knowledge of the network graph and its decomposition into tree and cotree branches. The *R-L-C* network is reduced to a resistive network with storage elements (*L* and *C*) represented as resistors in parallel with current sources which depend on the past history. This approach definitely simplifies the problem formulation and its solution. The simplicity and speed of solution (due to sparsity of the nodal conductance matrix) has resulted in its widespread application. However, problems can arise when switches are present and these turn on/off at regular intervals. For example, in a 12-pulse HVDC converter, each thyristor valve (represented as a switch) turns on/off in a cycle (50 or 60 times in a second) under normal conditions. Snubber circuits (of a resistor in series with a capacitor) are provided for voltage grading and protection during turn on and off of each thyristor device (several of them are connected in series to form a valve in HVDC converters). The transient analysis of an HVDC converter subjected to external faults does not require the modeling of snubber circuits (unless in the design stage) in the evaluation of HVDC system performance. However, it is common practice to represent a valve by a switch with a snubber circuit in parallel (see Figure 2.21). Here $R_v = 1\ \Omega$ during the on state and $R_v = R_d + \left(\frac{\Delta t}{2C}\right)$ during the off state.

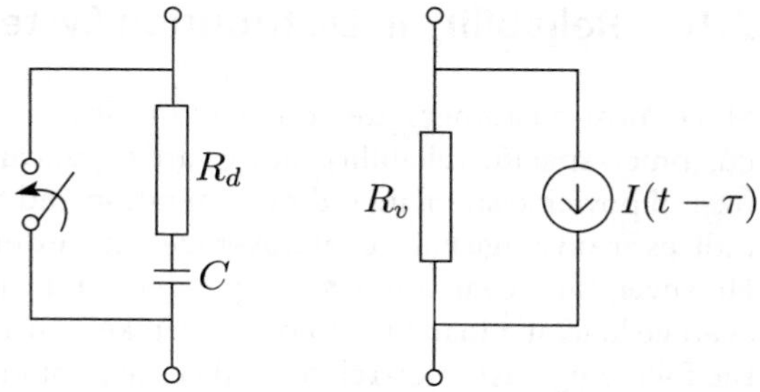

Figure 2.21 Representation of an HVDC valve.

2.9 Transmission Expansion Planning [18]

With load growth, there is a need for addition to the generation facilities and a plan for expanding transmission facilities. With vertically integrated utilities (VIU), the generation and transmission expansion is a combined exercise with the objective of minimizing the costs while maintaining the desired reliability indices. In the deregulated or restructured environment, the generation expansion is determined by market mechanism. However, the development of transmission facilities is aimed at ensuring market efficiency with benefits to customers. However, the introduction of renewable energy sources (RES) pose new challenges in transmission expansion. The environmental issues which lead to regulatory constraints also complicate and delay the addition of new lines.

The delivery of power to loads can be divided into two components: (i) bulk transmission at high or extra high voltages and (ii) distribution networks at medium or low voltages. The bulk power transmission network is a meshed network such that the disconnection of a line due to faults does not result in power blackouts. In contrast, distribution systems tend to be made of radial networks. The system reliability has two facets: (i) adequacy and (ii) security. The concept of adequacy implies the existence of sufficient facilities within the system to meet the consumer demand in steady-state operation under normal state. The security relates to the ability of the system to respond to various disturbances in the system such as loss of a line due to tripping caused by a fault. In a secure operating state, a contingency shall not result in the system transition to an emergency state and consequent power blackout leading to loss of load. Typically, the detection of an insecure condition can lead to preventive or emergency control to ensure a self-healing grid. A major control function in an energy control centre is to carry out contingency analysis that involves simulation of various contingencies that can result in loss of a transmission line and/or generator outage which can lead to an insecure operating state. A single major contingency during the system operation is an insecure state which can lead to an emergency that involves viability or stability crisis. A viability crisis involves the inability of the supply to serve the loads that can lead to frequency and/or voltage decay. A stability crisis is the result of one or more generators losing synchronism and being tripped by the protection system. Both crises can result in power blackouts. Static and dynamic security assessment during system operation is an essential control function to preserve the integrity of a power supply system.

2.10 Reliability in Distribution Systems [19]

Since most customers are connected to low or medium voltage distribution systems, customer-specific reliability indices are significant in evaluating the performance of the overall power system, including generation and transmission. Essentially, the reliability indices are average failure rate, average outage duration, and average annual outage time. However, their evaluation is independent of the number of customers connected or the average load at a load bus. In order to take into account the severity of a system outage, the following customer-related indices are commonly used.

(a) System average interruption index (SAIFI) This is defined as

$$\text{SAIFI} = \frac{\text{total number of customer interruptions}}{\text{total number of customers served}} = \frac{\sum \lambda_i N_i}{\sum N_i}$$

where λ_i is the failure rate and N_i is the number of customers served at load point i.

(b) Customer average interruption frequency index (CAIFI) This is defined as

$$\text{CAIFI} = \frac{\text{total number of customer interruptions}}{\text{total number of customers affected}} = \frac{\sum \lambda_i N_i}{\sum N_{if}}$$

The value of CAIFI is very useful in recognizing chronological (time-varying) trends in the reliability of a particular distribution system. Note that $N_{if} \leq N_i$.

(c) System average interruption duration index (SAIDI) This is defined as

$$\text{SAIDI} = \frac{\text{sum of customer interruption durations}}{\text{total number of customers}} = \frac{\sum U_i N_i}{\sum N_i}$$

where U_i is the annual outage time and N_i is the number of customers at load point i.

(d) Customer average interruption duration index (CAIDI) This is defined as

$$\text{CAIDI} = \frac{\text{sum of customer interruption durations}}{\text{total number of customer interruptions}} = \frac{\sum U_i N_i}{\sum \lambda_i N_i}$$

where λ_i is the failure rate, U_i is the annual outage time, and N_i is the number of customers at load point i.

(e) Average service availability index (ASAI) This is defined as

$$\text{ASAI} = \frac{\text{customer hours of available service}}{\text{customer hours demanded}} = \frac{\sum U_i N_i}{\sum N_i}$$

$$= \frac{\sum N_i \times 8760 - \sum U_i N_i}{\sum N_i \times 8760}$$

Note that 8760 is the number of hours in a calendar year. Also, CAIFI > SAIFI and CAIDI > SAIDI. Typically, in a well-designed distribution system (with adequate generation), ASAI is not more than one hour (of loss of service in a year).

2.11 Reliable Power Flows in a Transmission Network

The trend in the growth of a transmission network is due to the interconnection of individual power systems supplying a city or a state. The interconnection of AC systems (having common frequency) has several economic benefits that result from load diversity and sharing of generation reserves having different characteristics. For example, a nuclear plant can supply base load whereas hydro generation varies with the season. The wind energy may be abundant in one area whereas generation from fossil fuels is available in another area. Thus, the AC transmission networks are synchronized in most parts of the world. In North America there are Western and Eastern interconnected systems (apart from Texas), whereas in India five regional electricity supply systems have been synchronized to get an all-India grid. Even when two AC networks are not interconnected directly, they can be interlinked through HVDC links (typically back to back). In interconnected systems, it is necessary to determine the total transfer capability (TTC) and the available transfer capability (ATC) for power transfers from one area to another. The procedure to calculate the TTC is given below [20]:

1. Start with a base case power flow.
2. Increase generation in area A and increase demand in area B by the same amount.
3. Check the normal thermal, stabilty, and voltage constraints.
4. Evaluate the first contingency event and ensure that the emergency operating limits are met.
5. When the emergency limit is reached for a first contingency, the corresponding (pre-contingency) transfer of power from area A to B is the TTC.

It should be noted that it is necessary to define the critical single contingency before computing the TTC. The computation of the TTC is dependent on

(a) power system configurations in respective areas and the loading conditions
(b) critical contingencies
(c) parallel paths for power flow from one area to another
(d) non-simultaneous and simultaneous transfers (when several areas are interconnected)
(e) transmission line capacities.

In summary, it can be said that both TTC and ATC depend on system generation dispatch and load level, network topology, and the limits on the transmission network due to thermal, voltage, and stability constraints. ATC measures the residual transfer capability in the transmission network for the purpose of further power trading over the existing transmission commitments. The computation of ATC helps to allow trading of power among systems which can assist (i) delivery of reliable electrical power, (ii) provision of flexibility under changing system conditions, and (iii) reduction in the need for installed generating capacity. ATC is mathematically defined by the following equation:

$$\text{ATC} = \text{TTC} - \text{TRM} - \text{existing transmission commitments}$$

Here, TRM represents the transmission reliability margin (a certain percentage of the TTC). Existing transmission commitments include retail customer service and the capacity benefit margin (CBM). CBM is defined as the amount of TTC reserved by load-serving entities to ensure access to generation from interconnected systems to meet the generation reliability requirements. Its use is intended only for the duration of emergency generation deficiencies.

2.12 Reliability Analysis of Transmission Networks

Consider a system of n components. The status of a system can be described by the structure function ϕ of the system, defined as

$$\phi(x_1, x_2, \dots, x_i, \dots, x_n) = 0$$

if the system is not working and $\phi = 1$ if the system is working. x_i represents the status of component i. $x_i = 0$ if the component is not working and $x_i = 1$ if the component i is functioning. A system is called monotone if its structure function ϕ is increasing, that is,

$$\phi(x) \leq \phi(y) \text{ for all } x, y \text{ with } x \leq y$$

A component i is called irrelevant if the status of that component does not alter the value of the structure function. In other words, the value of ϕ is not dependent on the status (whether the component has failed or working) of the component.

A system is called coherent if it is monotone and none of its components are irrelevant [21].

Remarks

1. It should be noted that, typically, a system having n components can continue to work if some components have failed and, say, only k of its specified components are functioning. The repair of an additional component (say i) is not going to affect the functioning of the system.
2. The failure of any one of the (essential) specified components can result in failure of the system.

These are intuitive concepts and thus, in reliability theory, a coherent system represents a classical framework for describing the structure of the technical systems. However, it can be shown that a transmission network containing meshes is not a coherent system. This is illustrated using a simple three-bus system containing two generators and a load bus, as shown in Figure 2.22.

In the absence of line C, lines A and B carry the generated powers P_1 and P_2, respectively. If line C is switched on, the flow in lines A and B will change although the sum is $(P_1 + P_2 = P_A + P_B)$. Depending on the network parameters, line A or line B can carry increased power compared to the power carried in the absence of line C. Considering

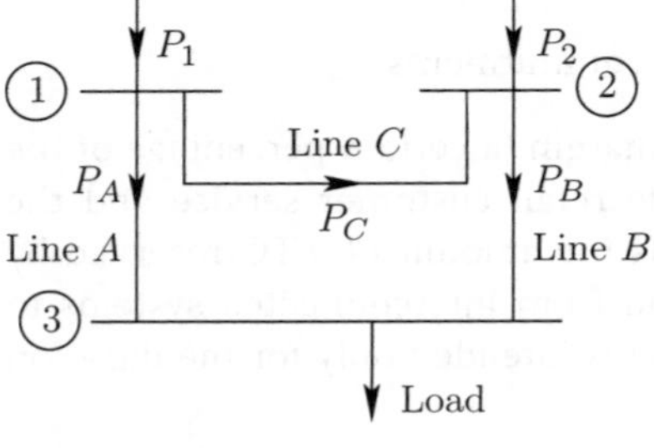

Figure 2.22 A three-bus system.

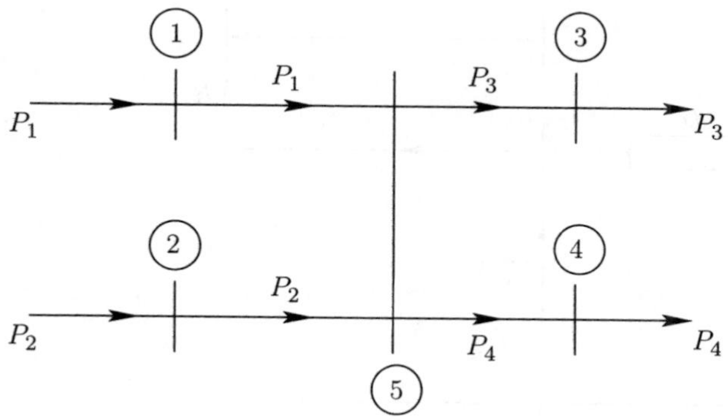

Figure 2.23 A radial network.

linear (DC) load flow, P_C is proportional to $(P_1b_2 - P_2b_1)$ where b_1 and b_2 are susceptances of lines A and B, respectively.

Proposition: A radial power transmission network is a coherent system.

Proof: The graph of the network consists only of tree branches, where any branch connecting nodes *i* and *j* is assumed to represent the path consisting of parallel connection of transmission lines connecting the nodes *i* and *j*. Since then are no loops in the graph, any specified transmission line is connected across adjacent nodes. The power flow (analogous to current) in the branch connecting nodes *i* and *j* is uniquely determined by the knowledge of injected powers at various nodes and the incidence matrix for the graph. Thus, addition of a line in parallel with the existing line(s) across nodes *i* and *j* will improve system reliability. Similarly, removal of a line will reduce system reliability.

Example 2.2 Consider a radial network with four lines, as shown in Figure 2.23. Here, $P_1 + P_2 = P_3 + P_4$. A new line can be added only in parallel with any one of the four lines without disturbing the radial nature of the network.

Remarks

1. Reference [22] describes a procedure of strategic switching in a network to achieve topology optimization with security constraints. Essentially, it is feasible to apply switching control for loss reduction, security improvement, and achieving optimal power flow.
2. With the availability of fast-acting static switches, it should be possible to achieve optimal operation of medium voltage distribution networks.

2.A Analysis of a Distributed Parameter Single-Phase Line in Steady State

Figure 2.A.1 shows a single-phase distributed parameter line. Considering a small element of the line length (dx) at a distance x from the receiving end, the following

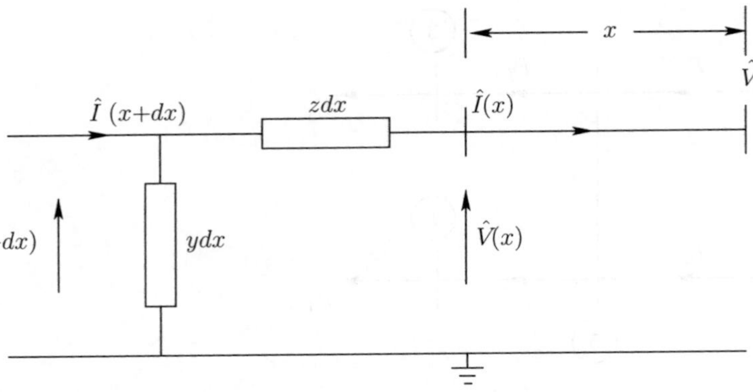

Figure 2.A.1 Single-phase distributed parameter transmission line.

equations apply

$$\hat{I}(x + dx) = \hat{I}(x) + (y\, dx)\hat{V}(x + dx) \tag{2.A.1}$$

$$\hat{V}(x + dx) = \hat{V}(x) + (zdx)\hat{I}(x) \tag{2.A.2}$$

where $y = g + jb, z = r + jx_l, b = \omega c, x_l = \omega l$. y and z are the shunt admittance and series impedance of the line per-unit length. It should be noted that both $\hat{V}$ and $\hat{I}$ are phasors that are functions of x.

From the above equations, we get the following differential equations for $\hat{V}(x)$ and $\hat{I}(x)$:

$$\frac{d\hat{V}}{dx} = z\hat{I}, \quad \frac{d\hat{I}}{dx} = y\hat{V} \tag{2.A.3}$$

Taking derivatives of the above equations, we can derive

$$\frac{d^2\hat{V}}{dx^2} = zy\hat{V}, \quad \frac{d^2\hat{I}}{dx^2} = zy\hat{I} \tag{2.A.4}$$

The solution of the above equations gives

$$\hat{V}(x) = A_1 e^{\gamma x} + A_2 e^{-\gamma x} \tag{2.A.5}$$

$$\hat{I}(x) = \frac{1}{Z_c}[A_1 e^{\gamma x} - A_2 e^{-\gamma x}] \tag{2.A.6}$$

where $Z_c = \sqrt{\frac{z}{y}}$ is termed the characteristic impedance and $\gamma = \sqrt{zy} = (\alpha + j\beta)$ is termed the propagation constant. α (real part of γ) is called the attenuation constant and β (imaginary part of γ) is called the phase constant. The constants A_1 and A_2 are determined from the boundary conditions:

$$\hat{V}_R = \hat{V}(x = 0) = A_1 + A_2 \tag{2.A.7}$$

$$\hat{I}_R = \hat{I}(x = 0) = \frac{(A_1 - A_2)}{Z_c} \tag{2.A.8}$$

Solving for A_1 and A_2 from the above and substituting in (2.A.5) and (2.A.6), we get

$$\hat{V}(x) = \hat{V}_R \cosh(\gamma x) + \hat{I}_R Z_c \sinh(\gamma x) \quad (2.A.9)$$

$$\hat{I}(x) = \frac{\hat{V}_R}{Z_c} \sinh(\gamma x) + \hat{I}_R \cosh(\gamma x) \quad (2.A.10)$$

where $\cosh(\gamma x) = \frac{e^{\gamma x}+e^{-\gamma x}}{2}$ and $\sinh(\gamma x) = \frac{e^{\gamma x}-e^{-\gamma x}}{2}$.

Normally, the conductance g of the line can be neglected. The series resistance r has only a secondary effect on the voltage and power flow in the line, and hence can be neglected for simplicity.

2.A.1 Expressions for a Lossless Line

Neglecting r and g, the propagation constant is purely imaginary and is given by

$$\gamma = j\beta = j\omega\sqrt{lc} \quad (2.A.11)$$

The characteristic impedance Z_c is purely resistive and defined as surge or natural impedance (Z_n), given by

$$Z_n = \sqrt{\frac{l}{c}} \quad (2.A.12)$$

We obtain the expressions for $\hat{V}(x)$ and $\hat{I}(x)$ for the lossless line as

$$\hat{V}(x) = \hat{V}_R \cos(\beta x) + j\hat{I}_R Z_n \sin(\beta x) \quad (2.A.13)$$

$$\hat{I}(x) = j\frac{\hat{V}_R}{Z_n} \sin(\beta x) + \hat{I}_R \cos(\beta x) \quad (2.A.14)$$

When $x = d$, where d is the length of the transmission line, we obtain

$$\begin{aligned} \hat{V}_S &= \hat{V}_R \cos\theta + j\hat{I}_R Z_n \sin\theta \\ \hat{I}_S &= \frac{j\hat{V}_R}{Z_n} \sin\theta + \hat{I}_R \cos\theta \end{aligned} \quad (2.A.15)$$

where $\theta = \beta d = \omega d\sqrt{lc} = \frac{2\pi}{\lambda} d$ is termed the electrical length of the line in radians. λ is the wavelength, which depends on frequency. $\lambda = \frac{u}{f}$ where u is the velocity of propagation of the voltage or current wave given by $u = \frac{1}{\sqrt{lc}}$. Typically, the value of u for the overhead high-voltage transmission lines is slightly less than the velocity of light (3×10^8 m/s).

Remarks

1. A transmission line can be considered as a two-port network defined by

$$\begin{bmatrix} \hat{V}_S \\ \hat{I}_S \end{bmatrix} = \begin{bmatrix} A & B \\ C & D \end{bmatrix} \begin{bmatrix} \hat{V}_R \\ \hat{I}_R \end{bmatrix} \quad (2.A.16)$$

Comparing with (2.A.15), we obtain

$$A = \cos\theta = D, \quad B = jZ_n \sin\theta, \quad C = j\frac{\sin\theta}{Z_n}$$

It should be noted that $AD - BC = 1$.

2.A.2 Performance of a Symmetrical Line

To control the receiving end voltage and increase the power transfer capability of the line, it is necessary to have a generator or a fast controlled reactive power source (such as SVC or STATCOM).

If the line is symmetrical ($|\hat{V}_S| = |\hat{V}_R| = V$), then from symmetry $Q_S = -Q_R$. The reactive power requirements of the line are shared equally at both ends of the line.

For a symmetrical lossless line, it can be shown that the power flow (P) in the line is given by

$$P = \frac{V^2 \sin\delta}{Z_n \sin\theta} = \frac{P_n \sin\delta}{\sin\theta} \tag{2.A.17}$$

where P_n is the SIL, defined by

$$P_n = \frac{V^2}{Z_n} \tag{2.A.18}$$

The reactive power at the two ends of the line is

$$Q_S = -Q_R = \frac{P_n}{\sin\theta}(\cos\theta - \cos\delta) \tag{2.A.19}$$

The voltage profile along the line (see Figure 2.A.2) varies as the line loading varies. For $P = P_n$, the voltage profile is flat. The voltage variation is a maximum at the midpoint of the line. The midpoint voltage $\hat{V}_m$ is given by

$$\hat{V}_m = \frac{V\cos\frac{\delta}{2}}{\cos\frac{\theta}{2}} \angle \frac{\delta}{2} \tag{2.A.20}$$

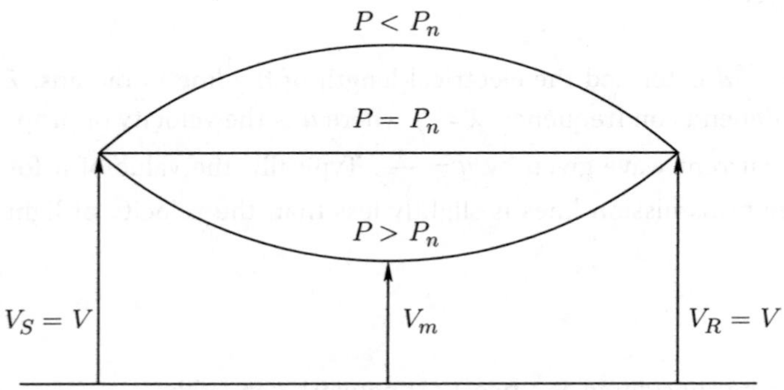

Figure 2.A.2 Voltage profile along the transmission line.

Remarks

1. The power transfer capability of the AC lines decreases as line length increases. Assuming that a line is operated with $\delta_{max} = 30°$, the maximum power transfer reduces below SIL for $\theta > 30°$ (approximately 500 km long line for systems operating at 50 Hz).
2. The problems of voltage control and charging reactive power (at no load) become severe as line length increases. Switchable shunt reactors are connected at the two ends of the line to regulate the voltage at loads below P_n.

2.B Computation of Electrical Torque

Assuming that the generator rotor oscillates (about a constant speed of ω_o) sinusoidally, the per-unit speed $\overline{\omega}$ is given by

$$\overline{\omega} = \overline{\omega_o} + A \sin \omega_m t \tag{2.B.1}$$

where ω_m is the oscillation frequency of the rotor. Since

$$\frac{d\delta}{dt} = (\overline{\omega} - \overline{\omega_o})\, \omega_B$$

We obtain

$$\delta = \delta_0 - \left(\frac{A\omega_o}{\omega_m}\right) \cos \omega_m t \tag{2.B.2}$$

The induced voltages ($e_{g\alpha}$ and $e_{g\beta}$) in the armature are given by

$$e_{g\alpha} = \overline{\omega} E' \sin(\omega_o t + \delta) \tag{2.B.3}$$

$$e_{g\beta} = \overline{\omega} E' \cos(\omega_o t + \delta)$$

For simplifying the notation, the subscript g is dropped in the following analysis.

If it is assumed that the amplitude (A) of the rotor oscillation is very small, the induced voltage in the stator (e_α) consists of three sinusoidal components, one of frequency f_o and other two components of frequencies $f_o \pm f_m$. This follows from substituting (2.B.1) and (2.B.2) in (2.B.3) and noting that

$$e_\alpha(t) = \overline{\omega}_o E' \sin(\omega_o t + \delta) + \frac{AE'}{2} \cos[(\omega_o - \omega_m)t + \delta] - \frac{AE'}{2} \cos[(\omega_o + \omega_m)t + \delta] \tag{2.B.4}$$

We also note that

$$\frac{AE'}{2} \cos[(\omega_o - \omega_m)t + \delta] \approx \frac{AE'}{2} \cos[(\omega_o + \omega_m)t + \delta_0] \tag{2.B.5}$$

since

$$\cos[(\omega_o - \omega_m)t + \delta] \approx \cos[(\omega_o - \omega_m)t + \delta_0] - \frac{A\omega_B}{\omega_m} \sin[(\omega_o - \omega_m)t + \delta_0] \cos \omega_m t$$

Also,

$$\begin{aligned}\overline{\omega}_0 E' \sin(\omega_o t + \delta) &\approx \omega_o E'[\sin(\omega_o t + \delta_0) + (\delta - \delta_0)\cos(\omega_o t + \delta_0)] \\ &= \overline{\omega}_o E' \sin(\omega_o t + \delta_0) - \frac{\omega_o A E'}{2\omega_m}\{\cos[(\omega_o - \omega_m)t + \delta_0] \\ &\quad + \cos[(\omega_o + \omega_m)t + \delta_0]\}\end{aligned} \tag{2.B.6}$$

Thus,

$$\begin{aligned}e_\alpha(t) = \overline{\omega}_o E' \sin(\omega_o t + \delta_0) &- \frac{AE'}{2\omega_m}(\omega_o - \omega_m)\cos[(\omega_o - \omega_m)t + \delta_0] \\ &- \frac{AE'}{2\omega_m}(\omega_o + \omega_m)\cos[(\omega_o + \omega_m)t + \delta_0]\end{aligned} \tag{2.B.7}$$

The subsynchronous frequency component of the voltage source, e_α is

$$e_\alpha^{sub} = \frac{AE'}{2\omega_m}(\omega_o - \omega_m)\cos[(\omega_o - \omega_m)t + \delta_0] \tag{2.B.8}$$

It can be easily derived that

$$e_\beta^{sub} = \frac{AE'}{2\omega_m}(\omega_o - \omega_m)\sin[(\omega_o - \omega_m)t + \delta_0] \tag{2.B.9}$$

When these voltages are applied to the α and β sequence networks, respectively, the subsynchronous frequency currents flow (in steady state) and are given by the expressions

$$(i_\beta^{sub} + j i_\alpha^{sub}) = Z^{-1}[j(\omega_o - \omega_m)](e_\beta^{sub} + j e_\alpha^{sub}) \tag{2.B.10}$$

If $Z(s) = R + Ls + \frac{1}{Cs}$

$$Z[j(\omega_o - \omega_m)] = R + j\left[(\omega_o - \omega_m)L - \frac{1}{(\omega_o - \omega_m)C}\right] = Z_{sub}\angle\phi_{sub}$$

If the resonance frequency f_{er} defined by $f_{er} = \frac{1}{2\pi\sqrt{LC}}$ is close to $(f_o - f_m)$, the impedance is small. It is resistive when $(f_o - f_m) = f_{er}$ and capacitive when $f_{er} > (f_o - f_m)$, otherwise it is inductive (when $f_{er} < (f_o - f_m)$). The d-q components of the currents are given by

$$(i_q + j i_d) = e^{-j(\omega_o t + \delta)}(i_\beta + j i_\alpha) \tag{2.B.11}$$

From (2.B.10) and (2.B.11), we can obtain

$$i_\alpha^{sub}(t) = -\frac{AE'}{2\omega_m Z_{sub}}(\omega_o - \omega_m)\cos[(\omega_o - \omega_m)t + \delta_0 - \phi_{sub}] \tag{2.B.12}$$

$$i_\beta^{sub}(t) = \frac{AE'}{2\omega_m Z_{sub}}(\omega_o - \omega_m)\sin[(\omega_o - \omega_m)t + \delta_0 - \phi_{sub}] \tag{2.B.13}$$

$$i_q^{sub}(t) = \frac{-AE'}{2\omega_m Z_{sub}}(\omega_o - \omega_m)\sin(\omega_m t + \phi_{sub}) \tag{2.B.14}$$

The component of the torque T_e^{sub} due to subsynchronous frequency currents is given by

$$T_e^{sub} = E' i_q^{sub} = -\frac{A(E')^2}{2\omega_m Z_{sub}}(\omega_o - \omega_m)\sin(\omega_m t + \phi_{sub}) \tag{2.B.15}$$

The supersynchronous frequency voltage components e_α^{sup} and e_β^{sup} result in supersynchronous frequency currents given by

$$i_\alpha^{sup}(t) = -\frac{AE'(\omega_o + \omega_m)}{2\omega_m Z_{sup}} \cos[(\omega_o + \omega_m)t + \delta_0 - \phi_{sup}] \tag{2.B.16}$$

$$i_\beta^{sup}(t) = \frac{AE'(\omega_o + \omega_m)}{2\omega_m Z_{sup}} \sin[(\omega_o + \omega_m)t + \delta_0 - \phi_{sup}] \tag{2.B.17}$$

where $Z_{sup}\angle\phi_{sup}$ is the network impedance at the supersynchronous frequency, viewed from generator (internal) bus. The magnitude of Z_{sup} and ϕ_{sup} is usually positive (the impedance is inductive).

The quadrature axis component, i_q^{sup}, due to the supersynchronous currents can be obtained as

$$i_q^{sup} = \frac{AE'}{2\omega_m Z_{sup}}(\omega_o + \omega_m)\sin(\omega_m t - \phi_{sup}) \tag{2.B.18}$$

The component of the torque due to the supersynchronous currents, T_e^{sup}, is given by

$$T_e^{sup} = E' i_q^{sup} = \frac{A(E')^2(\omega_o + \omega_m)}{2\omega_m Z_{sup}} \sin(\omega_m t - \phi_{sup}) \tag{2.B.19}$$

Note that both torque components T_e^{sub} and T_e^{sup} have same frequency ω_m, the frequency of oscillation of the generator rotor. The damping torque coefficient T_D is given by

$$T_D = -\frac{(E')^2}{2\omega_m}\left[\frac{(\omega_o - \omega_m)}{Z_{sub}}\cos\phi_{sub} - \frac{(\omega_o + \omega_m)}{Z_{sup}}\cos\phi_{sup}\right] \tag{2.B.20}$$

References

1 Kraus, J.D. (1991) *Electromagnetics*, McGraw Hill, New York.

2 Rothe, F.S. (1953) *An Introduction to Power System Analysis*, John Wiley & Sons, New York.

3 Hedman, D. (1965) Propagation on overhead transmission lines I – Theory of modal analysis. *IEEE Transactions on Power Apparatus and Systems*, **84** (3), 200–205.

4 Kron, G. (1965) *Tensor Analysis of Networks*, MacDonald, London.

5 Clarke, E. (1943) *Circuit analysis of AC power systems*, Vol. 1, John Wiley & Sons, New York.

6 Park, R.H. (1929) Two-reaction theory of synchronous machines generalized method of analysis – Part I. *Transactions of the American Institute of Electrical Engineers*, **48** (3), 716–727.

7 Padiyar, K.R. (2002) *Power System Dynamics – Stability and Control*, BS Publications, Hyderabad, 2nd edn.

8 Wood, A.J. and Wollenberg, B.F. (1996) *Power generation, operation, and control*, John Wiley & Sons, New York.

9 Kirchmayer, L.K. (1958) *Economic Operation of Power Systems*, John Wiley & Sons, New York.

10 deMello, F.P., Mills, R.J., and B'Rells, W.F. (1973) Automatic generation control, Parts I and II. *IEEE Transactions on Power Apparatus and Systems,* **PAS-92** (2), 710–724.
11 Cohn, N. (1971) *Control of Generation and Power Flow on Interconnected Systems,* John Wiley & Sons, New York.
12 Padiyar, K.R. (1999) *Analysis of Subsynchronous Resonance in Power Systems,* Kluwer, Boston.
13 Padiyar, K.R. and Kulkarni, A.M. (1997) Design of reactive current and voltage controller of static condenser. *International Journal of Electrical Power & Energy Systems,* **19** (6), 397–410.
14 Padiyar, K.R. (2007) *FACTS Controllers in Power Transmission and Distribution,* New Age, New Delhi.
15 McGillis, D., Huynh, N.H., and Scott, G. (1981) Reactive compensation. role of static compensation meeting AC system control requirements with particular to the James Bay system. *IEE Proceedings C – Generation, Transmission and Distribution,* **128** (6), 389–393.
16 Gutman, R., Marchenko, P.P., and Dunlop, R.D. (1979) Analytical development of loadability characteristics for EHV and UHV transmission lines. *IEEE Transactions on Power Apparatus and Systems,* **PAS-98** (2), 606–617.
17 Dommel, H.W. (1969) Digital computer solution of electromagnetic transients in single-and multiphase networks. *IEEE Transactions on Power Apparatus and Systems,* **PAS-88** (4), 388–399.
18 Hobbs, B.F., Xu, Q., Ho, J. *et al.* (2016) Adaptive transmission planning: Implementing a new paradigm for managing economic risks in grid expansion. *IEEE Power and Energy Magazine,* **14** (4), 30–40.
19 Billinton, R. and Allen, R.N. (1996) *Reliability Evaluation of Power Systems,* Plenum Press, New York, 2nd edn.
20 North American Electric Reliability Council Report (1996), Available Transfer Capability Definitions and Determination. Princeton, New Jersey.
21 Barlow, R.E. and Proschan, F. (1965) *Mathematical Theory of Reliability,* Wiley, New York.
22 Bacher, R. and Glavitsch, H. (1986) Network topology optimization with security constraints. *IEEE Transactions on Power Systems,* **1** (4), 103–111.

3

Modeling and Simulation of Synchronous Generator Dynamics

3.1 Introduction

Synchronous generators are extensively used in power stations, supplying bulk electric power over high-voltage (HV) or extra-high-voltage (EHV) transmission lines to load centres. Generally AC transmission is used, however, in some cases (particularly from remote hydropower stations), the generators may be transmitting power over bipolar HVDC lines. The nominal voltage (phase to phase) of generators is limited below 40 kV and a step-up generator transformer is used to increase the voltage to the transmission level. The generator may be driven by steam, hydro or gas turbines. In microgrids at lower power levels, the generators may be driven by diesel engines. The fuel used may be coal or gas in thermal stations whereas natural or enriched uranium is used in nuclear power plants. In hydro stations, the kinetic energy of water drives the hydro turbines.

In the early stages of power development, the only dynamic components in power systems were synchronous generators. The prime mover control was used to regulate speed (frequency) and power output. The fast-acting static excitation with high gain electronic voltage regulators were developed in the 1960s. For stability analysis, simple models based on the swing equation (3.1) were used. The swing equation describes the variation of rotors frequency as [1–5]:

$$\frac{2H}{\omega_B}\frac{d\omega}{dt} = T_m - T_e \approx (P_m - P_e) \tag{3.1}$$

where T_m and P_m are mechanical torque and power acting on the rotor, and T_e and P_e are electrical torque and power. With the small variations in rotor velocity, the torque and power (expressed per unit) can be assumed to be equal.

In this chapter, we consider detailed models of a generator including (two-axis) machine model, excitation, and prime mover controllers.

3.2 Detailed Model of a Synchronous Machine

The schematic of a three-phase synchronous machine is shown in Figure 3.1. Here, three-phase armature windings (a, b, and c) are on the stator and there are four

Dynamics and Control of Electric Transmission and Microgrids, First Edition.
K. R. Padiyar and Anil M. Kulkarni.

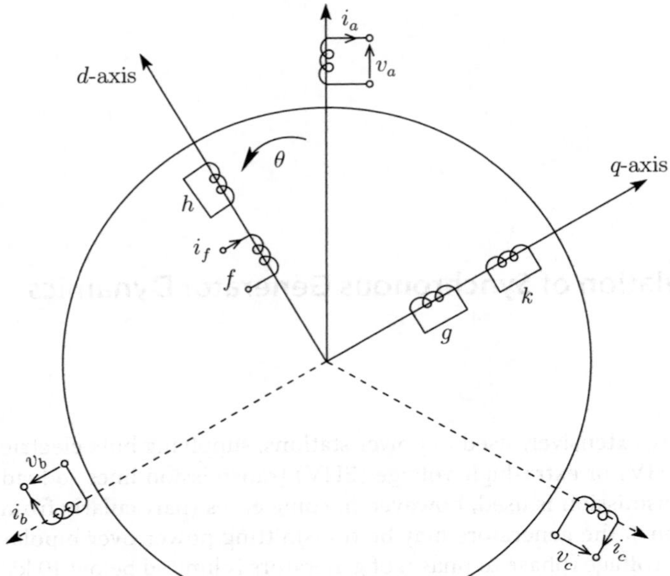

Figure 3.1 Synchronous machine.

windings on the rotor, including the field winding f. The amortisseur (damper) circuits in the salient pole machine or eddy current effects in the rotor are represented by a set of three closed coils (h in the direct axis, g and k in the quadrature axis). The number of damper coils can vary from zero (in the simplest model) to five or more in the Jackson–Winchester model [6]. However, from the point of availability of data, three coils, as shown in Figure 3.1, are considered adequate.

It is assumed that the rotor is rotating in the anticlockwise direction with angular velocity ω. The field winding f and damper winding h on the direct axis have relative motion with respect to the stationary winding a, b, and c on the stator. The angle between the d-axis and the phase a winding is θ.

3.2.1 Flux Linkage Equations

The stator and rotor flux linkages are given by

$$\psi_s = [L_{ss}]\bar{i}_s + [M]i_r \tag{3.2}$$

$$\psi_r = [M]^t\bar{i}_s + [L_{rr}]i_r \tag{3.3}$$

where $\bar{i}_s = -i_s$,

$$i_s^t = [i_a \quad i_b \quad i_c], \quad i_r^t = [i_f \quad i_h \quad i_g \quad i_k]$$

$$\psi_s^t = [\psi_a \quad \psi_b \quad \psi_c], \quad \psi_r^t = [\psi_f \quad \psi_h \quad \psi_g \quad \psi_k]$$

The matrices $[L_{ss}]$ and $[L_{rr}]$ are symmetric. From two-reaction theory, it is possible to express the inductance coefficients as follows:

$$[L_{ss}] = \begin{bmatrix} L_{aao} & L_{abo} & L_{abo} \\ L_{abo} & L_{aao} & L_{abo} \\ L_{abo} & L_{abo} & L_{aao} \end{bmatrix} + L_{aa2}\begin{bmatrix} \cos 2\theta & \cos\left(2\theta - \frac{2\pi}{3}\right) & \cos\left(2\theta + \frac{2\pi}{3}\right) \\ \cos\left(2\theta - \frac{2\pi}{3}\right) & \cos\left(2\theta + \frac{2\pi}{3}\right) & \cos 2\theta \\ \cos\left(2\theta + \frac{2\pi}{3}\right) & \cos 2\theta & \cos\left(2\theta - \frac{2\pi}{3}\right) \end{bmatrix} \quad (3.4)$$

$$[L_{rr}] = \begin{bmatrix} L_f & L_{fh} & 0 & 0 \\ L_{fh} & L_h & 0 & 0 \\ 0 & 0 & L_g & L_{gk} \\ 0 & 0 & L_{gk} & L_k \end{bmatrix} \quad (3.5)$$

$$[M] = \begin{bmatrix} M_{af}\cos\theta & M_{ah}\cos\theta & M_{ag}\sin\theta & M_{ak}\sin\theta \\ M_{af}\cos\left(\theta - \frac{2\pi}{3}\right) & M_{ah}\cos\left(\theta - \frac{2\pi}{3}\right) & M_{ag}\sin\left(\theta - \frac{2\pi}{3}\right) & M_{ak}\sin\left(\theta - \frac{2\pi}{3}\right) \\ M_{af}\cos\left(\theta + \frac{2\pi}{3}\right) & M_{ah}\cos\left(\theta + \frac{2\pi}{3}\right) & M_{ag}\sin\left(\theta + \frac{2\pi}{3}\right) & M_{ak}\sin\left(\theta + \frac{2\pi}{3}\right) \end{bmatrix} \quad (3.6)$$

Note that $[M]$ is a function of θ and is time varying if the rotor rotates at constant speed. $[L_{ss}]$ is also a function of θ if $L_{aa2} \neq 0$. This is true for salient pole machines.

3.2.2 Voltage equations

The voltage equations for the stator and rotor coils are as given below:

$$\frac{d\psi_s}{dt} + [R_s]\bar{i}_s = v_s \quad (3.7)$$

$$\frac{d\psi_r}{dt} + [R_r]i_r = v_r \quad (3.8)$$

where $v_s^t = \begin{bmatrix} v_a & v_b & v_c \end{bmatrix}$, $v_r^t = \begin{bmatrix} v_f & 0 & 0 & 0 \end{bmatrix}$

$$[R_s] = \begin{bmatrix} R_a & 0 & 0 \\ 0 & R_a & 0 \\ 0 & 0 & R_a \end{bmatrix} = R_a[U_3]$$

$$[R_r] = \begin{bmatrix} R_f & 0 & 0 & 0 \\ 0 & R_h & 0 & 0 \\ 0 & 0 & R_g & 0 \\ 0 & 0 & 0 & R_k \end{bmatrix}$$

and $[U_3]$ is a unit matrix of dimension 3.

It should be noted that we are using motor convention [4] although the stator currents are assumed to be flowing out of the generator, as is normally done.

Torque Equation

There is no loss of generality if we consider the machine has two poles as we use per-unit quantities. When the number of poles (P) is equal to 2, the mechanical angle θ_m and the electrical angle (θ) are identical. The equation of motion of the rotor is given by

$$J\frac{d^2\theta}{dt^2} + D\frac{d\theta}{dt} = T_m - T_e \tag{3.9}$$

J is the moment of inertia of the rotor and D is the damping (assumed to be viscous) coefficient. The equation for the electrical torque T_e is given by

$$T_e = -\frac{\partial W'}{\partial \theta} \tag{3.10}$$

where W' is the co-energy expressed as

$$W' = \frac{1}{2}\begin{bmatrix} \bar{i}_s^t & i_r^t \end{bmatrix}\begin{bmatrix} L_{ss} & M \\ M^t & L_{rr} \end{bmatrix}\begin{bmatrix} \bar{i}_s \\ i_r \end{bmatrix} \tag{3.11}$$

T_e can also be expressed as

$$T_e = -\frac{1}{2}\left[\bar{i}_s^t \frac{\partial \psi_s}{\partial \theta} + i_r^t \frac{\partial \psi_r}{\partial \theta}\right] \tag{3.12}$$

It can be shown that (3.9) can be expressed (using per-unit quantities) as (3.1), where

$$H = \frac{1}{2}\frac{J\omega_m^2}{S_B} = \frac{\text{kinetic energy stored in megajoules}}{\text{rating in MVA}}$$

and neglecting the damping term.

3.3 Park's Transformation [7]

The synchronous machine equations involve time-varying components even if the rotor speed is constant. It would be advantageous if the time-varying machine equations could be transformed to a time invariant set. This would result in the simplification of the calculations both for steady-state and transient conditions. R.H. Park (1929) introduced the following transformation:

$$\begin{bmatrix} f_a \\ f_b \\ f_c \end{bmatrix} = [C_P]\begin{bmatrix} f_d \\ f_q \\ f_0 \end{bmatrix} \tag{3.13}$$

where f_α can be stator voltage, current or flux linkage of the stator winding $\alpha(\alpha = a, b$ or $c)$. C_P is defined by

$$[C_P] = \begin{bmatrix} k_d \cos\theta & k_q \sin\theta & k_o \\ k_d \cos\left(\theta - \frac{2\pi}{3}\right) & k_q \sin\left(\theta - \frac{2\pi}{3}\right) & k_o \\ k_d \cos\left(\theta + \frac{2\pi}{3}\right) & k_q \sin\left(\theta + \frac{2\pi}{3}\right) & k_o \end{bmatrix} \quad (3.14)$$

where k_d, k_q, and k_o are appropriately chosen constants. In Park's original transformation $k_d = 1.0$, $k_q = -1.0$, and $k_o = 1$.

The inverse transformation is given by

$$\begin{bmatrix} f_d \\ f_q \\ f_0 \end{bmatrix} = [C_P]^{-1} \begin{bmatrix} f_a \\ f_b \\ f_c \end{bmatrix} \quad (3.15)$$

where

$$[C_P]^{-1} = \begin{bmatrix} k_1 \cos\theta & k_1 \cos(\theta - 2\pi/3) & k_1 \cos(\theta + 2\pi/3) \\ k_2 \sin\theta & k_2 \sin(\theta - 2\pi/3) & k_2 \sin(\theta + 2\pi/3) \\ k_3 & k_3 & k_3 \end{bmatrix}$$

$$k_1 = \frac{2}{3k_d}, \quad k_2 = \frac{2}{3k_q}, \quad k_3 = \frac{1}{3k_o}$$

Transformation of Flux Linkages

$$\begin{bmatrix} \psi_s \\ \psi_r \end{bmatrix} = \begin{bmatrix} C_P & 0 \\ 0 & U_4 \end{bmatrix} \begin{bmatrix} \psi_{dqo} \\ \psi_r \end{bmatrix} \quad (3.16)$$

where U_4 is a unit matrix of order 4 and $\psi_{dqo}^t = [\psi_d \ \ \psi_q \ \ \psi_o]$.

The left-hand side of (3.16) can be expressed as

$$\begin{bmatrix} \psi_s \\ \psi_r \end{bmatrix} = \begin{bmatrix} L_{ss} & M \\ M^t & L_{rr} \end{bmatrix} \begin{bmatrix} C_P & 0 \\ 0 & U_4 \end{bmatrix} \begin{bmatrix} -i_{dqo} \\ i_r \end{bmatrix} \quad (3.17)$$

where $i_{dqo}^t = [i_d \ \ i_q \ \ i_o]$. Substituting (3.17) in (3.16) we get

$$\begin{bmatrix} \psi_{dqo} \\ \psi_r \end{bmatrix} = \begin{bmatrix} C_P^{-1} & 0 \\ 0 & U_4 \end{bmatrix} \begin{bmatrix} L_{ss} & M \\ M^t & L_{rr} \end{bmatrix} \begin{bmatrix} C_P & 0 \\ 0 & U_4 \end{bmatrix} \begin{bmatrix} -i_{dqo} \\ i_r \end{bmatrix}$$

$$= \begin{bmatrix} C_P^{-1} L_{ss} C_P & C_P^{-1} M \\ M^t C_P & L_{rr} \end{bmatrix} \begin{bmatrix} -i_{dqo} \\ i_r \end{bmatrix}$$

$$= \begin{bmatrix} L'_{ss} & M' \\ M'' & L_{rr} \end{bmatrix} \begin{bmatrix} -i_{dqo} \\ i_r \end{bmatrix} \quad (3.18)$$

where

$$[L'_{ss}] = \begin{bmatrix} L_d & 0 & 0 \\ 0 & L_q & 0 \\ 0 & 0 & L_o \end{bmatrix}$$

$$\begin{aligned} L_d &= L_{aao} - L_{abo} + \frac{3}{2}L_{aa2} \\ L_q &= L_{aao} - L_{abo} - \frac{3}{2}L_{aa2} \\ L_o &= L_{aao} + 2L_{abo} \end{aligned} \tag{3.19}$$

$$[M'] = \begin{bmatrix} \frac{M_{af}}{k_d} & \frac{M_{ah}}{k_d} & 0 & 0 \\ 0 & 0 & \frac{M_{ag}}{k_q} & \frac{M_{ak}}{k_q} \\ 0 & 0 & 0 & 0 \end{bmatrix} \tag{3.20}$$

$$[M''] = \begin{bmatrix} \frac{3}{2}M_{af}k_d & 0 & 0 \\ \frac{3}{2}M_{ah}k_d & 0 & 0 \\ 0 & \frac{3}{2}M_{ag}k_q & 0 \\ 0 & \frac{3}{2}M_{ak}k_q & 0 \end{bmatrix} \tag{3.21}$$

Remarks

1) $[M''] \neq [M']^t$ unless

$$k_d^2 = \frac{2}{3}, k_q^2 = \frac{2}{3} \tag{3.22}$$

2) The mutual inductance terms between the stator and rotor coils in the q-axis are negative for $k_q < 0$ unless M_{ag} and M_{ak} are both negative. It should be noted that when the quadrature axis is lagging the direct axis (in the direction of rotation), as assumed in Figure 3.1, M_{ag} and M_{ak} are positive. These terms are negative only if the q-axis is assumed to be leading the d-axis.
   Hence, if the d-axis is assumed to lead the q-axis, it would be convenient to choose a positive value of k_q [5]. This is also recommended in [8].
3) Note that there is no transformation of the rotor currents and flux linkages. Hence the self inductance matrix of rotor coils is not altered.
4) Equation (3.18) shows that stator coils a, b, and c are replaced by fictitious d, q, and o coils from Park's transformation. Out of these, the o coil (in which zero-sequence current i_o flows) has no coupling with the rotor coils and may be neglected if $i_o = 0$. Since the (transformed) mutual inductance terms between the d and q coils and the rotor coils are constants, it can be interpolated that the d and q coils rotate at the same speed as the rotor. It can also be assumed that the d coil is aligned with the d-axis and the q coil with the q-axis. This is shown in Figure 3.2.

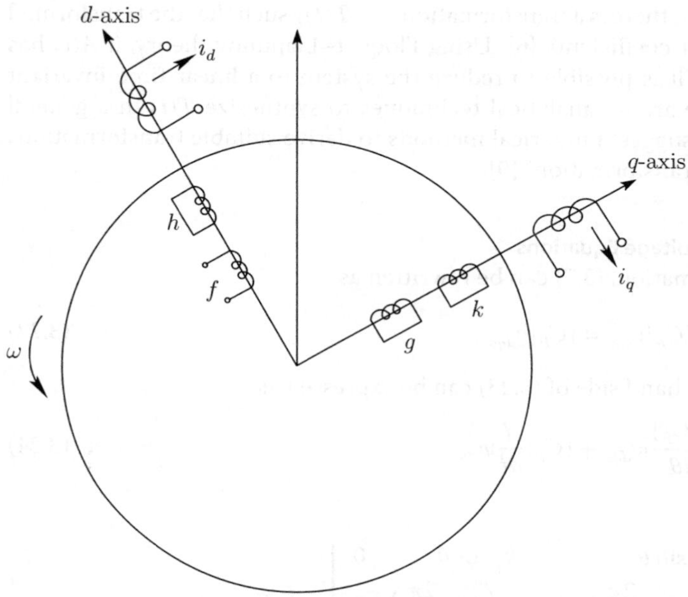

Figure 3.2 Synchronous machine with rotating armature windings.

5) The following trigonometric identities are useful in the derivation of the transformed equations:

$$\cos\theta + \cos\left(\theta - \frac{2\pi}{3}\right) + \cos\left(\theta + \frac{2\pi}{3}\right) = 0$$

$$\sin\theta + \sin\left(\theta - \frac{2\pi}{3}\right) + \sin\left(\theta + \frac{2\pi}{3}\right) = 0$$

$$\cos^2\theta + \cos^2\left(\theta - \frac{2\pi}{3}\right) + \cos^2\left(\theta + \frac{2\pi}{3}\right) = \frac{3}{2}$$

$$\sin^2\theta + \sin^2\left(\theta - \frac{2\pi}{3}\right) + \sin^2\left(\theta + \frac{2\pi}{3}\right) = \frac{3}{2}$$

Although the physical interpretation of Park's transformation is useful in gaining an intuitive understanding of its implications, it must be understood that it is not essential in the mathematical analysis of the synchronous machine. This is true of any mathematical transformation whose main objective is to simplify the analysis. From this point of view, the major benefit of Park's transformation is to obtain the machine equations in time-invariant form which simplifies the analysis. The transformation of stator voltage equations will clarify this point.

6) A general theory of transformations in rotating electric machines would be useful in situations where Park's transformation is not applicable. Park's transformation can be extended to electric machines connected to stationary symmetric networks (e.g., in the study of unsymmetrical faults and also for machines with magnetomotive force and permeance harmonics).

The flux linkages in a machine can be defined in terms of periodic functions of θ (rotor angle) and for constant speed, as periodic functions of time. Given a system described by $\dot{x} = A(t)x$, there is a transformation $x = T(t)y$ such that the transformed equation has constant coefficients [6]. Using Floquet–Liapunov theory, if $A(t)$ has periodic coefficients, it is possible to reduce the system to a linear time-invariant form. Although there are no analytical techniques to synthesize $T(t)$ in a general case, it is possible to suggest numerical methods to derive suitable transformations (analogous to Park's transformation) [9].

Transformation of Stator Voltage Equations

Applying Park's transformation, (3.7) can be rewritten as

$$\frac{d}{dt}[C_P\psi_{dqo}] - [R_s][C_P]i_{dqo} = [C_P]v_{dqo} \tag{3.23}$$

The first term on the left-hand side of (3.23) can be expressed as

$$\frac{d}{dt}[C_P\psi_{dqo}] = \dot{\theta}\frac{d[C_P]}{d\theta}\psi_{dqo} + [C_P]\frac{d}{dt}\psi_{dqo} \tag{3.24}$$

where $\dot{\theta} = \frac{d\theta}{dt}$ and

$$\frac{dC_P}{d\theta} = \begin{bmatrix} -k_d\sin\theta & k_q\cos\theta & 0 \\ -k_d\sin\left(\theta - \frac{2\pi}{3}\right) & k_q\cos\left(\theta - \frac{2\pi}{3}\right) & 0 \\ -k_d\sin\left(\theta + \frac{2\pi}{3}\right) & k_q\cos\left(\theta + \frac{2\pi}{3}\right) & 0 \end{bmatrix} = [C_P][P_1] \tag{3.25}$$

where

$$[P_1] = \begin{bmatrix} 0 & \frac{k_q}{k_d} & 0 \\ -\frac{k_d}{k_q} & 0 & 0 \\ 0 & 0 & 0 \end{bmatrix}$$

Substituting (3.24) in (3.23), we get

$$[C_P]\frac{d\psi_{dqo}}{dt} + \dot{\theta}[C_P][P_1]\psi_{dqo} - [R_s][C_P]i_{dqo} = [C_P]v_{dqo} \tag{3.26}$$

From (3.26), we obtain

$$\frac{d\psi_{dqo}}{dt} + \dot{\theta}[P_1]\psi_{dqo} - [R_a]i_{dqo} = v_{dqo} \tag{3.27}$$

after substituting for $[R_s] = R_a[U_3]$, where $[U_3]$ is the unit matrix of dimension 3. Equation (3.27) can be expanded as

$$\begin{aligned} \frac{d\psi_d}{dt} + \dot{\theta}\frac{k_q}{k_d}\psi_q - R_a i_d &= v_d \\ \frac{d\psi_q}{dt} - \dot{\theta}\frac{k_d}{k_q}\psi_d - R_a i_q &= v_q \\ \frac{d\psi_o}{dt} - R_a i_o &= v_o \end{aligned} \tag{3.28}$$

The rotor voltage equations are unchanged and can be written in the expanded form from (3.8) as

$$\begin{aligned}
&\frac{d\psi_f}{dt} + R_f i_f = v_f \\
&\frac{d\psi_h}{dt} + R_h i_h = 0 \\
&\frac{d\psi_g}{dt} + R_g i_g = 0 \\
&\frac{d\psi_k}{dt} + R_k i_k = 0
\end{aligned} \tag{3.29}$$

Transformation of the Torque Equation

After applying Park's transformation to (3.12) the electrical torque is expressed after some manipulation as (see Appendix 3.A)

$$\begin{aligned}
T_e &= \frac{3}{2} k_q k_d [\bar{i}_q \psi_d - \bar{i}_d \psi_q], \quad \bar{i}_d = -i_d \ , \ \bar{i}_q = -i_q \\
&= \frac{3}{2} k_q k_d [\psi_q i_d - \psi_d i_q]
\end{aligned} \tag{3.30}$$

Choice of Constants k_d, k_q, and k_o

The transformation $[C_P]$ defined by (3.14) is very general as no assumptions are made regarding the constants k_d, k_q, and k_o. However, the original Park's transformation used $k_d = 1$, $k_q = -1$, and $k_o = 1$.

Since the same transformation is applied for currents and voltages, it can be shown that the instantaneous power p is given by

$$p = v_a i_a + v_b i_b + v_c i_c = \frac{3}{2}(k_q^2 v_d i_d + k_q^2 v_q i_q) + 3k_o^2 v_o i_o \tag{3.31}$$

Power Invariant Transformation

A transformation $[C_P]$ is said to be power invariant if it is orthogonal, that is,

$$[C_P]^t = [C_P]^{-1} \tag{3.32}$$

The choice of k_d, k_q, and k_o to get power invariant transformation are

$$k_d = \pm\sqrt{\frac{2}{3}}, \quad k_q = \pm\sqrt{\frac{2}{3}}, \quad k_o = \pm\sqrt{\frac{1}{3}} \tag{3.33}$$

We normally assume only positive values of the constants, thereby defining a power invariant Park's transformation given by

$$[C_P] = \frac{1}{\sqrt{3}} \begin{bmatrix} \sqrt{2}\cos\theta & \sqrt{2}\sin\theta & 1 \\ \sqrt{2}\cos\left(\theta - \frac{2\pi}{3}\right) & \sqrt{2}\sin\left(\theta - \frac{2\pi}{3}\right) & 1 \\ \sqrt{2}\cos\left(\theta + \frac{2\pi}{3}\right) & \sqrt{2}\sin\left(\theta + \frac{2\pi}{3}\right) & 1 \end{bmatrix} \tag{3.34}$$

The major advantage of a power invariant transformation is that the mutual inductances in the transformed network are equal. For example

$$M_{df} = \frac{M_{af}}{k_d} \quad \text{and} \quad M_{fd} = \frac{3}{2} M_{af} k_d$$

for $k_d = \sqrt{\frac{2}{3}}$, $M_{df} = M_{fd} = \sqrt{\frac{3}{2}} M_{af}$

In what follows, we will use a power invariant transformation by selecting the following values:

$$k_d = \pm\sqrt{\frac{2}{3}} = k_q, \quad k_o = \sqrt{\frac{1}{3}}$$

We have assumed that the q-axis is lagging the d-axis (in the direction of rotation) and this is consistent with k_d and k_q having identical signs.

Although an IEEE Committee Report in 1969 [8] recommended a revision of the old convention (of the q-axis leading the d-axis), usage in power industry is often based on the old convention [4].

Example 3.1 A synchronous generator is connected to a balanced sinusoidal voltage source. Obtain the expression for the electrical torque in steady state.

Solution

Without loss of generality, the voltages at the generator terminals can be defined as

$$\begin{aligned} v_a &= \sqrt{\frac{2}{3}} V \sin \omega_o t \\ v_b &= \sqrt{\frac{2}{3}} V \sin \left(\omega_o t - \frac{2\pi}{3}\right) \\ v_c &= \sqrt{\frac{2}{3}} V \sin \left(\omega_o t + \frac{2\pi}{3}\right) \end{aligned} \tag{3.E.1}$$

Assuming $k_d = k_q = -\sqrt{\frac{2}{3}}$, we can obtain

$$v_{do} = V \sin \delta, \quad v_{qo} = -V \cos \delta \tag{3.E.2}$$

It is assumed that $\theta = \omega_o t + \delta$.

In steady state, the voltage equations in d-q variables (neglecting armature resistance) are obtained from (3.28) as

$$-\omega_o \Psi_{do} = v_{qo}, \quad \omega_o \Psi_{qo} = v_{do} \tag{3.E.3}$$

Since, $i_h = i_g = i_k = 0$ in steady state,

$$\omega_o \Psi_{do} = -x_d i_{do} + E_{fdo}, \quad \omega_o \Psi_{qo} = -x_q i_{qo} \tag{3.35}$$

Note that $E_{fd} = \frac{x_{df} v_f}{R_f}$. From (3.35), we obtain

$$i_{do} = \frac{E_{fdo} - \omega_o \Psi_{do}}{x_d}, \quad i_{qo} = \frac{-\omega_o \Psi_{qo}}{x_q} = -\frac{V \sin \delta}{x_q}$$

Since the electrical torque T_e is given by

$$T_e = \psi_q i_d - \psi_d i_q \quad \text{we can derive}$$

$$T_{eo} = \frac{V^2 \cos\delta_o \sin\delta_o}{\omega_o x_q} + \frac{V \sin\delta_o}{\omega_o x_d}[E_{fdo} - V\cos\delta_o]$$

The final expression for T_e is given by

$$T_{eo} = \frac{VE_{fdo}\sin\delta_o}{\omega_o x_d} + \frac{V^2 \sin 2\delta_o (x_d - x_q)}{2\omega_o x_d x_q}$$

Remarks

1) Since $T_e = T_m$ in steady state, the angle δ can be calculated from the knowledge of T_m, V, and E_{fd}.
2) The expression for torque in steady state (T_{eo}) is identical to the expression for power output (in steady state) divided by the operating frequency ω_o. Normally, $\omega_o = \omega_B$ where ω_B is the rated frequency.
3) The currents i_a, i_b, and i_c are sinusoidal and balanced as i_d and i_q are constants.
4) If the voltages at the terminal are unbalanced, v_d and v_q are no longer constants and contain second harmonic components. In such a case the currents i_d and i_q also contain second harmonic components and it can be shown that the phase currents i_a, i_b, and i_c contain third harmonic components.
5) In general, the harmonic currents flow in an unbalanced three-phase network connected to a synchronous generator.

3.4 Per-Unit Quantities

It is common to express voltages, currents, and impedances in per-unit quantities by choosing appropriate base quantities. The advantages of an appropriate per-unit system are as follows:

1) The numerical values of currents and voltages are related to their rated values irrespective of the size of the machine.
2) The per-unit impedances on the machine base lie in a narrow range for a class of machines of similar design.
3) The number of parameters required is minimized.

It should be noted that the base quantities for the stator and rotor circuits can be independently chosen with certain restrictions which result in per-unit mutual reactances being reciprocal. If the power-invariant Park's transformation is used, the constraints imply selecting the same base power for all the circuits.

Stator Base Quantities

The base quantities for the stator d-q windings are chosen as follows:

base power, S_B = three-phase rated power
base voltage, V_B = rated line-to-line voltage (rms)

base current, $I_B = \sqrt{3} \times$ rated line current

base impedance, $Z_B = \dfrac{V_B}{I_B} = \dfrac{\text{rated line to neutral voltage}}{\text{rated line current}}$

base flux linkages, $\psi_B = \dfrac{V_B}{\omega_B}$

base inductance, $L_B = \dfrac{\psi_B}{\omega_B} = \dfrac{Z_B}{\omega_B}$

ω_B is the base angular frequency in rad/s. (This is also the rated angular speed for a two-pole machine.)

The stator equations in put values, based on the quantities defined above are

$$\begin{aligned} \frac{1}{\omega_B}\frac{d\tilde{\psi}_d}{dt} + \frac{\omega}{\omega_B}\tilde{\psi}_q - \tilde{R}_a\tilde{i}_d = \tilde{v}_d \\ \frac{1}{\omega_B}\frac{d\tilde{\psi}_q}{dt} - \frac{\omega}{\omega_B}\tilde{\psi}_d - \tilde{R}_a\tilde{i}_q = \tilde{v}_q \end{aligned} \tag{3.36}$$

where the per-unit quantities are indicated by the ~ over the variables. For example,

$$\tilde{i}_d = \frac{i_d}{I_B}, \tilde{v}_d = \frac{v_d}{V_B}, \tilde{\psi}_d = \frac{\psi_d}{\psi_B}$$

Remarks

1) The base voltage and current used in the previous literature are

$$V_B = \text{peak rated voltage per phase}$$
$$I_B = \text{peak rated line current}$$

The choice of these base quantities is consistent with the original Park's transformation. This is because the per-unit voltage (or current) in the old system is identical to that in the revised system, defined above (which is consistent with the power-invariant version of Park's transformation)

2) The base impedances in both systems are identical. This fact combined with the identity of per-unit quantities implies that the equations (per unit) in both systems are identical (except for the differences in the orientation of d- and q-axes), thus eliminating the confusion about different versions of Park's transformation.
3) Anderson and Fouad [3] use different base quantities for voltage and current along with power-invariant Park's transformation. They define V_B as (rms) rated line to neutral voltage and I_B as rms line current. Although this results in identical base impedance as defined before, the per-unit voltages, currents, and fluxes are different, which leads to different equations using inconvenient factors (of $\sqrt{3}$).
4) The choice of base quantities for the rotor is related to the choice of stator base quantities. Although in the original Park's transformation the mutual inductances are not reciprocal, the selection of appropriate base quantities for the rotor will result in reciprocal per-unit mutual inductances.
5) The expression for the base three-phase power in the old system is

$$P_B = S_B = \frac{3}{2}V_B I_B$$

which leads to a per-unit power expression of

$$\tilde{P} = \frac{P}{P_B} = \tilde{v}_d \tilde{i}_d + \tilde{v}_q \tilde{i}_q$$

(It is assumed that the zero-sequence power is zero.) In the revised (new) system,

$$P_B = S_B = V_B I_B$$

which leads to the same per-unit power expression given above.

6) If the operating frequency is same as the base frequency, the per-unit inductances are identical to per-unit corresponding reactances. Then, $\tilde{x}_d = \tilde{L}_d$, $\tilde{x}_{df} = \tilde{M}_{df}$, etc.

Rotor Base Quantities

The base power and frequency are same as for the stator circuits. The base currents for the rotor circuits in the d-axis are chosen such that the base field current or base d-axis damper current (in the h coil) produce the same mutual flux (in the airgap) as produced by base current flowing in the stator d-axis coil. Similar conditions apply for q-axis coils. The mutual flux linkages in the d-axis are given by

$$\psi_{ad} = (L_d - L_{a\sigma})I_B = M_{df} I_{fB} = M_{dh} I_{hB} \tag{3.37}$$

where $L_{a\sigma}$ is the leakage inductance of the stator. I_{fB} and I_{hB} are the base currents in the field and damper windings (in the d-axis).

From (3.37),

$$I_{fB} = \frac{L_{ad}}{M_{df}} I_B, \quad I_{hB} = \frac{L_{ad}}{M_{dh}} I_B \tag{3.38}$$

where $L_{ad} = L_d - L_{a\sigma}$.

The base flux linkages for the rotor circuits are chosen such that

$$\psi_{fB} I_{fB} = \psi_B I_B = \psi_{hB} I_{hB} \tag{3.39}$$

Similar relations also apply for q-axis coils. The base currents and flux linkages for the g and k coils are given by

$$I_{gB} = \frac{L_{aq}}{M_{qg}} I_B, \quad I_{kB} = \frac{L_{aq}}{M_{qk}} I_B \tag{3.40}$$

$$\psi_{gB} = \frac{\psi_B}{I_{gB}} I_B, \quad \psi_{kB} = \frac{\psi_B}{I_{kB}} I_B \tag{3.41}$$

where $L_{aq} = L_q - L_{a\sigma}$.

Remarks

1) The per-unit system is chosen such that the per-unit mutual reactances M_{df} and M_{dh} are equal and can be expressed as

$$\tilde{M}_{df} = \tilde{M}_{dh} = \tilde{L}_{ad} = \tilde{L}_d - \tilde{L}_{a\sigma}$$

Similarly,

$$\tilde{M}_{qg} = \tilde{M}_{qk} = \tilde{L}_{aq} = \tilde{L}_q - \tilde{L}_{a\sigma}$$

2) The choice of base quantities is not unique. In general, the per-unit mutual inductances can be expressed as

$$\tilde{M}_{df} = \tilde{M}_{dh} = \tilde{L}_d - \tilde{L}_{c1}$$
$$\tilde{M}_{qg} = \tilde{M}_{qk} = \tilde{L}_q - \tilde{L}_{c2}$$

where $\tilde{L}_{c1}$ and $\tilde{L}_{c2}$ can be arbitrarily chosen (assuming magnetic linearity). It is only when representing saturation that it is convenient to define the per-unit mutual inductances as given earlier. The leakage inductance of the stator $L_{a\sigma}$ is normally assumed to be constant (unaffected by saturation) while mutual inductances are functions of mutual flux linkages (considering saturation).

3) The selection of the base quantities for the rotor circuits can be avoided if it is assumed that the rotor quantities can be referred to the stator, using appropriate turns ratios (which are not always uniquely defined). The equivalent circuit referred to the stator can then be described in per-unit quantities (on the stator base).

3.5 Equivalent Circuits of a Synchronous Machine

There are two equivalent circuits (in the d- and q-axes) which can be used to compute the transformer electromotive force (emf) ($p\psi_d$ and $p\psi_q$) in the d-axis and the q-axis, $\left(p = \frac{d}{dt}\right)$, that depend on the flux linkage associated with the stator d- and q-axis coils. For a machine with three damper windings (one in the d-axis and two in the q-axis) there are three coupled coils in each axis, (d, f, h) in the d-axis and (q, g, k) in the q-axis. It is common practice to assume a common inductance among the three coils and three leakage inductances that are associated with the three coils. Thus, the d- and q-axis equivalent circuits are shown in Figures 3.3 and 3.4, respectively.

It should be noted that $x_c \neq x_{a\sigma}$ where x_c is called the characteristic reactance [10]. Assuming $x_c = x_{a\sigma}$ (where $x_{a\sigma}$ is the armature leakage reactance) can give wrong results in predicting the field quantities. For turbo alternators, x_c is much larger than $x_{a\sigma}$ and for salient pole machines, x_c can be negative. Canay has also proposed a simple test for the determination of x_c.

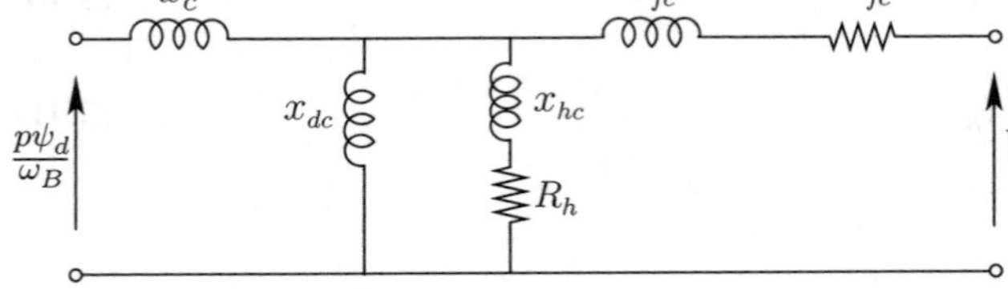

Figure 3.3 Equivalent circuit for the d-axis.

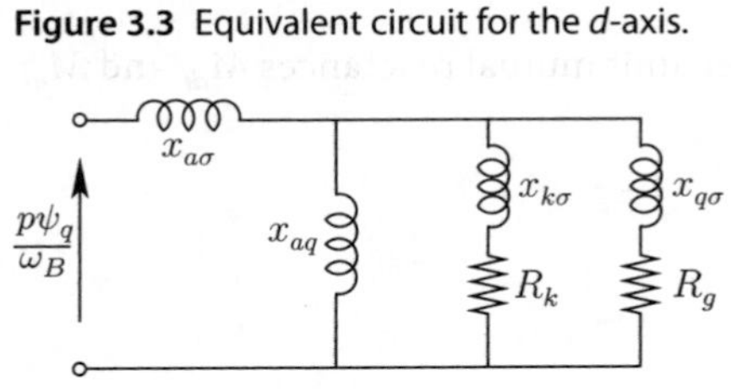

Figure 3.4 Equivalent circuit for the q-axis.

It should be noted that for the q-axis $x_{a\sigma}$ can be chosen to be any value if magnetic saturation is ignored. Generally, $x_{a\sigma}$ is assumed to be constant and x_{aq} is assumed to be affected by saturation (although this may not be accurate). It can be observed that

$$x_q = x_{a\sigma} + x_{aq},\ x_g = x_{g\sigma} + x_{aq},\ x_k = x_{k\sigma} + x_{aq},\ x_{aq} = x_{qk} = x_{gk} = x_{qg}$$

Similarly,

$$x_d = x_{dc} + x_c,\ x_f = x_{dc} + x_{fc},\ x_h = x_{dc} + x_{hc},\ x_{df} = x_{dh} = x_{fh} = x_{dc}$$

Taking Laplace transforms of the currents and flux linkages, we can define two equations as follows:

$$\psi_q(s) = -x_q(s)I_q(s) \tag{3.42}$$

$$\psi_d(s) = -x_d(s)I_d(s) + G(s)E_{fd}(s) \tag{3.43}$$

where

$$x_q(s) = \frac{x_q(1 + sT'_q)(1 + sT''_q)}{(1 + sT'_{qo})(1 + sT''_{qo})} \tag{3.44}$$

$$x_d(s) = \frac{x_d(1 + sT'_d)(1 + sT''_d)}{(1 + sT'_{do})(1 + sT''_{do})} \tag{3.45}$$

$$G(s) = \frac{(1 + sT''_{dc})}{(1 + sT'_{do})(1 + sT''_{do})},\ E_{fd} = \frac{x_{df}}{R_f}v_f \tag{3.46}$$

where T'_{do} and T''_{do} are the open circuit transient and subtransient time constants in the direct axis. Similarly T'_{qo} and T''_{qo} are the open circuit transient and subtransient time constants in the quadrature axis. T'_d and T''_d are the short-circuit transient and subtransient time constants in the d-axis. Similar definitions apply for T'_q and T''_q in the q-axis.

It can be shown that

$$T'_{do} > T'_d > T''_{do} > T''_d$$

Similarly,

$$T'_{qo} > T'_q > T''_{qo} > T''_q$$

Conceptually, the machine model can be viewed as a two-port network in the d-axis and a one-port model in the q-axis. Considering flux linkages analogous to voltages, it can be shown that the d-axis and q-axis equivalent circuits can be viewed as R-C networks which have real poles and zeros that have the interlacing property which gives rise to the above inequalities.

Determination of Parameters

Normally, the parameters R_a, $x_{a\sigma}$, x_c, x_d, and x_q are specified for a synchronous generator. The transient and subtransient quantities can be determined from measurements according to IEC or ASA recommendations. These are x'_d, x''_d, T'_d, and T''_d for the direct axis and x'_q, x''_q, T'_q, and T''_q for the quadrature axis.

Direct Axis Equivalent Circuit

The reciprocal of $x_d(s)$ can be expressed as (in accordance with IEC or ASA standards) as

$$\frac{1}{x_d(s)} = \frac{1}{x_d} + \left(\frac{1}{x'_d} - \frac{1}{x_d}\right)\frac{sT'_d}{1+sT'_d} + \left(\frac{1}{x''_d} - \frac{1}{x'_d}\right)\frac{sT''_d}{1+sT''_d} \tag{3.47}$$

From (3.45) and (3.47) we can obtain

$$T'_{do} + T''_{do} = \frac{x_d}{x'_d}T'_d + \left(1 - \frac{x_d}{x'_d} + \frac{x_d}{x''_d}\right)T''_d$$

$$T'_{do}\,T''_{do} = T'_d\,T''_d\frac{x_d}{x''_d}$$

These equations can be used to calculate T'_{do} and T''_{do} exactly if T'_d and T''_d are known or vice versa. It can be shown that

$$x_d(s) + x_e = \frac{x_{de}(1+sT'_{de})(1+sT''_{de})}{(1+sT'_{do})(1+sT''_{do})} = X_{de}(s) \tag{3.48}$$

where

$$x_{de} = x_d + x_e$$

$$T'_{de} + T''_{de} = \frac{x_d(T'_d + T''_d) + x_e(T'_{do} + T''_{do})}{x_d + x_e}$$

$$T'_{de}\,T''_{de} = T'_{do}\,T''_{do}\frac{x''_d + x_e}{x_d + x_e}$$

It is possible to calculate x'_{de} and x''_{de} from the expression for the inverse of $x_{de}(s)$:

$$x'_{de} = \frac{x_{de}(T'_{de} - T''_{de})}{T'_{do} + T''_{do} - \left(1 + \frac{x_{de}}{x''_{de}}\right)T''_{de}}$$

$$x''_{de} = x''_d + x_e$$

It should be noted that $x'_{de} \neq x'_d + x_e$.

Equation (3.47) represents the admittance of the equivalent circuit shown in Figure 3.5. An alternate representation is shown in Figure 3.6 where x_c is included in series and

$$x_{dc} = x_d - x_c,\ x''_{dc} = x''_d - x''_c$$

By using (3.48) where $x_e = -x_c$, we can obtain

$$x_{fo} = \frac{x_d x'_d}{x_d - x'_d},\ R_{fo} = \frac{x_{fo}}{\omega_B T'_d}$$

$$x_{h0} = \frac{x'_d x''_d}{x'_d - x''_d},\ R_{h0} = \frac{x_{h0}}{\omega_B T''_d}$$

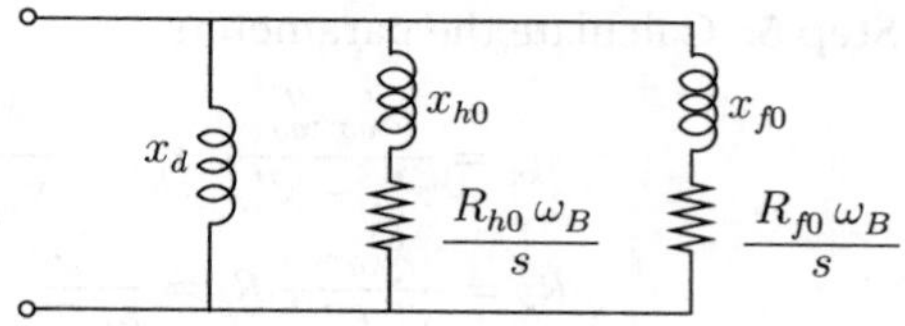

Figure 3.5 Direct axis equivalent circuit.

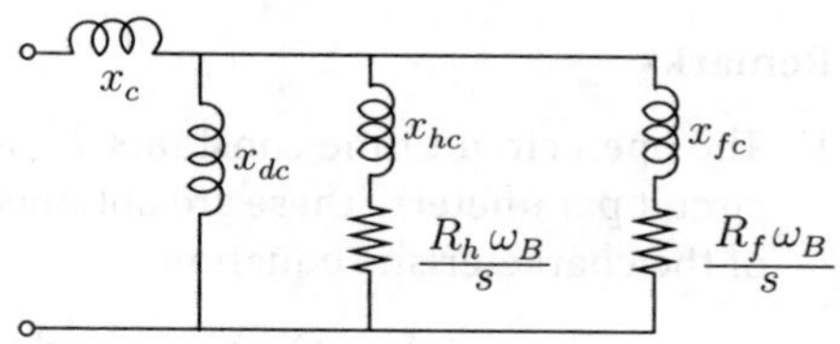

Figure 3.6 Alternative representation.

The expressions for x_{fc}, R_f, x_{hc}, and R_h are as given below:

$$x_{fc} = \frac{x_{de}x'_{de}}{x_{de} - x'_{de}}, \quad R_f = \frac{x_{fc}}{\omega_B T'_{de}}$$

$$x_{hc} = \frac{x'_{de}x''_{de}}{x'_{de} - x''_{de}}, \quad R_h = \frac{x_{hc}}{\omega_B T''_{de}}$$

Quadrature Axis Equivalent Circuit

The computation of circuit parameters is similar to what is given for the direct axis equivalent circuit. Hence one has to determine $x_{g\sigma}$, $x_{k\sigma}$, R_g, and R_k. It should be noted that x_q, x'_q, and x''_q replace x_d, x'_d, and x''_d, respectively. $x_{a\sigma}$ replaces x_c. The steps are given below:

Step 1: Compute $x_{aq} = x_q - x_{a\sigma}$ and $x''_{aq} = x''_q - x_{a\sigma}$.

Step 2: Compute T'_{qo} and T''_{qo} from

$$(T'_{qo} + T''_{qo}) = \frac{x_q}{x'_q}T'_q + \left(1 - \frac{x_q}{x'_q} + \frac{x_q}{x''_q}\right)T''_q$$

$$(T'_{qo}\ T''_{qo}) = T'_q\ T''_q\frac{x_q}{x''_q}$$

Step 3: Compute $T'_{q\sigma}$ and $T''_{q\sigma}$ from

$$(T'_{q\sigma} + T''_{q\sigma}) = (T'_q + T''_q)\frac{x_q}{x_{aq}} - (T'_{qo} + T''_{qo})\frac{x_{a\sigma}}{x_{aq}}$$

$$(T'_{q\sigma}\ T''_{q\sigma}) = T'_{qo}\ T''_{qo}\frac{x'_{qo}}{x_{aq}}$$

Step 4: Calculate the parameter

$$x'_{q\sigma} = \frac{x_{aq}(T'_{q\sigma} - T''_{q\sigma})}{T'_{qo} + T''_{qo} - \left(1 + \frac{x_{aq}}{x''_{q\sigma}}T''_{q\sigma}\right)}$$

Step 5: Calculate the parameters

$$x_{k\sigma} = \frac{x'_{q\sigma}x''_{q\sigma}}{x'_{q\sigma} - x''_{q\sigma}}, \quad x_{g\sigma} = \frac{x_{aq}x'_{q\sigma}}{x_{aq} - x'_{q\sigma}}$$

$$R_k = \frac{x_{k\sigma}}{\omega_B T''_{q\sigma}}, \quad R_g = \frac{x_{g\sigma}}{\omega_B T'_{q\sigma}}$$

Remarks

1) The open circuit time constants T'_{qo} and T''_{qo} can be determined from the equivalent circuit parameters. These are obtained as the negatives of the reciprocals of the roots of the characteristic equation

$$1 + \frac{s}{\omega_B}\left(\frac{x_g}{R_g} + \frac{x_k}{R_k}\right) + \frac{s^2}{\omega_B^2}\left[\frac{x_{aq}(x_{g\sigma} + x_{k\sigma}) + x_{g\sigma}x_{k\sigma}}{R_g R_k}\right] = 0$$

If $R_g \gg R_k$, T'_{qo} and T''_{qo} can be approximated as

$$T'_{qo} \approx \frac{x_g}{\omega_B R_g} = \frac{x_{g\sigma} + x_{aq}}{\omega_B R_g}$$

$$T''_{qo} \approx \frac{1}{\omega_B R_k}\left[x_{k\sigma} + \frac{x_{g\sigma}x_{a\sigma}}{(x_{aq} + x_{g\sigma})}\right]$$

2) If R_g and R_k are of similar magnitude the approximations given above are not valid.
3) Neglecting saturation, $x_{a\sigma}$ can be selected arbitrarily (even zero).

3.6 Synchronous Machine Models for Stability Analysis

The stability of power systems is affected by the rotor swings of the synchronous generators. Even in steady state, the system is subject to slow variations and perturbations in the connected loads. The small oscillations in the generator rotors are usually damped by the damper (amortisseur) windings. The frequency of the rotor oscillations does not exceed 5 Hz. Depending on the system loading and the network configuration, the oscillations can be undamped unless special controllers are employed to stabilize the system. Sometimes, due to faults, the system is subjected to large disturbances that can lead to the loss of synchronism and collapse of the system, leading to extensive blackouts. The prediction of system stability either by analysis or system simulation can be simplified as follows:

1) Neglecting the dynamics of the network and representing the network by bus admittance matrix in the complex domain.
2) Approximating the stator equations of the synchronous generator by ignoring transformer emfs and the variations in the rotor frequency.

The stator equations (3.36) in per-unit values are reproduced below for convenience:

$$\frac{1}{\omega_B}\frac{d\psi_d}{dt} + (1 + S_m)\psi_q - R_a i_d = v_d$$

$$\frac{1}{\omega_B}\frac{d\psi_q}{dt} - (1 + S_m)\psi_d - R_a i_q = v_q$$

where S_m is the slip defined by

$$S_m = \frac{\omega - \omega_B}{\omega_B} \tag{3.49}$$

Ignoring transformer emf implies assuming $\frac{d\psi_d}{dt} = 0$ and $\frac{d\psi_q}{dt} = 0$. Neglecting variations in the rotor frequency implies assuming $S_m = S_{mo}$ where

$$S_{mo} = \frac{\omega_o - \omega_B}{\omega_B}$$

Typically $S_{mo} \approx 0$. Thus we obtain the generator terminal voltages v_d and v_q as

$$v_d \approx \psi_q - R_a i_d \quad \text{and} \quad v_q \approx -\psi_d - R_a i_q \tag{3.50}$$

Stator Equations

The stator flux linkages are given by

$$\psi_d = -x_d i_d + x_{df} i_f + x_{dh} i_h \tag{3.51}$$

$$\psi_q = -x_q i_q + x_{qg} i_g + x_{qk} i_k \tag{3.52}$$

It is possible to eliminate i_f and i_h from (3.51) by expressing them in terms of ψ_f, ψ_h, and i_d. The expressions for i_f and i_h are obtained from

$$\begin{bmatrix} i_f \\ i_h \end{bmatrix} = \begin{bmatrix} x_f & x_{fh} \\ x_{fh} & x_h \end{bmatrix}^{-1} \left\{ \begin{bmatrix} \psi_f \\ \psi_h \end{bmatrix} + \begin{bmatrix} x_{df} \\ x_{dh} \end{bmatrix} i_d \right\} \tag{3.53}$$

Similarly i_g and i_k in (3.52) can be substituted from

$$\begin{bmatrix} i_g \\ i_k \end{bmatrix} = \begin{bmatrix} x_g & x_{gk} \\ x_{gk} & x_k \end{bmatrix}^{-1} \left\{ \begin{bmatrix} \psi_g \\ \psi_k \end{bmatrix} + \begin{bmatrix} x_{qg} \\ x_{qk} \end{bmatrix} i_q \right\} \tag{3.54}$$

Eliminating rotor currents, (3.51) and (3.52) reduce to

$$\begin{aligned} \psi_d &= -x''_d i_d - E''_q \\ \psi_q &= -x''_q i_q + E''_d \end{aligned} \tag{3.55}$$

where

$$\begin{aligned} -E''_q &= \frac{(x_{df}x_h - x_{dh}x_{fh})}{(x_f x_h - x_{fh}^2)}\psi_f + \frac{(x_{dh}x_f - x_{df}x_{fh})}{(x_f x_h - x_{fh}^2)}\psi_h \\ &= C_1\psi_f + C_2\psi_h \end{aligned} \tag{3.56}$$

$$\begin{aligned} E''_d &= \frac{(x_{qg}x_k - x_{qk}x_{gk})}{(x_g x_k - x_{gk}^2)}\psi_g + \frac{(x_{qk}x_g - x_{qg}x_{gk})}{(x_g x_k - x_{gk}^2)}\psi_k \\ &= C_3\psi_g + C_4\psi_k \end{aligned} \tag{3.57}$$

$$x''_d = x_d - C_1 x_{df} - C_2 x_{dh} \tag{3.58}$$

$$x''_q = x_q - C_3 x_{qg} - C_4 x_{qk} \tag{3.59}$$

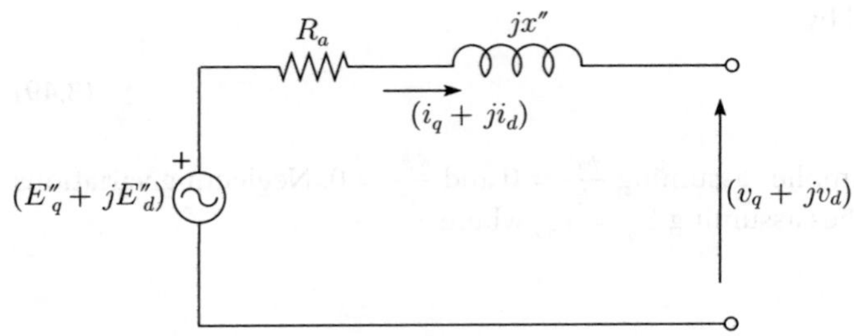

Figure 3.7 Stator equivalent circuit for Model (2.2).

Substituting the above in (3.55), we get

$$E''_q + x''_d i_d - R_a i_q = v_q \tag{3.60}$$

$$E''_d - x''_q i_q - R_a i_d = v_d \tag{3.61}$$

If subtransient saliency is neglected, that is, $x''_q = x''_d = x''$, then the above equations represent an equivalent circuit, shown in Figure 3.7.

Rotor Equations

The basic equations in the rotor flux linkages (using per-unit quantities) are

$$\frac{d\psi_f}{dt} = \omega_B \left[-R_f i_f + \frac{R_f}{x_{df}} E_{fd} \right]$$
$$\frac{d\psi_h}{dt} = -\omega_B R_h i_h \tag{3.62}$$

$$\frac{d\psi_g}{dt} = -\omega_B R_g i_g$$
$$\frac{d\psi_k}{dt} = -\omega_B R_k i_k \tag{3.63}$$

Eliminating the rotor currents using (3.53) and (3.54), we can express the above equations as

$$\frac{d\psi_f}{dt} = a_1 \psi_f + a_2 \psi_h + b_1 E_{fd} + b_2 i_d \tag{3.64}$$

$$\frac{d\psi_h}{dt} = a_3 \psi_f + a_4 \psi_h + b_3 i_d \tag{3.65}$$

$$\frac{d\psi_g}{dt} = a_5 \psi_g + a_6 \psi_k + b_4 i_q \tag{3.66}$$

$$\frac{d\psi_k}{dt} = a_7 \psi_g + a_8 \psi_k + b_5 i_q \tag{3.67}$$

where

$$a_1 = -\frac{\omega_B R_f x_h}{D_1}, a_2 = \frac{\omega_B R_f x_{fh}}{D_1}, \quad a_3 = \frac{\omega_B R_h x_{fh}}{D_1}, a_4 = -\frac{\omega_B R_h x_f}{D_1}$$

$$D_1 = x_f x_h - x_{fh}^2$$

$$a_5 = -\frac{\omega_B R_g x_k}{D_2}, a_6 = \frac{\omega_B R_g x_{gk}}{D_2}, \quad a_7 = \frac{\omega_B R_k x_{gk}}{D_2}, a_8 = -\frac{\omega_B R_k x_g}{D_2}$$

$$D_2 = x_g x_k - x_{gk}^2$$

$$b_1 = \frac{\omega_B R_f}{x_{df}}, \quad b_2 = \omega_B R_f C_1, \quad b_3 = \omega_B R_h C_2$$

$$b_4 = \omega_B R_g C_3, \quad b_5 = \omega_B R_k C_4$$

Electrical Torque

The new expression for electrical torque can be obtained by substituting (3.55) in the basic expression for the torque given by

$$T_e = \psi_q i_d - \psi_d i_q = E_q'' i_q + E_d'' i_d + (x_d'' - x_q'') i_d i_q \tag{3.68}$$

E_d'' and E_q'' can be substituted from expressions (3.56) and (3.57). The final expression for T_e can be written as

$$T_e = -(C_1 \psi_f + C_2 \psi_h) i_q + (C_3 \psi_g + C_4 \psi_k) i_d + (x_d'' - x_q'') i_q i_d \tag{3.69}$$

The last term is due to subtransient saliency.

Remarks

1. Since the stator equations (3.50) are algebraic (neglecting stator transients), it is not possible to choose stator currents i_d and i_q as state variables (state variables have to be continuous due to any sudden changes in the network). As rotor windings either remain closed (damper winding) or are closed through finite voltage source (field winding), the flux linkages of these windings cannot change suddenly. This implies that if i_d changes suddenly, the field and damper currents also change suddenly in order to maintain the field and damper flux linkages continuous. The flux linkage immediately after a disturbance remains constant at the value just prior to the disturbance. (This property is termed the theorem of constant flux linkages in the literature, see Kimbark [2].)
   The previous discussion shows that rotor winding currents cannot be treated as state variables when stator transients are neglected. The obvious choice of state variables are rotor flux linkages or transformed variables which are linearly dependent on the rotor flux linkages.
2. Depending on the degree of detail used, the number of rotor windings and corresponding state variables can vary from one to six. In a report published in 1986 by an IEEE Task Force [11], the following models were suggested based on varying degrees of complexity:
   (a) Classical model (Model (0.0))
   (b) Field circuit only (Model (1.0))
   (c) Field circuit with one equivalent damper on the q-axis (Model (1.1))
   (d) Field circuit with two damper windings, one each in the d- and q-axes (Model (2.1))
   (e) Field circuit with three damper windings, one in the d-axis and two in the q-axis (Model (2.2))

3.6.1 Application of Model (2.1)

We have assumed that the detailed synchronous machine is represented by Model (2.2). This model is widely used in the literature and data for it are supplied by manufacturers of machines or obtained by tests described in IEEE Standard No. 115. Model (2.1) and Model (1.1) are widely used for hydro generators. It should be noted that while higher-order models provide better results for special applications, they also require an exact determination of parameters. With constraints on data availability and for study of large systems, it may be adequate to use Model (1.1) if the data are correctly determined.

This model differs from Model (2.2) in that only one damper winding (say g) is considered in the q-axis. The equations are identical to those given in the previous section except for the following

$$E_d'' = C_3'\psi_g, \quad C_3' = \frac{x_{qg}}{x_g} \tag{3.70}$$

$$\frac{d\psi_g}{dt} = -\frac{\omega_B R_g \psi_g}{x_g} + b'_4 i_q, \quad b'_4 = \omega_B R_g C'_3 \tag{3.71}$$

$$C_4 = 0, \quad x_q'' = x_q - C_3' x_{qg} \tag{3.72}$$

It should be noted that in the classification of the machine models, the first number indicates the number of rotor windings on the d-axis while the second number indicates the number of rotor windings on the q-axis. (Alternately, the numbers represent the number of state variables considered in the d-axis and the q-axis.) Thus the classical model, which neglects damper circuits and field flux decay, ignores all state variables for the rotor coils and is termed Model (0.0).

3.6.2 Application of Model (1.1)

Here, neither ψ_h nor ψ_k is considered. Hence, the subtransient reactances are not relevant. We can show that

$$\begin{aligned} \psi_d &= -x_d' i_d - E_q' \\ \psi_q &= -x_q' i_q + E_d' \end{aligned} \tag{3.73}$$

where

$$E_q' = -\frac{x_{ad}\psi_f}{x_f}, \quad E_d' = \frac{x_{aq}\psi_g}{x_g} \tag{3.74}$$

The voltage equations for the rotor winding are

$$\frac{1}{\omega_B}\frac{d\psi_f}{dt} = -R_f i_f + v_f \tag{3.75}$$

$$\frac{1}{\omega_B}\frac{d\psi_g}{dt} = -R_g i_g \tag{3.76}$$

It is possible to express the above equations as

$$\frac{dE_q'}{dt} = \frac{1}{T_{do}'}[-E_q' + (x_d - x_d')i_d + E_{fd}] \tag{3.77}$$

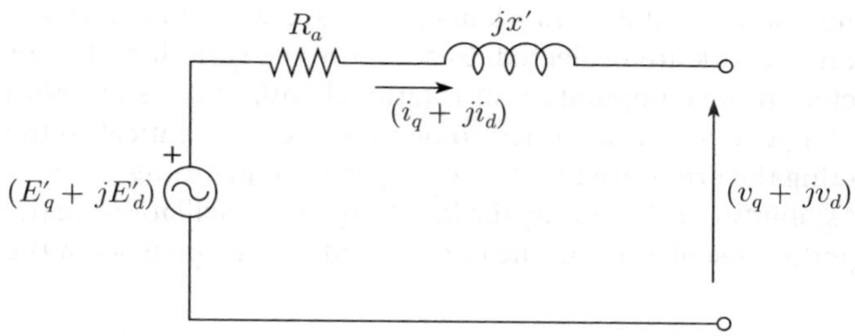

Figure 3.8 Stator equivalent circuit for Model (1.1).

$$\frac{dE'_d}{dt} = \frac{1}{T'_{qo}}[-E'_d - (x_q - x'_q)i_q] \tag{3.78}$$

where

$$T'_{do} = \frac{x_f}{\omega_B R_f} \quad \text{and} \quad T'_{qo} = \frac{x_g}{\omega_B R_g} \tag{3.79}$$

are the open circuit, d-axis and q-axis time constants, respectively. Note that these expressions are valid only for Model (1.1) and are not applicable for higher-order models.

Substituting (3.73) in (3.50), we obtain

$$E'_q + x'_d i_d - R_a i_q = v_q \tag{3.80}$$

$$E'_d - x'_q i_q - R_a i_d = v_d \tag{3.81}$$

These equations are similar to (3.60) and (3.61). If $x'_d = x'_q = x'$ (there is no transient saliency), then we can obtain an equivalent circuit shown in Figure 3.8, which is similar to that shown in Figure 3.7 except for the fact that subtransient reactances are replaced by transient reactances. Obviously, the emfs E'_q and E'_d are dependent only on the fluxes of field winding (on the d-axis) and damper winding (g) (on the q-axis), respectively.

Remarks

1) Typically, subtransient saliency is often neglected. However, transient saliency ($x'_d \neq x'_q$) is not normally neglected.
2) The transient saliency does not pose any problem for a single-machine system. However, for a multi-machine system, it can pose a problem.
3) It is convenient to introduce the term "dynamic saliency" [12] to take into account various levels of complexity in the representation of rotor windings. A simple approach of handling dynamic saliency to achieve a circuit model of the generator stator will be described next.
4) Models simpler than (1.1) can be derived from Model (1.1) by small adjustments in the data:
   (a) Model (1.0) can be obtained by letting $x'_q = x_q$ and $T'_{qo} \neq 0$.
   (b) Model (0.0) can be obtained from Model (1.0) by letting T'_{do} = large value (say 1000 s).

   It should be noted that Model (0.0), also called the classical model, assumes the field flux linkage remains constant (in the time interval of study involving first swing stability).

5) In transient stability simulation, the stator transients are neglected just as the transients in the transmission network are neglected. In addition, the variation of rotor frequency is also neglected in the computation of rotational emf. This also implies that the power output of a generator is assumed to be numerically identical to the electrical torque (neglecting the armature resistance) in per-unit quantities.
6) The analysis of transient stability considers only the low-frequency oscillations in the generator rotors which induce oscillations in the current and voltage phasors in the network.

3.6.3 Modeling of Saturation

Magnetic materials used in the stator and rotor of a synchronous machine exhibit saturation, which implies that the machine inductances are not constants but depend on the levels of flux in different parts of the machine. There are several saturation models proposed in the literature, but they tend to be highly empirical and lack theoretical justifications. Typically, the leakage reactances representing the flux paths in the air gap are assumed to be unaffected by saturation.

Accurate modeling of saturation is considered to be important for the determination of field excitation (current) required for a specified generator power output and operating rotor angle (δ). An important phenomenon resulting from saturation is the cross coupling between the d- and q-axis coils (termed the cross-magnetizing phenomenon). Due to saturation, the d-axis current i_d also produces q-axis flux linkages and vice versa.

From the point of view of conducting stability studies in multi-machine systems, what is required is the accuracy of a specific saturation model, which should be simple in application. In reference [13], the effects of various saturation models were investigated and the conclusions indicate that using saturated, but constant, values of reactances gives reasonably accurate results. While this result is significant particularly for large-scale system studies, it needs to be checked with experimental investigation. Unfortunately, there are no detailed reports of comparative studies on saturation models.

3.7 An Exact Circuit Model of a Synchronous Machine for Electromagnetic Transient Analysis [14]

Although the transient stability analysis involves low-frequency oscillations caused by rotor oscillations, the stability is also influenced by subsynchronous frequency rotor oscillations, which takes into account the fact that in a turbogenerator there are several masses due to several steam turbines (low, intermediate, and high pressure) connected to a long shaft which has elasticity (not rigid) and can be modeled by springs (connecting two masses). The torsional oscillations at frequency (f_r) result in voltages and currents oscillations in the network having frequencies $f_o \pm f_r$, which produce electrical torque oscillations of frequency f_r that can be sustained or tend to grow due to torsional interaction (TI). The analysis of subsynchronous resonance (SSR) requires a detailed model of a synchronous machine taking into account the stator transients along with the detailed modeling of torsional dynamics.

The simulation of SSR phenomena is usually performed using the Electromagnetic Transients Program (EMTP), which is briefly described in Chapter 2. Here, the stator transients also have to be modeled. Thus, there is a need for a detailed generator model. The system (involving the transmission network and the generator) can be easily handled if there is an exact circuit model that does not introduce any assumptions that cannot be justified.

3.7.1 Derivation of the Circuit Model

The objective is to obtain a circuit model which has constant parameters. It is assumed that the terminal voltages of the rotor coils are specified. This is a valid assumption as the field voltage is obtained as the output of the excitation control system and the damper windings are shorted. There is no restriction on the number of windings. Additional assumption are:

(a) saliency is restricted to the rotor
(b) magnetic saturation and hysteresis are ignored
(c) eddy current effects in the generator rotor are represented by a set of coils with constant parameters.

Because of the possibility that the stator currents are not treated as state variables in some cases, the rotor currents cannot be treated as state variables. Hence the rotor equations are expressed in terms of rotor flux linkages.

The circuit model with constant parameters is achieved mainly by transforming the time-varying coupling between the stator and rotor coils into dependent current sources. To handle *dynamic saliency* (defined later) the additional concept of a dummy coil on the rotor is essential. The machine equations (using flux linkages as state variables) are

$$p\psi_s + [R_s]\bar{i}_s = v_s \tag{3.82}$$

$$p\psi_r + [R_r]i_r = v_r \tag{3.83}$$

where $p = \dfrac{d}{dt}$ and

$$\psi_s = [L_{ss}]\bar{i}_s + [M]i_r \tag{3.84}$$

$$\psi_r = [M]^t\bar{i}_s + [L_{rr}]i_r \tag{3.85}$$

From the above two equations, ψ_s and i_r can be expressed as

$$\psi_s = [L''_s]\bar{i}_s + [M][L_{rr}]^{-1}\psi_r \tag{3.86}$$

$$i_r = [L_{rr}]^{-1}\psi_r - [L_{rr}]^{-1}[M]^t\bar{i}_s \tag{3.87}$$

where

$$[L''_s] = [L_{ss}] - [M][L_{rr}]^{-1}[M]^t \tag{3.88}$$

Let

$$I_s \stackrel{\Delta}{=} [L''_s]^{-1}[M][L_{rr}]^{-1}\psi_r \tag{3.89}$$

Then (3.86) can be written as

$$\psi_s = [L''_s](\bar{i}_s + I_s) \tag{3.90}$$

From (3.87) and (3.83), we get

$$p\psi_r + [R_r][L_{rr}]^{-1}\psi_r - [R_r][L_{rr}]^{-1}[M]^t\bar{i}_s = v_r \tag{3.91}$$

Since $[L_{rr}]$ is a constant matrix, (3.91) is an explicit state equation with the stator currents treated as inputs. The matrices $[L_{ss}]$ and $[M]$ are functions of the rotor angle θ (see (3.4) and (3.6)). Thus $[L_s'']$ is also dependent on θ, in general. However, it is possible by assuming the existence of additional coil(s) on the rotor to achieve the final value of $[L_s'']$ as constant. The additional coil(s) are termed dummy coil(s) and theoretically are assumed to be open. However, in a multi-machine system it becomes necessary to assume that the dummy coil(s) are closed through arbitrarily chosen resistances such that the flux linkage(s) of the dummy coil(s) are described by state equations that are introduced to account for dynamic saliency. Generally, a single dummy coil on the q-axis is adequate to approximate dynamic saliency. A convenient choice of the parameters of the dummy coil can be made by assuming that there is no coupling between the dummy coil(s) and the other rotor coils. Consequently $[L_{rr}]_{new}$ and $[M]_{new}$ can be written as

$$[L_{rr}]_{new} \stackrel{\Delta}{=} \begin{bmatrix} L_{rr} & 0 \\ 0 & L_c \end{bmatrix}, \quad [M]_{new} \stackrel{\Delta}{=} \begin{bmatrix} M & M_c \end{bmatrix}$$

The introduction of dummy coil modifies the value of $[L_s'']$ and the new value is

$$[L_s'']_{new} = [L_{ss}] - [M]_{new}[L_{rr}]_{new}^{-1}[M]_{new}^t \tag{3.92}$$

By choosing L_c and M_c appropriately, $[L_s'']_{new}$ can be arranged to be independent of the rotor angle θ. It should be noted that L_c is the self inductance of the dummy coil and M_c is the column vector of mutual inductances between the dummy coil and the stator coils. The current source vector I_s is also modified to

$$I_{s\ new} = [L_s'']_{new}^{-1}[ML_{rr}^{-1}\psi_r + M_cL_c^{-1}\psi_c] \tag{3.93}$$

The term "dynamic saliency", taken from [12], is defined by the condition that $[L_s'']$ is dependent on the rotor angle. A synchronous machine with dynamic saliency can be transformed to a machine with no dynamic saliency ($[L_s'']_{new}$ is a constant matrix) by the addition of a term in the expression for I_s that is proportional to the stator currents. The equivalent circuit for the stator coils of a synchronous machine is shown in Figure 3.9. It should be noted that $\bar{i}_s$ is defined as the armature (stator) currents flowing into the

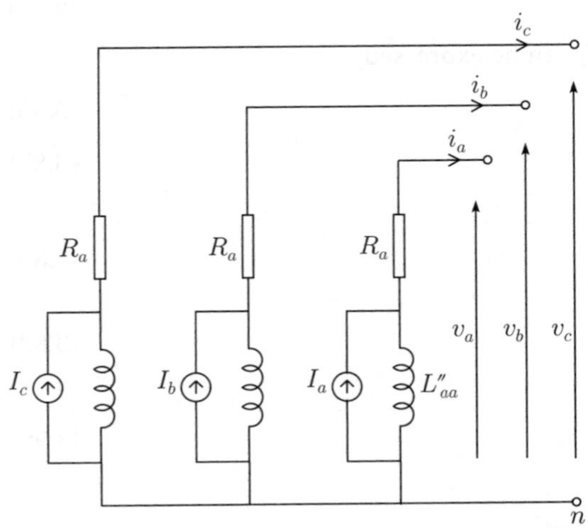

Figure 3.9 Circuit model for stator (phase variables).

generator (motor convention). The equation for the circuit model shown in Figure 3.11 is given by

$$[L_s'']p(\bar{i}_s + I_s) + [R_s]\bar{i}_s = v_s \tag{3.94}$$

For simplicity, the subscript *new* is dropped.

Torque Equation

The expression for the electrical torque is given by

$$\begin{aligned} T_e &= -\frac{1}{2}i^t\left[\frac{\partial L}{\partial \theta}\right]i = -\frac{1}{2}i^t\frac{\partial \psi}{\partial \theta} \\ &= -\frac{1}{2}\left[i_s^t\frac{\partial \psi_s}{\partial \theta} + i_r^t\frac{\partial \psi_r}{\partial \theta}\right] \end{aligned} \tag{3.95}$$

where

$$\begin{aligned} \frac{\partial \psi_s}{\partial \theta} &= [M][L_{rr}]^{-1}\frac{\partial \psi_r}{\partial \theta} + \left[\frac{\partial M}{\partial \theta}\right][L_{rr}]^{-1}\psi_r \\ &\quad + [M_c][L_c]^{-1}\frac{\partial \psi_c}{\partial \theta} + \left[\frac{\partial M_c}{\partial \theta}\right][L_c]^{-1}\psi_c \end{aligned}$$

$$\frac{\partial \psi_r}{\partial \theta} = \frac{\partial [M]^t}{\partial \theta}i_s, \qquad \frac{\partial \psi_c}{\partial \theta} = \frac{\partial [M_c]^t}{\partial \theta}i_s$$

The final torque expression is given as

$$\begin{aligned} T_e &= -i_s^t\left[\frac{\partial [M]^t}{\partial \theta}[L_{rr}]^{-1}\psi_r + \frac{\partial [M_c]^t}{\partial \theta}[L_c]^{-1}\psi_c\right] \\ &= -i_s^t[L_s'']I_t \end{aligned} \tag{3.96}$$

where

$$I_t = \frac{\partial I_s}{\partial \theta} = [L_s'']^{-1}\left\{\frac{\partial [M]}{\partial \theta}[L_{rr}]^{-1}\psi_r + \frac{\partial [M_c]}{\partial \theta}[L_{rr}]^{-1}\psi_c\right\} \tag{3.97}$$

Example 3.2 ***Model (2.2)*** For Model (2.2), $[L_{ss}]$, $[L_{rr}]$, and $[M]$ are defined in (3.4)–(3.6). The parameters $[L_s'']$ and I_s of the circuit model are obtained as

$$\begin{aligned} [L_s''] &= [L_{ss}] - [M][L_{rr}]^{-1}[M]^t \\ &= \begin{bmatrix} L_{aao} & L_{abo} & L_{abo} \\ L_{abo} & L_{aao} & L_{abo} \\ L_{abo} & L_{abo} & L_{aao} \end{bmatrix} - \frac{(s_1 + s_2)}{2}\begin{bmatrix} 1 & -\frac{1}{2} & -\frac{1}{2} \\ -\frac{1}{2} & 1 & -\frac{1}{2} \\ -\frac{1}{2} & -\frac{1}{2} & 1 \end{bmatrix} \\ &\quad + \left(L_{aa2} - \frac{(s_1 - s_2)}{2}\right)\begin{bmatrix} \cos 2\theta & \cos\left(2\theta - \frac{2\pi}{3}\right) & \cos\left(2\theta + \frac{2\pi}{3}\right) \\ \cos\left(2\theta - \frac{2\pi}{3}\right) & \cos\left(2\theta + \frac{2\pi}{3}\right) & \cos 2\theta \\ \cos\left(2\theta + \frac{2\pi}{3}\right) & \cos 2\theta & \cos\left(2\theta - \frac{2\pi}{3}\right) \end{bmatrix} \end{aligned} \tag{3.98}$$

where s_1 and s_2 are defined by

$$s_1 \stackrel{\Delta}{=} \frac{(M_{af}L_h - M_{ah}L_{fh})M_{af} + (M_{ah}L_f - M_{af}L_{fh})M_{ah}}{L_fL_h - L_{fh}^2} \tag{3.99}$$

$$s_2 \stackrel{\Delta}{=} \frac{(M_{ag}L_k - M_{ak}L_{gk})M_{ag} + (M_{ak}L_g - M_{ag}L_{gk})M_{ak}}{L_gL_k - L_{gk}^2} \tag{3.100}$$

Absence of Dynamic Saliency

Dynamic saliency is absent if

$$L_{aa2} - \frac{s_1 - s_2}{2} = 0 \tag{3.101}$$

In this case, $[L_s'']$ is reduced to a constant matrix and the current source I_s is directly obtained as

$$I_s = [L_s'']^{-1}[M][L_{rr}]^{-1}\psi_r$$
$$= I_1 \begin{bmatrix} \cos\theta \\ \cos\left(\theta - \frac{2\pi}{3}\right) \\ \cos\left(\theta + \frac{2\pi}{3}\right) \end{bmatrix} + I_2 \begin{bmatrix} \sin\theta \\ \sin\left(\theta - \frac{2\pi}{3}\right) \\ \sin\left(\theta + \frac{2\pi}{3}\right) \end{bmatrix} \tag{3.102}$$

where I_1 and I_2 are defined by

$$I_1 \stackrel{\Delta}{=} \frac{1}{L''} \frac{(M_{af}L_h - M_{ah}L_{fh})\psi_f + (M_{ah}L_f - M_{af}L_{fh})\psi_h}{L_fL_h - L_{fh}^2} \tag{3.103}$$

$$I_2 \stackrel{\Delta}{=} \frac{1}{L''} \frac{(M_{ag}L_k - M_{ak}L_{gk})\psi_g + (M_{ak}L_g - M_{ag}L_{gk})\psi_k}{L_gL_k - L_{gk}^2} \tag{3.104}$$

where $L'' = L_d'' = L_q''$. The rotor equations are given by

$$\frac{d\psi_r}{dt} = -[R_r][L_{rr}]^{-1}\psi_r + [R_r][M]^t i_s + v_r \tag{3.105}$$

where $\psi_r^t = \begin{bmatrix}\psi_f & \psi_h & \psi_g & \psi_k\end{bmatrix}$ and $v_r^t = \begin{bmatrix}v_f & 0 & 0 & 0\end{bmatrix}$. The electrical torque is given by

$$T_e = -i_s^t[L_s'']I_t = i_s^t[L_s'']\frac{\partial I_s}{\partial\theta} \tag{3.106}$$

$$\frac{\partial I_s}{\partial\theta} = I_1 \begin{bmatrix} -\sin\theta \\ -\sin\left(\theta - \frac{2\pi}{3}\right) \\ -\sin\left(\theta + \frac{2\pi}{3}\right) \end{bmatrix} + I_2 \begin{bmatrix} \cos\theta \\ \cos\left(\theta - \frac{2\pi}{3}\right) \\ \cos\left(\theta + \frac{2\pi}{3}\right) \end{bmatrix} \tag{3.107}$$

Substituting (3.107) in (3.106), we get

$$T_e = L''\sqrt{\frac{3}{2}}(I_1 i_q - I_2 i_d) \tag{3.108}$$

where i_d and i_q are defined by

$$i_d = \sqrt{\frac{2}{3}}\left\{i_a \cos\theta + i_b \cos\left(\theta - \frac{2\pi}{3}\right) + i_c \cos\left(\theta + \frac{2\pi}{3}\right)\right\}$$

$$i_q = \sqrt{\frac{2}{3}}\left\{i_a \sin\theta + i_b \sin\left(\theta - \frac{2\pi}{3}\right) + i_c \sin\left(\theta + \frac{2\pi}{3}\right)\right\}$$

Comparing (3.108) with (3.A.8), we note that

$$\psi_d = \sqrt{\frac{3}{2}}L''I_1, \quad \psi_q = \sqrt{\frac{3}{2}}L''I_2 \tag{3.109}$$

3.7.2 Transformation of the Circuit Model

We will consider the transformation of the circuit model (based on phase coordinates) to the model based on a rotating reference frame (Park's transformation) synchronized with the generator rotor. The transformation to *dqo* components is facilitated through $\alpha\beta o$ components. Hence, we can express

$$\begin{bmatrix} i_a \\ i_b \\ i_c \end{bmatrix} = [C_1]\begin{bmatrix} i_\alpha \\ i_\beta \\ i_o \end{bmatrix}; \begin{bmatrix} i_\alpha \\ i_\beta \\ i_o \end{bmatrix} = [C_2]\begin{bmatrix} i_d \\ i_q \\ i_o \end{bmatrix} \tag{3.110}$$

Evidently,

$$\begin{bmatrix} i_a \\ i_b \\ i_c \end{bmatrix} = [C_3]\begin{bmatrix} i_d \\ i_q \\ i_o \end{bmatrix}$$

where $[C_3] = [C_1][C_2]$. If $[C_1]$ and $[C_2]$ are chosen as orthogonal matrices, then $[C_3]$ is also orthogonal matrix and we can write

$$\begin{bmatrix} v_d \\ v_q \\ v_o \end{bmatrix} = [C_3]^t\begin{bmatrix} v_a \\ v_b \\ v_c \end{bmatrix} \tag{3.111}$$

(Note that $[C_3]^{-1} = [C_3]^t$). The matrices $[C_1]$ and $[C_2]$ are defined by

$$[C_1] = \begin{bmatrix} \sqrt{\frac{2}{3}} & 0 & \frac{1}{\sqrt{3}} \\ -\frac{1}{\sqrt{6}} & -\frac{1}{\sqrt{2}} & \frac{1}{\sqrt{3}} \\ -\frac{1}{\sqrt{6}} & \frac{1}{\sqrt{2}} & \frac{1}{\sqrt{3}} \end{bmatrix}, \quad [C_2] = \begin{bmatrix} \cos\theta & \sin\theta & 0 \\ -\sin\theta & \cos\theta & 0 \\ 0 & 0 & 1 \end{bmatrix}$$

If $\theta = \omega_o t + \delta$, then $\frac{d\delta}{dt} = \frac{d\theta}{dt} - \omega_o$. Since θ is the rotor angle w.r.t. a stationary reference, we obtain Park's reference frames. If $\theta = \omega_o t + \gamma$, with $\gamma =$ constant (chosen as zero) we obtain Kron's reference frame. First, applying the $\alpha\beta o$ transformation, (3.94) and (3.91) are transformed to

$$[C_1]^t[L''_s][C_1]\frac{di'_s}{dt} + [C_1]^t[L''_s][C_1]\frac{dI_s}{dt} + [C_1]^t[R_s][C_1]i'_s = v'_s \tag{3.112}$$

$$\frac{d\psi_r}{dt} + [R_r][L_{rr}]^{-1}\psi_r - [R_r][L_{rr}]^{-1}[M]^t[C_1]i'_s = v_r \tag{3.113}$$

where $i'_s = \begin{bmatrix} i_\alpha & i_\beta & i_o \end{bmatrix}^t$ and $v'_s = \begin{bmatrix} v_\alpha & v_\beta & v_o \end{bmatrix}^t$. Since $[L''_s]$ and its inverse are cyclic symmetric matrices, $[C_1][L''_s][C_1]$ and $[C_1][L''_s]^{-1}[C_1]$ will turn out to be diagonal matrices $[\Omega''_s]$ and $[\Omega''_s]^{-1}$. $[\Omega''_s]$ is defined as

$$[\Omega''_s] = \begin{bmatrix} L'' & 0 & 0 \\ 0 & L'' & 0 \\ 0 & 0 & L''_o \end{bmatrix} \tag{3.114}$$

where $L'' = L''_{aa} - L''_{ab}$ and $L''_o = L''_{aa} + 2L''_{ab}$.

I_s can be also expressed as

$$\begin{aligned} I_s &= [C_1][\Omega''_s]^{-1}[C_1]^t([M][L_{rr}]^{-1}\psi_r + [M_c][L_c]^{-1}\psi_c) \\ &= [C_1]I'_s \end{aligned} \tag{3.115}$$

where I'_s is defined by

$$I'_s = [\Omega''_s]^{-1}([M'][L_{rr}]^{-1}\psi_r + [M'_c][L_c]^{-1}\psi_c) \tag{3.116}$$

$$[M'] = [C_1]^t[M], \quad [M'_c] = [C_1]^t[M_c] \tag{3.117}$$

Substituting (3.115) in (3.112) and (3.113) gives, finally,

$$[\Omega''_s]\left(\frac{di'_s}{dt} + \frac{dI'_s}{dt}\right) + [R'_s]i'_s = v'_s \tag{3.118}$$

It should be noted that (3.115), (3.116), (3.118), and (3.113) apply to a machine when the stator phase windings are replaced by α, β, and o windings. The mutual inductances between these windings and the rotor windings are given by the matrix $[M']$.

It can be shown that I'_s can be expressed as

$$I'_s = I_d\begin{bmatrix} \cos\theta \\ -\sin\theta \\ 0 \end{bmatrix} + I_q\begin{bmatrix} \sin\theta \\ \cos\theta \\ 0 \end{bmatrix} \tag{3.119}$$

where

$$I_d = \sqrt{\frac{3}{2}}I_1, \quad I_q = \sqrt{\frac{3}{2}}I_2$$

The electrical torque is given by

$$\begin{aligned} T_e &= -i_s^t[L''_s]I_t \\ &= -i_s'^t[C_1]^t[L''_s][C_1]I'_t \\ &= -i_s'^t[\Omega''_s]I'_t \end{aligned} \tag{3.120}$$

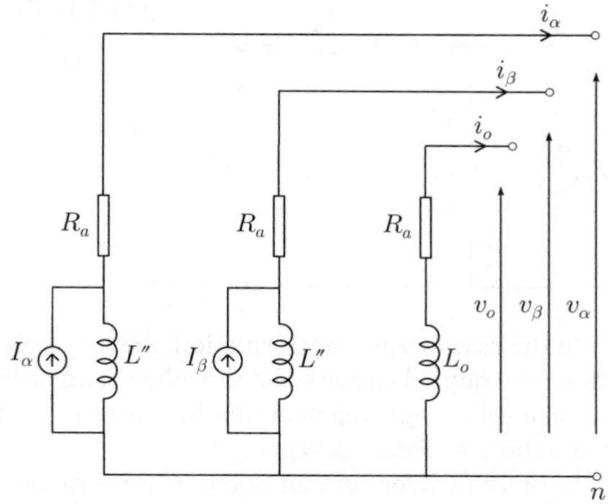

Figure 3.10 Circuit model for stator ($\alpha\beta o$ variables).

where I'_t is defined by

$$I'_t = \frac{\partial I'_s}{\partial \theta} = -I_d \begin{bmatrix} \sin\theta \\ -\cos\theta \\ 0 \end{bmatrix} + I_q \begin{bmatrix} \cos\theta \\ -\sin\theta \\ 0 \end{bmatrix} \tag{3.121}$$

Finally, the expression for T_e is obtained as

$$\begin{aligned} T_e &= -L''\{(i_\alpha \cos\theta - i_\beta \sin\theta)I_q - (i_\alpha \sin\theta + i_\beta \cos\theta)I_d\} \\ &= L''(I_d i_q - I_q i_d) \end{aligned} \tag{3.122}$$

The transformed circuit model is shown in Figure 3.10.

Transformation to Park's Variables

As mentioned earlier, the machine parameters are usually specified in terms of d-q parameters, determined from conventional tests. Neglecting dynamic saliency, $[L''_s]$ is expressed as

$$[L''_s] = \frac{L_o}{3}\begin{bmatrix} 1 & 1 & 1 \\ 1 & 1 & 1 \\ 1 & 1 & 1 \end{bmatrix} + \frac{2L''_d}{3}\begin{bmatrix} 1 & -\frac{1}{2} & -\frac{1}{2} \\ -\frac{1}{2} & 1 & -\frac{1}{2} \\ -\frac{1}{2} & -\frac{1}{2} & 1 \end{bmatrix}$$

When the above matrix is diagonalized, we get the matrix shown in (3.114), with $L'' = L''_d = L''_q$. Further, by defining,

$$c^t = \sqrt{\frac{2}{3}}\left[\ \cos\theta \quad \cos\left(\theta - \frac{2\pi}{3}\right) \quad \cos\left(\theta + \frac{2\pi}{3}\right)\ \right]$$

$$s^t = \sqrt{\frac{2}{3}}\left[\ \sin\theta \quad \sin\left(\theta - \frac{2\pi}{3}\right) \quad \sin\left(\theta + \frac{2\pi}{3}\right)\ \right]$$

The current source vector I_s can be expressed as

$$I_s = I_d c + I_q s \tag{3.123}$$

where I_d and I_q are defined in (3.119).

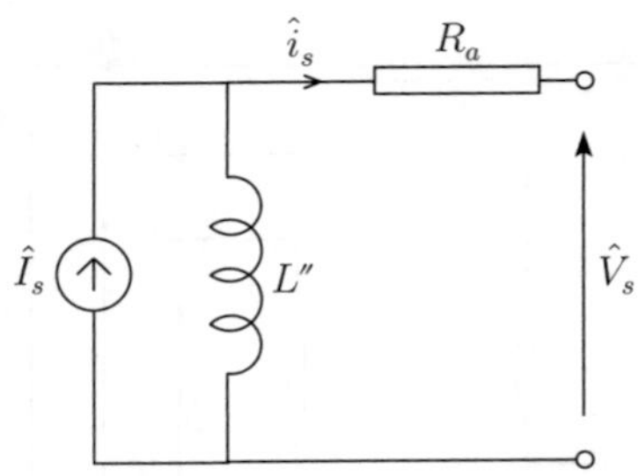

Figure 3.11 Circuit model for stator (*d-q* variables).

In the case of $\alpha\beta o$ transformation, the three-phase coupled circuit is transformed into three uncoupled circuits. On the other hand, using *d-q* variables, it is possible to derive a single-phase equivalent circuit for the stator of a synchronous machine using complex variables (see Figure 3.11).

In this equivalent circuit, $\hat{I}_s$ and $\hat{i}_s$ are complex variables defined by

$$\begin{aligned}\hat{I}_s &= (I_d - jI_q)e^{j\theta}, \hat{i}_s = (i_d - ji_q)e^{j\theta}\\ \hat{V}_s &= (v_d - jv_q)e^{j\theta}\end{aligned} \tag{3.124}$$

The complex equation for the equivalent circuit in Figure 3.11 is given by

$$L''\frac{d}{dt}(\hat{I}_s - \hat{i}_s) - R_s\hat{i}_s = \hat{V}_s \tag{3.125}$$

Substituting (3.124) in (3.125) and separating the real and imaginary parts, we get

$$\begin{aligned}\frac{d\psi_d}{dt} + \dot{\theta}\psi_q + R_a i_d = v_d\\ \frac{d\psi_q}{dt} - \dot{\theta}\psi_d + R_a i_q = v_q\end{aligned} \tag{3.126}$$

where

$$\psi_d = L''(I_d + i_d), \psi_q = L''(I_q + i_q)$$

$$\dot{\theta} = \frac{d\theta}{dt} = \omega_o + \frac{d\delta}{dt}$$

A major feature of this complex equivalent circuit is that the approximation of neglecting stator transients and variation in the rotor frequency can be easily implemented by replacing the inductor in Figure 3.11 by the reactance (x'') computed at the operating or nominal frequency (as the operating frequency (ω_o) is normally assumed to be identical to the nominal (ω_B) frequency). The equivalent circuit for this case is shown in Figure 3.7 when a complex voltage source $\hat{E}''$ (in series with impedance $Z = R_a + jx''$) replaces the shunt current source $\hat{I}_s$ shown in Figure 3.11. The voltage source $\hat{E}''$ is defined by

$$\hat{E}'' = E''_d - jE''_q \tag{3.127}$$

where

$$E''_d = x''I_q, \quad E''_q = -x''I_d \tag{3.128}$$

The equivalent circuit shown in Figure 3.7 represents the generator stator in stability analysis where the external network is represented by algebraic equations based on the admittance matrix. It is interesting to note that while Figure 3.11 is exact, Figure 3.7 is an approximation.

Remarks

1) The three-phase circuit model derived in Section 3.7 is the most general and appropriate to use in the analysis of electromagnetic transients in an unsymmetrical network subjected to any disturbance such as single phase to ground fault. There is no restriction on the machine details, for example it can be applied even for machine Model (3.3).
2) The transformations of the model to the one based on $\alpha\beta o$ components is applicable when the synchronous generator is connected to any symmetric three-phase network. Here the complexity of the model is reduced by ignoring the zero-sequence network. Also, the α and β sequence networks become identical (although excited by α and β sequence voltage or currents).
3) In the analysis of the low-frequency (below 5 Hz) response of the system in simulations involving the study of system stability, the machine model can be simplified to what is shown in Figure 3.11 based on the *d-q* transformation. It should be noted that while application of the model shown in Figure 3.11 is exact, the stator equivalent circuit (for Model (2.2)) shown in Figure 3.7 is not exact. The model shown in Figure 3.7 assumes that (i) stator transients are neglected and (ii) the variations in frequency are neglected.
4) The derivation of the detailed model assumes that dynamic saliency is not present. However, it is possible to introduce a dummy rotor coil (typically in the quadrature axis) with appropriate choice of its parameters. This is explained in Section 3.7.

3.7.3 Modeling of a Synchronous Generator in the Simulation of Electromagnetic Transients

The EMTP described in Chapter 2 has emerged as a powerful tool for the transient analysis of power systems. For the study of lightning and switching surges which persist for a short duration, a simple model of a synchronous generator (of a voltage source behind a subtransient reactance) is adequate. However, studies involving subsynchronous resonance (SSR) require detailed generator models. The solution of the synchronous generator equations with the network requires a suitable interfacing technique. There are several interfacing techniques reported in the literature, one set based on the prediction of voltage or current at the generator bus while others do not require prediction.

The application of the circuit model given in Section 3.7 is ideally suited for the equivalent circuit representation of the stator (Padiyar *et al.*, unpublished). Here, the prediction of the equivalent current sources is required, which are the functions of the rotor flux linkages and δ. Since rotor windings are closed, rotor flux linkages are continuous functions of time. Hence the interface technique proposed by Padiyar (unpublished) is robust with better numerical stability. With the help of a case study based on the First Benchmark model, it was shown that there is no numerical instability even with a step size of 1 ms, although for accuracy, a step size of 100 or 200 μs was selected. The time constant of the dummy coil (to represent dynamic saliency) can be chosen to be not less than 0.1 ms, although 1 ms may suffice. The application of the circuit model of the synchronous machine given in [14] has been applied for the study of the control of torsional interactions in HVDC turbine- generators [15].

3.7.4 Treatment of Dynamic Saliency

If $L_{aa2} - (s_1 - s_2)/2 \neq 0$, then $[L_s'']$ as defined by (3.98) is not a constant matrix. However, it is possible to adjust s_1 or s_2 such that the new value of $[L_s'']$ becomes a constant matrix (independent of θ). Adjustment of s_1 or s_2 is physically explained by considering a dummy (open) rotor coil on the d- or q-axis. One can choose the parameters of the dummy coil such that the new value of s_1 or s_2 satisfies (3.101) and the new value of $[L_s'']$ is a constant matrix.

To illustrate these ideas, let us assume that $L_{aa2} - (s_1 - s_2)/2 < 0$ (which happens in all practical cases). We increase the modified value of s_2 such that

$$L_{aa2} - \frac{s_1 - s_2'}{2} = 0$$

where s_2' is the new value of s_2 in the presence of the dummy coil on the q-axis. The new value of $[L_s'']$ is given by

$$[L_s'']_{new} = [L_s'']_{old} - \frac{M_{ax}^2}{2L_x}\begin{bmatrix} 1 & -\frac{1}{2} & -\frac{1}{2} \\ -\frac{1}{2} & 1 & -\frac{1}{2} \\ -\frac{1}{2} & -\frac{1}{2} & 1 \end{bmatrix} +$$

$$\frac{M_{ax}^2}{2L_x}\begin{bmatrix} \cos 2\theta & \cos\left(2\theta - \frac{2\pi}{3}\right) & \cos\left(2\theta + \frac{2\pi}{3}\right) \\ \cos\left(2\theta - \frac{2\pi}{3}\right) & \cos\left(2\theta + \frac{2\pi}{3}\right) & \cos 2\theta \\ \cos\left(2\theta + \frac{2\pi}{3}\right) & \cos 2\theta & \cos\left(2\theta - \frac{2\pi}{3}\right) \end{bmatrix} \tag{3.129}$$

L_x is the self inductance of the dummy coil and M_{ax} is the maximum mutual inductance between the dummy coil and any stator coil. The matrix $[M_c]^t$ for a single dummy coil is defined by

$$[M_c]^t = \left[\; M_{ax}\sin\theta \quad M_{ax}\sin\left(\theta - \frac{2\pi}{3}\right) \quad M_{ax}\sin\left(\theta + \frac{2\pi}{3}\right) \;\right] \tag{3.130}$$

In obtaining the expression for $[L_s'']_{new}$ given in (3.129), we have arbitrarily chosen the mutual inductances between the dummy coil and other rotor coils as zero. This results in

$$s_2' = s_2 + \frac{M_{ax}^2}{L_x} \tag{3.131}$$

If L_x is chosen arbitrarily (say $L_x = L_q$), M_{ax} is calculated from the relationship

$$s_2' = s_2 + \frac{M_{ax}^2}{L_x} = s_1 - 2L_{aa2} \tag{3.132}$$

The expressions for I_1 and I_2 (given in (3.103) and (3.104)) are changed due to replacement of L'' by L''_d and adding a term to I_2 such that

$$I_{2new} = I_{2old} + \frac{M_{ax}}{L''_d L_x}\psi_c \tag{3.133}$$

where ψ_c is the flux linkage of the dummy coil. If it is assumed to be open (ideal case), then

$$\psi_c = M_{ax}\left[i_a \sin\theta + i_b \sin\left(\theta - \frac{2\pi}{3}\right) + i_c \sin\left(\theta + \frac{2\pi}{3}\right)\right] \tag{3.134}$$

Remarks

1) The dynamic saliency introduces an additional term to the current source in the equivalent circuit that depends on the stator (armature) currents. This representation is exact and it does not pose any numerical problems in a single-machine system.
2) In multi-machine systems, it would be advantageous to treat ψ_c (flux linkage of the dummy coil) as a state variable by considering the dummy coil being shorted through a high resistance. This introduces approximation, but the degree of approximation can be arbitrarily chosen by selecting an arbitrarily small time constant for the dummy coil. For stability studies, the choice of 10 ms as the time constant is adequate whereas for the analysis of electromagnetic transients, the time constant can be chosen as 1 ms.

3.8 Excitation and Prime Mover Controllers

The synchronous generator is provided with two automatic (feedback) controllers for the regulation of the terminal voltage and frequency. These controllers indirectly influence the reactive power and active power outputs of the generator, respectively. The regulation of the voltage is the faster of the two controllers and has bearing on the system stability much more than the regulation of speed.

In this section, we will look at the modeling of the excitation and prime mover controllers for the purposes of stability analysis of power systems. For each control system, the models are grouped into a few standard types which are conveniently handled in computer simulation and analysis. The block diagram structure of each standard type is well defined such that an equipment belonging to that type is characterized by a set of parameters.

3.8.1 Excitation Systems [16, 17]

The main objective of the excitation system is to control the field current of the synchronous machine. The field current is controlled to regulate the terminal voltage of the machine. As the field circuit time constant is high (of the order of a few seconds), fast control of the field current requires field forcing. Thus, the exciter should have a high ceiling voltage which enables it to operate transiently with voltage levels that are three to four times the normal. The rate of change of voltage should also be fast. Because of the high reliability required, a unit exciter scheme is prevalent where each generating unit has its individual exciter.

There are three distinct types of excitation systems based on the power source for the exciter:

1) DC excitation systems (DC), which utilize a DC generator with commutator
2) AC excitation systems (AC), which use alternators and either stationary or rotating rectifiers to produce the direct current needed
3) static excitation systems (ST) in which the power is supplied through transformers and rectifiers.

The first two types of exciters are also called rotating exciters and are mounted on the same shaft as the generator and driven by the prime mover.

Modeling of Excitation Systems

The general functional block diagram of an excitation system (for all three types defined earlier) is shown in Figure 3.12. The modeling of the various components of the excitation system is described below.

Terminal Voltage Transducer and Load Compensation

The transducer and load compensation are shown in Figure 3.13. The terminal voltage of the generator is sensed and transformed to a DC quantity. Although the filtering associated with the voltage transducer may be complex, it is usually modeled as a single time constant T_R. In many systems, T_R is very small and can be assumed to be zero for simplicity.

The purpose of the load compensation is to synthesize a voltage which differs from the terminal voltage by the voltage drop in an impedance $(R_c + jX_c)$. Both voltage and

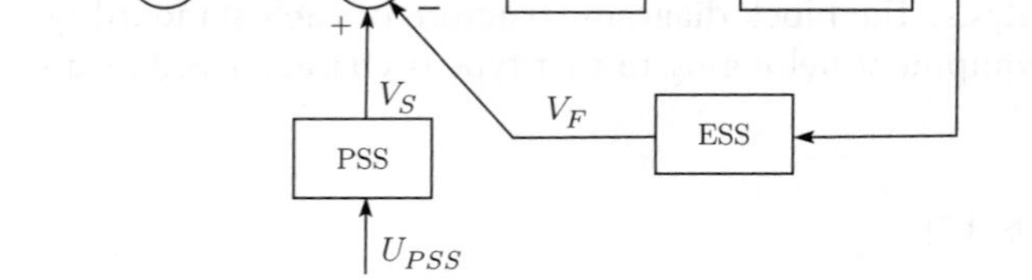

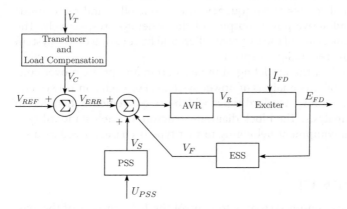

Figure 3.12 Functional block diagram of an excitation control.

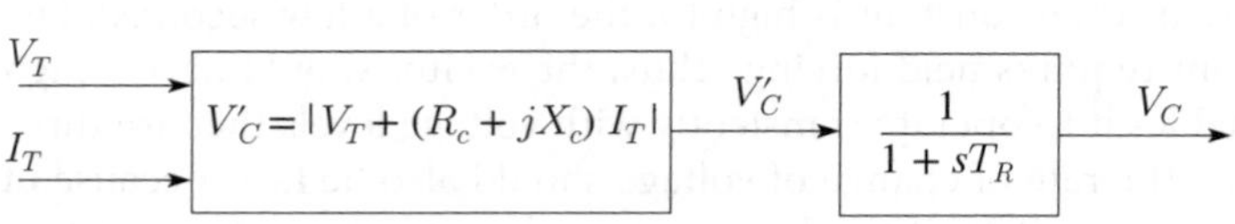

Figure 3.13 Transducer and load compensation.

current phasors must be used in computing V_C. The objectives of the load compensation are as follows:

a) Sharing of reactive power among units which are bussed together with zero impedance between them. In this case, R_c and X_c are positive and the voltage is regulated at a point internal to the generator.
b) When the generating units are connected in parallel through unit transformers, it may be desirable to regulate voltage at a point beyond the machine terminals to compensate for a portion of the transformer impedance. In this case both R_c and X_c are negative values.

In most cases, R_c is negligible and can be ignored.

Exciters and Voltage Regulators

The modeling of various excitation systems has been reported in two IEEE Committee Reports [16, 17]. Modern automatic voltage regulators (AVRs) are continuously acting electronic regulators with high gain and small time constants. The exciters can be of the following types:

1) Field-controlled DC generator-commutator.
2) (a) Field-controlled alternator with non-controlled rectifier (using diodes): (i) with slip rings and brushes (stationary rectifier) or (ii) brushless, without sliprings (rotating rectifier).
   (b) Alternator with controlled rectifier.
3) Static exciter with:
   (a) potential source-controlled rectifier in which the excitation power is supplied through a potential transformer connected to generator terminals.
   (b) compound source (using both current and voltage transformers at the generator terminals) with (i) a non-controlled rectifier (control using magnetic elements such as saturable reactors) or (ii) a controlled rectifier (for controlling the voltage).

Historically, DC generator-commutator exciters were first to be used. The DC generator may be self-excited or separately excited (using a pilot exciter). The voltage regulator for DC excitation systems was based on a rotating amplifier (amplidyne) or magnetic amplifiers.

AC and static excitation systems invariably use electronic regulators that are fast acting and result in the phase control of the controlled rectifiers using thyristors. In type 2(a) exciters, field control of the alternator is achieved using a controlled rectifier with a power source derived from the alternator output. With brushless exciters, the field circuit of the alternator is mounted in the stationary member and supplied through a controlled rectifier from a permanent magnet generator. The armature of the alternator is on the rotor and connected directly to a rotating diode rectifier and thus sliprings are eliminated.

The performances of the type 2(b) and 3(a) exciters are expected to be similar as in both systems the generator field is directly supplied through controlled rectifiers which have fast response. The only difference is the power source for the rectifiers (and the generator field): in 2(b) it comes from the alternator (hence is part of the AC excitation systems) and in 3(a) it comes from static elements (potential transformer) and thus belongs to the static excitation systems.

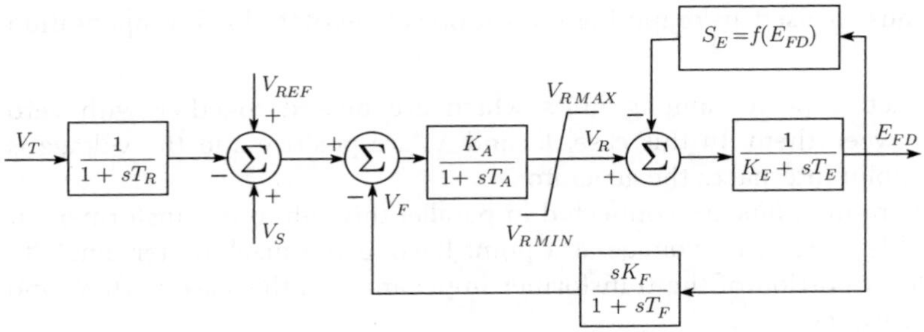

Figure 3.14 IEEE Type 1 excitation system. Source: [16] Fig. 1, p. 1461, reproduced with permission ©IEEE 1968.

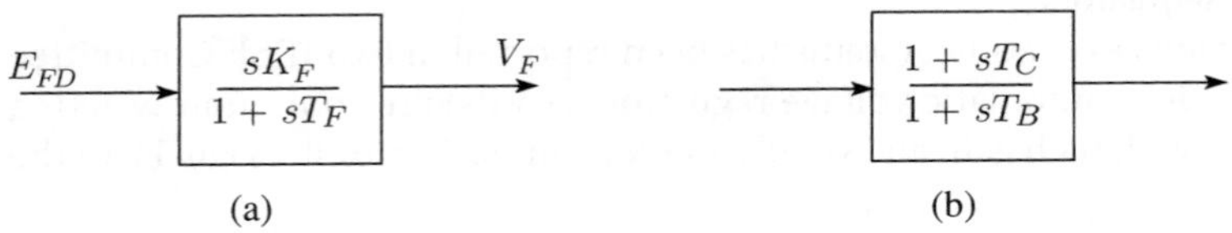

Figure 3.15 (a) Excitation system stabilizer and (b) transient gain reduction.

In the first IEEE Committee Report, published in 1968 [17], excitation systems were classified not according to their power source but in an arbitrary manner. It essentially represents rotating exciters, but with some modifications can also represent static exciters. This is shown in Figure 3.14. Here, V_R is the output of the regulator, which is limited. The regulator transfer function has single time constant T_A and a positive gain of K_A. The saturation function $S_E = f(E_{FD})$ represents the saturation of the exciter. IEEE Type 1 can also represent the static excitation system 3(a) by specifying the following parameters:

$$K_E = 1, \quad T_E = 0, \quad S_E = 0, \quad \text{and} \quad V_{RMAX} = K_P V_T \tag{3.135}$$

Excitation System Stabilizer and Transient Gain Reduction

An excitation system stabilizer (ESS) and transient gain reduction (TGR) block are used to increase the stable region of operation of the excitation system and permit higher regulator gains. It should be noted that feedback control systems, of which the excitation system is an example, often require lead/lag compensation or derivative (rate) feedback. The transfer function for ESS is shown in Figure 3.15a.

The time constant (T_F) is usually taken as 1 s. Instead of feedback compensation for ESS, a series-connected lead/lag circuit can also be used, as shown in Figure 3.15b. Here T_C is usually less than T_B. Hence, this means of stabilization is termed TGR. The objective of TGR is to reduce the transient gain or gain at higher frequencies, thereby minimizing the negative contribution of the regulator to system damping. However, if the power system stabilizer (PSS) is specifically used to enhance system damping, TGR may not be required. A typical value of the transient gain reduction factor (T_B/T_C) is 10.

Power System Stabilizer

The stabilization provided by PSS is not to be confused with that obtained by ESS. While ESS is designed to provide effective voltage regulation under open or short-circuit conditions, the objective of PSS is to provide damping of the rotor oscillations whenever there is a transient disturbance. The damping of these oscillations (whose frequency varies from 0.2 to 2.0 Hz) can be impaired by the provision of high gain AVR, particularly at high loading conditions when a generator is connected through a high external impedance (due to weak transmission network).

While detailed discussion of PSS is given in Chapter 6, it is worth noting here that the input signal for PSS is derived from speed/frequency, or accelerating power. The PSS design in a multi-machine environment can be complex, as several rotor oscillation frequencies have to be considered. In any case, the stabilizer is designed to have zero output in steady state. The output is also limited in order not to adversely affect the voltage control. The stabilizer output V_S is added to the terminal voltage error signal.

Excitation Systems: Standard Block Diagram

The second IEEE Committee Report, published in 1981 [17], distinguished between the excitation systems based on their power source. This classification is more logical and can avoid gross approximations in the representation of different excitation systems. Here, we will consider only the static excitation system for illustration.

Static Excitation System

In these systems, transformers are used to convert voltage (and also current in compounded systems) to the required level of the field voltage. Controlled or uncontrolled rectifiers are then used to provide the DC voltage for the generator field. Although negative field voltage forcing is used, many of the excitation systems do not permit negative field current. This aspect is normally ignored in the computer simulation, but can be significant sometimes (particularly in asynchronous operation). As the exciter ceiling voltage tends to be high in static exciters, field current limiters are used to protect the exciter and the field circuit. However, this protection is also not modeled except in special cases.

The block diagram of the potential source, the controlled rectifier excitation system, is shown in Figure 3.16. The internal limiter following the summing junction can be neglected, but the field voltage limits which are dependent on both V_T and I_{FD} must be considered. For transformer fed systems, K_C is small and can be neglected. The block diagram given in Figure 3.16 is also similar to that of the alternator supplied, the controlled rectifier excitation system. The only difference is that the field voltage limits are not dependent on the generator terminal voltage V_T in the case of the alternator supplied systems.

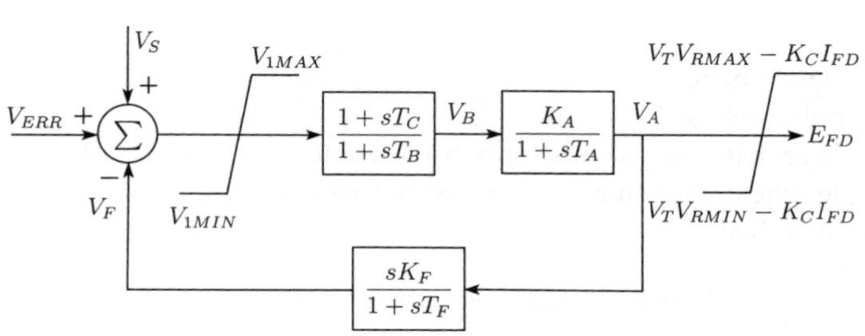

Figure 3.16 Static excitation system (Type STI). Source: [17] Fig. 10, p. 499, reproduced with permission ©IEEE 1981.

3.8.2 Modeling of Prime-Mover Control Systems [18]

The regulation of frequency in the system requires the speed control of a prime-mover using governor. However, parallel operation of generators requires a droop characteristic incorporated in the speed-governing system to ensure stability and proper division of load. Hence, to maintain constant frequency, an auxiliary control is required which responds to a load unbalance. It is also necessary for the prime-mover control to adjust the generation according to economic dispatch.

Thus, different prime-mover controls are classified as (i) primary (speed governor), (ii) secondary (load frequency control), and (iii) tertiary (involving economic dispatch). With increase in the system size due to interconnections, the frequency variations (in normal conditions) become less and less, and load frequency control assumes importance. However, the role of speed governors in rapid control of frequency cannot be underestimated.

In stability studies, the secondary and tertiary controls are usually neglected. Only speed-governing systems, including turbines, need to be represented. In this section, both turbine and governor models are presented based on the IEEE report published in 1973 [18].

Hydraulic Turbine

The hydraulic turbine is approximately represented by the block diagram shown in Figure 3.17. The time constant T_W is called the water starting time or water time constant. The equation for T_W is

$$T_W = \frac{LV}{H_T g} \tag{3.136}$$

where L is the length of the penstock, V is the water velocity, H_T is the total head, and g is the acceleration due to gravity. For more accurate models, traveling wave phenomena in penstock need to be considered. However, this is not required in stability studies.

The input P_{GV} for the turbine comes from the speed-governor. It is the gate opening expressed in per unit. Values for T_W lie in the range 0.5–5.0 s, with the typical value around 1.0 s. It should be noted that a hydro-turbine has a non-minimum phase characteristic that results in a slower response. The response to a unit step input is shown in Figure 3.18 and is compared to the response if the zero in the right half of the s plane (RHP) for the transfer function shown in Figure 3.17 did not exist.

Steam Turbine

There are six common steam turbine systems:

(a) nonreheat
(b) tandem compound, single reheat
(c) tandem compound, double reheat
(d) cross compound, single reheat with two low pressure turbines
(e) cross compound, single reheat with single low pressure turbine
(f) cross compound, double reheat

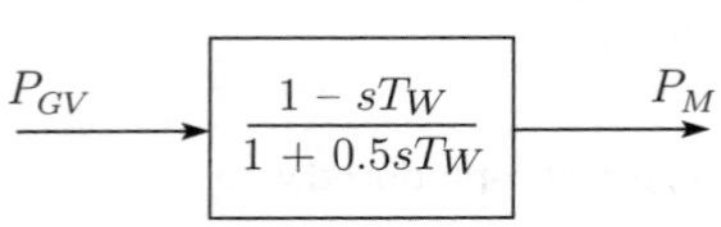

Figure 3.17 Hydraulic turbine model.

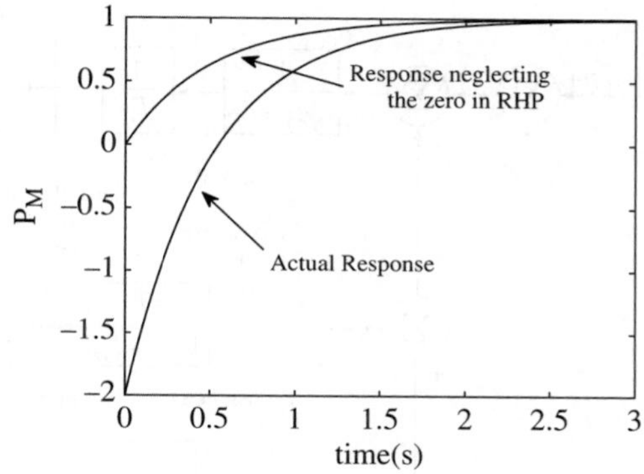

Figure 3.18 Response to a unit step.

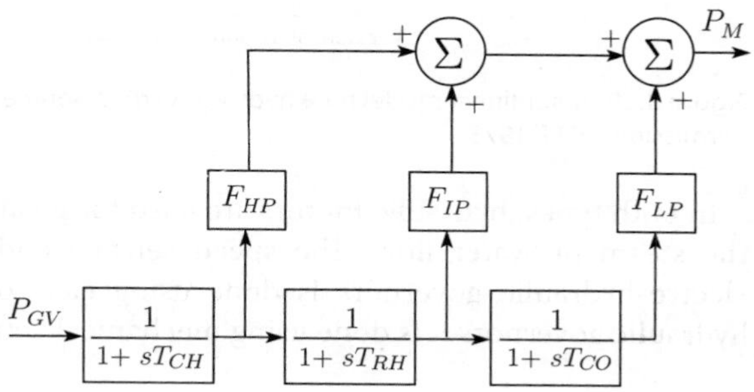

Figure 3.19 Steam turbine model. Source: [18] Fig. 7(B), p. 1908, reproduced with permission ©IEEE 1973.

The tandem compound has only one shaft on which all the turbines, high pressure (HP), intermediate pressure (IP) and low pressure (LP), are mounted. Sometimes there is also a very-high-pressure (VHP) turbine. Cross-compound systems have two shafts driving two independent generators. The configuration corresponding to tandem compound, single reheat is shown in Figure 3.19. This does not show the extraction of steam taken at various turbine stages to heat feedwater, as this has no major significance in stability studies. All compound steam turbines use governor- controlled valves at the inlet to the high pressure turbine to control the steam flow. The steam chest, reheater, and crossover piping all introduce delays. The time constants T_{CH}, T_{RH}, and T_{CO} represent these delays. The fractions F_{HP}, F_{IP}, and F_{LP} represent fractions of the total turbine power developed in the HP, IP, and LP turbines, respectively. Typical values for T_{CH}, T_{RH}, and T_{CO} are

$$T_{CH} = 0.1 - 0.4\text{s}$$
$$T_{RH} = 4 - 11\text{s}$$
$$T_{CO} = 0.3 - 0.5\text{s}$$

The typical values of F_{HP}, F_{IP}, and F_{LP} are 0.3, 0.3, and 0.4, respectively, the sum adding to unity.

Speed-Governing Systems

There are two types of speed-governing systems, namely

a) mechanical-hydraulic
b) electro-hydraulic.

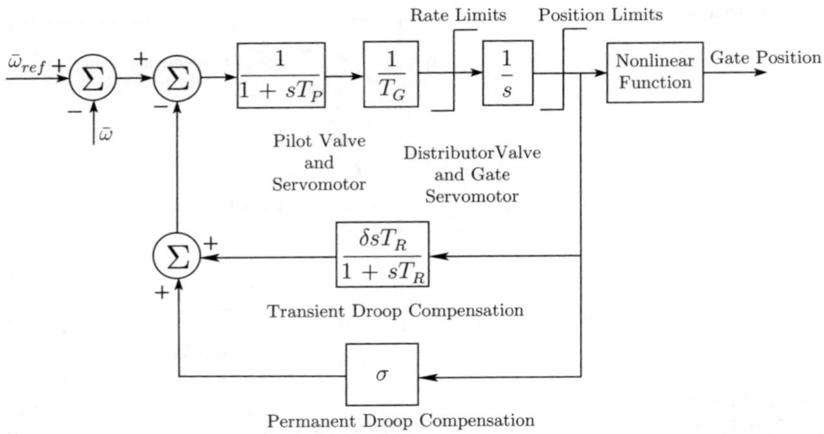

Figure 3.20 A nonlinear model for a hydro-governor. Source: [18] Fig. 9(B), p. 1909, reproduced with permission ©IEEE 1973.

In both types, hydraulic motors are used for positioning the valve or gate controlling the steam or water flow. The speed sensing and conditioning (at low power) for electro-hydraulic governors is done using electronic circuits while for mechanical hydraulic governors it is done using mechanical components.

Hydroturbines

An approximate nonlinear model for the hydro-governing system is shown in Figure 3.20. It should be noted that T_R and δ are computed from

$$T_R = 5T_W, \quad \delta = 1.25T_W/H$$

where H is the generator inertia constant and T_W is the water starting time.

Steam Turbines

A simplified, general model for the speed-governing systems for a steam turbine is shown in Figure 3.21. Typical values of time constants (in second) are given below.

$$\text{Mechanical-hydraulic governor}: T_1 = 0.2\text{–}0.3\text{ s}, \quad T_2 = 0\text{ s}, \quad T_3 = 0.1\text{ s}$$

$$\text{Electro-hydraulic governor}: T_1 = T_2, \quad T_3 = 0.025\text{–}0.15\text{ s}$$

Note that when $T_1 = T_2$, the value of T_1 or T_2 has no effect as there is pole-zero cancellation. For studies involving the dynamic performance for the midterm and long term, it is essential also to model boiler controls in addition to automatic generation control (AGC). In this context, it is relevant to note that there are two basic modes of energy control in fossil-fueled steam generator units:

1) boiler following (or turbine leading) mode
2) turbine following (or boiler leading) mode.

In the first mode of control, which is applicable to many drum-type boilers (and also a few once-through boilers), changes in the power are initiated by turbine control valves and boiler controls respond to changes in steam flow and pressure. The response to small

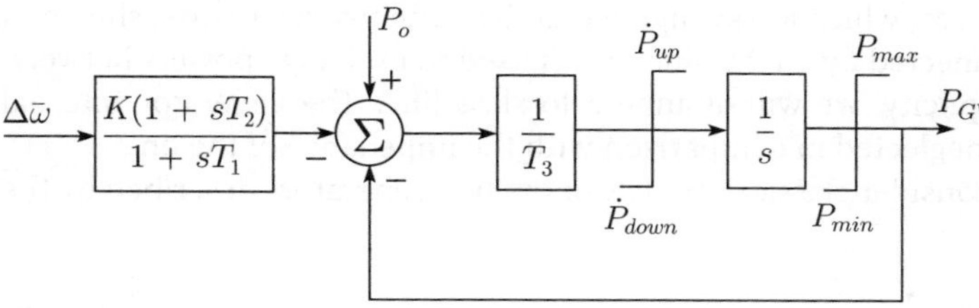

Figure 3.21 General model for the speed-governor system for a steam turbine. Source: [18] Fig. 4(A), p. 1906, reproduced with permission ©IEEE 1973.

changes in power demand are rapid as the turbine utilizes the stored energy in the boiler. However, large changes can be detrimental to the boiler operation as large excursions in steam pressure and flow result following changes in the valve position.

In the second mode of control (boiler leading turbine), the turbine control valves are made to regulate boiler pressure and changes in generation are made through boiler controls. The fast action of the turbine control valves can be utilized to keep the boiler pressure almost constant. However, in this mode, the response of the turbine power to a change in load demand is slow as the lags in the fuel system and boiler storage affect the response. A compromise between the two desired objectives, (i) fast response and (ii) constraints of regulating pressure and temperatures for maintaining boiler safety, is achieved by the adoption of control modes termed coordinated, integrated or direct energy balance. The need for close coordination between boiler and turbine controls is more crucial for once-through boilers than drum-type boilers. A recent development is to use turbine leading mode for small changes in steam pressure and switch to boiler leading mode for large changes.

3.9 Transient Instability due to Loss of Synchronism

In steady state (under normal conditions) the average electrical frequency of all synchronous generators connected to an AC grid must remain identical. This condition is termed "synchronous operation". Any disturbance, small or large, can affect the synchronous operation and lead to loss of synchronism. For example, there can be a sudden increase in the load or loss of generation. Another type of disturbance is the switching out of a transmission line, which may occur due to overloading or a fault. The transient stability of a system determines whether the system can settle down to a new or original steady state after the transients disappear.

The simplest system that can exhibit the loss of synchronism is the two-generator system shown in Figure 3.22. We can consider each generator as an equivalent representing

Figure 3.22 A two-generator system.

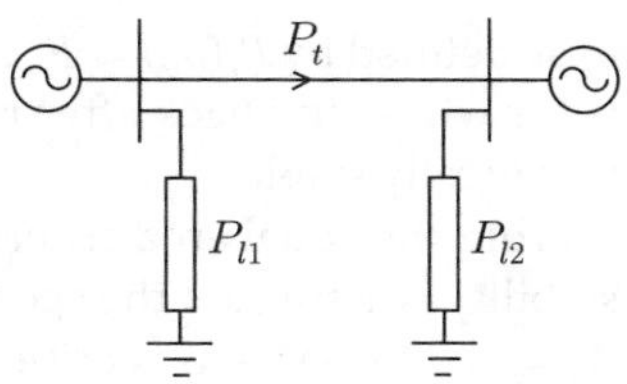

all the generators in an area, which are swinging together. The two generators shown in Figure 3.22 are interconnected by an AC line that is used to exchange powers between the two areas. For simplicity, we will assume a lossless line. The generator internal impedance can also be neglected in comparison with the impedance of the line.

For simplicity, let us consider classical models of the two generators described by the swing equations

$$M_1 \frac{d^2\delta_1}{dt^2} = P_{a1} - P_t \tag{3.137}$$

$$M_2 \frac{d^2\delta_2}{dt^2} = P_{a2} + P_t \tag{3.138}$$

where $P_{a1} = P_{m1} - P_{l1}$ and $P_{a2} = P_{m2} - P_{l2}$ are the accelerating powers of the generators 1 and 2, respectively, in the absence of power flow in the line. Note that $M_1 = \frac{2H_1}{\omega_B}$ and $M_2 = \frac{2H_2}{\omega_B}$, if δ is expressed in radians and time in seconds. H_1 and H_2 are inertia constants expressed on a common base.

Multiplying (3.137) by M_2 and (3.138) by M_1 and subtracting (3.138) from (3.137), we get

$$M_1 M_2 \frac{d^2\delta_{12}}{dt^2} = M_2 P_{a_1} - M_1 P_{a2} - (M_2 + M_1) P_t$$

We can rewrite the above equation as

$$M_{eq} \frac{d^2\delta_{12}}{dt^2} = P_m^{eq} - P_e^{eq} \tag{3.139}$$

where $M_{eq} = \frac{M_1 M_2}{(M_1+M_2)}$, $P_m^{eq} = \frac{M_2 P_{a1} - M_1 P_{a2}}{(M_1+M_2)}$, $P_e^{eq} = P_t$ and $\delta_{12} = \delta_1 - \delta_2$.

Equation (3.139) shows that a two-machine system can be reduced to an equivalent single-machine system with the equivalent inertia of $\frac{M_1 M_2}{M_1+M_2}$. Note that $M_{eq} \approx M_1$ if $M_2 \gg M_1$. In steady state, $P_m^{eq} = P_t$.

If there is a major disturbance in area 1 or 2 such that P_m^{eq} suddenly increases to a value (P_{mf}^{eq}) which is greater than P_{to} (initial value of P_t), the angular difference δ_{12} increases as the equivalent machine accelerates. When $\delta_{12} = \delta_{cl}$, the P_m^{eq} assumes the nominal value of P_{to}. The power angle curve is defined by

$$P_t = \frac{E_1 E_2}{X} \sin \delta_{12} \tag{3.140}$$

where X is the reactance of the tie line. At $\delta_{12} = \delta_{cl}$, $\frac{d\delta_{12}}{dt} > 0$ and the equivalent generator angle δ_{12} continues to increase until $\frac{d\delta_{12}}{dt} = 0$. The rotor of the machine continues to decelerate from $\delta_{12} = \delta_{cl}$ until $\delta_{12} = \delta_{max}$. It can be shown that the area A_1 in Figure 3.23 represents the kinetic energy $\left(\frac{1}{2} M_{eq} \left[\frac{d\delta_{12}}{dt}\right]^2\right)$ gained by the rotor. If $\delta_{max} < \delta_u$, where δ_u is defined by $P_t(\delta_u) = P_{to}$, the system is transiently stable as the equivalent generator rotor will swing back after reaching a maximum value of δ_{max}. If $\delta_{max} = \delta_u$, the system is critically stable.

From the equal area criterion (EAC) (see Appendix A), it can be shown that transient stability is assured if the accelerating area A_1 is less than the decelerating area A_2. When $A_1 = A_2$, the system is critically stable. δ_{cl} represents the clearing angle of the fault and

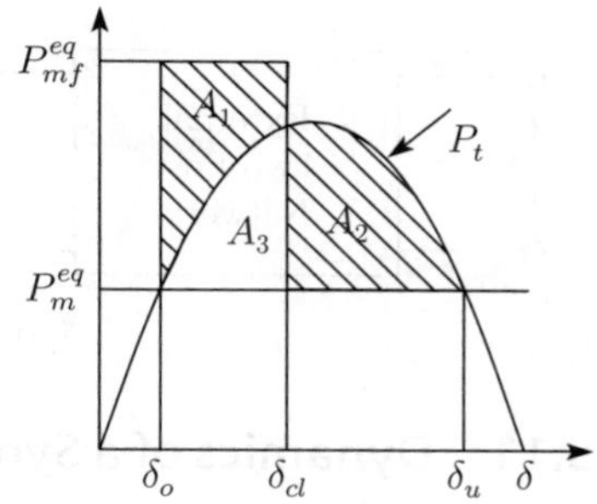

Figure 3.23 Application of the equal area criterion.

from the EAC we can determine the critical (fault) clearing angle (CCA). If we know the fault dynamics, we can determine the critical clearing time (CCT) of the specified fault.

A major assumption in this analysis is that the system remains stable if it is stable in the first swing. This assumption may not be true when the generator is equipped with a fast-acting exciter with high gain AVR. The oscillatory instability is associated with the instability of the final equilibrium point (EP). This will be discussed in Chapter 6.

3.10 Extended Equal Area Criterion

The EAC is applicable for a single- or two-machine system [1, 2, 5]. However, it is feasible to extend the EAC to multi-machine systems [19, 20]. The justification for the extended equal area criterion (EEAC) is based on the following proposition [21].

Proposition: However complex, the mechanism of loss of synchronism (LoS) in a power system originates from the irrevocable separation of its synchronous generators into two groups.

This proposition has been used extensively to determine the mode of instability (MoI) that characterizes the loss of synchronous stability when a disturbance occurs. MoI refers to the determination of the number and identity of generators that belong to the accelerating group (with respect to the remaining group of generators). The knowledge of MoI has been used to compute the controlling UEP and the correct value of kinetic energy that leads to system separation. However, the determination of MoI is not straightforward and several empirical approaches have been suggested and tried.

A new proposition [22] states "There is a unique cutset (called as the 'critical cutset') consisting of transmission lines, transformers (or series elements) connecting the two groups of machines that separate. The cutset angles (difference between the angles of the terminal buses of each member of the cutset) become unbounded when there is a loss of synchronism".

It is relatively simple to identify the critical cutset by just monitoring the angles across the critical elements (say by PMUs or other means). Applying prediction techniques can speed up the on-line detection of loss of synchronism [23]. The knowledge of the critical cutset helps to apply controlled system separation by adaptive out-of-step protection. The network analogy (see Appendix A) where a lossless transmission line is analogous to a nonlinear inductor (through which the power flows) and the bus (radian) frequency is analogous to the bus voltage, helps in applying network theorems to develop algorithms for on-line determination of LoS.

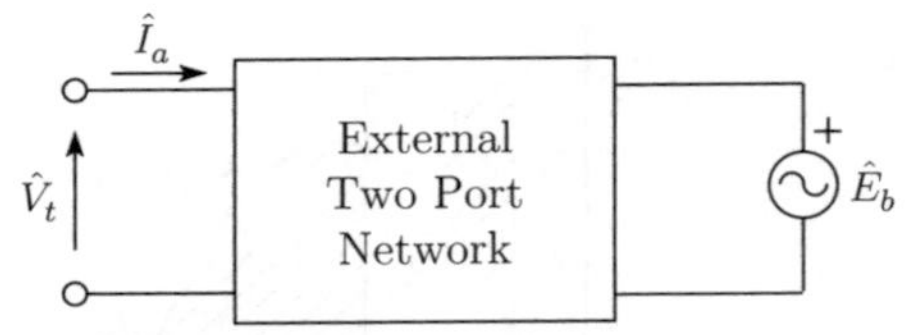

$\hat{V}_t = (v_q + jv_d)e^{j\delta}$, $\hat{I}_a = (i_q + ji_d)e^{j\delta}$, $\hat{E}_b = E_b\angle\delta_b$

Figure 3.24 A two-port network connecting a generator to the external bus.

3.11 Dynamics of a Synchronous Generator

Here we consider the simulation of a synchronous generator connected to an external bus (with variable voltage and frequency) through a linear two-port network (see Figure 3.24). A major assumption in the analysis of the dynamic performance involving the low-frequency (<5 Hz) behavior of the system is to neglect the transients in the external network. This implies that the stator transients are also neglected. This simplifies the analysis as both the stator windings (d and q) and the external network are modeled by algebraic equations based on single-phase (positive sequence) representation.

Model (2.2) is usually applicable to the turbo alternators, while Models (2.1) and (1.1) are widely used for hydro generators. Although higher-order models provide better results for special applications, they require exact determination of parameters. With constraints on data availability for the study of large systems, it may be adequate to use Model (1.1) or even (1.0) if the data are correctly determined.

Network Equations

A two-port network representing a transmission line can be characterized by either a symmetric admittance or an impedance matrix of order two. Thus, there are only three parameters that are required to determine the terminal voltage (V_t) of the generator along with the current (I_b) injected by the external bus. If the bus voltage (E_b) and phase (δ_b) are constants, the bus is labeled as an infinite bus. (Note that the variation in the frequency of the external bus can also be modeled by the variation in the bus angle.)

Assuming that the first port, connected to the generator terminals, is of interest, the voltage $\hat{V}_t$ can be expressed as

$$\hat{V}_t = \frac{\hat{I}_a}{y_{11}} + h_{12}\hat{E}_b \tag{3.141}$$

where y_{11} is the short-circuit self-admittance of the network, measured at the generator terminals, and h_{12} is a hybrid parameter (open circuit voltage gain). In general, both y_{11} and h_{12} are complex quantities. For a simple network consisting of only series impedance ($R_e + jx_e$), it is easy to obtain

$$y_{11} = \frac{1}{R_e + jx_e}, \quad h_{12} = 1.0 + j0.0 \tag{3.142}$$

In the general case, let

$$\frac{1}{y_{11}} = z_R + jz_I, h_{12} = h_1 + jh_2$$

$\hat{V}_t$ can be expressed as

$$\hat{V}_t = (v_q + jv_d)e^{j\delta} = (z_R + jz_I)(i_q + ji_d)e^{j\delta} + (h_1 + jh_2)\hat{E}_b \tag{3.143}$$

Multiplying both sides of the above equation by $e^{-j\delta}$, we get

$$(v_q + jv_d) = (z_R + jz_I)(i_q + ji_d) + (h_1 + jh_2)E_b e^{-j(\delta-\delta_b)} \tag{3.144}$$

where δ_b is the phase of $\hat{E}_b$. Equating real and imaginary parts we get

$$\begin{aligned} v_q &= z_R i_q - z_I i_d + h_1 E_b \cos(\delta - \delta_b) + h_2 E_b \sin(\delta - \delta_b) \\ v_d &= z_I i_q + z_R i_d + h_2 E_b \cos(\delta - \delta_b) - h_1 E_b \sin(\delta - \delta_b) \end{aligned} \tag{3.145}$$

The above equations can be substituted in (3.80) and (3.81) and solved for i_d and i_q in terms of state variables E'_d, E'_q, and $(\delta - \delta_b)$.

Example 3.3 ***Solution of Network Equations*** The simplest external network is a series impedance $(R_e + jx_e)$. If $R_e = 0$ and $\delta_b = 0$ then

$$z_R = 0, \quad z_I = x_e, \quad h_1 = 1.0, \quad h_2 = 0$$

Substituting these values in (3.145) we get

$$\begin{aligned} v_q &= -x_e i_d + E_b \cos\delta \\ v_d &= x_e i_q - E_b \sin\delta \end{aligned} \tag{3.146}$$

The expressions for i_d and i_q are obtained as

$$i_d = \frac{E_b \cos\delta - E'_q}{(x_e + x'_d)} \tag{3.147}$$

$$i_q = \frac{E_b \sin\delta + E'_d}{(x_e + x'_q)} \tag{3.148}$$

By eliminating the non-state variables, we can express the equations in the form

$$\dot{x}_m = f_m(x_m, u_m) \tag{3.149}$$

where

$$x^t_m = [\ \delta \quad S_m \quad E'_q \quad E'_d\ ]$$

$$u^t_m = [\ E_{fd} \quad T_m\ ]$$

It should be noted that E_b is treated as a parameter. E_{fd} and T_m are inputs from the excitation and turbine-governor system, respectively. If the dynamics of the controllers are ignored, then E_{fd} and T_m are also treated as parameters. Otherwise E_{fd} and T_m are treated as outputs of dynamic systems represented by differential equations which are to be appended to (3.149).

Calculation of Initial Conditions

The system equations in (3.149) are nonlinear and have to be solved numerically. In solving these equations it is assumed that the system is at a stable equilibrium point (SEP) till time $t = 0$, and a disturbance occurs at $t = 0$ or later. It is necessary to calculate the initial conditions x_o at time $t = 0$ based on the system operating point determined from load (power) flow.

From power flow calculations in steady state, we get the real and reactive power (P_t and Q_t), the voltage magnitude (V_t), and angle (θ) at the generator terminals. Here θ is the angle with respect to the slack (infinite) bus.

In steady state, the derivatives of all state variables $\dot{x} = 0$. From this condition, we get

$$\begin{aligned} E'_{qo} &= E_{fdo} + (x_d - x'_d)i_{do} \\ E'_{do} &= -(x_q - x'_q)i_{qo} \end{aligned} \tag{3.150}$$

$$T_{mo} = T_{eo} = E'_{qo}i_{qo} + E'_{do}i_{do} + (x'_d - x'_q)i_{do}i_{qo} \tag{3.151}$$

In the above equations, the subscript o indicates the operating values. It should be noted that, in general, the initial slip S_{mo} cannot be determined. It has to be specified separately. It can be taken as zero.

Substituting (3.151) in (3.80), we get

$$\begin{aligned} E_{fdo} + x_d i_{do} - R_a i_{qo} &= v_{qo} \\ -x_q i_{qo} - R_a i_{do} &= v_{do} \end{aligned} \tag{3.152}$$

From the above, we can obtain

$$\begin{aligned} E_{fdo} + (x_d - x_q)i_{do} &= (v_{qo} + jv_{do}) + (R_a + jx_q)(i_{qo} + ji_{do}) \\ &= \hat{V}_{to}e^{-j\delta_o} + (R_a + jx_q)\hat{I}_{ao}e^{-j\delta_o} \end{aligned} \tag{3.153}$$

Defining

$$E_{qo}\angle\delta_o = \hat{V}_{to} + (R_a + jx_q)\hat{I}_{ao} \tag{3.154}$$

we can express

$$E_{fdo} = E_{qo} - (x_d - x_q)i_{do} \tag{3.155}$$

Equation (3.154) can be used to fix the position of the q-axis. The phasor diagram shown in Figure 3.25 represents (3.154) and (3.155). The reference coincides with the position of $\hat{E}_b$ in steady state. This implies that $\delta_b(t = 0) = 0$.

The procedure for the computation of the initial conditions is given below.

1. Compute $\hat{I}_{ao}$ from

$$\hat{I}_{ao} = I_{ao}\angle\phi_o = \frac{P_t - jQ_t}{V_{to}\angle - \theta_o}$$

2. Compute E_{qo} and δ_o from

$$E_{qo}\angle\delta_o = V_{to}\angle\theta_o + (R_a + jx_q)I_{ao}\angle\phi_o$$

(Note that P_t, Q_t, V_{to}, and θ_o are obtained from the power flow analysis in steady state.)

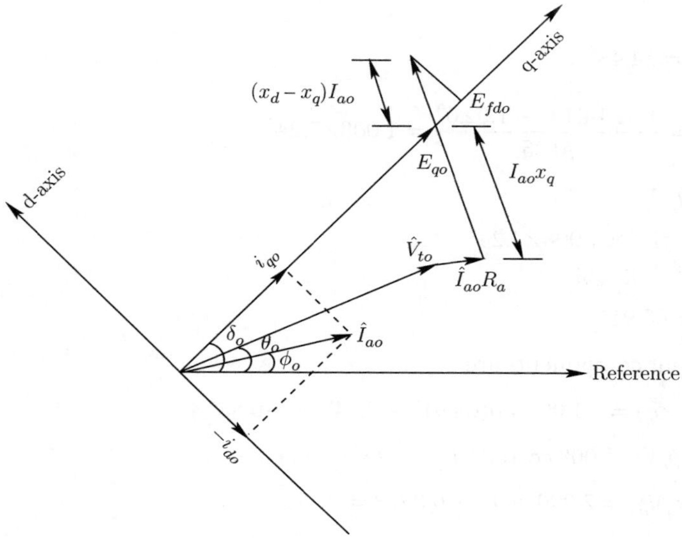

Figure 3.25 Phasor diagram.

3. Compute

$$i_{do} = -I_{ao}\sin(\delta_o - \phi_o)$$
$$i_{qo} = I_{ao}\cos(\delta_o - \phi_o)$$
$$v_{do} = -V_{to}\sin(\delta_o - \theta_o)$$
$$v_{qo} = V_{to}\cos(\delta_o - \theta_o)$$

4. Compute

$$E_{fdo} = E_{qo} - (x_d - x_q)i_{do}$$
$$E'_{qo} = E_{fdo} + (x_d - x'_d)i_{do}$$
$$E'_{do} = -(x_q - x'_q)i_{qo}$$
$$T_{eo} = E'_{qo}i_{qo} + E'_{do}i_{do} + (x'_d - x'_q)i_{do}i_{qo} = T_{mo}$$

Example 3.4 A generator is connected to an infinite bus through an external impedance of jx_e. If $E_b = V_{to} = 1.0$ pu and the power delivered is 1 pu, find the initial conditions. Assume $x_e = 0.25$pu. The generator data in per-unit quantities is:

$$x_d = 1.8, x_q = 1.7, x'_d = 0.23, R_a = 0.0$$
$$T'_d = 0.4\,\text{s}, T'_q = 0.1\text{ s}, H = 4\,\text{s}, f_B = 60\text{ Hz}$$

Solution

The generator terminal bus angle θ_o is found from

$$P_{to} = \frac{V_{to}E_b\sin\theta_o}{x_e}$$

Substituting values,

$$\sin\theta_o = 0.25, \quad \theta_o = 14.48^\circ$$

$$\hat{I}_{ao} = \frac{\hat{V}_{to} - E_b\angle 0}{jx_e} = \frac{1.0\angle 14.48^\circ - 1.0\angle 0^\circ}{j0.25} = 1.008\angle 7.24^\circ$$

$$\begin{aligned}\hat{E}_{qo} &= V_{to}\angle\theta_o + jx_q\hat{I}_{ao}\\ &= 1.0\angle 14.48^\circ + j1.7\times 1.008\angle 7.24^\circ\\ &= 2.090\angle 68.91^\circ = E_{qo}\angle\delta_o\end{aligned}$$

$$E_{qo} = 2.090, \quad \delta_o = 68.91^\circ$$

The initial armature current components are

$$i_{do} = -I_{ao}\sin(\delta_o - \phi_o) = -1.008\sin(68.91^\circ - 7.24^\circ) = -0.8873$$

$$i_{qo} = I_{ao}\cos(\delta_o - \phi_o) = 1.008\cos(68.91^\circ - 7.24^\circ) = 0.4783$$

$$E_{fdo} = E_{qo} - (x_d - x_q)i_{do} = 2.090 + 0.1\times 0.8873 = 2.179$$

$$E'_{qo} = E_{fdo} + (x_d - x'_d)i_{do} = 0.7327$$

$$E'_{do} = -(x_q - x'_q)i_{qo} = -0.7031$$

$$T_{eo} = E'_{qo}i_{qo} + E'_{do}i_{do} + (x'_d - x'_q)i_{do}i_{qo} = 1.0$$

(Check: $T_{eo} = P_{to} + I_{ao}^2 R_a$)
The initial slip, S_{mo}, can be assumed to be equal to zero.

System Simulation

The synchronous machine is represented by Model (1.1). Magnetic saturation is either neglected or considered by using saturated values or mutual inductances, x_{dc} (or x_{aq}).

The machine equations are

$$\begin{aligned}\frac{d\delta}{dt} &= \omega_B(S_m - S_{mo})\\ \frac{dS_m}{dt} &= \frac{1}{2H}[-D(S_m - S_{mo}) + T_m - T_e]\end{aligned} \tag{3.156}$$

$$\begin{aligned}\frac{dE'_q}{dt} &= \frac{1}{T'_{do}}[-E'_q + (x_d - x'_d)i_d + E_{fd}]\\ \frac{dE'_d}{dt} &= \frac{1}{T'_{qo}}[-E'_d - (x_q - x'_q)i_q]\end{aligned} \tag{3.157}$$

The electrical torque, T_e, is expressed in terms of state variables E'_d and E'_q and non-state variables i_d and i_q. The expression for T_e is

$$T_e = E'_d i_d + E'_q i_q + (x'_d - x'_q)i_d i_q \tag{3.158}$$

The non-state variables i_d and i_q can be obtained from the stator algebraic equations and the network equations and i_q.

Simulation of a Single-Machine Finite Bus System

Figure 3.24 shows a single generator connected to a voltage source ($E_b \angle \delta_b$) through a two-port network. Typically E_b and δ_b are assumed to be constants (one can even assume $\delta_b = 0$) in which case it is called an infinite bus. However, realistically, E_b and δ_b can be functions of time as they are influenced by the remaining generators in a large system, whose dynamics is ignored in the simulation. As a matter of fact, even in multi-machine power systems it is possible to represent each generator connected to a finite bus (with variable voltage magnitude and frequency) through a two-port network. This approach can simplify the simulation of a large practical system.

Example 3.5 A synchronous generator is connected to a finite bus through an external reactance of 0.4 pu. The operating data in pu is as follows:

$$P_g = 0.5, \ V_t = 1.0, \ E_b = 1.0$$

The generator is represented by Model (1.0) with the following data:

$$x_d = 1.6, \ x_q = 1.55, \ x'_d = 0.32, \ T'_{do} = 6.05 \text{ s}, \ H = 5, \ D = 0, \ F_B = 60 \text{ Hz}$$

The AVR data are

$$T_E = 0.05 \text{ s}, \ K_e = 200 \text{ (static exciter)}$$

It can be shown that the operating point is stable and the small signal analysis gives the eigen values as $-0.1512 \pm j5.5407$ and $-10.0803 \pm j14.3810$. It should be noted that the first complex pair of eigenvalues is related to the *swing mode* and the second pair is related to the *exciter model.*

The disturbance considered is perturbations in E_b or δ_b. It is assumed that

$$E_b = E_{bo} + A\sin(2\pi f't) \text{ or } \delta_b = \delta_{bo} + B\sin(2\pi f't)$$

There are two cases considered:

i) $B = 0, A = 0.1$
ii) $A = 0, B = 0.1$

In both the cases nine different frequencies of oscillation (in magnitude or angle of the finite bus) are considered, namely

$$f' = 0.7, \ 0.8, \ 0.9, \ 1.0, \ 1.1, \ 1.6, \ 1.7, \ 1.8, \text{ and } 1.9$$

The variations in the rotor angle δ are shown in Figure 3.26 for different values of f'.

Remarks It is interesting to note that when f' is around f_n or $2f_n$, the system can be unstable. Note that f_n is the natural frequency of oscillation (which in this example is 0.9 Hz).

This example illustrates the complexity of dynamic response in a power system which needs to be considered in the design of damping controllers.

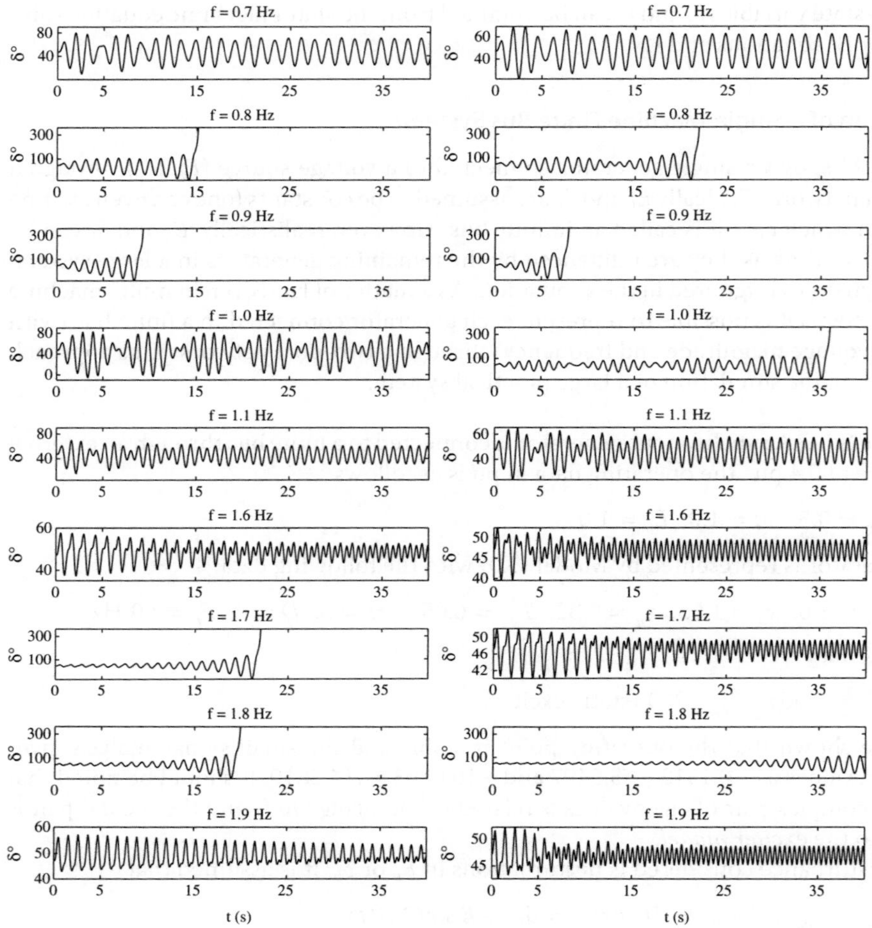

(a) $B = 0, A = 0.1$

(b) $B = 0.1, A = 0$

Figure 3.26 Variation of rotor angle δ in degrees with respect to time.

3.A Derivation of Electrical Torque

Applying Park's transformation, we can express (3.12) as

$$T_e = -\frac{1}{2}\left[\bar{i}_{dqo}^{t}[C_P]^t\left[\frac{\partial L_{ss}}{\partial \theta}\right][C_P]\bar{i}_{dqo} + 2\bar{i}_{dqo}^{t}[C_P]^t\left[\frac{\partial L_{sr}}{\partial \theta}\right]i_r\right] \tag{3.A.1}$$

$$\left[\frac{\partial L_{ss}}{\partial \theta}\right] = -2L_{aa2}\begin{bmatrix} \sin 2\theta & \sin\left(2\theta - \frac{2\pi}{3}\right) & \sin\left(2\theta + \frac{2\pi}{3}\right) \\ \sin\left(2\theta - \frac{2\pi}{3}\right) & \sin\left(2\theta + \frac{2\pi}{3}\right) & \sin 2\theta \\ \sin\left(2\theta + \frac{2\pi}{3}\right) & \sin 2\theta & \sin\left(2\theta - \frac{2\pi}{3}\right) \end{bmatrix} \tag{3.A.2}$$

$$\left[\frac{\partial L_{sr}}{\partial \theta}\right] = \begin{bmatrix} -M_{af}\sin\theta & -M_{ah}\sin\theta & M_{ag}\cos\theta & M_{ak}\cos\theta \\ -M_{af}\sin\left(\theta-\frac{2\pi}{3}\right) & -M_{ah}\sin\left(\theta-\frac{2\pi}{3}\right) & M_{ag}\cos\left(\theta-\frac{2\pi}{3}\right) & M_{ak}\cos\left(\theta-\frac{2\pi}{3}\right) \\ -M_{af}\sin\left(\theta+\frac{2\pi}{3}\right) & -M_{ah}\sin\left(\theta+\frac{2\pi}{3}\right) & M_{ag}\cos\left(\theta+\frac{2\pi}{3}\right) & M_{ak}\cos\left(\theta+\frac{2\pi}{3}\right) \end{bmatrix} \tag{3.A.3}$$

$$\left[\frac{\partial L_{ss}}{\partial \theta}\right][C_P] = -3L_{aa2}[C_P][P_2]$$

where

$$[P_2] = \begin{bmatrix} 0 & \frac{k_q}{k_d} & 0 \\ \frac{k_d}{k_q} & 0 & 0 \\ 0 & 0 & 0 \end{bmatrix} \tag{3.A.4}$$

$$[C_P]^t\left[\frac{\partial L_{sr}}{\partial \theta}\right] = \begin{bmatrix} 0 & 0 & \frac{3}{2}k_dM_{ag} & \frac{3}{2}k_dM_{ak} \\ -\frac{3}{2}k_qM_{af} & -\frac{3}{2}k_qM_{ah} & 0 & 0 \\ 0 & 0 & 0 & 0 \end{bmatrix} \tag{3.A.5}$$

After some manipulations, the expression for electrical torque reduces to

$$T_e = \frac{3}{2}k_dk_q\left[\bar{i}_q\left(\frac{M_{af}}{k_d}i_f + \frac{M_{ah}}{k_d}i_h + \frac{3}{2}L_{aa2}\bar{i}_d\right) - \bar{i}_d\left(\frac{M_{ag}}{k_q}i_g + \frac{M_{ak}}{k_q}i_k - \frac{3}{2}L_{aa2}\bar{i}_q\right)\right] \tag{3.A.6}$$

Simplifying (3.A.6), we obtain

$$T_e = \frac{3}{2}k_dk_q[i_d\psi_{qx} - i_q\psi_{dx}] \tag{3.A.7}$$

where

$$\psi_{dx} = \frac{M_{af}}{k_d}i_f + \frac{M_{ah}}{k_d}i_h$$

$$\psi_{qx} = \frac{M_{ag}}{k_q}i_g + \frac{M_{ak}}{k_q}i_k$$

Since $\psi_d = \psi_{dx} - L_d i_d$ and $\psi_q = \psi_{qx} - L_q i_q$, we get

$$T_e = \frac{3}{2}k_dk_q[\psi_q i_d - \psi_d i_q] \tag{3.A.8}$$

If $k_d = k_q = \sqrt{\frac{2}{3}}$, we finally obtain

$$T_e = \psi_q i_d - \psi_d i_q \tag{3.A.9}$$

References

1 Crary, S. (1945, 1947) *Power System Stability, Vol. I: Steady-State Stability, Vol. II: Transient Stability*, Wiley, New York.

2 Kimbark, E. (1948, 1956) *Power System Stability, Vol. I: Elements of Stability Calculations, Vol. III, Synchronous Machine*, Wiley, New York.

3 Anderson, P.M. and Fouad, A.A. (1977) *Power System Control and Stability*, Iowa State University Press, Ames.

4 Kundur, P. (1994) *Power System Stability and Control*, EPRI power system engineering series, McGraw-Hill, New York.

5 Padiyar, K.R. (1996) *Power System Dynamics: Stability and Control*, John Wiley & Sons, Singapore.

6 Jackson, W.B. and Winchester, R.L. (1969) Direct- and quadrature-axis equivalent circuits for solid-rotor turbine generators. *IEEE Transactions on Power Apparatus and Systems*, **88** (7), 1121–1136.

7 Park, R.H. (1929) Two-reaction theory of synchronous machines – Part I: Generalized method of analysis. *AIEE Transactions*, **48** (7), 716–730.

8 IEEE Committee Report (1969) Recommended phasor diagram for synchronous machines. *IEEE Transactions on Power Apparatus and Systems*, **88** (11), 1593–1610.

9 Padiyar, K.R. and Ramshaw, R.S. (1975) Transformations in rotating electric machines based on Floquet–Liapunov theory. *Seventh Annual Southeastern Symposium on Circuit Theory, Auburn, Alabama.*

10 Canay, I.M. (1969) Causes of discrepancies on calculation of rotor quantities and exact equivalent diagrams of the synchronous machine. *IEEE Transactions on Power Apparatus and Systems*, **7** (88), 1114–1120.

11 IEEE Task Force (1986) Current usage and suggested practices in power system stability simulations for synchronous machines. *IEEE Transactions on Energy Conversion*, **EC-1** (1), 77–93.

12 Undrill, J.M. (1969) Structure in the computation of power-system nonlinear dynamical response. *IEEE Transactions on Power Apparatus and Systems*, **PAS-88** (1), 1–6.

13 Harley, R.G., Limebeer, D.J., and Chirricozzi, E. (1980) Comparative study of saturation methods in synchronous machine models. *IEE Proceedings B Electric Power Applications*, **1** (127-B), 1–7.

14 Ramshaw, R.S. and Padiyar, K.R. (1973) Generalised system model for slipring machines. *Proceedings of the Institution of Electrical Engineers*, **120** (6), 647–658.

15 Padiyar, K.R., Sachchidanand and Senthil, J. (1994) Digital computer study of the control of torsional interactions in HVDC turbine-generators, *Electric Machines and Power Systems*, **22** (1), 87–103.

16 IEEE Committee Report (1968) Computer representation of excitation systems. *IEEE Transactions on Power Apparatus and Systems*, **PAS-87** (6), 1460–1464.

17 IEEE Committee Report (1981) Excitation system models for power system stability studies. *IEEE Transactions on Power Apparatus and Systems*, **PAS-100** (2), 494–509.

18 IEEE Committee Report (1973) Dynamic models for steam and hydro turbines in power system studies. *IEEE Transactions on Power Apparatus and Systems*, **PAS-92** (6), 1904–1915.

19 Xue, Y., Van Custem,T., and Ribbens-Pavella, M. (1989) Extended equal area criterion justifications, generalizations, applications. *IEEE Transactions on Power Systems*, **4** (1), 44–52.

20 Pavella, M. and Murthy, P.G. (1994) *Transient Stability of Power Systems: Theory and Practice*, John Wiley & Sons, New York.

21 Pavella, M., Ernst, D., and Ruiz-Vega, D. (2000) *Transient Stability of Power Systems: A Unified Approach to Assessment and Control*, Kluwer, Boston.

22 Padiyar, K.R. (2013) *Structure Preserving Energy Functions in Power Systems: Theory and Applications*, CRC Press, Boca Raton.

23 Padiyar, K.R. and Krishna, S. (2006) On-line detection of loss of synchronism using energy function criterion. *IEEE Transactions on Power Delivery*, **21** (1), 46–55.

4

Modeling and Simulation of Wind Power Generators

4.1 Introduction

Wind power is now a major source of energy in power grids and its share of overall energy production has grown rapidly in the past two decades. It is reported in [1] that the installed wind power generation world-wide, which is currently more than 440 GW, is expected to exceed 760 GW by 2020. Large wind turbines with individual capacities of up to 6–8 MW are now available and large wind farms (onshore and offshore) having overall ratings of hundreds of megawatts are in operation. The availability of wind energy is variable, intermittent, and location dependent. This poses a challenge in grid operation if a large amount of wind generation is integrated into the grid. The injection of wind power into the grid has to be coordinated with the prevailing demand, available generation, storage, and transmission capacity so that the objectives of economy and reliability are met during power system operation.

While wind power is extracted using an electro-mechanical system (a wind turbine coupled to a generator), the electrical configuration is different from conventional power plants, which use synchronous machines directly connected to the grid. The intermittent/varying nature of wind power availability and the operation at variable turbine speeds (to extract maximum power) favor the use of power-electronic converters to interface generators to the electrical grid. Squirrel-cage/wound-rotor induction machines, wound-rotor synchronous machines, or permanent magnet synchronous machines may be used for electro-mechanical energy conversion. Therefore, the behavior of wind turbine-generator systems can be quite different from the behavior of conventional generation systems. Importantly, the use of power electronic converters at the generator-grid interface facilitates the controlled injection of active power and reactive power during steady-state and transient conditions.

This chapter presents the various configurations of wind energy systems and the modeling of these systems. Particular emphasis is given to the modeling of the Type III wind-turbine system, which uses a variable speed wind turbine coupled to a doubly-fed wound rotor induction machine, and a partial-rating frequency converter implemented using back-to-back connected voltage source converters.

Dynamics and Control of Electric Transmission and Microgrids, First Edition.
K. R. Padiyar and Anil M. Kulkarni.

4.2 Power Extraction by Wind Turbines

A wind turbine can be thought of as intercepting a moving tube of wind which has a cross-sectional area A_w in m^2. A_w is the area swept by the blades of the turbine and is given by $A_w = \pi \times \mathcal{R}^2$, where $\mathcal{R}$ is the radius of the blades. The mass of air flowing through this cross-section in h seconds is given by

$$M = A_w \rho \, v_w \, h \tag{4.1}$$

where ρ is the density of air in kg/m^3 and v_w is the wind velocity in m/s. The density of air is a function of the height above sea level and the temperature. The kinetic energy contained in this mass of wind is $\frac{1}{2} A_w \rho v_w^3 \times h$. If this entire kinetic energy was to be extracted by the wind turbine then its power output would be $\frac{1}{2} A_w \rho v_w^3$. However, this is not feasible as the wind has to continuously flow past the turbine, and cannot be abruptly halted. In practice only a fraction C_p (also called the turbine power coefficient) of this energy can be extracted. The theoretical maximum value of C_p, which is also called the Betz limit, is approximately 0.59 [2]. Thus, the power output of a wind turbine P_m in W can be expressed as follows:

$$P_m = \frac{1}{2} \rho A_w C_p v_w^3 \tag{4.2}$$

C_p is a function of the tip-speed ratio λ, and the pitch angle of the turbine blades β. The tip speed ratio is given by $\lambda = \frac{\omega'_m \mathcal{R}}{v_w}$, where ω'_m is the speed of the turbine in rad/s. The following expression [3, 4] may be used for power system studies:

$$C_p = c_1 \left(\frac{c_2}{\lambda_i} - c_3 \beta - c_4 \beta^{c_5} - c_6 \right) e^{\frac{-c_7}{\lambda_i}} \tag{4.3}$$

where

$$\frac{1}{\lambda_i} = \frac{1}{\lambda + c_8 \beta} - \frac{c_9}{\beta^3 + 1} \tag{4.4}$$

The pitch angle β is expressed in degrees in these equations. The coefficients c_1 to c_9 can be determined by using a numerical optimization procedure which minimizes the error between the power curve obtained from these equations and the one obtained from the manufacturer's documentation.

Figure 4.1 shows the plots of C_p versus λ for various values of β, evaluated using the parameters shown in the figure. For a given β, there is an optimum value of tip speed ratio (λ_{opt}) for which C_p is maximum ($C_{p_{opt}}$). It is seen that for $\beta = 0°$, the approximate optimum values are $\lambda_{opt} = 7.2$ and $C_p = 0.44$.

The power versus turbine-speed characteristics of a 2 MW wind turbine for $\beta = 0°$ are shown in Figure 4.2 for various wind speeds. The parameters of the turbine are the same as those used for the computation of C_p earlier. The diameter of the turbine blades, $2 \times \mathcal{R}$, is 75 m, while the air density (ρ) is taken to be 1.225 kg/m^3.

The maximum power that can be extracted as a function of turbine speed is given by

$$P_{m_{opt}} = \left[\frac{1}{2} \frac{\rho A_w \mathcal{R}^3 C_{p_{opt}}}{\lambda_{opt}^3} \right] (\omega'_m)^3 = k_{opt} \, (\omega'_m)^3 \tag{4.5}$$

The locus of the maximum power that can be extracted for different wind speeds is indicated in Figure 4.2.

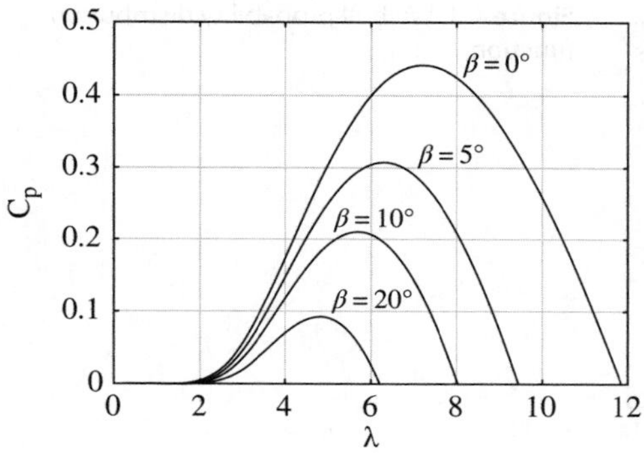

Turbine parameters [5]

c_1	0.73
c_2	151
c_3	0.58
c_4	0.002
c_5	2.14
c_6	13.2
c_7	18.4
c_8	–0.02
c_9	–0.003

Figure 4.1 Variation of C_p with respect to λ.

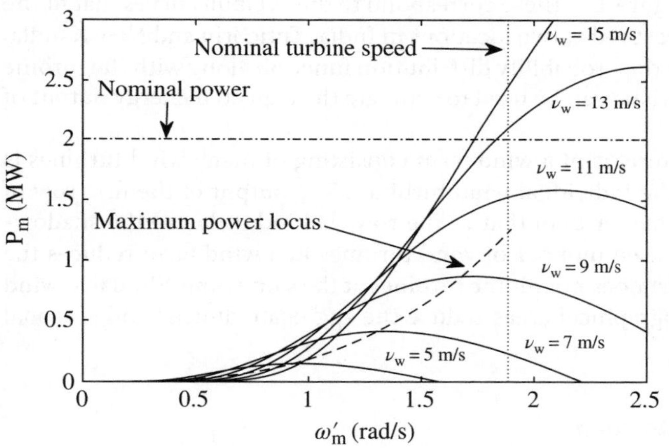

Figure 4.2 Power-speed characteristics of a wind turbine for different wind speeds and a constant pitch angle $\beta = 0°$. $\mathcal{R} = 37.5$ m, $\rho = 1.225$ kg/m³.

4.2.1 Wind Speed Characteristics

The wind speed at a given height varies continuously as a function of time. The wind speed cannot be predicted precisely and can only be analyzed statistically. While wind speed exhibits diurnal and seasonal variations, there may also be turbulent periods caused by gusts of wind in the subsecond to minute range.

The probability density distribution of the wind speed at a given location $p(v_w)$ gives us the proportion of time spent by the wind within narrow bands of wind speed. This distribution is obtained from measurements of wind speeds taken over a long period of time. The Weibull distribution function gives a reasonable fit to the wind speed data in a large number of cases [6]:

$$p(v_w) = \frac{k}{c}\left(\frac{v_w}{c}\right)^{k-1} e^{[-(\frac{v_w}{c})^k]} \tag{4.6}$$

where c is the scale factor having the unit of speed and k is a dimensionless shape factor. The probability distributions obtained for $k = 1.96$ and $c = 5.21$ m/s, and $k = 1.99$ and

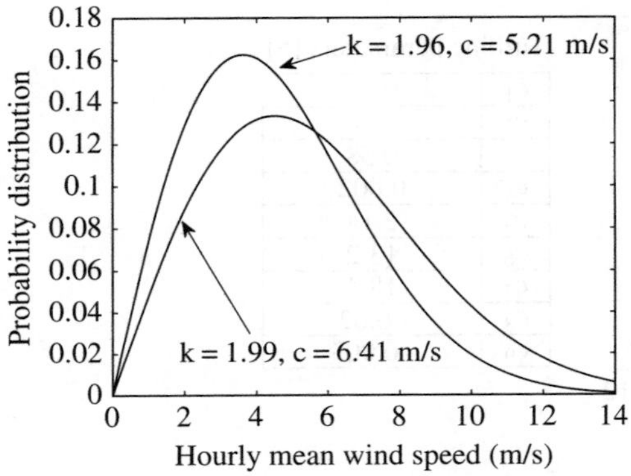

Figure 4.3 Weibull probability distribution function.

$c = 6.41$ m/s are shown in Figure 4.3. These correspond to the Weibull curves that fit the hourly mean wind speed measured at two locations in India, Tuticorin and New Kandla, as reported in [7]. Note that the probability distribution function along with the turbine power-wind speed characteristic can be used to estimate the expected energy output of a wind turbine in a year.

The overall power characteristic of a wind farm consisting of many wind turbines is not the scaled-up value of the individual wind turbines. The output of the first row of turbines facing the wind is higher than that of the rows behind it due to the shadowing or wake effect. An increased number of wind turbines in a wind farm reduces the impact of wind gusts as a gust does not hit the turbines at the same time. Similarly, wind farms spread over large geographical areas reduce the aggregate diurnal and seasonal variations.

4.2.2 Control of Power Extraction

The wind speed at which a wind turbine starts to operate is called the cut-in speed (typically around 3 m/s). As wind speed increases, the power that can be captured by the wind turbine also increases. At wind speeds greater than the rated wind speed (typically around 12 to 16 m/s), the available power may exceed the rated power. The turbine speed will then tend to increase beyond its limit. To prevent the power and turbine speed from exceeding the limits, the power extracted from the wind is limited by pitch control or stall control [2].

Pitch (β) control involves turning the blades around their longitudinal axis. This reduces C_p, which is a function of β, as shown in Figure 4.1, thereby reducing the power output. A proportional control-based pitch angle controller is shown in Figure 4.4. The controller increases the pitch angle when the turbine speed increases beyond the

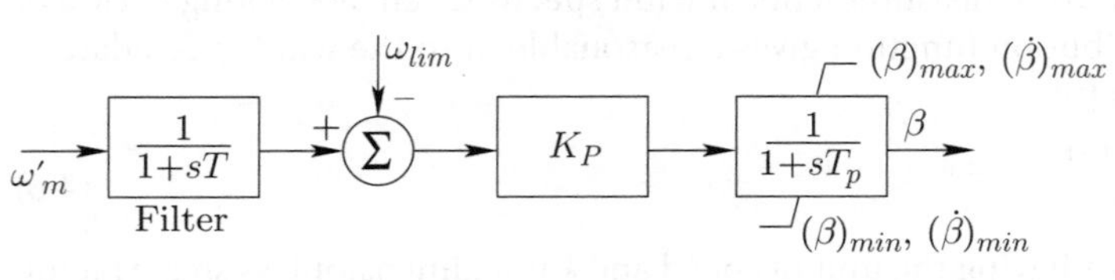

Figure 4.4 Pitch angle control.

turbine speed limit ω_{lim}, which may be set to be equal to or slightly above the nominal turbine speed. The rate of increase in pitch angle is dependent on the blade-drive system. The maximum ramp-up rate of the pitch angle is in the order of 3 to 10°/s and depends on the size of the wind turbine [5]. The maximum ramp-down rate may be different from the maximum ramp-up rate. The speed will exceed the value ω_{lim} during high-speed conditions since the rate of change of β is limited. Hence some headroom may be provided for this [8].

Even when the blades of the turbine are bolted to the hub at a fixed angle (no pitch control), the rotor aerodynamics may be designed to stall (lose power) when the wind speed exceeds a certain level. This is known as stall control.

Beyond a certain speed (cut-out wind speed), which is typically in the range of 20–25 m/s, the turbine stops power generation and turns out of the main wind direction. To avoid a sudden drop of power, a gradual reduction may be implemented. The subsequent restart of the wind turbine usually requires a reduction in wind speed by 3–4 m/s. The restart may involve a time delay depending on the type of turbine-generator configuration that is used.

Since C_p is also a function of tip-speed ratio, the mechanical power P_m extracted by the turbine (for a given wind speed) can be changed by controlling ω'_m. Since the turbine is coupled to the rotor of a electric generator (usually through a gearbox) the turbine speed and generator speed are related by the gear ratio. Therefore, for a given wind speed, the turbine speed is determined by the point of intersection of the mechanical power–speed characteristic of the wind turbine at that wind speed, and the electrical power–speed characteristic of the generator.

For example, if the generator is a squirrel cage induction machine connected to the grid, then the speed of the generator is almost constant and nearly equal to the synchronous speed (the slip is usually small over the entire operating power range). Therefore the turbine speed is nearly constant over the entire power range when this electrical configuration is used. In this situation, the turbine speed is practically not controllable by the electrical system.

The control of the turbine speed by the electrical system (for a given wind speed) is feasible if the electrical power speed characteristics of the generator can be controlled. This can be achieved using power electronic converters. The turbine speed may be controlled so that the tip speed ratio is maintained at λ_{opt}, that is, maximum power is extracted. This is indicated by the dashed line shown in Figure 4.2. However, once the turbine speed limit is reached, the turbine speed may be regulated at this value by the electrical system, as long as the power extracted is less than the rated power.

To summarize, a wind-turbine system has the following modes of operation:

1) Below cut-in wind speed: No power extraction from the wind.
2) Between cut-in wind speed and rated wind speed: The turbine speed and the extracted power depend on the characteristics and/or control of the electrical system (generator, power electronic interfaces, and grid). The electrical system may be controlled so that maximum power is extracted from the wind, subject to the power rating of the equipment and the turbine speed limit.
3) Between rated wind speed and cut-out wind speed: The power extracted from the wind is regulated at the rated value or reduced by pitch angle control or stall control.
4) Above cut-out wind speed: No power extraction from the wind.

4.3 Generator and Power Electronic Configurations

Wind turbines may be coupled to squirrel cage induction generators (SCIGs) which are directly connected to the grid. This configuration is known as a *fixed-speed wind turbine* or a *Type I wind-turbine generator (WTG)* (see Figure 4.5). The speed is nearly constant because the slip of an induction generator is generally small in its normal operating range, and the grid frequency also normally remains within a small range around the nominal value. SCIGs always draw reactive power from the grid. The amount of reactive power drawn depends on the power output and the grid voltage. Hence, SCIGs require reactive power compensation in the form of shunt-connected switched capacitor banks and/or controlled reactive volt-ampere (VAr) sources like Static VAr Compensator (SVC) or STATic (Synchronous) COMpensators (STATCOMs), which are described in Chapter 5. A soft-starting arrangement is generally provided to ensure a graceful grid connection (i.e., lower transient inrush currents when connecting to the grid). A disadvantage of fixed speed turbines is that at low wind speeds the power output is lower than the maximum power that may be extracted at that wind speed (see Figure 4.2).

To facilitate the control of the turbine speed so that maximum power can be extracted from the wind, as discussed in the previous section, variable speed turbines may be used. These turbines are coupled to generators whose electrical output is controlled using power electronic controllers. Variable-speed turbines may be used in conjunction with the following generator/power electronic configurations:

1) **Wound rotor induction generators (WRIGs) with variable rotor resistance (Type II WTG):**
   In this configuration (see Figure 4.6), a variable resistance is connected externally to the rotor windings of the WRIG. The resistance is varied using power electronic control. This enables control of the generator speed and electrical power over a limited range. The slip power is dissipated in the rotor resistance as losses. The configuration requires both the soft-starter as well as reactive power compensation.
2) **Doubly-fed induction generators with partial-rating frequency converter (Type III WTG):**
   By using a partial-rating AC-AC frequency converter to inject a voltage into the rotor windings of a WRIG, as shown in Figure 4.7, it is possible to control the generator speed. The converter rating is typically about 30% of the generator power rating. The speed variation of approximately ±30% around the synchronous speed is achievable with this, and is normally sufficient to obtain the benefits of maximum power extraction. The WRIG used in such a configuration is known as a doubly-fed

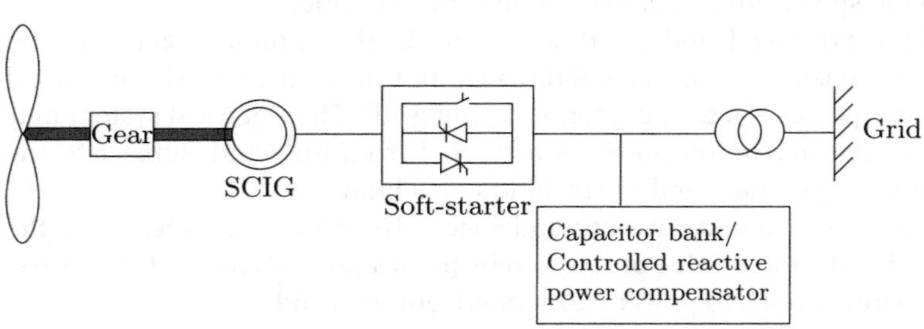

Figure 4.5 Type I WTG: SCIG with fixed-speed wind turbine.

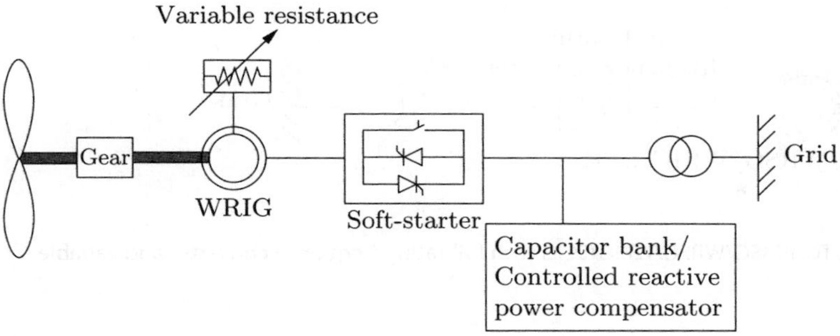

Figure 4.6 Type II WTG: WRIG with variable rotor resistance and variable speed wind turbine.

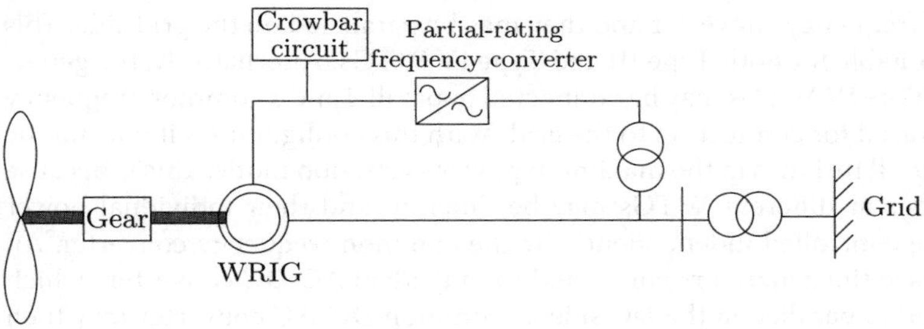

Figure 4.7 Type III WTG: DFIG with partial-rating frequency converter and variable speed wind turbine.

induction generator (DFIG). In this configuration, the need for a soft-starter is also obviated. A separate reactive power compensator may not be necessary as the converters can facilitate a controlled exchange of reactive power with the grid (subject to their overall volt-ampere ratings).

The AC-AC frequency converter is typically implemented using two stages of conversion: an AC-DC converter and a DC-AC converter, with a DC link between them. Since the stator of a DFIG is directly connected to the grid, a fault results in high currents being induced in the rotor circuit, which may damage the converter. If DFIGs are to remain connected and smoothly recover following a fault then a protection system such as a crowbar circuit is necessary. In this arrangement, the rotor windings of a wound-rotor machine are connected to a resistance during fault conditions to divert the high currents which would otherwise flow into the rotor-side converter. The crowbar is deactivated when the rotor current falls below a threshold value, and the converter can resume operation thereafter.

3) **Permanent magnet or wound rotor synchronous generator (PMSG/WRSG) or WRIG/SCIG with a full-rating frequency converter (Type IV WTG):**
   In this configuration the generator stator is connected to the grid via a full-rating AC-AC converter (see Figure 4.8). Both active and reactive power can be controlled and a smooth connection to the grid is possible in this configuration. A gearbox may not be necessary if a generator with a large number of poles is used.

Note that pitch control is used in both fixed- and variable-speed turbines, while stall control is used only with fixed-speed wind turbines. In [2] it is reported that the largest turbines from most manufacturers use variable-speed turbines with a pitch control mechanism.

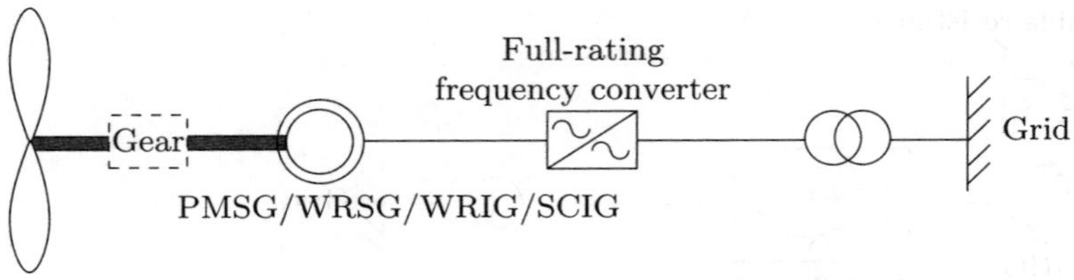

Figure 4.8 Type IV WTG: PMSG/WRSG/WRIG/SCIG with full-rating frequency converter and variable speed wind turbine.

4.3.1 Wind Farm Configurations

A wind farm may consist of a large number of wind turbines. Each variable-speed WTG may have its own frequency converter and they may be paralleled on the grid side. This configuration is suitable for both Type III and Type IV WTGs. Alternatively, the generators of different Type IV WTGs may be connected in parallel and a common frequency converter may be used for connection to the grid. With this configuration it may not be possible to operate all turbines in the maximum power extraction mode. This is because the local conditions at different WTGs may be different and their individual power outputs cannot be controlled independently by the common frequency converter. An alternative is to have the generators connected to individual AC-DC converters which are then connected in parallel on the DC side. A common DC-AC converter may then be used to connect to the AC grid. Electrical storage devices like batteries (if present) can be connected in shunt at the DC bus. For offshore wind farms, the output can be pooled on the DC side of the AC-DC converters and DC transmission from the offshore platforms to the mainland can be used for power evacuation. The DC-AC conversion is then done on the mainland.

Small WTGs may also be used in stand-alone mode (not connected to a grid, feeding a local load). However, these will require devices to regulate the frequency and voltage magnitude. Small WTGs may also be integrated with other energy sources (solar and/or diesel generators) and storage systems to form microgrids. These systems are considered in Chapter 12.

4.4 Modeling of the Rotating System

If a wind turbine is connected to a generator via a gear assembly, as is generally the case, then the mechanical torque T'_m and speed (ω'_m) on the turbine side are related to the generator side mechanical torque (T_m) and speed (ω_m), respectively, as follows: $T'_m = \frac{n_g}{n_t} T_m$ and $\omega_m = \frac{n_g}{n_t}\omega'_m$, where ($\frac{n_g}{n_t}$) denotes the teeth ratio of the gear. The mechanical power P_m is given by:

$$P_m = T'_m \, \omega'_m = T_m \, \omega_m \tag{4.7}$$

All the mechanical system parameters may be referred to one side of the gearbox (as is done with transformers). The inertia of the turbine, when referred to the generator side, has to be multiplied by $(\frac{n_t}{n_g})^2$. These transformations can be avoided by using a per-unit system with appropriate mechanical speed and torque base values on either side of the gearbox.

The dynamical equation of the rotor electrical speed ω in rad/s is given by:

$$\frac{2H}{\omega_B}\frac{d\omega}{dt} = \frac{2H}{\omega_{mB}}\frac{d\omega_m}{dt} = T_m - T_e. \tag{4.8}$$

In this equation, the mechanical input torque T_m and the electrical torque T_e are per unit. ω_B denotes the base electrical frequency in rad/s. ω_{mB} is the base mechanical frequency in rad/s on the generator side of the gearbox. The mechanical speed of the generator ω_m in rad/s is related to the electrical frequency in rad/s by the equation $\omega_m = \frac{2}{P}\omega$, where P is the number of poles. This also holds true for the corresponding base values ω_{mB} and ω_B. If S_B denotes the volt-ampere (VA) base then the torque base on the generator side T_B is related to S_B by $T_B = \frac{S_B}{\omega_{mB}}$.

Note that the rotor is modeled in this equation as a single mass; the inertia constant H (in MJ/MVA) is that of the entire rotating system, that is, $H = H_t + H_g$, where H_t and H_g denote the inertia constants of the turbine and generator, respectively. Note that

$$H_t = \frac{\frac{1}{2}J_t\omega_{mB}^2}{S_B} \quad \text{and} \quad H_g = \frac{\frac{1}{2}J_g\omega_{mB}^2}{S_B}$$

J_t denotes the turbine inertia in kg-m^2, referred to the generator side of the gearbox as discussed earlier. J_g denotes the inertia of the generator in kg-m^2.

For Type I and II WTGs, it may be necessary to model the shaft torsional dynamics (the relative motion between the generator and the turbine) to estimate the impact of disturbances on the shaft. This modeling is also needed to study small-signal oscillatory instabilities due to interactions between the torsional system and the electrical network and power electronic controllers. A two-mass-shaft model of the system is described by the following equations.

$$\frac{2H_t}{\omega_B}\frac{d\omega_t}{dt} = T_m - K_s(\theta_t - \theta) \tag{4.9}$$

$$\frac{2H_g}{\omega_B}\frac{d\omega}{dt} = K_s(\theta_t - \theta) - T_e \tag{4.10}$$

$$\frac{d(\theta_t - \theta)}{dt} = \omega_t - \omega \tag{4.11}$$

ω_t and ω are the electrical speeds in rad/s of the turbine and generator rotors respectively, and K_s is the shaft stiffness in per-unit torque/electrical radians. $\theta_t - \theta$ denotes the relative angular displacement of the turbine and generator in electrical radians.

For a Type IV WTG, the fast power-electronic control of the electrical power fed to the grid using the full-rating converter effectively shields the electrical network from the torsional dynamics of the rotor. In a Type III system, although electrical power is controlled by the rotor-side converter under normal conditions, crowbar protection is activated during faults, which diverts the current from the rotor-side converter. During this period, the DFIG behaves like a conventional fixed-speed generator and the multi-mass model may be needed to examine the torsional dynamics.

4.5 Induction Generator Model

Induction machines are commonly used in wind-turbine generator systems. This section presents the per-unit (pu) form of the stator and rotor equations of an induction machine. A wound-rotor induction machine is shown in Figure 4.9. The model presented here is based on the following assumptions:

1) The magnetomotive force in the airgap is distributed sinusoidally and the harmonics are neglected.
2) The rotor is cylindrical (no saliency). The effect of slots on the stator and rotor is neglected.
3) Magnetic saturation and hysteresis are ignored.

The machine shown in Figure 4.9 has two poles, but since the equations are finally written in pu form, they are valid for a machine with a higher number of pole pairs as well. The direction of currents with respect to the voltages, torques with respect to the direction of rotation, and the flux-current relationships follow the *generator convention*. In the following equations, $\psi, v,$ and i represent the fluxes, voltages, and currents respectively, all expressed in pu form. The subscript r denotes the rotor quantities. Time-variant transformations are applied to the stator *and* the rotor three-phase variables in order to obtain a time-invariant set of equations.

Stator Equations

$$-\frac{1}{\omega_B}\frac{d\psi_D}{dt} - \frac{\omega_s}{\omega_B}\psi_Q - R_s i_D = v_D \tag{4.12}$$

$$-\frac{1}{\omega_B}\frac{d\psi_Q}{dt} + \frac{\omega_s}{\omega_B}\psi_D - R_s i_Q = v_Q \tag{4.13}$$

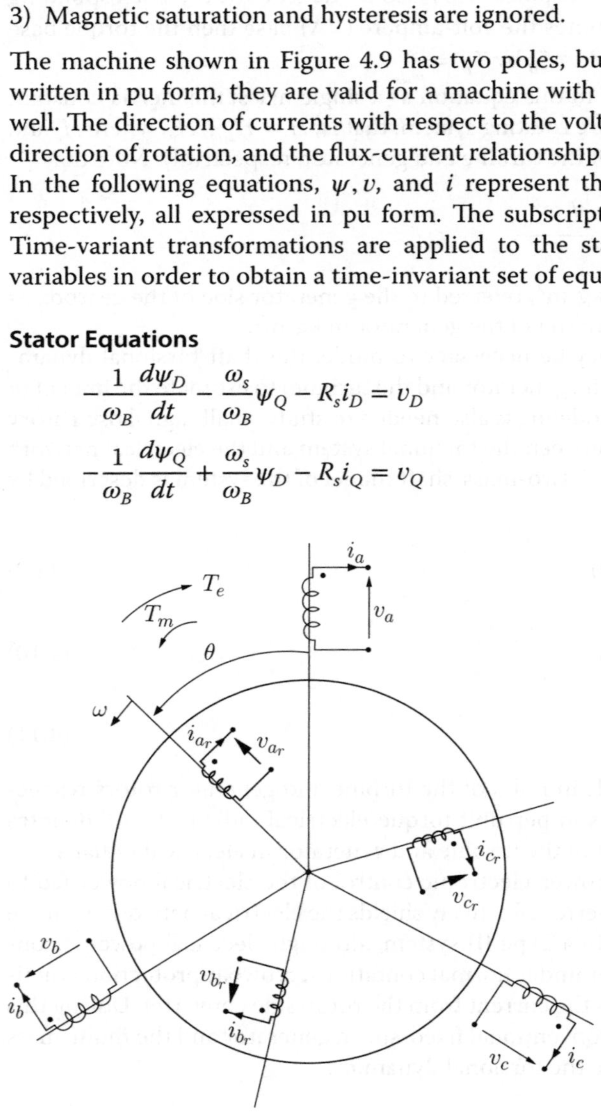

Figure 4.9 Induction machine windings.

$$-\frac{1}{\omega_B}\frac{d\psi_o}{dt} - R_s i_o = v_o \tag{4.14}$$

where ω_s is the frequency of the electrical grid in rad/s (synchronous frequency). For a machine which is not connected to a grid, this could be taken to be the base speed, ω_B. R_s is the stator winding resistance. Note that $\omega_s = \frac{d\theta_s}{dt}$, where θ_s is the angle (in radians) used in the matrix C_S which is used to transform the DQo variables to the abc variables. The matrix C_S is

$$C_S = \sqrt{\frac{2}{3}}\begin{bmatrix} \cos\theta_s & \sin\theta_s & \sqrt{\frac{1}{2}} \\ \cos(\theta_s - 2\pi/3) & \sin(\theta_s - 2\pi/3) & \sqrt{\frac{1}{2}} \\ \cos(\theta_s + 2\pi/3) & \sin(\theta_s + 2\pi/3) & \sqrt{\frac{1}{2}} \end{bmatrix} \tag{4.15}$$

Rotor Equations

$$-\frac{1}{\omega_B}\frac{d\psi_{D_r}}{dt} - \frac{(\omega_s - \omega)}{\omega_B}\psi_{Q_r} - R_r i_{D_r} = v_{D_r} \tag{4.16}$$

$$-\frac{1}{\omega_B}\frac{d\psi_{Q_r}}{dt} + \frac{(\omega_s - \omega)}{\omega_B}\psi_{D_r} - R_r i_{Q_r} = v_{Q_r} \tag{4.17}$$

$$-\frac{1}{\omega_B}\frac{d\psi_{o_r}}{dt} - R_r i_{o_r} = v_{o_r} \tag{4.18}$$

where $\omega = \frac{d\theta}{dt}$ is the electrical speed of the rotor in rad/s. θ is the rotor electrical position in radians. R_r is the rotor winding resistance. The transformation matrix C_R that is used to obtain the a_r, b_r, c_r variables from the $D_r Q_r o_r$ variables is given below.

$$C_R = \sqrt{\frac{2}{3}}\begin{bmatrix} \cos(\theta_s - \theta) & \sin(\theta_s - \theta) & \sqrt{\frac{1}{2}} \\ \cos(\theta_s - \theta - 2\pi/3) & \sin(\theta_s - \theta - 2\pi/3) & \sqrt{\frac{1}{2}} \\ \cos(\theta_s - \theta + 2\pi/3) & \sin(\theta_s - \theta + 2\pi/3) & \sqrt{\frac{1}{2}} \end{bmatrix} \tag{4.19}$$

Flux-Current Relationship

The flux-current relationship in pu form is given by the following equation:

$$\begin{bmatrix} \psi_D \\ \psi_Q \\ \psi_o \\ \psi_{D_r} \\ \psi_{Q_r} \\ \psi_{o_r} \end{bmatrix} = \begin{bmatrix} X_{ss} & 0 & 0 & X_m & 0 & 0 \\ 0 & X_{ss} & 0 & 0 & X_m & 0 \\ 0 & 0 & X_o & 0 & 0 & 0 \\ X_m & 0 & 0 & X_{rr} & 0 & 0 \\ 0 & X_m & 0 & 0 & X_{rr} & 0 \\ 0 & 0 & 0 & 0 & 0 & X_{o_r} \end{bmatrix}\begin{bmatrix} i_D \\ i_Q \\ i_o \\ i_{D_r} \\ i_{Q_r} \\ i_{o_r} \end{bmatrix} \tag{4.20}$$

where X_m, X_{ss}, and X_{rr} are the magnetizing reactance, stator, and rotor reactances, respectively. Note that $X_m = X_{ss} - X_s = X_{rr} - X_r$, where X_s and X_r are the stator and rotor leakage reactances. X_o and X_{o_r} are the zero-sequence reactances of the stator and rotor windings.

Electrical Torque

The per-unit electrical torque is given by

$$T_e = \psi_D i_Q - \psi_Q i_D = -(\psi_{D_r} i_{Q_r} - \psi_{Q_r} i_{D_r}) = \frac{\omega_B}{\omega} P_e \tag{4.21}$$

The torque expression per unit is not dependent on the number of poles and the zero-sequence variables. P_e is the electrical power at the shaft per unit.

Steady-State Real Power Relationships

In steady state, the DQ and D_rQ_r variables are constant. Therefore, the steady-state relationships between the flux, current, and voltages can be obtained by setting the derivative terms in the differential equations (4.12)–(4.13) and (4.16)–(4.17) to zero. From the resulting equations, we obtain the following:

$$\frac{\omega_s}{\omega_B}(-\psi_Q i_D + \psi_D i_Q) = (v_D i_D + v_Q i_Q) + R_s(i_D^2 + i_Q^2) \tag{4.22}$$

$$\frac{(\omega_s - \omega)}{\omega_B}(\psi_{D_r} i_{Q_r} - \psi_{Q_r} i_{D_r}) = (v_{D_r} i_{D_r} + v_{Q_r} i_{Q_r}) + R_r(i_{D_r}^2 + i_{Q_r}^2) \tag{4.23}$$

The right-hand sides of these equations denote the electrical power available at the stator and rotor windings respectively. In steady state, $T_e = T_m$ and $P_m = P_e = \frac{\omega}{\omega_B} T_m = \frac{\omega}{\omega_B} T_e$. Using (4.21), (4.22), and (4.23), we obtain the following equations:

$$(v_D i_D + v_Q i_Q) + R_s(i_D^2 + i_Q^2) = \frac{\omega_s}{\omega_B} T_e = \frac{\omega_s}{\omega} P_m = \frac{P_m}{1+S} \tag{4.24}$$

$$(v_{D_r} i_{D_r} + v_{Q_r} i_{Q_r}) + R_r(i_{D_r}^2 + i_{Q_r}^2) = -\frac{\omega_s - \omega}{\omega_B} T_e = -\frac{\omega_s - \omega}{\omega} P_m = \frac{P_m S}{1+S} \tag{4.25}$$

where $\omega = (1 + S)\omega_s$, S being the slip of the induction machine (generator convention). In steady state,

$$P_m = P_e = \frac{P_m S}{1+S} + \frac{P_m}{1+S} \tag{4.26}$$

This means that the fraction of power available at the stator and rotor windings depends on the slip. This is depicted in Figure 4.10. The total electrical power injected into the grid is equal to $P_e - (i_D^2 + i_Q^2)R_s - (i_{D_r}^2 + i_{Q_r}^2)R_r$.

Remarks

1) The model of an SCIG can be obtained by setting v_{a_r}, v_{b_r}, and v_{c_r} to zero, or equivalently by setting v_{D_r}, v_{Q_r}, and v_{o_r} to zero.
2) A wound-rotor machine with resistances R_{ext} externally connected to the rotor windings can be modeled by setting $v_{D_r} = R_{ext} i_{D_r}$, $v_{Q_r} = R_{ext} i_{Q_r}$, and $v_{o_r} = R_{ext} i_{o_r}$.

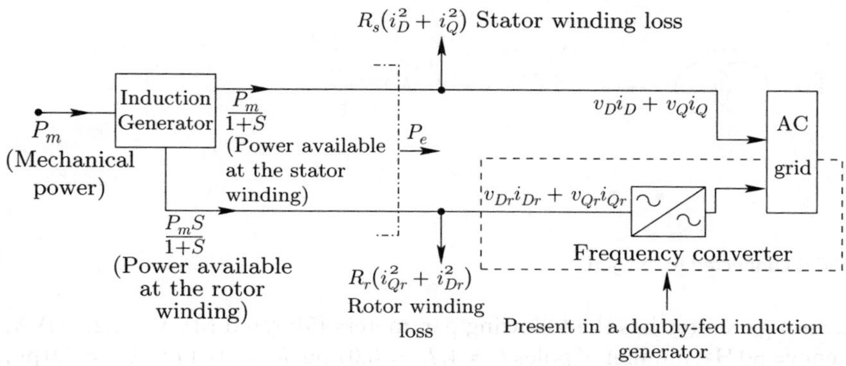

Figure 4.10 Steady-state real power flow in an induction generator.

3) In the case of a DFIG, the voltage applied to the rotor windings (v_{D_r} and v_{Q_r}) may be used to control the slip (or speed) of the generator, and therefore the turbine speed. As discussed previously, this allows us to extract maximum power from the wind by adjusting the tip-speed ratio. The control of a DFIG is discussed in detail in Section 4.6.
4) A SCIG that is not connected to a grid or voltage source can nonetheless be *self-excited* (build up AC voltage at its terminals and feed a load) if it is connected to a three-phase bank of capacitors. For self-excitation to happen, the per-phase capacitance of the bank has to be above a critical value, which is dependent on the rotational speed. Self-excitation occurs if the system is small-signal unstable at the zero-flux, zero-voltage equilibrium state [9], thereby causing a voltage buildup if some initial flux or charge is present. The voltage buildup continues until it is limited by the nonlinear magnetic characteristics of the generator.
5) The zero-sequence variables are decoupled from the *D*-*Q* variables and are zero under balanced conditions, therefore these equations may be neglected in the analysis if the system is balanced.

4.5.1 Rotor Speed Instability

In the Type I WTG system, which uses an SCIG directly connected to a grid, a dynamic phenomenon of interest is the loss of stability of the generator if it is subjected to large disturbances like faults. This is also called speed instability [10, 11] and is the generator counterpart of the stalling of induction motors under severe voltage sags. This is illustrated in the following example.

Example 4.1 ***Rotor Speed Instability*** The induction generator of a Type I WTG system is connected to the grid by a transformer and a transmission line, as shown in Figure 4.11. The WTG system uses a 2 MW wind turbine with the following parameters [3]: $c_1 = 0.5$, $c_2 = 116$, $c_3 = 0.4$, and $c_4 = 0.0$. c_5 is not required as $c_4 = 0.0$, $c_6 = 5$, $c_7 = 21$, $c_8 = 0.08$, and $c_9 = 0.035$. $\mathcal{R} = 37.5$ m. The pitch angle is assumed to be constant at $\beta = 0°$ and $\rho = 1.225$ kg/m^3. The gear-ratio of the turbine-generator system ($\frac{n_g}{n_t}$) is 79.

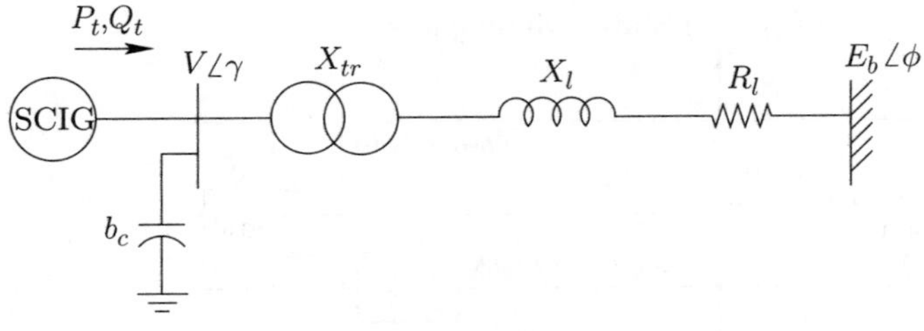

Figure 4.11 SCIG connected to the grid.

The induction generator has the following parameters [5]: rated MVA = 2.27 MVA, rated frequency= 50 Hz, number of poles $P = 4$, $R_s = 0.01$ pu, $R_r = 0.01$ pu, $X_m = 3.0$ pu, $X_{ss} = 3.1$ pu, and $X_{rr} = 3.08$ pu. Note that $X_s = X_{ss} - X_m = 0.1$ pu and $X_r = X_{rr} - X_m = 0.08$ pu.

The inertia of the generator is $H_g = 0.5$ MJ/MVA and that of the turbine is $H_t = 2.5$ MJ/MVA. The generator and turbine rotors are represented as one lumped rotor mass with inertia $H = H_t + H_g = 3$ MJ/MVA.

The network parameters are $R_l = 0.02$ pu, $X_l = 0.2$ pu, $X_{tr} = 0.08$ pu, and $b_c = 0.5$ pu.

The MVA base S_B is the rated three-phase MVA of the induction generator. The base electrical and mechanical speeds are $\omega_B = 2\pi \times 50$ rad/s and $\omega_{mB} = \frac{2}{P}\omega_B$, respectively.

The initial (steady-state) torque of the generator T_e is 0.85 pu and the initial terminal voltage of the induction generator V is 1.0 pu. The grid frequency ω_b is equal to the base frequency ω_B.

Investigate the rotor speed instability phenomenon by evaluating the response of the system to a three-phase fault in the grid with different fault-clearing times.

Solution

The electrical and mechanical equations of the induction generator and wind turbine are given in the previous sections. The *D-Q* equations of the shunt capacitor and the network are obtained by transforming the equations in the phase variables using the same transformation matrix (C_S) that is used for transforming the stator quantities of the generator. The pu forms of these equations for the shunt capacitor are:

$$\frac{1}{\omega_B}\frac{dv_D}{dt} = -\frac{\omega_s}{\omega_B}v_Q + \frac{1}{b_c}(i_D - i_{Dl}) \tag{4.27}$$

$$\frac{1}{\omega_B}\frac{dv_Q}{dt} = \frac{\omega_s}{\omega_B}v_D + \frac{1}{b_c}(i_Q - i_{Ql}) \tag{4.28}$$

where i_{Dl} and i_{Ql} denote the currents in the transmission line. The equations of the transmission line are

$$\frac{1}{\omega_B}\frac{di_{Dl}}{dt} = -\frac{R_l}{X_l + X_{tr}}i_{Dl} - \frac{\omega_s}{\omega_B}i_{Ql} + \frac{1}{X_l + X_{tr}}(v_D - e_D) \tag{4.29}$$

$$\frac{1}{\omega_B}\frac{di_{Ql}}{dt} = -\frac{R_l}{X_l + X_{tr}}i_{Ql} + \frac{\omega_s}{\omega_B}i_{Dl} + \frac{1}{X_l + X_{tr}}(v_Q - e_Q) \tag{4.30}$$

where e_D and e_Q denote the *D-Q* components of the grid voltage. Note that in the pre-fault steady state, $e_Q + je_D = E_b\angle\phi$ and $v_Q + jv_D = V\angle\gamma$.

Table 4.1 Initial conditions

Variables	Value	Variables	Value	Variables	Value
T_e	0.85 pu	V	1.0 pu	γ	0°
T_m	0.85 pu	E_b	1.0015 pu	ϕ	−13.56°
P_m	0.8578 pu	ω	317.03 rad/s	ω_m	158.52 rad/s
P_e	0.8578 pu	$\frac{\omega}{\omega_B}$	1.009 pu	$\frac{\omega_m}{\omega_{mB}}$	1.009 pu
P_t	0.8408 pu	ω'_m	2.0066 rad/s	v_w	14 m/s
Q_t	−0.4657 pu	λ	5.3713	C_p	0.2621

The system is assumed to be in steady state prior to the disturbance. The initial conditions of all the variables for the given torque and the terminal voltage are given in Table 4.1. Note that P_t and Q_t are the real and reactive power injected by the generator at its terminals, respectively. θ_s, which is used in the transformation C_S, is chosen such that the initial value of $v_D = 0$, and $\frac{d\theta_s}{dt} = \omega_s = \omega_b$. The wind-speed ($v_w$), grid voltage phase angle (ϕ), and the grid frequency (ω_b) are assumed to be constant during transient conditions. The system behavior is simulated by numerically integrating the differential equations of the system, as given in (4.8), (4.12)–(4.21) and (4.27)–(4.30). The three-phase fault in the grid is simulated by setting E_b equal to zero for the fault duration. E_b is constant at its initial value except during the fault. The response of the speed of the induction generator for different fault durations is shown in Figure 4.12a. The generator is unable to come back to an acceptable equilibrium speed when the fault clearing time is more than 0.25 s (the critical clearing time). This happens because the electrical torque starts decreasing as the speed crosses a certain value. The change in the mechanical torque due to the change in the tip-speed ratio is relatively smaller. The accelerating torque (mechanical minus electrical) may then become positive, causing the machine to accelerate. This phenomenon is called rotor speed instability. The mechanical and electrical torques for an unstable case are shown in Figure 4.12b. The generator terminal voltage magnitude $\left(\sqrt{v_D^2 + v_Q^2}\right)$ and generator current magnitude $\left(\sqrt{i_D^2 + i_Q^2}\right)$ are shown for the unstable case in Figure 4.13. The unstable situation is accompanied by a large generator current and a low voltage, and cannot be tolerated for long (the induction generator has to be disconnected from the grid). Even in the marginally stable case (clearing time = 0.24 s), the voltage remains depressed and the current magnitude is high for a prolonged period before they settle to the equilibrium.

Speed instability is affected by the quiescent power output of the generator and the external reactance. If the external reactance X_l is doubled, the critical clearing time decreases to approximately 0.1 s, while if the quiescent power output of the generator is reduced to 0.74 pu, the critical clearing time increases to approximately 0.36 s.

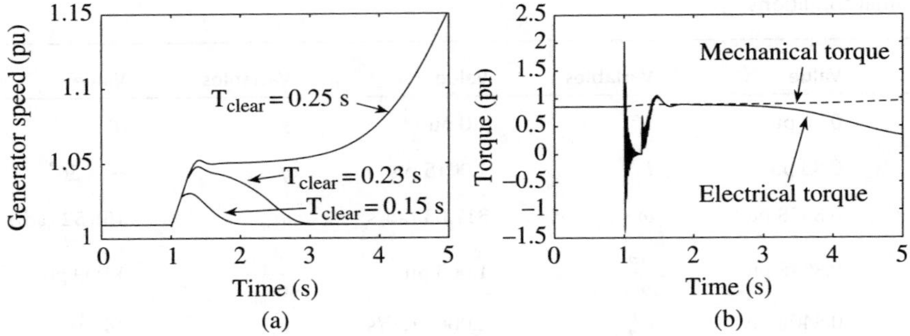

Figure 4.12 Response of generator speed and torque for a fault in the grid. (a) Generator speed ($\frac{\omega}{\omega_B}$) (b) Torques T_m and T_e ($T_{clear} = 0.25$ s).

4.5.2 Modeling Issues

The responses of the electrical torque, current, and voltage in Figures 4.12 and 4.13 exhibit two dominant patterns:

1) A high-frequency, decaying oscillatory transient which is excited when the fault occurs and when it is cleared.
2) A relatively slow transient which corresponds to the electro-mechanical interaction.

The high-frequency oscillatory pattern is primarily due to electromagnetic interactions, as it is almost unobservable in the speed. The slower component of the response is of interest in the study of rotor speed instability. For this component, the terms $\frac{1}{\omega_B}\frac{d\psi_D}{dt}$, $\frac{1}{\omega_B}\frac{d\psi_Q}{dt}$, $\frac{1}{\omega_B}\frac{dv_D}{dt}$, $\frac{1}{\omega_B}\frac{dv_Q}{dt}$, $\frac{1}{\omega_B}\frac{di_{Dl}}{dt}$, and $\frac{1}{\omega_B}\frac{di_{Ql}}{dt}$ are small relative to the other terms in the corresponding differential equations. Therefore, an approximate model may be constructed by neglecting these terms in the differential equations, thereby converting them into algebraic equations, which are given below.

$$-\frac{\omega_s}{\omega_B}\psi_Q - R_s i_D - v_D = 0$$

$$\frac{\omega_s}{\omega_B}\psi_D - R_s i_Q - v_Q = 0$$

$$-\frac{R_l}{X_l + X_{tr}} i_{Dl} - \frac{\omega_s}{\omega_B} i_{Ql} + \frac{1}{X_l + X_{tr}}(v_D - e_D) = 0$$

$$-\frac{R_l}{X_l + X_{tr}} i_{Ql} + \frac{\omega_s}{\omega_B} i_{Dl} + \frac{1}{X_l + X_{tr}}(v_Q - e_Q) = 0$$

$$-\frac{\omega_s}{\omega_B} v_Q + \frac{1}{b_c}(i_D - i_{Dl}) = 0$$

$$\frac{\omega_s}{\omega_B} v_D + \frac{1}{b_c}(i_Q - i_{Ql}) = 0$$

These algebraic equations are combined with the differential equations of the rotor fluxes (4.16), (4.17) and the speed (4.8) to obtain a differential-algebraic model of the system. This model has three differential equations. The model is expected to capture the slow response with good accuracy, but will exclude the faster electromagnetic transients.

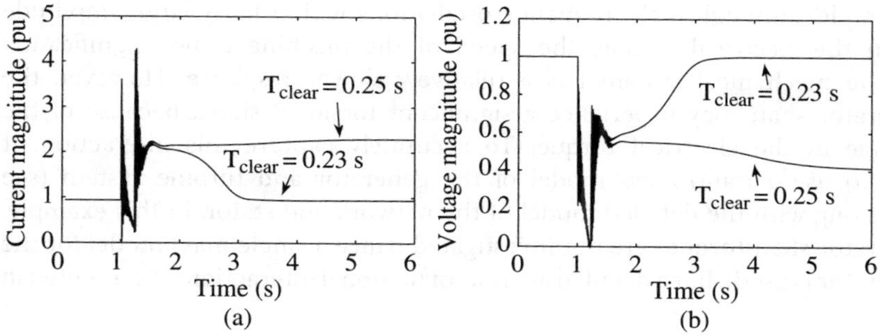

Figure 4.13 Generator current and voltage magnitudes for the unstable and marginally stable condition. (a) Generator current magnitude (b) Generator voltage magnitude.

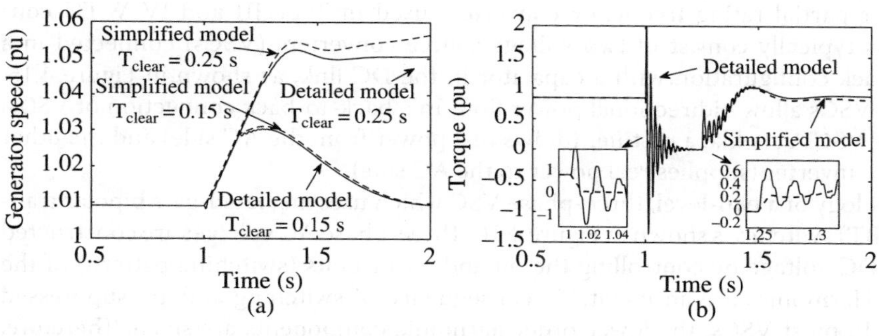

Figure 4.14 Generator speed and electrical torque response: simplified and detailed model. (a) Generator speed ($\frac{\omega}{\omega_B}$) (b) Electrical torque (T_{clear} = 0.25 s).

A comparison of the responses obtained by using this simplified model and the detailed model (in which the differential equations of the stator fluxes and the network are considered) is given in Figure 4.14. It can be observed that the speed responses are nearly the same for both the models. The electrical torque responses obtained using the simplified model exclude the high-frequency oscillations but track the slower transient quite accurately. The simplified model gives slightly pessimistic results, that is, the speed deviation is larger and the critical clearing is smaller when the simplified model is used.

Since the high-frequency transients are effectively excluded from the response when the simplified model is used, easy-to-implement explicit numerical integration methods with relatively larger time steps can be used for simulating the response. This dramatically improves the efficiency of the simulation without a significant loss of accuracy. In this example, it is possible to use the Runge–Kutta fourth order explicit method with a time step of 10 ms for the simulation of the simplified model. If both slow and fast transients are to be captured (the detailed model) then the time step has to be much smaller (50 μs is used here).

The neglect of the time-derivative terms in the differential equations of the stator fluxes and the network states in the DQ frame is a common approximation in the study of slow electro-mechanical transients. This approximation will also be used in the study of power swings and loss of synchronism in later chapters.

In this example, although high-frequency oscillations with a fairly large amplitude are visible in the electrical torque, the speed of the machine is not significantly affected as the mechanical system has a relatively slower response. However, the turbine-generator shaft may experience a significant torsional shock because of the sudden change in the electrical torque. To accurately capture this interaction, it is necessary to use a multi-mass model of the generator and turbine system (see Section 4.4) along with the detailed model of the network and stator. In this example, turbine-generator shaft torques are not investigated (since a single mass model for the turbine-generator is used), but a detailed analysis of torsional interactions is presented in Chapter 10.

4.5.3 Frequency Conversion Using Voltage Source Converters

The full or partial rating frequency converters used in Type III and IV WTG configurations typically consist of two voltage source converters (VSCs) connected in a back-to-back configuration with a capacitor in the DC link, as shown in Figure 4.15. Note that VSCs allow bidirectional power flow. In a back-to-back connection of VSCs, one of the VSCs acts like a rectifier (draws real power from the AC side) and the other acts like an inverter (supplies real power to the AC side).

The topology of a two-level, three-phase VSC which uses insulated-gate bipolar transistor (IGBT) switches is shown in Figure 4.16. Three-phase AC voltages are constructed from the DC voltage by controlling the on and off instants (switching pattern) of the switches. Harmonics are an inevitable consequence of switching and are suppressed by filters. In most VSCs, the lower-order harmonic components are small. Therefore, small-sized filters are needed only to remove the higher-order harmonics.

The magnitude and phase of the fundamental component of the voltages at the AC terminals can be changed by changing the switching pattern (pulse-width modulation, PWM) and the switching instants. In other words, the modulation-index (m) and the phase α (see Figure 4.17) can be controlled. Controllability of the AC voltage allows us to control the real and reactive power output of the converter, since the current flowing into the grid can be controlled by the converter AC voltage.

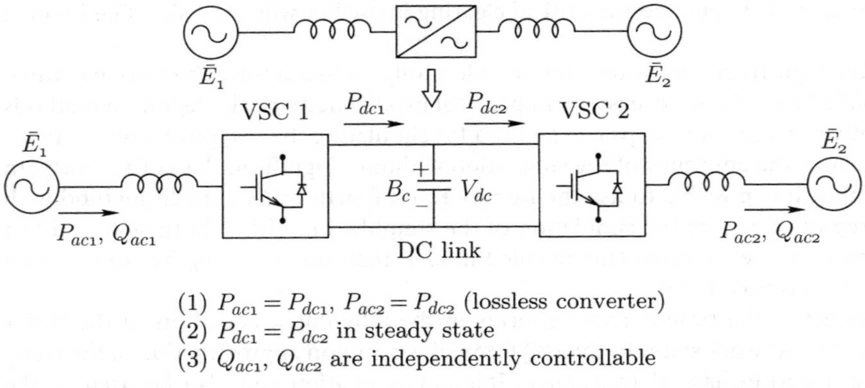

Figure 4.15 Back-to-back connection of VSCs (converters are assumed to be lossless).

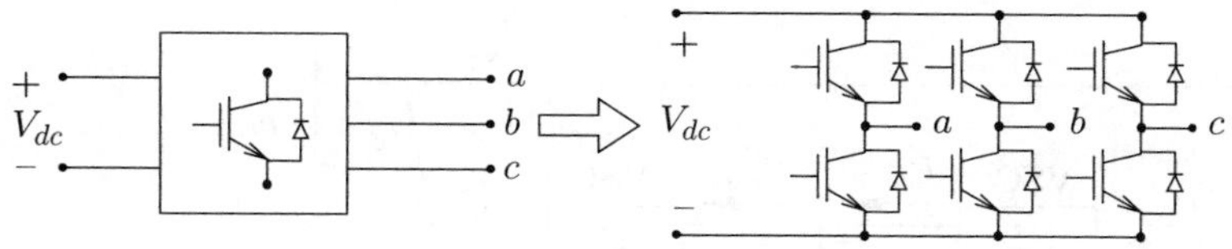

Figure 4.16 A three-phase, two-level VSC using IGBTs.

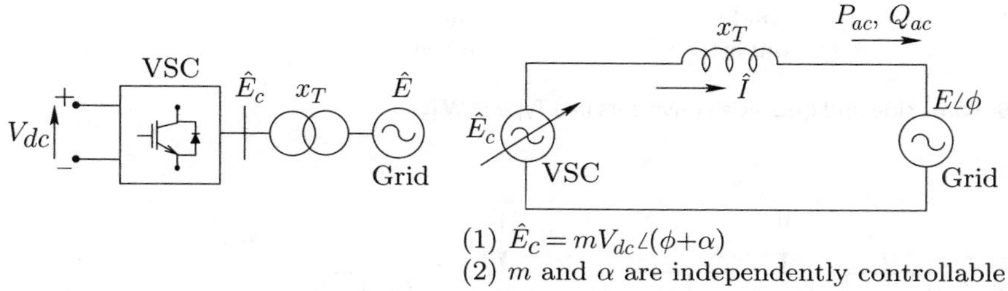

Figure 4.17 AC voltage control using a VSC.

Since the converter itself is (ideally) lossless, the real power supplied at the AC terminals has to come from the DC link. If the DC link consists of a capacitor (which can store very little power), this means that any power demanded by the AC side of VSC 2 in the back-to-back configuration of Figure 4.15, has to be met from the AC side of VSC 1. However, each VSC can independently exchange reactive power with the respective AC sides. For most power system studies, it is usually adequate to model a VSC as a sinusoidal source with controllable m and α. Representation of switching action and harmonics are required only for filter design and high-frequency interaction studies.

While the back-to-back connection of VSCs is the most widely used configuration for frequency conversion [2], alternative power electronic topologies like matrix converters [12], which avoid the use of a DC link capacitor, may also be considered. If steady-state power flow is unidirectional, as in a Type IV WTG system, then a diode bridge can be used as a rectifier while a VSC may be used as an inverter.

4.6 Control of Type III WTG System

A Type III WTG uses a DFIG and a partial-rating frequency converter, as shown in Figure 4.7. The frequency conversion achieved using back-to-back VSC converters is depicted in Figure 4.18. The converters are called rotor-side converters (RSCs) and grid-side converters (GSCs) based on the circuit they are connected to on the AC side. The control of these converters is described in the following subsections, followed by the overall control strategy for a Type III WTG.

4.6.1 Rotor-Side Converter Control

The RSC applies an appropriate voltage (v_{D_r} and v_{Q_r}) so that the desired level of real and reactive power is injected into the electrical grid. The control strategy to determine

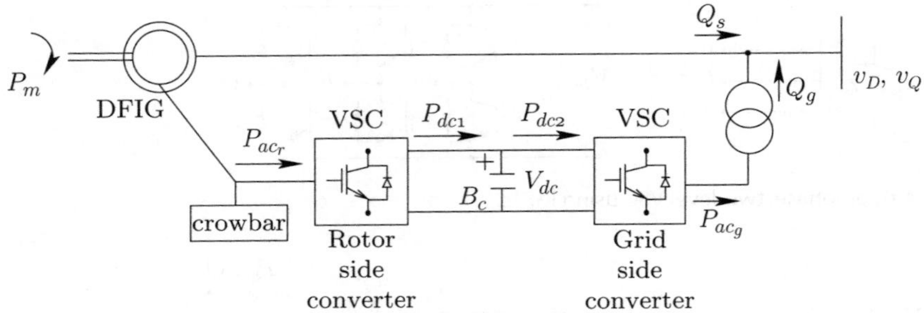

Figure 4.18 Rotor-side and grid-side converters in a Type III WTG.

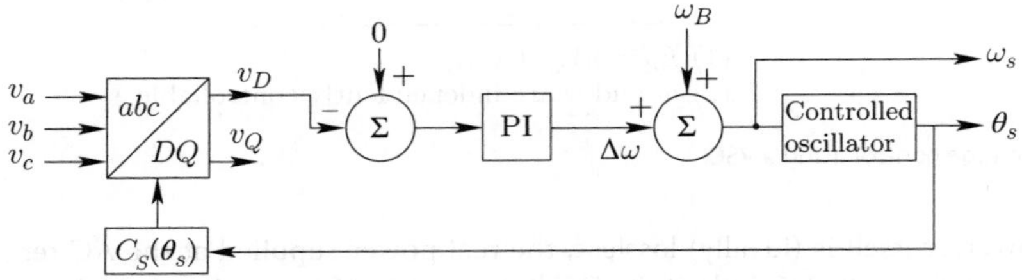

Figure 4.19 A phase locked loop.

these voltages can be derived from the equations of the induction generator described in Section 4.5 as given below.

1) If $\sin\theta_s$, which is used in the transformation matrix given in (4.15), is in phase with the machine terminal voltage $v_a(t)$ (phase a), then we obtain

$$v_D = 0,\ v_Q = V$$

where $\sqrt{\frac{2}{3}}\,V$ is the peak value of the phase voltage $v_a(t)$. In practice, the locking of $\sin\theta_s$ with $v_a(t)$ is achieved using a phase locked loop (PLL), such as the one shown in Figure 4.19 [13]. The proportional-integral (PI) controller in the PLL drives v_D to zero by adjusting the frequency of the controlled oscillator.

2) If the voltages v_{a_r}, v_{b_r}, and v_{c_r} are balanced and are of the slip frequency $\omega - \omega_s$, then v_{D_r} and v_{Q_r} are constants as well. The flux linkages and currents in the DQo and $D_rQ_ro_r$ variables are also constant in steady state.

3) If stator resistance is neglected, then in steady state

$$\psi_D = \frac{\omega_B}{\omega_s}v_Q = \frac{\omega_B}{\omega_s}V \text{ and } \psi_Q = -\frac{\omega_B}{\omega_s}v_D = 0 \tag{4.31}$$

4) Therefore, the steady-state torque T_e is proportional to i_Q. The steady-state reactive power injected into the grid via the stator windings is proportional to i_D, since $Q_s = v_D i_Q - v_Q i_D = -V i_D$.

5) It follows from (4.31) and the flux-current relationship (4.20) that in steady state

$$i_Q = -\frac{1}{X_{ss}}\frac{\omega_B}{\omega_s}v_D - \frac{X_m}{X_{ss}}i_{Q_r} \text{ and } i_D = \frac{1}{X_{ss}}\frac{\omega_B}{\omega_s}v_Q - \frac{X_m}{X_{ss}}i_{D_r} \tag{4.32}$$

6) Therefore, the control of i_Q and i_D (which translates to the control of T_e and Q_s) can be indirectly achieved by the control of i_{Q_r} and i_{D_r}, respectively. These rotor currents are controlled using the controllable rotor voltages (v_{D_r} and v_{Q_r}).

Decoupled Control of i_{Q_r} and i_{D_r}

Consider the differential equations describing the rotor currents as given below. These equations are derived from differential equations of the rotor flux given in Section 4.5 and the flux-current relationship (4.20). The stator flux differential equations are converted to algebraic equations by assuming that $\frac{1}{\omega_B}\frac{d\psi_D}{dt} \approx 0$ and $\frac{1}{\omega_B}\frac{d\psi_Q}{dt} \approx 0$, since the transients that can be affected by the controller are expected to be relatively slow. Stator resistances are also neglected.

$$\frac{di_{D_r}}{dt} = -\frac{R_r\omega_B}{\sigma X_{rr}} i_{D_r} + \frac{\omega_B}{\sigma X_{rr}} v'_{D_r} \tag{4.33}$$

$$\frac{di_{Q_r}}{dt} = -\frac{R_r\omega_B}{\sigma X_{rr}} i_{Q_r} + \frac{\omega_B}{\sigma X_{rr}} v'_{Q_r} \tag{4.34}$$

where $\sigma = (1 - \frac{X_m^2}{X_{ss}X_{rr}})$ and

$$\begin{aligned} v_{D_r} &= -v'_{D_r} - \sigma X_{rr}\frac{(\omega_s - \omega)}{\omega_B} i_{Q_r} + \frac{(\omega_s - \omega)}{\omega_s}\frac{X_m}{X_{ss}} v_D \\ &= -v'_{D_r} - \frac{(\omega_s - \omega)}{\omega_B}[X_m i_Q + X_{rr} i_{Q_r}]. \end{aligned} \tag{4.35}$$

$$\begin{aligned} v_{Q_r} &= -v'_{Q_r} + \sigma X_{rr}\frac{(\omega_s - \omega)}{\omega_B} i_{D_r} + \frac{(\omega_s - \omega)}{\omega_s}\frac{X_m}{X_{ss}} v_Q \\ &= -v'_{Q_r} + \frac{(\omega_s - \omega)}{\omega_B}[X_m i_D + X_{rr} i_{D_r}] \end{aligned} \tag{4.36}$$

If v'_{D_r} and v'_{Q_r} are determined independently, then the control of i_{D_r} and i_{Q_r} can be achieved in a decoupled fashion since there is no coupling between (4.33) and (4.34). In particular, the following PI control laws, which are expressed in the Laplace domain (s is the Laplace variable), may be used to obtain v'_{D_r} and v'_{Q_r}:

$$V'_{D_r}(s) = \left(k_p + \frac{k_i}{s}\right)[I^*_{D_r}(s) - I_{D_r}(s)] \tag{4.37}$$

$$V'_{Q_r}(s) = \left(k_p + \frac{k_i}{s}\right)[I^*_{Q_r}(s) - I_{Q_r}(s)] \tag{4.38}$$

The required converter output voltages v_{D_r} and v_{Q_r} are *subsequently* calculated from v'_{D_r} and v'_{Q_r} using the algebraic relationships given in (4.35) and (4.36). The modulation index m_r and phase angle α_r are then computed as follows:

$$m_r = \frac{\sqrt{v_{D_r}^2 + v_{Q_r}^2}}{V_{dc}} \quad \text{and} \quad \alpha_r = \tan^{-1}\frac{v_{D_r}}{v_{Q_r}} \tag{4.39}$$

m_r and α_r are used to generate the switching pattern for the RSC. Note that the resulting fundamental frequency waveform (for phase a_r) is given by $v_{a_r} = \sqrt{\frac{2}{3}} m_r V_{dc} \sin(\theta_s - \theta + \alpha_r)$.

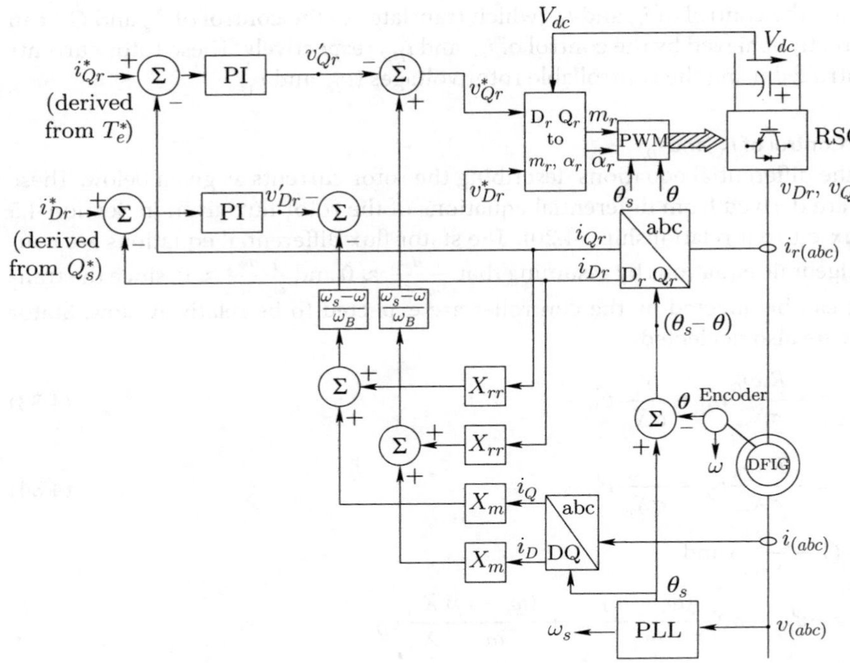

Figure 4.20 DFIG RSC control.

The controller structure for the RSC is shown in Figure 4.20. The set-points of the current controller $i_{Q_r}^*$ can be calculated directly from the torque set-point T_e^* by using the steady-state relationship $T_e^* = -\frac{X_m}{X_{ss}}\frac{\omega_B}{\omega_s}Vi_{Q_r}^*$. Similarly, $i_{D_r}^*$ can be obtained from the desired value of the stator reactive power Q_s^* from the relationships $i_D^* = -\frac{Q_s^*}{V}$ and $i_{D_r}^* = \frac{1}{X_m}\frac{\omega_B}{\omega_s}V - \frac{X_{ss}}{X_m}i_D^*$.

4.6.2 Grid-Side Converter Control

The per-unit equations describing the GSC currents in the DQ frame are (the matrix C_S is used for the transformation):

$$\begin{aligned}\frac{di_{D_g}}{dt} &= -\frac{R_g\omega_B}{X_{gl}}i_{D_g} - \omega_s i_{Q_g} + \frac{\omega_B}{X_{gl}}(v_{D_g} - v_D)\\ &= -\frac{R_g\omega_B}{X_{gl}}i_{D_g} + \frac{\omega_B}{X_{gl}}v'_D\end{aligned} \tag{4.40}$$

$$\begin{aligned}\frac{di_{Q_g}}{dt} &= -\frac{R_g\omega_B}{X_{gl}}i_{Q_g} + \omega_s i_{D_g} + \frac{\omega_B}{X_{gl}}(v_{Q_g} - v_Q)\\ &= -\frac{R_g\omega_B}{X_{gl}}i_{Q_g} + \frac{\omega_B}{X_{gl}}v'_{Q_g}\end{aligned} \tag{4.41}$$

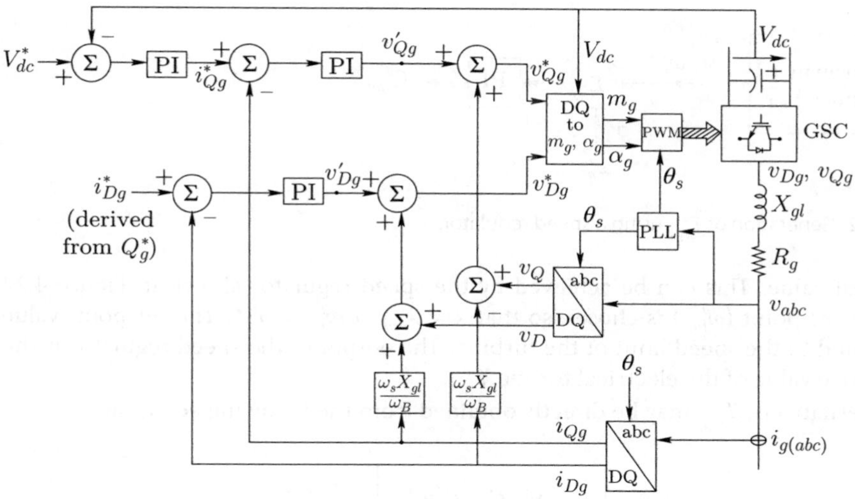

Figure 4.21 DFIG GSC control.

where R_g and X_{gl} are the per-unit resistance and leakage reactance of the coupling transformer. i_{D_g} and i_{Q_g} are the currents that are injected by the GSC into the grid while v_{D_g} and v_{Q_g} are the inverter output voltages. The form of these differential equations is similar to (4.33) and (4.34). A current control strategy similar to the one used for the RSC may therefore be used for GSC control, as shown in Figure 4.21. The dynamics of the DC-link capacitor can be described by the following equation (refer to Figure 4.18):

$$\frac{dV_{dc}}{dt} = \frac{\omega_B}{B_c}(i_{dc1} - i_{dc2}) = \frac{\omega_B}{B_c}\frac{P_{dc1} - P_{dc2}}{V_{dc}} = \frac{\omega_B}{B_c}\frac{P_{ac_r} - P_{ac_g}}{V_{dc}} \tag{4.42}$$

where B_c is the susceptance of the DC-link capacitor and P_{ac_r} is the power injected into the RSC by the rotor windings. In steady state, the power injected by the GSC is given by $P_{ac_g} = Vi_{Q_g}$. The GSC acts like a slave to the RSC, that is, it ensures that the power supplied to (drawn from) the induction generator by the RSC is drawn from (supplied to) the grid. It does this indirectly by ensuring that V_{dc} is regulated. Therefore, the set-point $i^*_{Q_g}$ is set by a capacitor voltage regulator, as shown in Figure 4.21.

In addition to maintaining the DC-link capacitor voltage, the GSC can also independently exchange reactive power with the grid (provided that the overall converter current rating is not exceeded). Since in steady state the reactive power injected by the GSC into the grid is given by $Q_g = -Vi_{D_g}$, the set-point $i^*_{D_g}$ may be set according to the reactive power requirements of the grid and is discussed in the next section.

4.6.3 Overall Control Scheme for a Type III WTG system

The RSC and GSC current references are set based on the torque and reactive power requirements. The set-point values of the torque and reactive power are usually decided based on the following considerations.

1) Torque reference (T_e^*) generation [14, 15]: If maximum power is to be extracted from wind then the turbine speed (ω'_m) should be such that the tip-speed ratio is at the

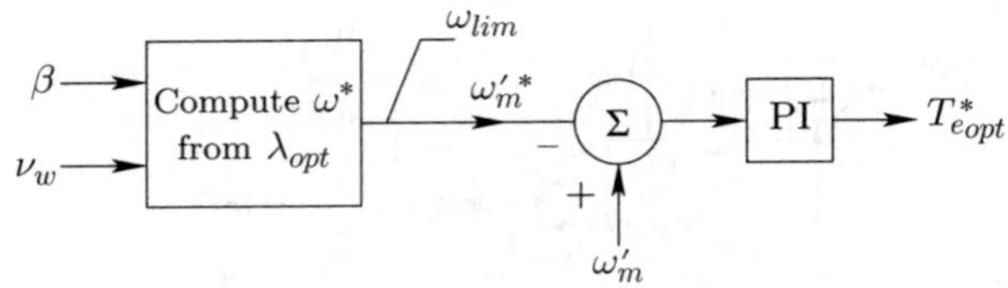

Figure 4.22 Generation of $T^*_{e_{opt}}$ using a speed regulator.

optimal value. This can be achieved by the speed regulator shown in Figure 4.22 whose set-point ($\omega_m'^*$) is chosen so that $\lambda = \lambda_{opt}$ at $\omega_m' = \omega_m'^*$. The set-point value is limited to the speed limit of the turbine. The output of the speed regulator is the reference value of the electrical torque $T^*_{e_{opt}}$.

Alternatively, $T^*_{e_{opt}}$ may be directly obtained from the following equation:

$$T^*_{e_{opt}} = K_{opt}\left(\frac{\omega}{\omega_B}\right)^2 = \left[\frac{1}{2}\,\frac{\rho A_w C_{p_{opt}} R^3 \omega_B^3}{S_B \lambda_{opt}^3 \left(\frac{P}{2} \times \frac{n_g}{n_t}\right)^3}\right]\left(\frac{\omega}{\omega_B}\right)^2 \tag{4.43}$$

Note that K_{opt} is per unit and is derived from (4.5).

2) If there is a generation surplus in the grid, then the power injected by the WTG may need to be *curtailed* so as to maintain the grid frequency. In such a case, the torque set-point is determined from the power order (received from a system operator) or from a power-frequency droop controller [16], and not from the maximum power extraction considerations. Alternatively, the wind energy system may extract maximum power from the wind and store it if it is not required at that point of time. The stored energy may be used later to support the grid when wind is not available.
3) An additional signal (P_{aux}) from an auxiliary controller may also be used to augment the torque set-point, as shown in Figure 4.23. This controller is used to improve the electro-mechanical stability of a power system. The auxiliary controller may consist of a power swing damping controller as well as a controller that rapidly increases (decreases) power injection into the grid when frequency suddenly tends to fall (rise) due to a load–generation imbalance. The latter controller reduces the rate of change of frequency. In this way the WTG system can contribute to the

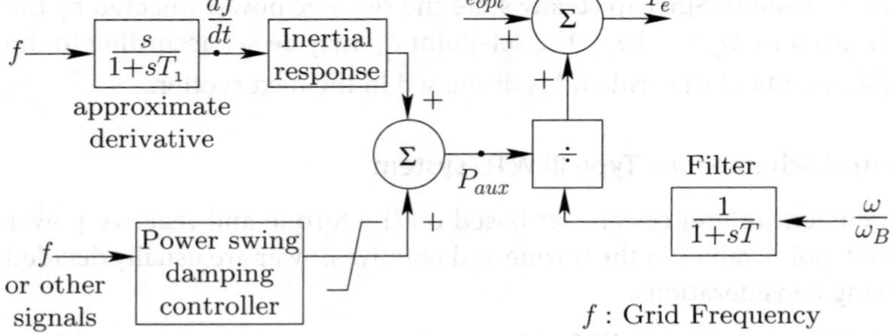

Figure 4.23 Modulation of torque reference with an auxiliary input.

system inertia. The power swing damping controller is used to contribute to the damping of low-frequency oscillations (0.2–2 Hz) which are generally present in a synchronous grid. This aspect is discussed in detail in Chapter 6. P_{aux} is non-zero during transient conditions but is practically zero otherwise. Therefore, providing this capability is not expected to affect annual energy production [17].

The set-points of the current regulators of the RSC and GSC controllers are limited to prevent currents from exceeding the ratings of the machine and converters. Therefore, the electrical torque is effectively limited by the current limits of the electrical system. If these limits are hit and wind speed is high, then the turbine speed cannot be controlled by the electrical system. The pitch angle controller then has to reduce the power extracted from the wind in order to limit the turbine speed, as discussed in Section 4.2.

4) A terminal voltage regulator with the set-point denoted by V^* may be used to determine the overall reactive power exchange between the WTG system and the grid. Reactive power exchange with the system can happen via the GSC as well as the stator of the induction machine. The corresponding set-points Q_s^* and Q_g^* may be determined in a coordinated fashion, while ensuring that the converter and stator winding currents are within their respective ratings.

Case Study: Power Injection Control in a Type III WTG

Using the auxiliary power signal P_{aux}, a Type III WTG can modulate the power injected to damp power swings in the electrical grid or transiently increase power injection if there is a sudden drop in grid frequency. The power boost can be for a short while before alternative frequency control mechanisms in the power system respond. Power swing damping is normally a small-amplitude low-frequency modulation of power. The power boost/modulation is easily achievable since the torque controller is expected to have a fast response time.

Let us consider a 2 MW Type III WTG which uses the control scheme given in Figure 4.24. The torque reference $T^*_{e_{opt}}$ is derived from the speed regulator shown in Figure 4.22. The turbine parameters are as shown in Figure 4.1. The pitch angle β is assumed to be zero. The corresponding turbine characteristics are shown in Figure 4.2. The induction machine and external system parameters are assumed to be the same as those in Example 4.1 except that the turbine-generator coupling gear ratio is 1:100, as the generator is a DFIG which can operate at a larger slip. The reactive power references are $Q_s^* = 0$ and $Q_r^* = 0$. The initial operating power $P_m = 0.4795$ pu (1.089 MW) and the initial generator speed $\frac{\omega}{\omega_B} = 1.1867$ pu. This corresponds to a turbine speed $\omega'_m = 1.8641$ rad/s (17.8 rpm). The WTG is operating at the maximum power extraction mode at a wind speed of 9.7 m/s.

To test the capability of the Type III WTG system to transiently boost power injection, the response to a pulse change in P_{aux} of 0.4 pu at 1 s and lasting for 2 s is simulated and shown in Figure 4.25. Since the wind speed is assumed to be constant and the WTG system is operating at the maximum power extraction mode to begin with, the additional power cannot be extracted from the wind. This means that the additional power is essentially drawn from the kinetic energy of the rotor, thereby slowing it down. This change in the turbine speed also results in a slight reduction in the power extracted from the wind. The speed regulator attempts to reaccelerate the rotor (slowly) back to

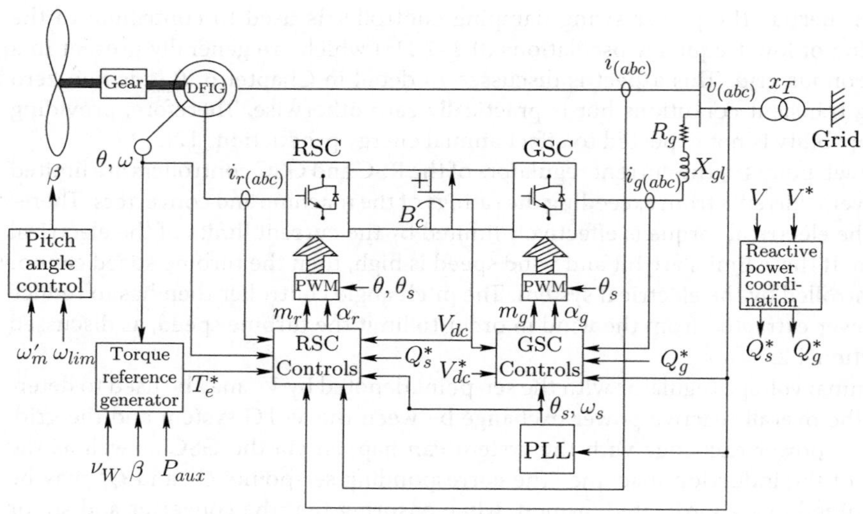

Figure 4.24 Overall DFIG converter control.

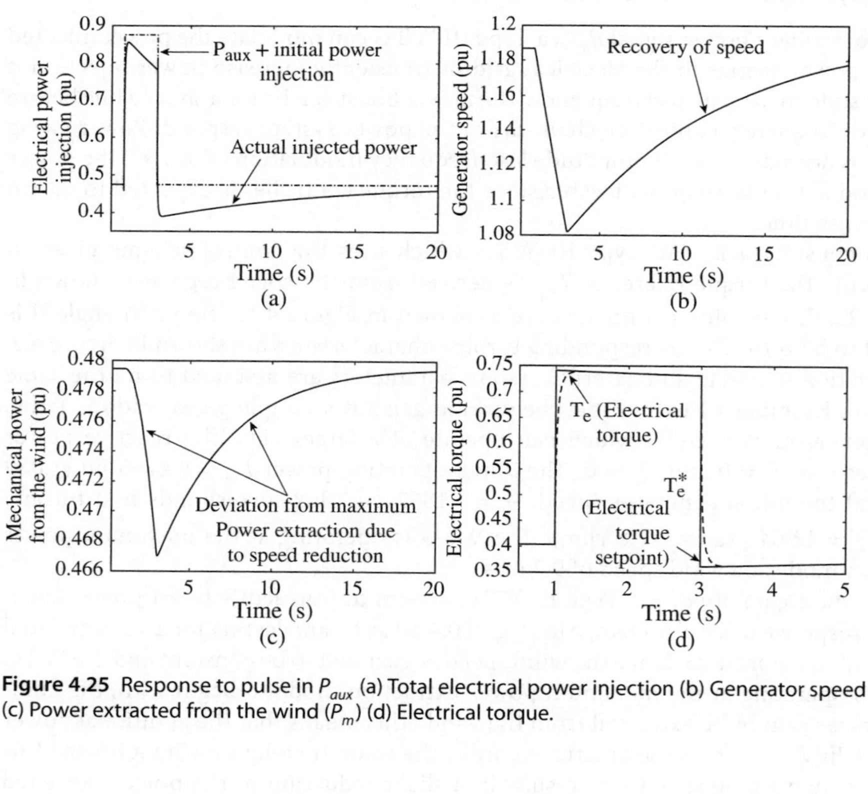

Figure 4.25 Response to pulse in P_{aux} (a) Total electrical power injection (b) Generator speed (ω/ω_B) (c) Power extracted from the wind (P_m) (d) Electrical torque.

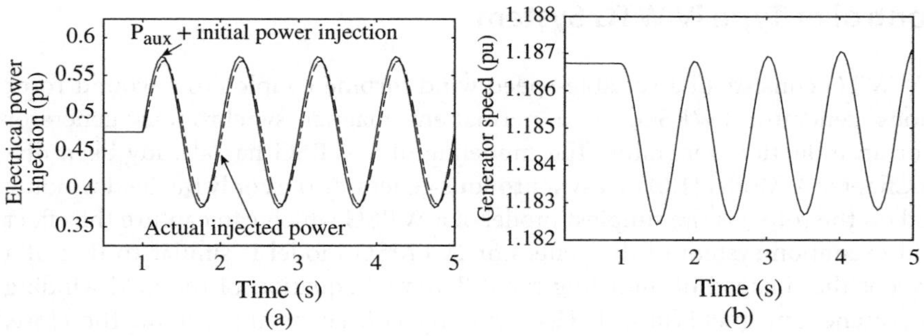

Figure 4.26 Response to power modulation. (a) Electrical power injection (b) Generator speed ($\frac{\omega}{\omega_B}$)

the point of maximum power extraction, but this is feasible only by reducing the electrical power injected into the grid. This explains the slight reduction in electrical power injected into the grid after the pulse change. However, the system eventually settles back to the original equilibrium.

To test the response of the system to a power modulation, let P_{aux} be a sinusoidal signal of 1 Hz with an amplitude of 0.1 pu. The response of the generator speed and the total power injected into the system is shown in Figure 4.26. Note that the power injected follows the P_{aux} signal quite closely because of the fast response of the RSC and GSC controllers. The speed regulator of Figure 4.22 is designed to have a slower bandwidth so that it does not negate the 1 Hz variations. Note that the average power remains practically at the original value (the maximum power point).

4.6.4 Simplified Modeling of the Controllers for Slow Transient Studies

It is evident from the previous case study that the inner current regulators of the RSC and GSC converters can respond quickly to variations in P_{aux}. Therefore, a simplified first-order current injection model may be used to model the net effects of the current regulators, VSC, and the PLL instead of representing each of these components in detail. For example, i_{D_r} and i_{Q_r} may be obtained from the following dynamic model:

$$I_{D_r}(s) = \frac{1}{1+sT_c} I^*_{D_r}(s), \quad I_{Q_r}(s) = \frac{1}{1+sT_c} I^*_{Q_r}(s) \tag{4.44}$$

The time constant T_c depends on the response of the controller and is typically about 20–30 ms. Similar models may be derived for the other currents i_{D_g} and i_{Q_g}. Note that the rotor currents i_{D_r} and i_{Q_r} are related algebraically to the stator currents i_D and i_D and the stator fluxes. It is necessary, however, to model the the higher level WTG controllers (e.g., the speed regulator) from which the current references are derived. The simplified modeling is suitable for the studies of slow electro-mechanical transients which include power swings and system frequency dynamics. The derivative terms of the stator fluxes and network voltages/currents in the DQ frame can also be neglected in such studies, as discussed in Example 4.1.

4.7 Control of Type IV WTG System

A Type IV WTG consists of a variable-speed wind turbine coupled to a wound-rotor synchronous generator (WRSG) or a permanent magnet synchronous generator (PMSG) or an induction generator. The modeling of a WRSG has already been discussed in Chapter 3. Model (1.0) of a synchronous generator (i.e., only the field winding is modeled on the rotor) is the simplest model of a WRSG which can capture the effect of the field excitation system of the generator. A PMSG model is similar to that of a WRSG except that instead of modeling the differential equation of the field winding flux, the permanent magnet is modeled as a winding with a constant current. Therefore, the per-unit model of a PMSG in Park's reference frame (and using the generator convention) can be written as follows.

$$-\frac{1}{\omega_B}\frac{d\psi_d}{dt} - \frac{\omega}{\omega_B}\psi_q - R_s i_d = v_d \tag{4.45}$$

$$-\frac{1}{\omega_B}\frac{d\psi_q}{dt} + \frac{\omega}{\omega_B}\psi_d - R_s i_q = v_q \tag{4.46}$$

where

$$\psi_d = L_d i_d + \psi_{fp} \tag{4.47}$$

$$\psi_q = L_q i_q \tag{4.48}$$

ψ_{fp} is a constant flux linkage, which represents the effect of the permanent magnet. The model does not consider any other rotor currents (eddy currents or damper windings). Therefore, the only differential equations of the generator are those corresponding to the stator (d–q) fluxes. The expression for electrical torque is:

$$T_e = \psi_d i_q - \psi_q i_d \tag{4.49}$$

A Type IV WTG system uses a full-rating frequency converter which may be implemented using back-to-back connected VSCs. The two converters are designated as the stator-side converter (SSC) and the grid-side converter (GSC). As in a Type III WTG, a Type IV WTG may be controlled to extract maximum power from the wind. To achieve this, the turbine speed is controlled by controlling the electrical torque of the generator. The set-point derivation for the electrical torque may be similar to that of a Type III WTG.

The electrical torque of the generator may be controlled by the SSC, while the GSC regulates the DC-link capacitor voltage by pushing the power drawn by the SSC into the grid. The roles may also be reversed, that is, the SSC regulates the capacitor voltage while the electrical torque control is indirectly done by controlling the power output of the GSC. The GSC can independently control the reactive power injected into the grid as per the grid requirements, subject to the converter current rating. The SSC can also control the reactive power drawn from the generator. The full frequency converter may employ a diode rectifier on the generator side instead of a VSC. In such a case, the GSC controls the power output, while the capacitor voltage at the DC link is determined by the generator AC voltage magnitude.

References

1 Blaabjerg, F. and Ma, K. (2017) Wind energy systems. *Proceedings of the IEEE*, **105** (11), 2116–2131.
2 Ackermann, T. (2005) *Wind power in power systems*, John Wiley & Sons, Chichester.
3 Heier, S. (2014) *Grid integration of wind energy: onshore and offshore conversion systems*, John Wiley & Sons, Chichester.
4 Slootweg, J.G., de Haan, S.W.H., Polinder, H., and Kling, W.L. (2003) General model for representing variable speed wind turbines in power system dynamics simulations. *IEEE Transactions on Power Systems*, **18** (1), 144–151.
5 Slootweg, J.G., Polinder, H., and Kling, W.L. (2003) Representing wind turbine electrical generating systems in fundamental frequency simulations. *IEEE Transactions on Energy Conversion*, **18** (4), 516–524.
6 Justus, C.G., Hargraves, W.R., Mikhail, A., and Graber, D. (1978) Methods for estimating wind speed frequency distributions. *Journal of Applied Meteorology*, **17** (3), 350–353.
7 Sarkar, A., Gugliani, G., and Deep, S. (2017) Weibull model for wind speed data analysis of different locations in India. *Korean Society of Civil Engineers (KSCE) Journal of Civil Engineering*, **21** (7), 2764–2776.
8 Muljadi, E. and Butterfield, C.P. (2001) Pitch-controlled variable-speed wind turbine generation. *IEEE Transactions on Industry Applications*, **37** (1), 240–246.
9 Grantham, C., Sutanto, D., and Mismail, B. (1989) Steady-state and transient analysis of self-excited induction generators. *IEE Proceedings B – Electric Power Applications*, **136** (2), 61–68.
10 Samuelsson, O. and Lindahl, S. (2005) On speed stability. *IEEE Transactions on Power Systems*, **20** (2), 1179–1180.
11 Grilo, A.P., Mota, A.d.A., Mota, L.T.M., and Freitas, W. (2007) An analytical method for analysis of large-disturbance stability of induction generators. *IEEE Transactions on Power Systems*, **22** (4), 1861–1869.
12 Wheeler, P. and Grant, D. (1993) A low loss matrix converter for AC variable-speed drives, in *Fifth European Conference on Power Electronics and Applications*, IET, Brighton, pp. 27–32.
13 Gole, A., Sood, V.K., and Mootoosamy, L. (1989) Validation and analysis of a grid control system using d-q-z transformation for static compensator systems, in *Canadian Conference on Electrical and Computer Engineering*, Montreal, pp. 745–758.
14 Pena, R., Clare, J.C., and Asher, G.M. (1996) Doubly fed induction generator using back-to-back PWM converters and its application to variable-speed wind-energy generation. *IEE Proceedings – Electric Power Applications*, **143** (3), 231–241.
15 Buehring, I.K. and Freris, L.L. (1981) Control policies for wind-energy conversion systems. *IEE Proceedings C – Generation, Transmission and Distribution*, **128** (5), 253–261.
16 Buckspan, A., Aho, J., Fleming, P. *et al.* (2012) Combining droop curve concepts with control systems for wind turbine active power control, in *IEEE Power Electronics and Machines in Wind Applications*, Denver, pp. 1–8.
17 Singh, M., Allen, A., Muljadi, E., and Gevorgian, V. (2014) Oscillation damping: A comparison of wind and photovoltaic power plant capabilities, in *IEEE Symposium on Power Electronics and Machines for Wind and Water Applications*, Milwaukee, pp. 1–7.

5
Modeling and Analysis of FACTS and HVDC Controllers

5.1 Introduction

As mentioned in Chapter 2, the AC transmission network has no inherent power controllability. The flow in a line depends on the power (and reactive power) injections at the generation and load nodes in radial networks. In a mesh-type network (with more than one path between any two nodes) the power flow is also dependent on Kirchoff's voltage law (KVL). Transmission planning was a major exercise in the vertically integrated utilities (VIU). However, with the advent of deregulation or restructuring, transmission expansion to accommodate load growth becomes an ad hoc exercise. There are two issues: one involves the need to increase power flows in the existing right of way (RoW) and the second relates to making transmission "flexible" to accommodate modifications of power flow pattern during system contingencies. It should be noted that the maximum power that can be carried in a line depends on the maximum temperature of the conductor. This depends on the power losses in the line and the ambient temperature. The dynamic rating of a line takes into account the temperature. The power losses include both I^2R losses and the corona losses, which depend on weather conditions.

The maximum power flow that can be permitted in an AC line also depends on its length. Steady-state stability issues limit the power flow in a line below the thermal limit. Series compensation using fixed capacitors is an economic solution to enhance power transfer capability in long lines. There are problems of protection (against faults) that have to be tackled in the application of series capacitors. The series capacitor has to be bypassed to protect against fault currents and reinserted as soon as the fault is cleared by the operation of circuit breakers. Since the early 1970s, the problems of subsynchronous resonance (SSR) also have to be considered in the application of series capacitors.

High-voltage direct current (HVDC) transmission was first applied in 1954 using mercury arc valves which had problems of arc-backs that resulted in loss of rectification. Nevertheless, 6000 MW of DC power transfer was used until 1972, when thyristor valves replaced mercury arc valves. A thyristor valve is made up of a number thyristor devices in series. Over the years, the thyristor valve-based line commutated (current source) converters (LCC) have been used up to ±800 kV, 6400 MW.

HVDC power transmission is primarily applied for (i) long distance bulk power transmission from remote hydro and thermal stations, (ii) asynchronous interconnections, and (iii) sea-crossing underwater cables. In recent times, HVDC transmission has also been considered for control and stabilization of power flows in AC tie lines in integrated power systems. The first application (for long distance, bulk power transmission) is due

Dynamics and Control of Electric Transmission and Microgrids, First Edition.
K. R. Padiyar and Anil M. Kulkarni.

to economic and ROW considerations as two conductors in a bipolar HVDC line can carry about the same amount of power (for specified voltage and current ratings) as a three-phase AC line. A bipolar HVDC line is also more reliable than a single circuit AC line, as one healthy pole can carry more than half the rated power flow during the failure of one pole due to a fault.

There is no alternative to HVDC when an asynchronous interconnection is required to connect two systems operating at different frequencies (e.g., in Japan). In such cases, back-to-back (BTB) converter stations are used with no HVDC line. Sometimes even systems having the same nominal frequency are interconnected using BTB converter stations (e.g., in the USA, Canada and India).

The thyristor valve technology developed for HVDC power transmission has been applied for fast voltage and reactive power control in AC transmission systems by developing Static VAr Compensators (SVCs). These are shunt-connected controllers. Subsequently, thyristor valve technology was applied to develop thyristor-controlled series capacitors (TCSCs). The basic idea is to apply rapid adjustment of network impedance (RANI), a concept proposed by Dr. John Vithayathil [1]. This concept was generalized by Dr. N.G. Hingorani in 1988, who proposed the flexible AC transmission system (FACTS) [2].

The research on FACTS controllers led to the second generation of controllers based on voltage source converters (VSCs). For example, the static synchronous compensator (STATCOM) is also applied for fast reactive power control and has superior characteristics to SVCs.

Interestingly, in 1997, ABB in Sweden installed an experimental VSC-based HVDC transmission. Several VSC-HVDC systems (called HVDC light) were installed by ABB. The development of the modular multilevel (voltage source) converter (MMC) by Siemens (first applied in 2010) has given fresh impetus to scaling up the power levels and novel applications, such as transmission of off-shore wind power to land-based converter stations. A major advantage of VSC-HVDCs is the ability to operate inverters with no voltage sources, thus providing black-start capability to off-shore wind generators. VSC-HVDCs are also more suited for multi-terminal operation as power reversal does not require voltage reversal.

5.2 FACTS Controllers [3–5]

5.2.1 Description

As mentioned earlier, stability and security issues require the maintenance of adequate margins in power transfer in transmission networks. From economic considerations, it is desirable to minimize the margins without compromising system security. The difficulties in the expansion of the transmission network are also due to environmental issues. The required operating margin can be substantially reduced by the introduction of fast dynamic control over active and reactive power by the application of power electronic controllers. This can make the AC transmission network "flexible" enough to adapt to changing conditions caused by contingencies and load variations. The flexible AC transmission system (FACTS) is defined as "alternating current transmission systems incorporating power electronic based and other static controllers to enhance controllability and increase power transfer capability" [2, 3]. The FACTS controller is defined as "a power electronics based system and other static equipment that provide control of one or more AC transmission system parameters".

The FACTS controllers can be classified as:

1) shunt-connected controllers
2) series-connected controllers
3) combined series–series controllers
4) combined shunt–series controllers.

Depending on the power electronic devices used, FACTS controllers can be classified as:

(a) thyristor valve-based variable impedance type
(b) gate turn-off thyristor (GTO) or insulated-gate bipolar transistor (IGBT) valve-based VSCs.

The variable impedance-type controllers include:

(a) SVCs (shunt-connected)
(b) TCSCs (series-connected)
(c) thyristor-controlled phase shifting transformers (TCPSTs) or static phase-shifting transformers (PSTs) (combined shunt and series).

The VSC-based FACTS controllers are:

(a) STATCOM (shunt-connected)
(b) static synchronous series compensators (SSSCs) (series-connected)
(c) interline power flow controllers (IPFCs) (combined series–series)
(d) unified power flow controllers (UPFCs) (combined shunt–series).

The FACTS controllers based on VSC have several advantages over the variable impedance type. For example, a STATCOM is much more compact than a SVC for similar rating and is technically superior. It can supply the required reactive current even at low values of the bus voltage and can be designed to have in-built short-term overload capability. Also, a STATCOM can supply active power if it has an energy source or large energy storage at its DC terminals.

It is interesting to note that while SVC was the first FACTS controller (which utilized the thyristor valves developed in connection with HVDC line commutated converters) several new FACTS controllers based on VSC have been developed. This has led to the introduction of VSC in HVDC transmission for ratings up to 1000 MW.

5.2.2 A General Equivalent Circuit for FACTS Controllers

The UPFC (shown in Figure 5.1) is the most versatile FACTS controller, with three control variables (the magnitude and phase angle of the series-injected voltage in addition to the reactive current drawn by the shunt-connected VSC). The equivalent circuit of a UPFC on a single-phase basis is shown in Figure 5.2. The current i is drawn by the shunt-connected VSC while the voltage e is injected by the series-connected VSC. Neglecting harmonics, both the quantities can be represented by phasors $\hat{I}$ and $\hat{E}$.

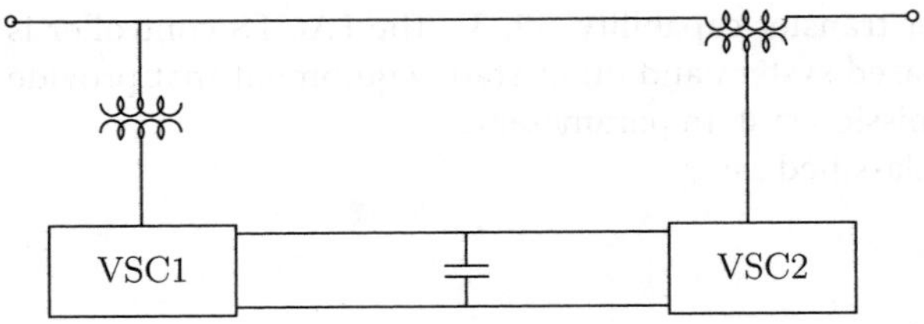

Figure 5.1 Unified power flow controller.

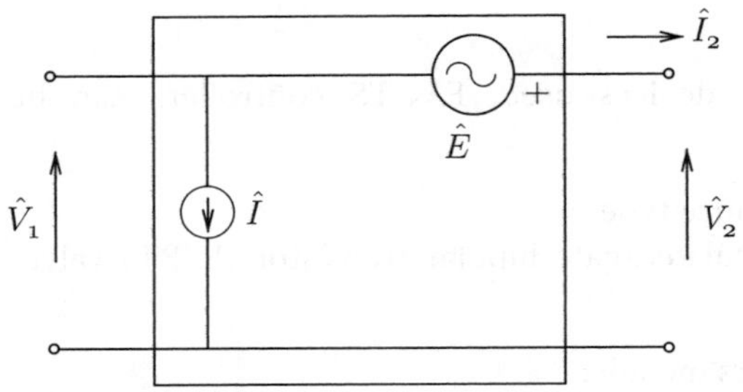

Figure 5.2 Equivalent circuit of UPFC.

Neglecting power losses in the UPFC, the following constraint equation applies:

$$\Re[\hat{V}_1\hat{I}^*] = \Re[\hat{E}\hat{I}_2^*] \tag{5.1}$$

Assuming that $\hat{V}_1 = V_1e^{j\theta_1}$ and $\hat{I}_2 = I_2e^{j\phi_2}$, $\hat{I}$ and $\hat{E}$ can be expressed as

$$\hat{I} = (I_p - jI_r)e^{j\theta_1} \tag{5.2}$$

$$\hat{E} = (V_p + jV_r)e^{j\phi_2} \tag{5.3}$$

where I_p and I_r are "real" and "reactive" components of the current drawn by the shunt-connected VSC. Similarly V_p and V_r are the "real" and "reactive" voltages injected by the series-connected VSC. Positive I_p and V_p indicate positive "real" (active) power flowing into the shunt-connected VSC and flowing out of the series-connected VSC. The positive values of I_r and V_r indicate reactive power drawn by the shunt converter and supplied by the series converter.

Using (5.2) and (5.3), (5.1) can be expressed as

$$V_1I_p = I_2V_p \tag{5.4}$$

The remaining shunt- and series-connected FACTS controllers can be viewed as special cases of a UPFC. For example, in an SVC,

$$V_p = 0, V_r = 0, I_p = 0, I_r = -B_{SVC}V_1 \tag{5.5}$$

There are three constraint equations and one control variable (B_{SVC}) in an SVC. In a STATCOM, I_r is the control variable. Table 5.1 gives the constraint equations and control variables for all the FACTS controllers. Note that in a STATCOM or SSSC with an energy source at the DC terminals, there are two control variables as I_p or V_p is non-zero.

5.2.3 Benefits of the Application of FACTS Controllers

Primarily, FACTS controllers provide voltage support at critical buses in the system (shunt-connected controllers) and regulate power flow in critical lines (series-connected

Table 5.1 Constraint equations and control variables for FACTS controllers.

Controller	Constraint equations	Control variable(s)
SVC	$V_p = 0, V_r = 0, I_p = 0$ $I_r = -B_{SVC}V_1$	B_{SVC}
TCSC	$I_p = 0, I_r = 0, V_p = 0$ $V_r = X_{TCSC}I_2$	X_{TCSC}
SPST (TCPAR)	$\hat{E} = V_1(e^{j\phi} - 1) \approx jV_1\phi$ $V_1I_p = V_pI_2, \quad V_1I_r = I_2V_r$	ϕ
STATCOM	$V_p = 0, V_r = 0, I_p = 0$	I_r
STATCOM with energy source	$V_p = 0, Vr = 0$	I_p, I_r
SSSC	$I_p = 0, I_r = 0, V_p = 0$	V_r
SSSC with energy source	$I_p = 0, I_r = 0$	V_p, V_r

controllers). Both voltage and power flow are controlled by the combined series and shunt controller (UPFC). The power electronic control is quite fast and this enables regulation under both steady-state and dynamic conditions (when the system is subjected to disturbances). The benefits due to FACTS controllers are listed below.

1. They contribute to optimal system operation by reducing power losses and improving voltage profile.
2. The power flows in critical lines can be enhanced as the operating margins can be reduced due to fast controllability. In general, the power-carrying capacity of lines can be increased to values up to the thermal limits (imposed by the current-carrying capacity of the conductors).
3. The transient stability limit is increased, thereby improving the dynamic security of the system and reducing the incidence of blackouts caused by cascading outages.
4. The steady-state or small signal stability region can be increased by providing auxiliary stabilizing controllers to damp low-frequency oscillations.
5. FACTS controllers such as TCSCs can counter the problem of SSR experienced with fixed-series capacitors connected in lines evacuating power from thermal power stations (with turbogenerators).
6. The problem of voltage fluctuations, and in particular dynamic overvoltages, can be overcome by FACTS controllers.

The capital investment and the operating costs (essentially the cost of power losses and maintenance) are offset against the benefits provided by FACTS controllers and the payback period is generally used as an index in the planning. The major issues in the deployment of FACTS controllers are (i) location, (ii) ratings (continuous and short term), and (iii) control strategies required for the optimal utilization. Here, both steady-state and dynamic operating conditions have to be considered. Several systems studies involving power flow, stability, and short-circuit analysis are required to prepare the specifications. The design and testing of the control and protection equipment is based on real-time digital simulators (RTDSs) or physical simulators. It should be noted that a series-connected FACTS controller (such as a TCSC) can control power flow not only in the line in which it is connected, but also in the parallel paths (depending on the control strategies).

5.2.4 Application of FACTS Controllers in Distribution Systems

Although the concept of FACTS was developed originally for transmission networks, this has been extended in the last 10 years to the improvement of power quality (PQ) in distribution systems operating at low or medium voltages.

In the early days, PQ referred primarily to the continuity of power supply at acceptable voltage and frequency. However, the prolific increase in the use of computers, microprocessors, and power electronic systems has resulted in PQ issues involving transient disturbances in voltage magnitude, waveform, and frequency. The nonlinear loads not only cause PQ problems but are also very sensitive to voltage deviations.

In the modern context, the PQ problem is defined as "Any problem manifested in voltage, current or frequency deviations that results in failure or misoperation of customer equipment" [6].

Hingorani (1995) [7] was the first to propose FACTS controllers for improving PQ. He termed them custom power devices (CPDs). These are based on VSC and exist in three types [7]:

1) shunt-connected distribution STATCOM (DSTATCOM)
2) series-connected dynamic voltage restorer (DVR)
3) combined shunt and series, unified power quality conditioner (UPQC).

The DVR is similar to SSSC while UPQC is similar to UPFC. In spite of the similarities, the control strategies are quite different for improving PQ. A major difference involves the injection of harmonic currents and voltages to isolate the source from the load. For example, a DVR can work as a harmonic isolator to prevent the harmonics in the source voltage reaching the load in addition to balancing the voltages and providing voltage regulation. A UPQC can be considered to be the combination of DSTATCOM and DVR. A DSTATCOM is utilized to eliminate the harmonics from the source currents and also balance them in addition to providing reactive power compensation (to improve the power factor or regulate the load bus voltage).

The terminology is yet to be standardized. The terms *active filters* and *power conditioners* are also employed to describe the custom power devices.

5.3 Reactive Power Control [5, 8]

Reactive power control applied to a transmission line or a network can result in several benefits involving voltage regulation, enhancement of power flow, and stability improvement. The reactive power control can involve the injection of (i) variable reactive voltage in series with a line or (ii) variable reactive current in shunt at suitable location(s) in critical transmission lines and/or suitable buses in the network. The injection of reactive

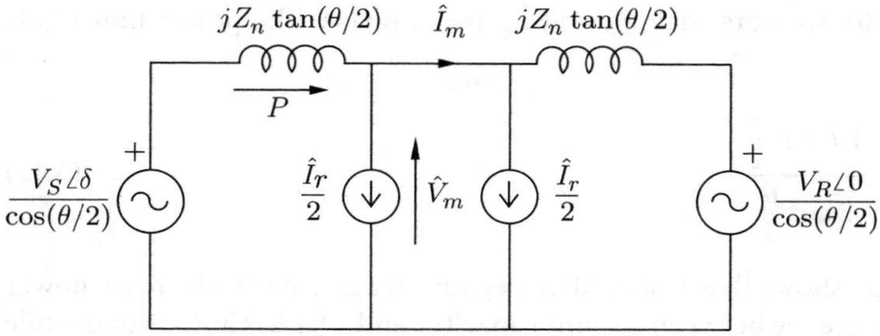

Figure 5.3 Equivalent circuit of a line with STATCOM at the midpoint.

voltage using a TCSC will be discussed in the next section whereas the role of injection of reactive current by shunt-connected controllers will be discussed here. It should be noted that both SVC and STATCOM are shunt devices that can control injected reactive current. To illustrate the role of the shunt reactive power control, let us consider a STATCOM connected at the midpoint of a lossless transmission line, as shown in Figure 5.3. There is no loss of generality if we consider the injected reactive current as the sum of two equal components ($\frac{\hat{I}_r}{2}$). This enables the marking of the current $\hat{I}_m$ at the midpoint of the line.

Note that $\hat{I}_r$ is shown as flowing away from the midpoint as a common convention when I_r is assumed to be positive when the current is lagging the bus voltage. This implies that the controller is operating in the inductive region. In the capacitive region, I_r is negative.

In Figure 5.3, the two halves of a transmission line are represented by Thevenin equivalents as viewed from the midpoint of the line. Assuming $V_S = V_R = V$, the following equations apply

$$\frac{V\angle\delta}{\cos\frac{\theta}{2}} - jZ_n \tan\frac{\theta}{2}\left(\hat{I}_m + \frac{\hat{I}_r}{2}\right) = \hat{V}_m \tag{5.6}$$

$$\frac{V\angle 0}{\cos\frac{\theta}{2}} + jZ_n \tan\frac{\theta}{2}\left(\hat{I}_m - \frac{\hat{I}_r}{2}\right) = \hat{V}_m \tag{5.7}$$

where

$$\hat{I}_r = -j\frac{\hat{V}_m}{V_m} I_r \tag{5.8}$$

From the above equations, we can derive

$$\hat{V}_m = \left(\frac{V\cos\frac{\delta}{2}}{\cos\frac{\theta}{2}} - \frac{I_r Z_n}{2}\tan\frac{\theta}{2}\right)\angle\frac{\delta}{2} \tag{5.9}$$

$$\hat{I}_m = \frac{V\sin\frac{\delta}{2}}{Z_n \sin\frac{\theta}{2}}\angle\frac{\delta}{2} \tag{5.10}$$

From (5.9) and (5.10), we note that $\hat{V}_m$ and $\hat{I}_m$ are in phase. The power flow (P) is given by

$$P = \frac{V^2 \sin\delta}{Z_n \sin\theta} - \frac{VI_r \sin\frac{\delta}{2}}{2\cos\frac{\theta}{2}} \tag{5.11}$$

The above expression shows that I_r should be negative (capacitive) to increase power transfer. The major difference between a shunt capacitor and a STATCOM is that while the maximum power for the former occurs when $\delta = 90°$, it occurs when $90° < \delta < 180°$ for a STATCOM.

Remarks

1) In general, when the power angle relationship is expressed as

$$P = A\sin\delta + B\sin\frac{\delta}{2}$$

the angle δ_p at which the power is maximum is defined by

$$\frac{dP}{d\delta} = 0 = A\cos\delta_p + \frac{1}{2}B\cos\frac{\delta_p}{2}$$

Defining $x = \cos\frac{\delta_p}{2}$, the above equation can be expressed as a quadratic equation in x as

$$A(2x^2 - 1) + \frac{1}{2}Bx = 0$$

The solution of this equation is given by

$$x = \frac{-B}{8A} \pm \frac{\sqrt{\frac{B^2}{4} + 8A^2}}{4A}$$

Only the positive solution for x is relevant.
2) The injection of reactive current is also feasible by connecting a variable susceptance. This is achieved by connecting a fixed capacitor (FC) in parallel with a thyristor-controlled reactor (TCR). This is a FC-TCR type SVC. If thyristor valves are also applied for switching in or out a set of capacitors connected in parallel we obtain a TSC-TCR-type SVC (TSC denotes thyristor switched capacitor).
3) The effectiveness of a STATCOM is reduced if the location of the reactive current source (SVC or STATCOM) is not exactly at the midpoint of the transmission line.
4) The voltage regulation can be applied at a bus which can be considered as the electrical centre of the system. Reactive power control at such a bus can increase the power transfer capability of the network.

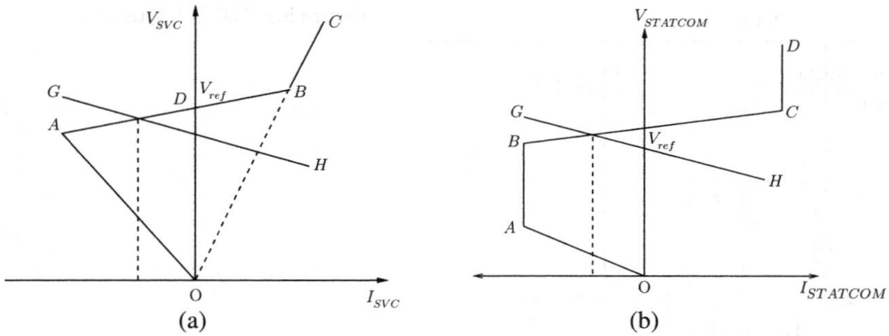

Figure 5.4 Steady-state control characteristics of a reactive power compensator (a) SVC (b) STATCOM.

Control Characteristics

The steady-state control characteristics of a reactive power compensator are shown in Figure 5.4. AB on Figure 5.4a and BC on Figure 5.4b are the control ranges of SVC and STATCOM, respectively. The range is defined by

$$V_s = V_{ref} + X_s I_s \tag{5.12}$$

where X_s is the slope of the control characteristics (in the control range). V_{ref} is the bus voltage when I_s (reactive current drawn by the compensator) is zero. A positive slope (in the range of 1–5%) is given in the control range to:

1) enable parallel operation of more than one reactive power compensator (RPC) connected at the same or neighboring buses and
2) prevent the compensator from hitting the limits frequently.

The system characteristics GH shown in Figure 5.4 represent the variation of RPC voltage (V_s) as a function of the current drawn by the RPC. The intersection of the system and control characteristics defines the operating voltage (V_s) at the terminals of the RPC. The control parameters of the RPC are V_{ref} and X_s. As the system operating conditions change, it becomes necessary to modify V_{ref} in order to ensure that the full control range of RPC is available whenever there is a disturbance.

The SVC is the first-generation shunt FACTS controller based on thyristor valves. There are two types of SVC: (i) fixed capacitor, thyristor-controlled reactor (FC-TCR) and (ii) thyristor switched capacitor, thyristor-controlled reactor (TSC-TCR). A typical TSC-TCR type SVC schematic diagram is shown in Figure 5.5. The STATCOM is described in Section 5.5.

5.4 Thyristor-Controlled Series Capacitor

Series capacitors have been applied in long distance transmission lines for increasing power transfer. The series capacitors provide the most economic solution for enhancing power transfer. However, the problem of SSR has deterred system planners from going in a big way for series compensation. While the use of shunt capacitors does not have the problem of SSR, these capacitors have the drawbacks as their effectiveness is dependent

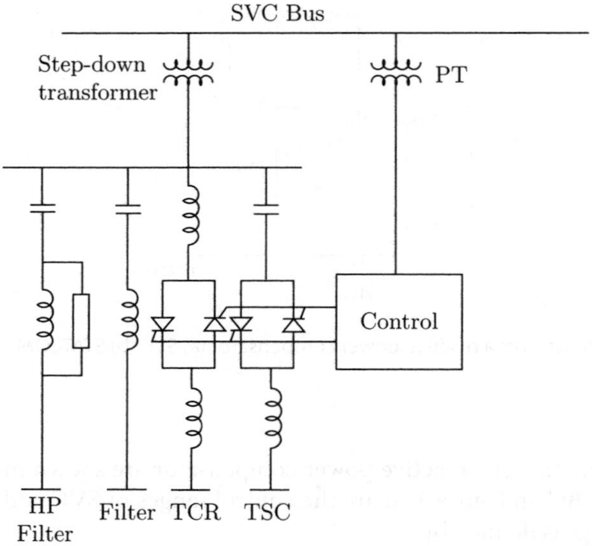

Figure 5.5 TSC-TCR type SVC.

largely on their location. Even when a shunt capacitor is located at the midpoint of a line, it requires much larger rating to achieve the same level of increase in the power transfer as a series capacitor. The ratio of the two ratings is given by

$$\frac{Q_{se}}{Q_{sh}} = \tan^2\left(\frac{\delta_{\max}}{2}\right) \tag{5.13}$$

where Q_{se} and Q_{sh} are the ratings of the series and shunt capacitors, respectively, and $\delta_{\max}$ is the maximum operating angular difference between the two ends of the line. For $\delta_{\max}$ in the range of 30–40°, Q_{se} varies from 7% to 13% of Q_{sh}. Although series capacitors tend to be twice as costly as shunt capacitors (per-unit var), they are still cheaper to use. In addition, the location of series capacitor is not critical.

The application of thyristor control to provide variable series compensation is a viable means to provide "flexibility" in AC transmission networks. A major advantage is that the SSR problem is significantly reduced. The feasibility of fast control of thyristor valves enables the improvement in stability and damping of oscillations using suitable control strategies.

The first demonstration project of a TCSC was commissioned in 1991 at the 345 kV Kanawha River substation in West Virginia in the USA by the American Electric Power company. This was a test installation of thyristor switches in one phase for rapid switching of series capacitor segment and was supplied by ABB, Sweden.

In October 1992, the first three-phase TCSC was installed at 230 kV Kayenta Substation in Arizona by the Western Area Power Administration (WAPA) [9]. Here a 15 Ω capacitor bank is connected in parallel with a TCR and permits a smooth and rapid control of (capacitive) reactance between 15 and 60 Ω through phase control of TCR.

A larger prototype three-phase TCSC was installed in 1993 at 500 kV Slatt Substation in Oregon by the Bonneville Power Administration (BPA). The project was sponsored by the Electric Power Research Institute (EPRI) and the equipment was developed by

General Electric in the USA. Here, six modules of TCSCs are connected in series and controlled to provide a variation in impedance from +1.4 Ω to −16 Ω.

In Sweden, a long, series-compensated 400 kV transmission line connected to a nuclear power station was chosen to install a TCSC by splitting the existing series capacitor into two parts: a fixed capacitor and a TCSC at Stode Station. This was necessitated by an SSR problem that caused repeated triggering of protections. The TCSC allowed the existing level of compensation to be continued without any SSR problems.

At present, there are several installations of TCSCs all over the world, including India and China.

5.4.1 Basic Concepts of Controlled Series Compensation [5]

By controlled series compensation, we imply dynamic control of the degree of series compensation in a long line. This can be achieved in two ways:

1. discrete control using a thyristor switched series capacitor (TSSC)
2. continuous control using a TCSC.

The configuration using a TSSC is shown in Figure 5.6a. Here, the thyristor switch is either off or on continuously. To insert the capacitor, the switch is turned off while it is turned on to bypass the capacitor. A major problem with this configuration is that the SSR characteristics of the TSSC are no different from those of the fixed capacitor. Also, the full line current flows through the thyristor switch when it is on. Thus, this configuration is not common.

The configuration using a TCSC is shown in Figure 5.6b. Here, a TCR is used in parallel with a fixed capacitor to enable continuous control over the series compensation. Although harmonics are present in steady state with partial conduction of thyristor switches, the TCSC can be used to mitigate SSR. In addition, the TCSC provides inherent protection against over voltages.

The TCSC was also labeled as rapid adjustment of network impedance (RANI) in a paper by Dr. John Vithayathil *et al.* [1]. TCSCs are a mature technology available for application in AC lines of voltage up to 500 kV.

Consider the equivalent circuit of the TCSC modeled as a capacitor in parallel with a variable inductor (as shown in Figure 5.7). The impedance of the TCSC (Z_{TCSC}) is given by

$$Z_{TCSC} = \frac{-jX_C \cdot jX_{TCR}}{j(X_{TCR} - X_C)} = \frac{-jX_C}{\left(1 - \dfrac{X_C}{X_{TCR}}\right)} \tag{5.14}$$

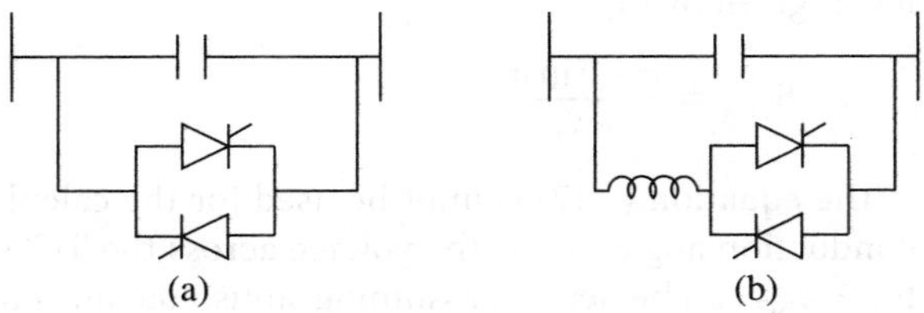

Figure 5.6 Controlled series compensation (a) TSSC (b) TCSC.

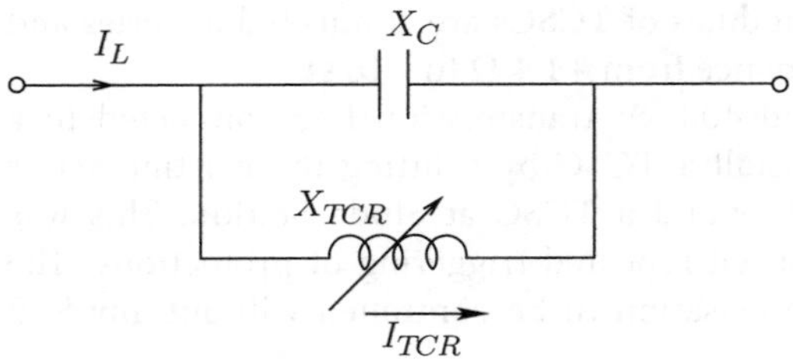

Figure 5.7 TCSC equivalent circuit.

The current through the TCR (I_{TCR}) is given by

$$\hat{I}_{TCR} = \frac{-jX_C}{j(X_{TCR} - X_C)}\hat{I}_L = \frac{\hat{I}_L}{\left(1 - \frac{X_{TCR}}{X_C}\right)} \tag{5.15}$$

If $X_{TCR} > X_C$ then $I_{TCR} < 0$, which implies that the current through the capacitor is greater than I_L.

Since the losses are neglected, the impedance of the TCSC is purely reactive. The capacitive reactance of the TCSC is obtained from (5.14) as

$$X_{TCSC} = \frac{X_C}{\left(1 - \frac{X_C}{X_{TCR}}\right)} \tag{5.16}$$

Note that X_{TCSC} is capacitive as long as $X_C < X_{TCR}$. $X_{TCR} = \infty$ when the thyristors are blocked since $I_{TCR} = 0$. For the condition when $X_C < X_{TCR}$, $\hat{I}_{TCR}$ is 180° out of phase with the line current $\hat{I}_L$. In other words, $\hat{I}_L$ is in phase with $-\hat{I}_{TCR}$.

For the condition where $X_C > X_{TCR}$, the effective reactance of TCSC (X_{TCSC}) is negative, implying that it behaves like an inductor. In this case, $\hat{I}_L$ and $\hat{I}_{TCR}$ are in phase.

It can be shown that although the magnitude of the TCSC reactance is same in both (capacitive and inductive) cases, the current through the TCR is more for the inductive operation. This indicates that thyristor ratings determine the maximum limit on the X_{TCSC} for the inductive operation.

The TCR also injects harmonics in addition to the fundamental frequency current. This distorts the capacitor voltage, which does not remain sinusoidal even though the line current remains approximately sinusoidal. Note that the presence of harmonics in the capacitor voltage can result in harmonics in the line current, but for long lines these can be neglected. It is also obvious that inductive operation of the TCSC results in higher voltage harmonics across the capacitor.

The expression for the effective susceptance of the TCR (neglecting voltage harmonics) is given by [5]

$$B_{TCR} = \frac{\sigma - \sin\sigma}{\pi X_L} \tag{5.17}$$

The equation (5.17) cannot be used for the calculation of X_{TCSC} as a function of the conduction angle (σ), as the voltage across the TCSC is not sinusoidal. The expression for X_{TCSC} can be derived assuming sinusoidal line current.

It should be noted that $X_L < X_C$ and a typical value of $\frac{X_L}{X_C}$ is 0.16. This results in protection against overvoltages due to fault currents by increasing the conduction angle to 180°. It is obvious that there is a value of the conduction angle (σ_r) for which $X_{TCR} = X_C$ and the TCSC reactance is a maximum (theoretically infinite when losses are neglected). It is necessary to ensure that the conduction angle stays within limits even during transient operation. The ratio of $\frac{X_{TCSC}}{X_C}$ is kept below a limit (say 3.0 in the capacitive region). The actual value of the limit is also a function of the line current to ensure that the capacitor voltage rating is not exceeded. In general, there are three voltage ratings based on the current ratings:

1) V_{rated} is the maximum continuous voltage across the TCSC. This must be more than $X_C \times I_{rated}$ to allow for continuous modulation with vernier operation (typically with 15% margin).
2) V_{temp} is the maximum temporary voltage ($X_C \times I_{temp}$) where I_{temp} is the temporary line current (typically $1.35 \times I_{rated}$) that has a typical duration of 30 minutes.
3) V_{tran} is the maximum voltage required during transient swings. This must be equal to or more than $X_C \times I_{tran}$, where I_{tran} is the maximum line current for which the thyristor control is not affected by protective considerations. Typically, $I_{tran} = 2 \times I_{rated}$ and lasts for a duration of 3–10 seconds.

5.4.2 Operation of a TCSC

A single line diagram of a TCSC is shown in Figure 5.8, which shows two modules connected in series. There can be one or more modules depending on the requirement. To reduce the costs, a TCSC may be used in conjunction with fixed series capacitors. Each module has three operating modes.

1) **Bypassed**
   Here the thyristor valves are gated for 180° conduction (in each direction) and the current flow in the reactor is continuous and sinusoidal. The net reactance of the module is slightly inductive as the susceptance of the reactor is larger than that of the capacitor. During this mode most of the line current is flowing through the reactor and thyristor valves, with some current flowing through the capacitor. This mode is used mainly for protecting the capacitor against overvoltages (during transient overcurrents in the line). This mode is also termed the thyristor switched reactor (TSR) mode.

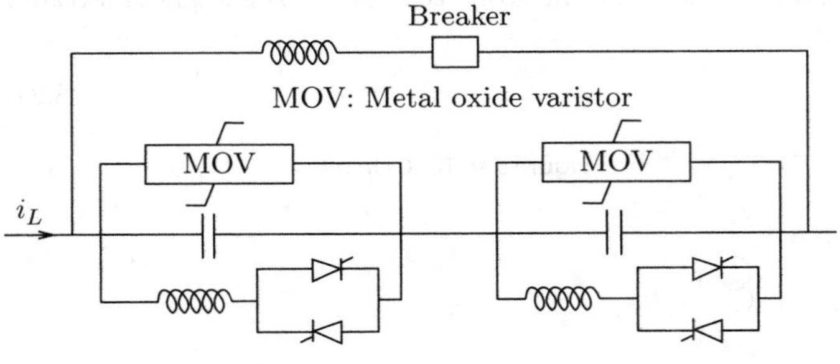

Figure 5.8 Single line diagram of a TCSC.

2) **Inserted with thyristor valve blocked**
   In this operating mode no current flows through the valves with the blocking of gate pulses. Here, the TCSC reactance is same as that of the fixed capacitor and there is no difference in the performance of the TCSC in this mode with that of a fixed capacitor. Hence this operating mode is generally avoided. This mode is also termed the waiting mode.
3) **Inserted with vernier control**
   In this operating mode the thyristor valves are gated in the region of ($\alpha_{min} < \alpha < 90°$) such that they conduct for a part of the cycle. The effective value of TCSC reactance (in the capacitive region) increases as the conduction angle increases from zero (α_{min} is above the value corresponding to the parallel resonance of the TCR and the capacitor, at fundamental frequency). In the inductive vernier mode, the TCSC (inductive) reactance increases as the conduction angle is reduced from 180°.

Generally, vernier control is used only in the capacitive region and not in the inductive region.

5.4.3 Analysis of a TCSC [10]

To understand the vernier control operation of a TCSC, it is necessary to analyze the TCSC circuit (see Figure 5.9). For simplicity, it is assumed that the line current i_s is specified and can be viewed as a current source. The equations are

$$C\frac{dv_C}{dt} = i_s(t) - i_T \tag{5.18}$$

$$L\frac{di_T}{dt} = v_C u \tag{5.19}$$

where $u = 1$ when the switch is closed and $u = 0$ when it is open. The current in the thyristor switch and the reactor (i_T) is zero at the instant when the switch is opened. Note that when $u = 0$, and the initial current $i_T = 0$, it remains at the zero value until S is turned on and $u = 1$. The line current i_s is defined by

$$i_s(t) = I_m \cos \omega t \tag{5.20}$$

It is convenient to measure the firing angle (α) from the zero crossing instant of the line current. It can be shown that the zero crossing of the capacitor voltage (v_C) coincides with the peak value of the line current in steady state. The range of α is from 0° to 90°, corresponding to the conduction angle varying from 180° to 0°. The angle of advance (β) is defined as

$$\beta = 90° - \alpha \tag{5.21}$$

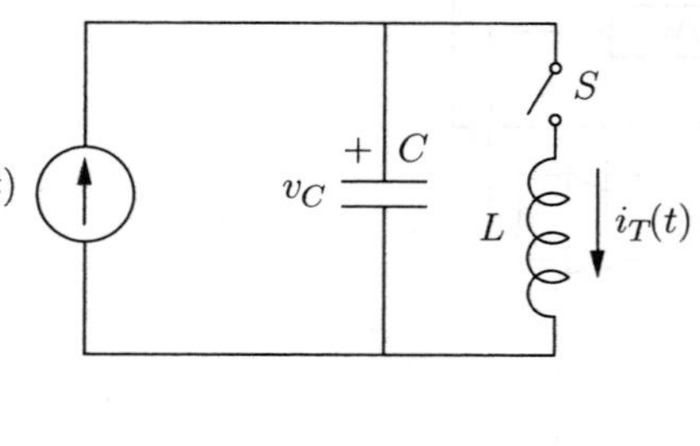

Figure 5.9 TCSC circuit.

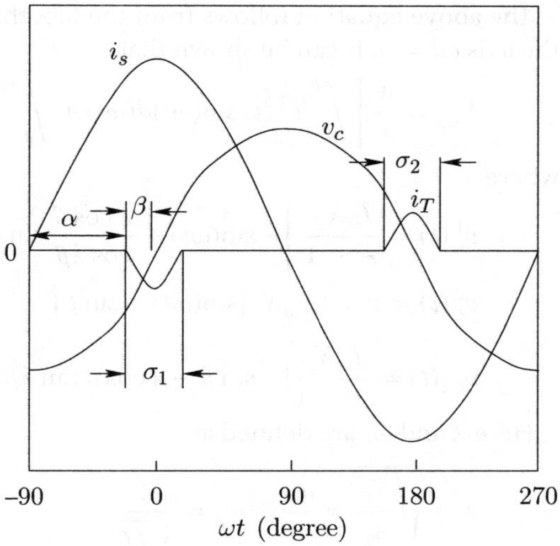

Figure 5.10 Waveforms of $i_s(t)$, $i_T(t)$, and $v_c(t)$.

which also varies from 0° to 90°. Figure 5.10 shows the waveforms of $i_s(t)$, $i_T(t)$, and $v_C(t)$ with delay angle (α), angle of advance (β), and conduction angle (σ) indicated.

The equations (5.18) and (5.19) can be solved if the switching instants are known. The switch S is turned on twice in a cycle (of the line current) at the following instants (assuming equidistant gating pulses)

$$t_1 = \frac{-\beta}{\omega}$$
$$t_3 = \frac{\pi - \beta}{\omega} \tag{5.22}$$

where $0 < \beta < \beta_{max}$. The thyristor switch turns off at the instants t_2 and t_4 given by

$$t_2 = t_1 + \frac{\sigma_1}{\omega}$$
$$t_4 = t_2 + \frac{\sigma_2}{\omega} \tag{5.23}$$

where σ_1 and σ_2 are the conduction angles in the two halves of the cycle. In steady state, $\sigma_1 = \sigma_2 = \sigma$ with half wave symmetry and

$$\sigma = 2\beta \tag{5.24}$$

5.4.4 Computation of the TCSC Reactance (X_{TCSC})

The TCSC reactance corresponding to the fundamental frequency is obtained by taking the ratio of the peak value of the fundamental frequency component (V_{C1}) to the peak value of the sinusoidal line current. From Fourier analysis, the fundamental frequency component (V_{C1}) is calculated from

$$V_{C1} = \frac{4}{\pi} \int_0^{\pi/2} v_C(t) \sin(\omega t) d(\omega t) \tag{5.25}$$

The above equation follows from the fact that v_C has half-wave odd symmetry about the axis $\omega t = 0$. It can be shown that

$$V_{C1} = \frac{4}{\pi}\left[\int_0^{\beta} v_C^1(t)\sin(\omega t)d(\omega t) + \int_{\beta}^{\pi/2} v_C^2(t)\sin(\omega t)d(\omega t)\right]$$

where

$$v_C^1(t) = \frac{I_m X_C}{\lambda^2 - 1}\left[-\sin(\omega t) + \frac{\lambda\cos\beta}{\cos\lambda\beta}\sin\omega_r t\right]$$

$$v_C^2(t) = v_{C2} + I_m X_C[\sin(\omega t) - \sin\beta]$$

$$v_{C2}(t) = \frac{I_m X_C}{\lambda^2 - 1}[-\sin\beta + \lambda\cos\beta\tan\lambda\beta]$$

Here λ and ω_r are defined as

$$\lambda = \sqrt{\frac{X_C}{X_L}} = \frac{\omega_r}{\omega}, \quad \omega_r = \frac{1}{\sqrt{LC}} \tag{5.26}$$

X_C and X_L are reactances calculated at the fundamental frequency (ω). The reactance X_{TCSC} is usually expressed in terms of X_C. By defining

$$X_{TCSC} = \frac{V_{C1}}{I_m} \tag{5.27}$$

we can derive the ratio $\frac{X_{TCSC}}{X_C}$ as

$$\frac{X_{TCSC}}{X_C} = 1 + \frac{2}{\pi}\frac{\lambda^2}{(\lambda^2 - 1)}\left[\frac{2\cos^2\beta}{(\lambda^2 - 1)}(\lambda\tan\lambda\beta - \tan\beta) - \beta - \frac{\sin 2\beta}{2}\right] \tag{5.28}$$

The above expression can be simplified as

$$\frac{X_{TCSC}}{X_C} = 1 + \frac{2}{\pi}\frac{\lambda^2}{(\lambda^2 - 1)}\left[-\left(\frac{\lambda^2 + 1}{\lambda^2 - 1}\right)\frac{\sin 2\beta}{2} - \beta + \frac{2\lambda\cos^2\beta\tan\lambda\beta}{(\lambda^2 - 1)}\right]$$

The capacitor voltage also contains odd harmonics of the order

$$n = 2k - 1, \quad k = 1, 2, 3, \ldots$$

Remarks

1) The fundamental frequency component and the harmonics in the capacitor voltage can also be obtained from the calculation of harmonics in the TCR current $i_T(t)$. The peak value of the fundamental component I_{T1} can be found from

$$\begin{aligned} I_{T1} &= \frac{4}{\pi}\int_0^{\pi/2} i_T(t)\cos\omega t\, d(\omega t) \\ &= \frac{4}{\pi}\left[\int_0^{\beta}\frac{\lambda^2}{\lambda^2 - 1}I_m\left[\cos\omega t - \frac{\cos\beta}{\cos\lambda\beta}\cos\lambda\omega t\right]d(\omega t)\right] \end{aligned} \tag{5.29}$$

The fundamental component of the capacitor voltage is obtained from

$$V_{C1} = (I_m - I_{T1})X_C \tag{5.30}$$

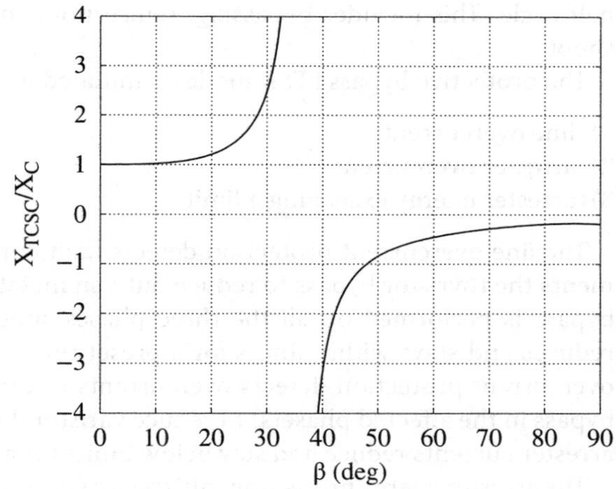

Figure 5.11 Variation of X_{TCSC}/X_C as a function of β.

2) $\frac{X_{TCSC}}{X_C} \to \infty$ at $\beta_{res} = \frac{(2k-1)\pi}{2\lambda}$, where k is an integer. Since the range of β is limited to 90°, if $\lambda < 3$, then there will be only one value of β_r at which the TCR and the capacitor will be in resonance at the fundamental frequency. Typically, $\lambda = 2.5$. The variation of $\frac{X_{TCSC}}{X_C}$ as a function of β for $\lambda = 2.5$ is shown in Figure 5.11. The negative value of $\frac{X_{TCSC}}{X_C}$ indicates that the reactance is inductive.
3) As the TCSC impedance is very high at resonance β_r given by

$$\beta_r = \frac{\pi}{2\lambda} \tag{5.31}$$

the operation of the TCSC must ensure that the upper limit on β ($\beta_{\max}$) should be strictly enforced even under transient conditions. Generally $\frac{X_{TCSC}}{X_C} < 4$ to ensure that the voltage ratings on the capacitor are not exceeded.

5.4.5 Control of the TCSC

The control of the TCSC also includes protective functions (protective bypass). The control functions are partitioned into two levels: common (to all modules) and module (level). Commands for the control flow from the common level to the module levels while the status information is sent back from each module level.

Module Control Functions

There are three basic functions at each module level:

1) reactance control
2) SSR damping control (involving modulation of the reactance)
3) bypass (for protection).

The controller also ensures that the transients associated with mode transitions are minimized. The module controller executes the ordered change to reactance within one half cycle. This includes bypassing, reinsertion, and setting the vernier without overshoot.

The protective bypass (TSR mode) is initiated in response to

1) line overcurrent
2) arrester overcurrent
3) arrester energy exceeding a limit.

The line overcurrent protection detects fault currents in the line and rapidly implements the thyristor bypass to reduce duty on metal oxide varistors and capacitors. The bypass is performed on all the three phases simultaneously. When the line current reduces and stays within limits for a preset time, the bypass is removed. The arrester overcurrent protection detects overcurrents in the arrester and implements thyristor bypass in the affected phase(s) to reduce varistor duty. The bypass is removed when the arrester currents reduce and stay below limits for a preset time.

The arrester energy protection initiates a bypass when $\int i^2 dt$ in the arrester exceeds its rating. In this case, the bypass is not removed automatically because of the long thermal time constants associated with the decay of excessive arrester energy.

Common Control Functions

The common level receives signals of line current and TCSC voltage to generate feedback signals for closed-loop control functions. It also receives commands from energy management centre for setting power order. The major control functions in a TCSC are described below.

Power Scheduling Control [11]

The simplest type of power scheduling control adjusts the reactance order (or setpoint) slowly to meet the required steady-state power flow requirements of the transmission network. The adjustment may be done manually or by a slow-acting feedback loop.

An alternative approach is to use a closed-loop current control in which the measured line current is compared to a reference current (which may be derived from the knowledge of the required power level).

An interesting approach to power scheduling is one where during a transient, the line in which the TCSC is situated carries the required power so that the power flow in parallel paths is kept constant. This is equivalent to maintaining the angular difference across the line constant and has been termed constant angle (CA) control. Assuming the voltage magnitudes at the two ends of the line are regulated, maintaining CA is equivalent to maintaining constant voltage difference between the two ends of the line.

Both constant current (CC) and CA controllers can be of PI type with dynamic compensation for improving the response. The steady-state control characteristics of both CC and CA control are shown in Figures 5.12a and 5.12b, respectively. Assuming V_{TCSC} to be positive in the capacitive region, the characteristics have three segments: OA, AB, and BC. The control range is AB. OA and BC correspond to the limits on X_{TCSC}. In Figure 5.12b the control range AB is described by the equation

$$V_{TCSC} = I_L X_{LR} - V_{Lo} \tag{5.32}$$

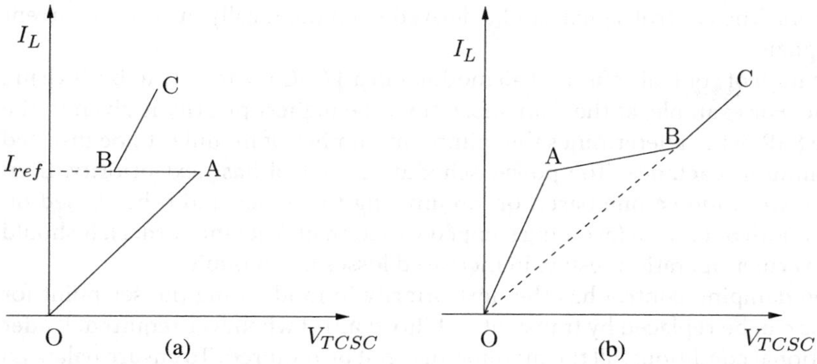

Figure 5.12 Control characteristics (a) CC control (b) CA control.

where I_L is the magnitude of the line current, X_{LR} is the net line reactance (taking into account the fixed-series compensation if any), and V_{Lo} is the constant (regulated) voltage drop across the line (including TCSC). Thus, the slope of the line AB for CA control is the reciprocal of X_{LR}. OA in Figure 5.12b corresponds to the lower limit on TCSC reactance while BC corresponds to the higher limit on TCSC reactance.

Power Swing Damping Control [12]

This is designed to modulate the TCSC reactance in response to an appropriately chosen control signal derived from local measurements. The objective is to damp low-frequency swing modes (corresponding to the oscillation of generator rotors) of frequencies in the range of 0.2 to 2.0 Hz. One of the signals that is easily accessible is the line current magnitude. Alternatively, the signal corresponding to the frequency of Thevenin (equivalent) voltage of the system across the TCSC can be used. This signal can be synthesized from the knowledge of voltage and current measurements [12].

Transient Stability Control

This is generally a discrete control in response to the detection of a major system disturbance. The discrete or bang-bang control of TCSC in response to signals from locally measured variables is described in [13]. The controller is activated immediately after a major disturbance such as clearing of a fault and is deactivated when the magnitude of frequency deviation is below threshold. This type of control is beneficial not only in reducing the first swing but also for damping subsequent swings.

Subsynchronous Damping Control [14, 15]

The use of vernier control mode at the module level by setting the reactance set-point at the requisite (minimum) level is often adequate to damp subsynchronous oscillations caused by series resonance in the line and sustained due to torsional interaction. However, in some cases the constant reactance control may not be adequate. In such cases, a damping control is added. The control signal is based on the synthesis of speed of remote turbo-generators. The control signal can be derived from the locally measured current and voltage signals.

The coordination of control actions of all modules in a TCSC is carried out by devising a suitable logic. For example, at the Slatt substation, the highest priority is given to the need to tackle SSR, which determines the minimum number of modules to be inserted and their minimum reactance. The power scheduling control has next priority. Even here there are two options, one based on minimizing the losses and other based on maximizing smooth operation (avoiding stepped variation in reactance order). It should be noted that vernier operation results in increased losses in a module.

Power swing damping control has the next priority in modulating the set-point for reactance. This can be replaced by transient stability control whenever required. Under normal operational conditions, all the modules may not be required. To ensure balanced long-term duty for each module in a TCSC, the control logic also incorporates a rotation feature, according to which the modules in series are rotated if no insert/bypass operations occur for some preset time (say an hour). The rotation is performed without changing the net reactance.

Modeling of a TCSC for Stability Studies [5]

For stability studies it is not necessary to model the gate pulse unit and the generation of gate pulses. It is adequate to assume that the desired value of TCSC reactance is implemented within a well-defined time frame. This delay can be modeled by first-order lag, as shown in Figure 5.13. The value of T_{TCSC} is typically 15–20 ms. X_{ref} is determined by the power scheduling controller or, in its absence, by manual control based on order from load dispatch control.

The block diagram of a CC or CA controller is shown in Figure 5.14. T_m is the time constant of a first-order low pass filter associated with the measurement of line current I_L and the TCSC voltage. $S = 0$ for CC control and $S = \frac{1}{X}$ for CA control. X is the net reactance of the line given by

$$X = X_{line} - X_{FC} \tag{5.33}$$

where X_{line} is the line reactance and X_{FC} is the reactance of the fixed-series capacitor if any. Generally, a TCSC will be used in conjunction with a fixed-series capacitor to

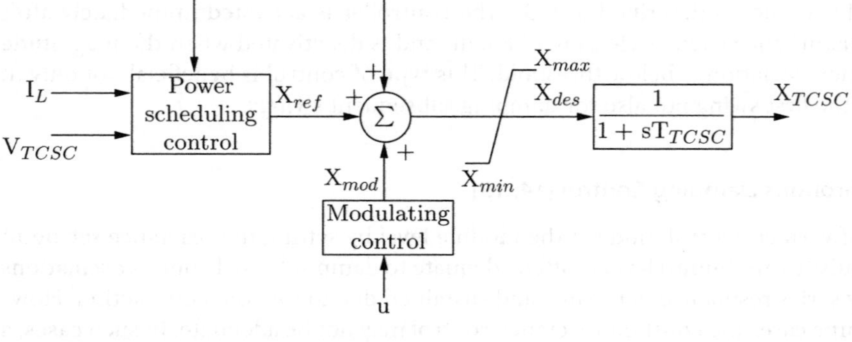

Figure 5.13 Block diagram of TCSC control.

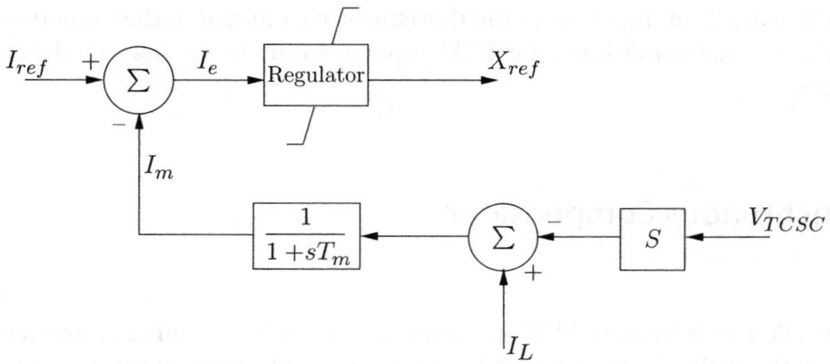

Figure 5.14 Block diagram of CC or CA controller.

minimize the overall cost of compensation while providing effective control for stability improvement.

The modulation controller is designed to damp power swings in the line by using a control signal derived from local measurement. A convenient signal to use is the magnitude of line current. The control configuration is similar to the supplementary modulation controller (SMC) used in an SVC. X_{aux} could represent the increase in the reactance order required for mitigating SSR or improving transient stability.

The limits on the total reactance order X_{des} are based on the TCSC capability. For a single module, the operational regions of the TCSC in the (V_{TCSC}–I_L) plane and the ($\frac{X_{TCSC}}{X_C}$)–I_L plane are shown in Figure 5.15. The line AB corresponds to the maximum limit $X_{\max 0}$ and is selected based on the TCSC design. It should be such that the TCSC does not operate close to the resonant point, which would be inherently unstable. A typical value of $X_{\max 0}$ is 3.0 pu. The line AB also corresponds to the maximum voltage rating of the TCSC. The reactance of the TCSC in the bypass mode (X_{bypass}) is inductive. The line DE corresponds to the voltage limit of the TCSC in the inductive region. The

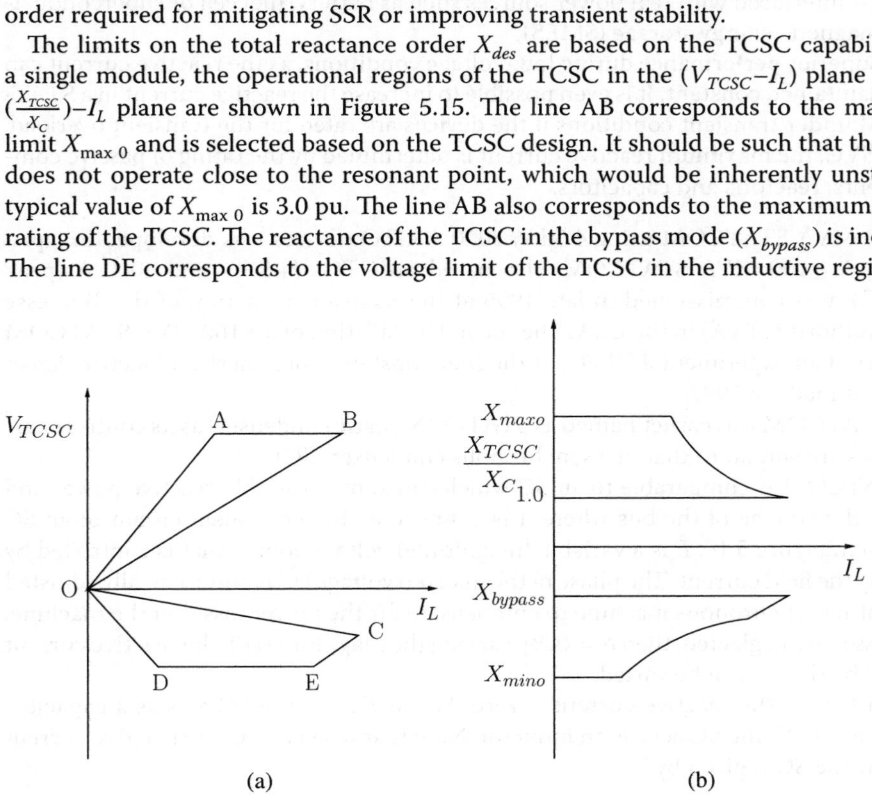

Figure 5.15 TCSC capability curves for a module (a) V-I plane (b) X-I plane.

line EC corresponds to the limit imposed on the thyristor valve current in the inductive vernier mode. Under normal conditions, the TCSC operates only in the first quadrant of the X_{TCSC}–I_L plane.

5.5 Static Synchronous Compensator

5.5.1 General

STATCOM is also called an *advanced SVC* and is based on a voltage source converter (VSC) that injects a voltage that is synchronized to the voltage at the bus where it is connected. Although VSCs require self-commutating power semiconductor devices (such as GTO or IGBT), with higher costs and losses (compared to the variable impedance type SVCs, which employ thyristor devices), there are many technical advantages of a STATCOM over a SVC:

1) faster response independent of the effective short-circuit ratio (ESCR)
2) requires less space as bulky components, such as reactors, are not required
3) inherently modular and relocatable
4) can be interfaced with real power sources such as battery, fuel cell or superconducting magnetic energy storage (SMES)
5) has superior performance during low-voltage conditions, as the reactive current can be maintained constant. It is even possible to increase the reactive current in a STATCOM under transient conditions if the devices are rated for the transient overload. In a SVC, the maximum reactive current is determined by the rating of passive components, reactors, and capacitors.

A ±80 MVA STATCOM using 4.5 kV, 3000 A GTO devices was developed in Japan in 1991. A ±100 MVA STATCOM, also based on GTOs (4.5 kV and 4000 A (peak turn-off)), was commissioned in late 1995 at the Sullivan substation of the Tennesse Valley Authority (TVA) in the USA. The second installation of a ±160 MVA STATCOM (as a part of an experimental UPFC) at the Inez substation of American Electric Power (AEP) was made in 1997.

The STATCOM was earlier named as STATCON (static condenser) as its control characteristics are similar to that of a synchronous condenser (SC).

A STATCOM is comparable to an SC which can supply variable reactive power and regulate the voltage of the bus where it is connected. The equivalent circuit of an SC is shown in Figure 5.16. $\hat{E}$ is a variable (magnitude) voltage source that is controlled by adjusting the field current. The phase of the injected voltage ($\hat{E}$) is automatically adjusted in a rotating synchronous machine to compensate for the power losses in the machine. If the losses are neglected, then $\delta = 0$. By varying the magnitude (E), the reactive current supplied by the SC can be varied.

When $E = V$, the reactive current is zero. When $E > V$, the SC acts as a capacitor and when $E < V$, the SC acts as an inductor. Note that when $\delta = 0$, the reactive current drawn by the SC is given by

$$I_r = \frac{V - E}{X'} \tag{5.34}$$

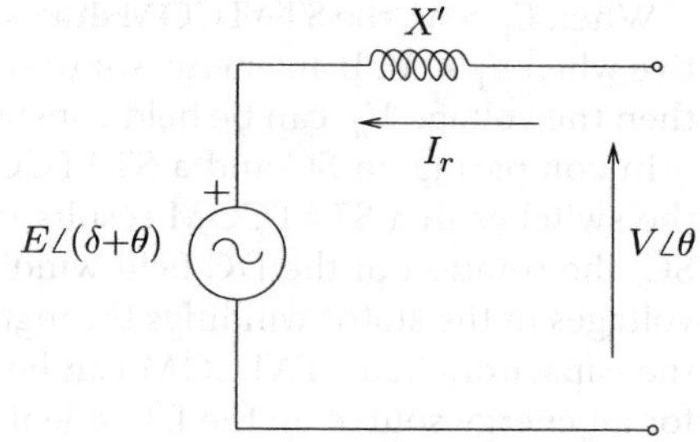

Figure 5.16 Synchronous condenser.

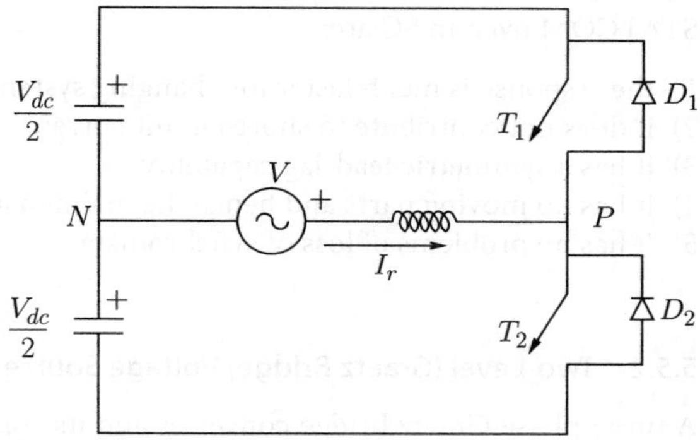

Figure 5.17 A single-phase STATCOM.

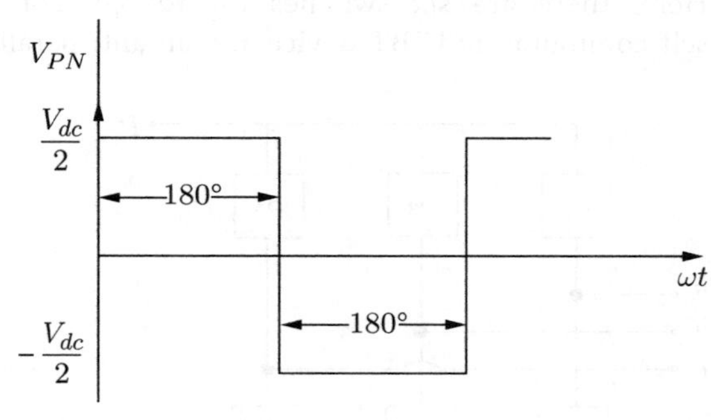

Figure 5.18 The waveform of V_{PN}.

The circuit of a single-phase STATCOM is shown in Figure 5.17. The voltage (V_{PN}) developed across the terminals P and N is shown in Figure 5.18. The self-commutating switches T_1 and T_2 (based on GTO or IGBT) are switched on and off once in a cycle. The conduction period of each switch is 180° and care has to be taken to ensure that T_1 is "off" when T_2 is "on" and vice versa. The diodes D_1 and D_2 enable the conduction of currents in the reverse direction. Note that $V_{PN} = \frac{V_{dc}}{2}$ when T_1 is conducting (T_2 is off) and $V_{PN} = -\frac{V_{dc}}{2}$ when T_2 is conducting (and T_1 is off). The switches are synchronized with the supply voltage (V), which is assumed to be sinusoidal of frequency ω. The rms value of the fundamental component (E_1) of V_{PN} is obtained as

$$E_1 = \frac{\sqrt{2}}{\pi}\int_0^{\pi} \frac{V_{dc}}{2} \sin\theta \, d\theta = \frac{\sqrt{2}}{\pi} V_{dc} = 0.45 V_{dc} \tag{5.35}$$

When $E_1 > V$, the STATCOM draws a capacitive reactive current, whereas it is inductive when $E_1 < V$. If an energy source (a battery or a rectifier) is present on the DC side, then the voltage V_{dc} can be held constant.

In comparing an SC and a STATCOM, we note that the synchronous operation of the switches in a STATCOM results in the AC voltages at the output, whereas in the SC, the rotation of the DC field winding on the rotor results in the generation of AC voltages in the stator windings through magnetic induction. Unlike in the case of a SC, the capacitors in a STATCOM can be charged from the AC side and there is no need for an energy source on the DC side if only reactive current is to be provided in steady state. The losses in the STATCOM can be met from the AC source. The advantages of a STATCOM over an SC are:

1) the response is much faster for changing system conditions
2) it does not contribute to short-circuit current
3) it has a symmetric lead-lag capability
4) it has no moving parts and hence the maintenance is easier
5) it has no problems of loss of synchronism.

5.5.2 Two-Level (Graetz Bridge) Voltage Source Converter

A three-phase Graetz bridge converter and its components are shown in Figure 5.19a. Here, there are six switches (S_1 to S_6). For a VSC, the switch consists of a self-commutating IGBT device and an anti-parallel connected diode. The DC side of

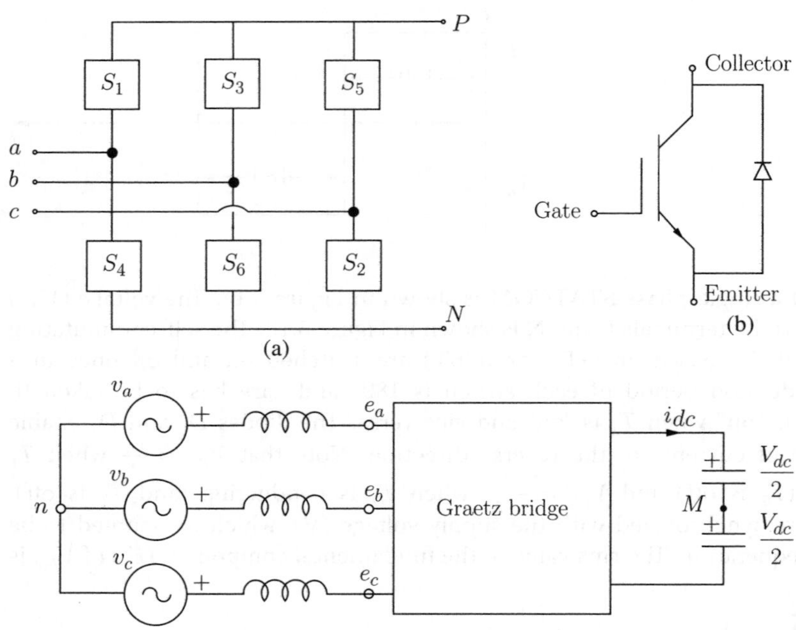

Figure 5.19 A two-level VSC and its components (a) Graetz bridge (b) Switch consisting of IGBT and diode.

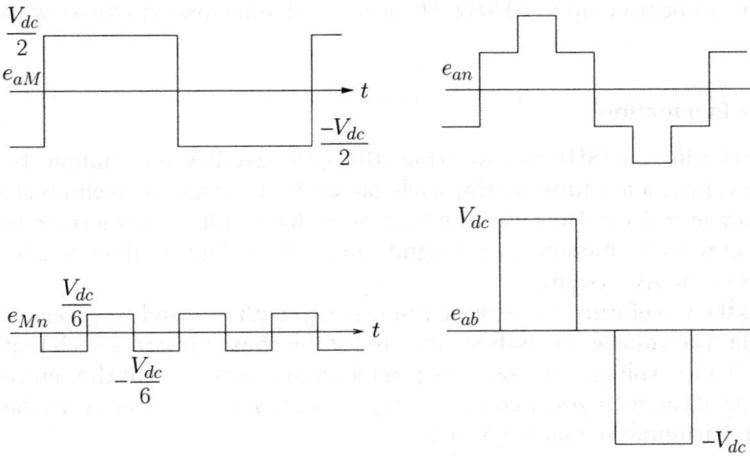

Figure 5.20 Waveforms of e_{aM}, e_{Mn}, e_{an}, and e_{ab}.

the converter is connected to the capacitor and the AC side is connected to the source voltage through reactors. The voltages e_a, e_b, and e_c injected by the VSC are dependent on V_{dc} (the voltage across the capacitor). The current i_{dc} flowing in the capacitor is a function of the source currents i_a, i_b, and i_c. In each leg of the converter, only one switch can be turned on at any given time to avoid a short circuit across the capacitor. The waveform of the voltage e_a depends on the reference point. The waveforms of e_{aM}, e_{Mn}, e_{an}, and e_{ab} are shown in Figure 5.20. The voltage of the midpoint M w.r.t the supply neutral, e_{Mn}, consists of only a zero-sequence voltage with triplen harmonics. Note that e_{Mn} is obtained by subtracting e_{aM} from e_{an}.

It can be shown that both e_{an} and e_{ab} contain harmonics of the order $h = 6k \pm 1$, where k is an integer. The magnitude of the fundamental frequency voltage E_1 is given in (5.35). The harmonic voltage component is given by

$$E_n = \frac{E_1}{n} \tag{5.36}$$

5.5.3 Pulse Width Modulation

In order to control the reactive power injected by a STATCOM, it is essential to control the magnitude of the injected voltage. If there is an energy source on the DC side of the VSC, it is also possible to control the power output by changing the phase angle of the injected voltage. It is also desirable to regulate the DC bus voltage and minimize the injection of low order harmonics.

The pulse width modulation (PWM) is utilized to achieve the following objectives:

1) to control the output AC voltage for a constant DC voltage (across the capacitor)
2) to minimize the harmonics subject to the constraints on the switching losses and generation of noise.

PWM is not feasible in a GTO-based VSC. Multipulse or multilevel converters which require one switching (on and off) in a cycle use GTO devices. On the other hand, IGBT devices can be used up to 10 kHz. However, switching losses increase with frequency.

Selective Harmonic Elimination

Selective harmonic elimination (SHE) is also termed the optimized PWM technique. By reversing the phase voltage a few times during each half cycle, it is possible to eliminate low order harmonics selectively. However, the high order harmonics may increase in magnitude, but the current harmonics are not significantly affected due to the low-pass filter characteristics of the AC system.

Figure 5.21 shows the waveforms of the phase voltage (e_{aM}) with four and six reversals within one half cycle. The voltage reversals are affected at the chosen instants such that the notches (caused by the voltage reversals) are placed symmetrically about the centre line of each half cycle. If there are p switchings (voltage reversals) in a quarter cycle, the rms value of the nth harmonic voltage is given by

$$E_n = \pm\frac{1}{\sqrt{2}}\frac{4}{\pi}\frac{V_{dc}}{2}\left[-\int_0^{\alpha_1}\sin n\theta\, d\theta + \int_{\alpha_1}^{\alpha_2}\sin n\theta\, d\theta \ldots \pm \int_{\alpha_p}^{\pi/2}\sin n\theta\, d\theta\right]$$

$$= \pm\frac{0.45}{n}V_{dc}[2(\cos n\alpha_1 - \cos n\alpha_2 + \cos n\alpha_3 - \ldots) - 1]$$

where $\alpha_1, \alpha_2 \ldots \alpha_p$ are switching angles within each quarter cycle. There are p degrees of freedom and these are used to cancel $(p-1)$ harmonic components in the voltage and control the fundamental voltage. For example, if there are three switchings at α_1, α_2, and α_3, then we can eliminate the fifth and seventh harmonics in addition to controlling the fundamental voltage. We get three transcendental equations given by

$$\overline{m} = 2[\cos\alpha_1 - \cos\alpha_2 + \cos\alpha_3] - 1 \tag{5.37}$$

$$0 = 2[\cos 5\alpha_1 - \cos 5\alpha_2 + \cos 5\alpha_3] - 1 \tag{5.38}$$

$$0 = 2[\cos 7\alpha_1 - \cos 7\alpha_2 + \cos 7\alpha_3] - 1 \tag{5.39}$$

where $\overline{m}$ is the normalized modulation index defined by

$$\overline{m} = \frac{E_{mod1}}{0.45V_{dc}} \tag{5.40}$$

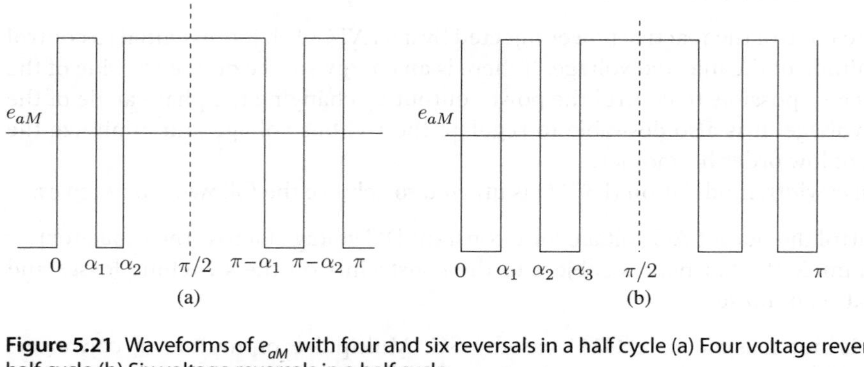

Figure 5.21 Waveforms of e_{aM} with four and six reversals in a half cycle (a) Four voltage reversals in a half cycle (b) Six voltage reversals in a half cycle.

E_{mod1} is the rms value of the fundamental component of the modulated output voltage. Note that although the phase voltage contains triplen harmonics, the line to line voltages are free from them.

Example 5.1 If there are two voltage reversals in a quarter cycle, the values of α_1 and α_2 are 16.2° and 22.0° such that the fifth and seventh harmonics are zero. The modulation index ($\overline{m}$) is given by

$$\overline{m} = 1 - 2[\cos \alpha_1 - \cos \alpha_2] = 0.934$$

Sinusoidal PWM

The sinusoidal PWM technique is widely used for industrial converters. The method is also known as the suboscillation method and uses a triangular carrier wave of frequency F to generate a sinusoidal reference voltage of frequency f.

The sine PWM technique can be compared to the SHE PWM discussed earlier. Here, the instants of voltage reversals are determined by the intersection between a reference sinusoidal voltage of peak value $\overline{m}$ and a triangular voltage of amplitude 1. Figure 5.22 shows the operation of sine PWM for $p = \frac{F}{f} = 9$ with $\overline{m} = 0.5$.

The modulation is said to be in the linear phase range if $0 < \overline{m} < 1$. The same carrier wave can be used for all three phases with the reference voltages in each phase shifted from the other phases by 120°. The harmonics present in the output of sinusoidal PWM have frequencies given by

$$f_n = MF \pm Nf \tag{5.41}$$

where M and N are integers. The harmonics are significant only around the frequencies MF.

5.5.4 Analysis of a Voltage Source Converter

Here, we consider a three-phase VSC using six switches made up of IGBT valves with antiparallel diodes (see Figure 5.23).

The following assumptions are made in the analysis:

1) The supply voltages are sinusoidal and balanced (contain only fundamental frequency, positive sequence components).
2) The losses in the switches are ignored and hence the switches are assumed to be ideal.

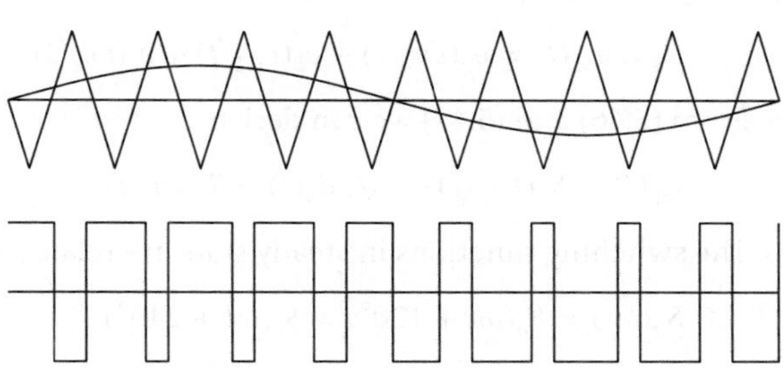

Figure 5.22 Sinusoidal PWM.

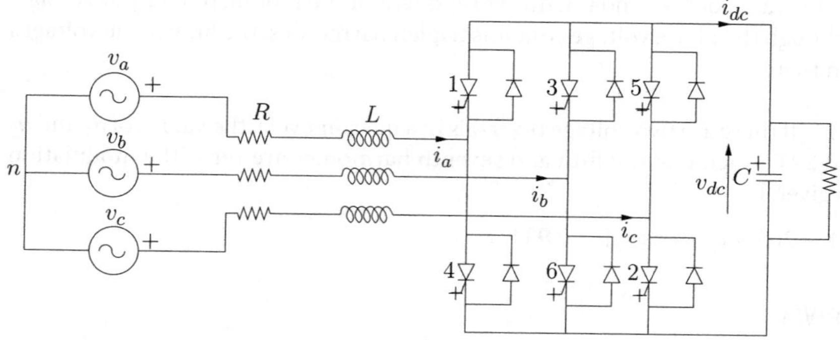

Figure 5.23 A six-pulse VSC.

3) The inductance L and the resistance R are the parameters of the series reactor connected between the supply and the VSC (or represent the leakage reactance and ohmic losses of the interfacing transformer). G represents the losses in the DC capacitor.

The VSC shown in the figure is described by the following differential equations:

$$L\frac{di_a}{dt} + Ri_a = v_a - e_a \tag{5.42}$$

$$L\frac{di_b}{dt} + Ri_b = v_b - e_b \tag{5.43}$$

$$L\frac{di_c}{dt} + Ri_c = v_c - e_c \tag{5.44}$$

$$C\frac{dv_{dc}}{dt} + Gv_{dc} = i_{dc} \tag{5.45}$$

where e_a, e_b, and e_c are the output voltages of the VSC. All the voltages are with reference to the source neutral n.

We can express the VSC output voltages as

$$e_a(t) = S_a(t)v_{dc}(t), e_b(t) = S_b(t)v_{dc}(t), e_c(t) = S_c(t)v_{dc}(t) \tag{5.46}$$

where S_a, S_b, and S_c are the switching functions. For square wave operation (without any modulation), they are periodic in steady state and are shown in Figure 5.24. From the principle of energy conservation, we get

$$v_{dc}(t)i_{dc}(t) = e_a(t)i_a(t) + e_b(t)i_b(t) + e_c(t)i_c(t) \tag{5.47}$$

From (5.46) and (5.47) we can derive

$$i_{dc}(t) = S_a(t)i_a(t) + S_b(t)i_b(t) + S_c(t)i_c(t) \tag{5.48}$$

The switching functions in steady state are related by

$$S_a(\omega t) = S_b(\omega t + 120°) = S_c(\omega t + 240°) \tag{5.49}$$

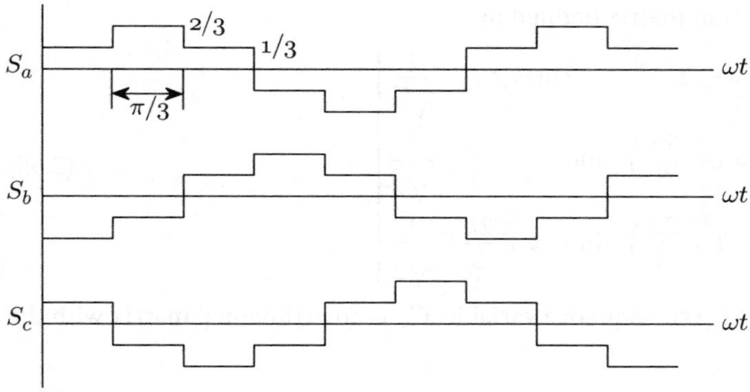

Figure 5.24 Switching function for a six-pulse VSC controller.

Substitution of (5.46) and (5.48) in (5.42) to (5.45) makes the system time varying. If switching functions are approximated by their fundamental components only, we obtain for a two-level converter

$$e_a \approx \overline{m} v_{dc} \frac{2}{\pi} \sin(\omega_o t + \theta + \alpha) \tag{5.50}$$

$$e_b \approx \overline{m} v_{dc} \frac{2}{\pi} \sin\left(\omega_o t + \theta + \alpha - \frac{2\pi}{3}\right) \tag{5.51}$$

$$e_b \approx \overline{m} v_{dc} \frac{2}{\pi} \sin\left(\omega_o t + \theta + \alpha - \frac{4\pi}{3}\right) \tag{5.52}$$

where $\overline{m}$ is the modulation index (normalized).

In deriving the above equations, we have assumed that the injected voltages lead the source voltages by angle α, which can be controlled. θ is the angle of the source voltage whose line to line magnitude is V_s. We can express $v_a(t)$ as

$$v_a(t) = \sqrt{\frac{2}{3}} V_s \sin(\omega_o t + \theta) \tag{5.53}$$

The expressions for $v_b(t)$ and $v_c(t)$ are obtained from the fact that the source voltages are balanced and contain only positive sequence components. The system frequency is assumed to be ω_o and a constant (or it varies very slowly).

Equations in the *D-Q* Reference Frame

The voltages and currents in the AC circuit can be transformed to a synchronously rotating reference frame by Kron's transformation defined by

$$f_{abc} = [C_K]\, f_{DQo} \tag{5.54}$$

where f_{abc} can be a voltage or current vector defined by

$$f_{abc} = [f_a \quad f_b \quad f_c]^t$$

Similarly,

$$f_{DQo} = [f_D \quad f_Q \quad f_o]^t$$

and C_K is the transformation matrix defined by

$$C_K = \sqrt{\frac{2}{3}} \begin{bmatrix} \cos\omega_o t & \sin\omega_o t & \frac{1}{\sqrt{2}} \\ \cos\left(\omega_o t - \frac{2\pi}{3}\right) & \sin(\omega_o t - \frac{2\pi}{3}) & \frac{1}{\sqrt{2}} \\ \cos\left(\omega_o t + \frac{2\pi}{3}\right) & \sin(\omega_o t + \frac{2\pi}{3}) & \frac{1}{\sqrt{2}} \end{bmatrix} \tag{5.55}$$

The subscript o indicates a zero-sequence variable. C_K is an orthogonal matrix with the property

$$[C_K]^{-1} = [C_K]^t \tag{5.56}$$

It should be noted that the transformation is power invariant. Thus,

$$v_a i_a + v_b i_b + v_c i_c = v_D i_D + v_Q i_Q + v_o i_o \tag{5.57}$$

By applying Kron's transformation, we can derive the circuit equations in D-Q variables:

$$L\frac{di_D}{dt} + Ri_D + \omega_o L i_Q = v_D - \overline{m}k\sin(\alpha + \theta)v_{dc} \tag{5.58}$$

$$L\frac{di_Q}{dt} + Ri_Q - \omega_o L i_D = v_Q - \overline{m}k\cos(\alpha + \theta)v_{dc} \tag{5.59}$$

$$C\frac{dv_{dc}}{dt} + Gv_{dc} = \overline{m}k[\sin(\alpha + \theta)i_D + \cos(\alpha + \theta)i_Q] \tag{5.60}$$

where $k = \frac{\sqrt{6}}{\pi}$ for a two-level converter. $\overline{m}$ is the normalized modulation index that can vary from 0 to 1. The zero-sequence current is zero as $i_a + i_b + i_c = 0$.

Remarks

1) In a high power VSC with GTO thyristor switches, a switch is turned on and off only once in a cycle to minimize switching losses. Here, $\overline{m}$ is a constant and is equal to 1. The magnitude of the injected voltage can be controlled by PWM. There are many variations in PWM, but all PWM converters require several switchings in a cycle.
2) In a STATCOM with an energy source on the DC side, it is advisable to control both the magnitude and phase angle of the injected voltage by the VSC in order to control the power and reactive power output.
3) It is possible to express (5.58) and (5.59) in terms of current and voltage phasors. By multiplying (5.58) by j on both sides and adding it to (5.59), we get

$$L\frac{d\hat{I}}{dt} + (R + j\omega_o L)\hat{I} = \hat{V} - \hat{E} \tag{5.61}$$

where

$$\hat{I} = i_Q + ji_D,\ \hat{V} = v_Q + jv_D,\ \hat{E} = \overline{m}kV_{dc}\angle(\theta + \alpha) \tag{5.62}$$

Note that, since v_D and v_Q are the components of the (dynamic) phasor, they become constants in steady state (corresponding to an operating point). Generally, we use

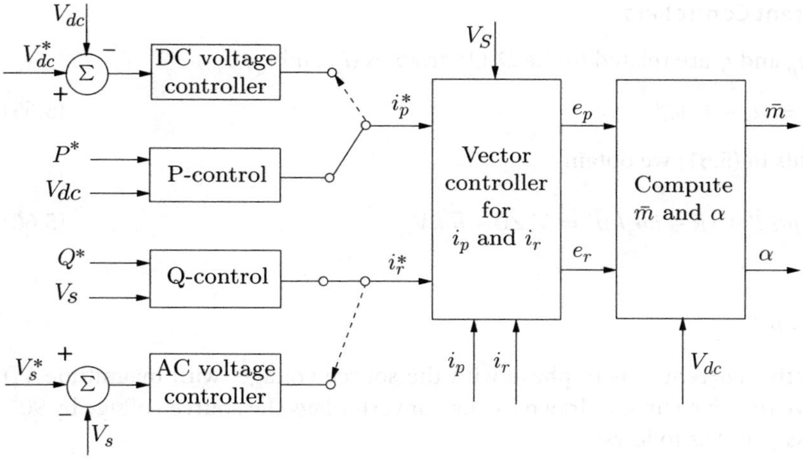

Figure 5.25 Block diagram of a VSC controller.

upper case to denote phasors. On the other hand, the *D*-*Q* components are also instantaneous values. Thus, to simplify the notation, we will use both lower and upper case symbols interchangeably.

5.5.5 Control of VSC

A VSC injects a controllable voltage at the AC terminals given by

$$\hat{E} = \overline{m}E_{ao}e^{j(\theta+\alpha)}, E_{ao} = kV_{dc}$$

where k is a constant depending on the converter configuration. For a two-level converter, $k = \frac{\sqrt{6}}{\pi}$. $\overline{m}$ (normalized modulation index) and α (phase angle of the injected voltage leading the AC bus voltage) are the control variables used to control the active power and reactive power output of the VSC. Note that $\overline{m} < 1$, and α can be positive or negative depending on whether the converter is operating as an inverter or a rectifier.

The reactive power injected by the converter is a function of V_s, the AC bus voltage magnitude. The block diagram of the VSC controller is shown in Figure 5.25. Here i_s and i_r are the active and reactive components of the AC current flowing into the converter. That is $i_p = \frac{P}{V_s}$ and $i_r = \frac{Q}{V_s}$. The reference value for the active current (i_p^*) is obtained either from the power controller or the DC voltage controller. The reference value for the reactive current (i_r^*) is obtained from the specified reactive power of the AC voltage controller. The outputs of the current controllers (active and reactive current) are the desired converter voltage components e_p and e_r, which are defined as

$$e_p = \overline{m}kV_{dc}\cos\alpha, \quad e_r = -\overline{m}kV_{dc}\sin\alpha \tag{5.63}$$

From the above equation, $\overline{m}$ and α can be calculated as

$$\overline{m} = \frac{\sqrt{e_p^2 + e_r^2}}{kV_{dc}}, \quad \alpha = -\tan^{-1}\frac{e_r}{e_p} \tag{5.64}$$

Design of Current Controllers

The variables i_p and i_r are related to the D-Q variables (i_D and i_Q) by

$$i_Q + ji_D = (i_p - ji_r)e^{j\theta} \tag{5.65}$$

Substituting this in (5.61) we obtain

$$L\frac{d\hat{I}'}{dt} + j\dot{\theta}L\hat{I}' + (R + j\omega_o L)\hat{I}' = V_s\angle 0 - \overline{m}kV_{dc}e^{j\alpha} \tag{5.66}$$

where

$$\hat{I}' = i_p - ji_r$$

Note that active current i_p is in phase with the source voltage (with magnitude V_s) and the positive reactive current drawn by the converter lags the source voltage by 90°. We can express (5.66) as follows:

$$\frac{d}{dt}\begin{bmatrix} i_p \\ i_r \end{bmatrix} = \begin{bmatrix} \frac{-R\omega_B}{X} & 0 \\ 0 & \frac{-R\omega_B}{X} \end{bmatrix}\begin{bmatrix} i_p \\ i_r \end{bmatrix} + \frac{\omega_B}{X}\begin{bmatrix} V_s - e_p - \frac{X\omega}{\omega_B}i_r \\ -e_r + \frac{X\omega}{\omega_B}i_p \end{bmatrix} \tag{5.67}$$

where $X = \omega_B L$, $\omega = \omega_o + \frac{d\theta}{dt}$, and ω_B is the rated (nominal) frequency. Note that we normally assume $\omega = \omega_B$. If we arrange

$$u_p = V_s - e_p - \frac{X\omega}{\omega_B}i_r \tag{5.68}$$

$$u_r = -e_r + \frac{X\omega}{\omega_B}i_p \tag{5.69}$$

where u_p and u_r are obtained as the outputs of the active and reactive current controllers, then we get decoupled control (also called vector control) of the two current components, as shown in Figure 5.26. u_p and u_r are obtained as the outputs of controllers and

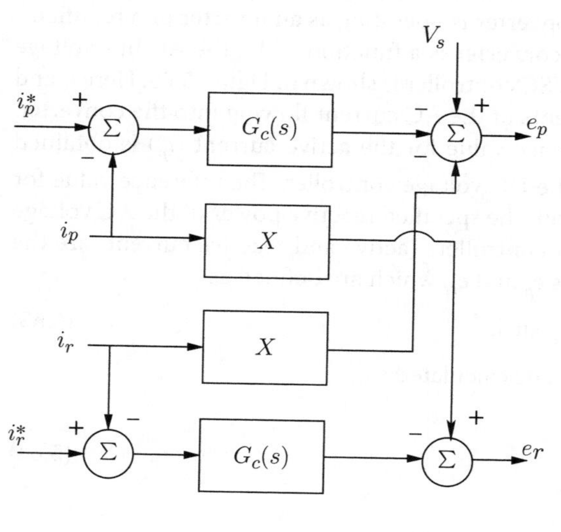

Figure 5.26 Decoupled current controllers.

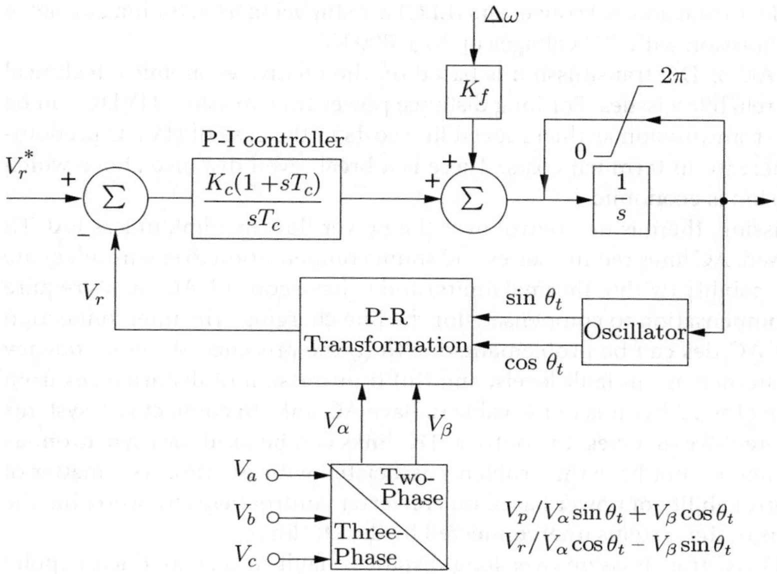

Figure 5.27 Block diagram of PLL.

defined by

$$u_p = G_c(s)(i_p^* - i_r), u_r = G_c(s)(i_r^* - i_r) \tag{5.70}$$

where $G_c(s) = \frac{K_c(1+sT_c)}{sT_c}$ for the P-I controller.

It should be noted that the decoupled structure of the current controllers is obtained by the feed-forward terms involving i_p, i_r, and V_s (see (5.68) and (5.69)). The P-I controller time constant T_c can be chosen to cancel the poles introduced by the current flow in the inductive circuit. If $T_c = \frac{L}{R}$ and K_c is chosen as $K_c = \frac{L}{T_i}$, then it can be shown that

$$i_p = \frac{1}{1+sT_i} i_p^*, \quad i_r = \frac{1}{1+sT_i} i_r^* \tag{5.71}$$

The value of T_i can be chosen to provide a specified response time for the current controllers. In general, the response time is a fraction of a cycle. The implementation of the vector control scheme involves computation of i_p and i_r, which requires a phase locked loop (PLL) synchronized to the (AC) bus voltage of the converter. A typical PLL is shown in Figure 5.27 [16].

5.6 HVDC Power Transmission [17–21]

Since 1972 when the first HVDC link using thyristor valves was installed at Eel River, Canada, there has been significant growth in the application of HVDC links. The application of IGBT valves in VSC-based HVDC links since 1997 has expanded the scope of HVDC transmission. The total installed capacity of HVDC links exceeds 130 GW. The line commutated (current source) converters (LLC) are still being used for long distance bulk power transmission with DC voltages up to $\pm$ 800 kV.

The choice of AC or DC transmission is based on the relative economics, technical advantages, and reliability issues. For long distance power transmission, HVDC can be cheaper than AC transmission as the reduced line costs in the case of HVDC predominate over the increase in terminal costs. There is a break-even distance above which HVDC transmission is economical.

In AC transmission, there is no control over the power flow in a link unless FACTS controllers are used. AC lines require series and shunt compensation to ensure adequate power transfer capability (within thermal limits) and voltage control. AC cables require shunt reactive compensation to compensate for the line charging. The interconnection of systems using AC ties can be problematic due to (i) the presence of low-frequency oscillations, (ii) an increase in fault levels, and (iii) transmission of disturbances from one system to the other. Also, it is not feasible to have AC links to connect two systems operating at different frequencies. In contrast, DC links can be used for asynchronous interconnection and do not have the problems associated with AC ties. As a matter of fact, the fast controllability of power based on converter control helps to overcome the stability problems of the systems interconnected by the DC links.

In a bipolar HVDC transmission over long distances, fault in one conductor (pole) does not lead to stoppage of power flow over the link as monopolar operation with ground return is feasible. Thus it is claimed that a bipolar HVDC line with two conductors can be as reliable as a double circuit AC line with six conductors.

5.6.1 Application of DC Transmission

Based on a detailed comparison of AC and DC transmission in terms of economics and technical performance, the following areas of application are suggested:

1) long distance bulk power transmission
2) underground or underwater cables
3) asynchronous interconnections of AC systems operating at different frequencies or where independent control of systems is desired
4) control and stabilization of power flows in AC ties in an integrated power system.

The asynchronous DC links are typically back-to-back (BTB) links without any HVDC line. The provision of HVDC links and FACTS controllers (like TCSC) has enabled the development of an all-India power grid interconnecting five regional grids.

5.6.2 Description of HVDC Transmission Systems

Typically, HVDC links are bipolar (see Figure 5.28). A bipolar link has two conductors, one positive and the other negative (each may be a bundled conductor in EHV lines). Each terminal has two sets of converters of identical ratings. The junction between the two sets of converters is grounded at one or both ends. Normally, both poles operate at equal currents and hence there is no ground current flowing under normal conditions.

A schematic diagram of a converter station is shown in Figure 5.29. A converter station can operate as a rectifier station (power conversion from AC to DC) or inverter station (power conversion from DC to AC). Typically, a converter station has one or

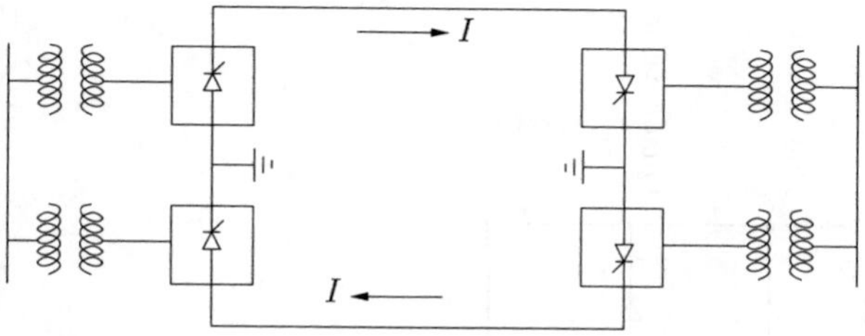

Figure 5.28 A bipolar HVDC link.

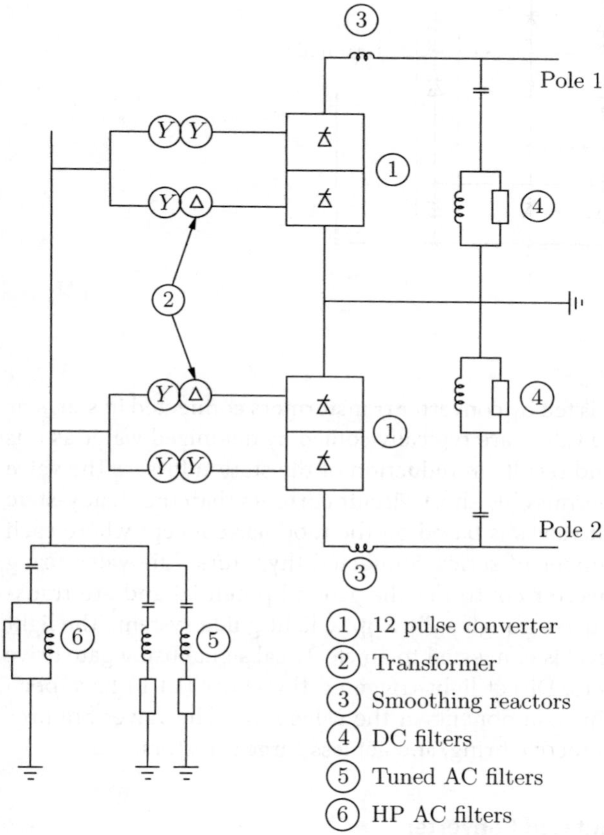

Figure 5.29 Schematic diagram of a HVDC converter station.

two 12-pulse converter units per pole. In addition to the converter units, a station has converter transformers, AC and DC filters, reactive power sources, a smoothing reactor (SR), and switchgear (not shown in the figure).

A 12-pulse converter is shown in Figure 5.30. This consists of two three-phase converter (Graetz) bridges connected in series to form a 12-pulse converter unit. The total number of thyristor valves is 12. The valve can be packaged as a single valve, double valve or quadri valve arrangement. Each valve is used to switch in a segment of an AC

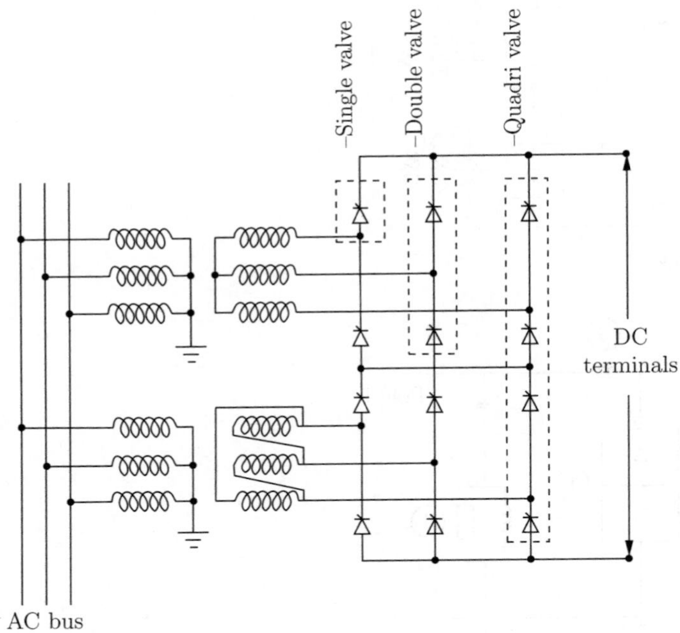

Figure 5.30 A 12-pulse converter unit.

voltage waveform. The converter is fed by converter transformers connected in star/star and star/delta arrangements. The valves are typically cooled by deionized water as it is more efficient than air cooling and results in reduction of the station losses. The valve ratings are limited more by the permissible short-circuit currents than the steady-state load requirements. The design of valves is based on the modular concept where each module consists of a limited number of series connected thyristors. The valve firing signals are generated in the converter control at the ground potential and are transmitted to each thyristor in the valve through a fibre optic light guide system. The light signal received at the thyristor level is converted to an electrical signal using gate drive amplifiers with pulse transformers. Direct light triggered thyristors (LTT) have been developed to reduce the electronic components at the valve level. The valves are protected using snubber circuits, protective firing, and gapless surge arrestors.

5.6.3 Analysis of a Line Commutated Converter

A line commutated (Graetz bridge) current source converter is shown in Figure 5.31a. The six switches in the converter are made of thyristor valves which can turn on when a firing pulse is applied at the gate terminal during the interval (in a cycle) when they are forward biased. The thyristor symbol is shown in Figure 5.31b. The AC current magnitude depends on the DC current and the average DC voltage across the converter depends on the AC voltage magnitude and the delay angle (α).

The operation of the converter can be explained from the fact that it is equivalent to a series connection of the two valve groups (see Figure 5.32). A valve (commutation) group is a group of valves which commutate the DC current from one valve to the next

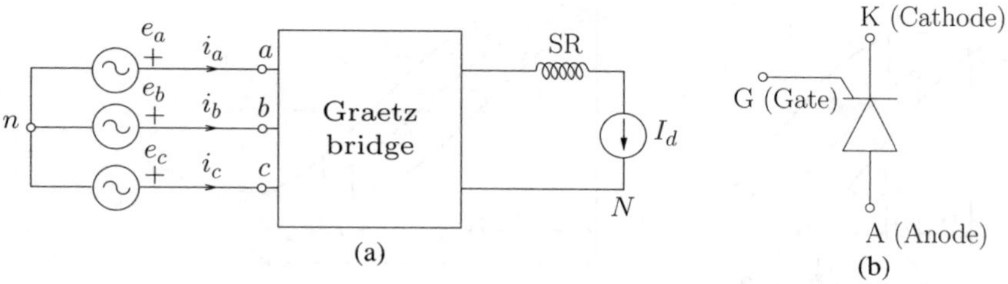

Figure 5.31 A LCC-HVDC converter (a) A line commutated converter (b) Thyristor symbol.

Figure 5.32 A six-pulse converter as a series connection of two valve groups.

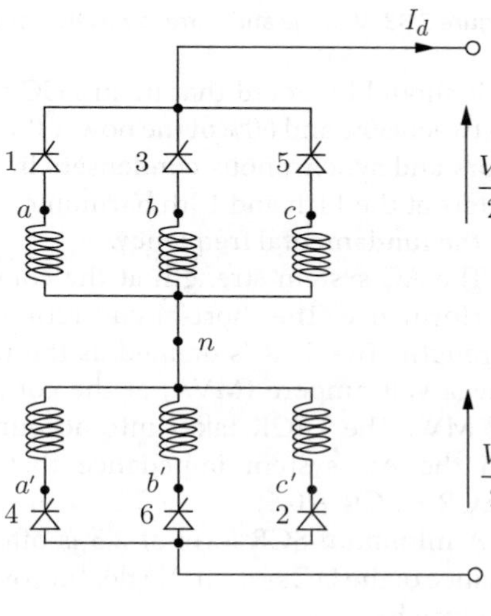

in the group during a cycle. Note that the valves are numbered in the sequence in which they are fired.

It should be noted that a valve cannot be turned on earlier than the instant of zero crossing of the commutation voltage (it should become positive). The delay angle (α) is the angle when the valve is fired with respect to the zero crossing of the commutation voltage. Normally, the rectifier station controls the DC current (and power) while the inverter station regulates the DC voltage, ensuring the extinction angle does not reduce below a minimum. The DC current in the HVDC link is increased by reducing α. However, there is a minimum limit on α (say 5°) and a further increase in current is effected by reducing α in the inverter. To raise the voltage, α at the inverter is increased subject to the limit determined by the extinction angle (γ). This limit is required for the outgoing valve to regain its current blocking capability, which depends on its recovery charge (Q_{rr}). The DC voltage and valve current waveforms during the firing of (i) a rectifier and (ii) an inverter are shown in Figure 5.33. The figures also define α, μ, and γ, where μ and γ are the overlap and extinction angles, respectively. Apart from the converter control, the tap changer control on the converter transformers also acts in a slow fashion to ensure that the α and γ are within the limits of normal operation. It is also possible to modulate α and γ (at the rectifier and inverter, respectively) to help in damping low-frequency oscillations in the transmission network.

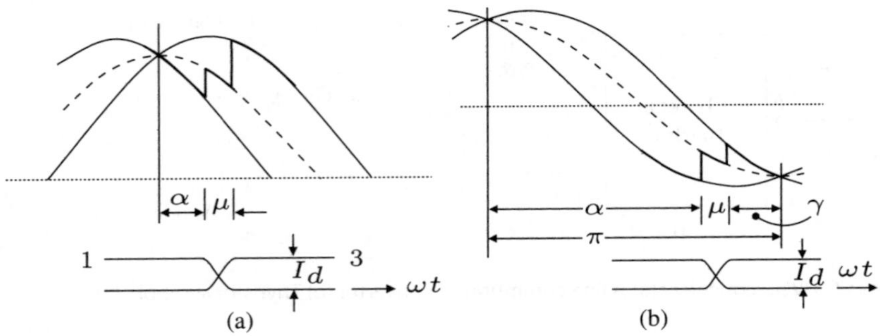

Figure 5.33 Voltage and current waveforms during firing of a valve (a) Rectifier (b) Inverter.

It should be noted that in an LCC the converter requires reactive power that varies between 50% and 60% of the power flow in the link. This is met by the provision of capacitors and synchronous condensers in addition to the AC filters. The AC filters (tuned filters at the 11th and 13th harmonics, and the high-pass filter) generate reactive power at the fundamental frequency.

The AC system strength at the converter bus has a significant effect on the system performance. The short-circuit ratio (SCR) or ESCR is an indicator of the AC system strength. The SCR is defined as the ratio of the AC system three-phase short-circuit mega volt-ampere (MVA) at the converter AC bus to the capacity of the DC station in MW. The ESCR takes into account the effect of AC filters and shunt capacitors on the AC system impedance that determines the short-circuit MVA. Typically ESCR = SCR − 0.5.

A minimum SCR level of 2.5 is often used as the lower limit for acceptable performance of the DC system. Neglecting overlap, it can be shown that the average DC voltage is given by

$$V_d = V_{do} \cos \alpha \tag{5.72}$$

$$V_{do} = \frac{3\sqrt{2}}{\pi} E_{LL} = 1.35 E_{LL}$$

where E_{LL} is the magnitude of line-to-line voltage appearing across a valve. For $0^\circ < \alpha < 90^\circ$, V_d is positive (rectifier mode) and for $90^\circ < \alpha < 180^\circ$, V_d is negative (inverter mode).

Since a thyristor valve is not self-commutating, care has to be taken to ensure that it turns off (when reverse biased) with a minimum margin angle of 8–10°. Considering overlap (when three valves conduct as the current through the outgoing valve cannot be reduced to zero instantaneously), the extinction angle γ defined by

$$\gamma = 180^\circ - \alpha - \mu$$

(where μ is the overlap angle) is maintained at a value around 15° to avoid commutation failures. It can be shown that the (rms) fundamental component of the AC current (I_1) is given by (assuming $\mu \approx 0$)

$$I_1 = \frac{\sqrt{6}}{\pi} I_d \tag{5.73}$$

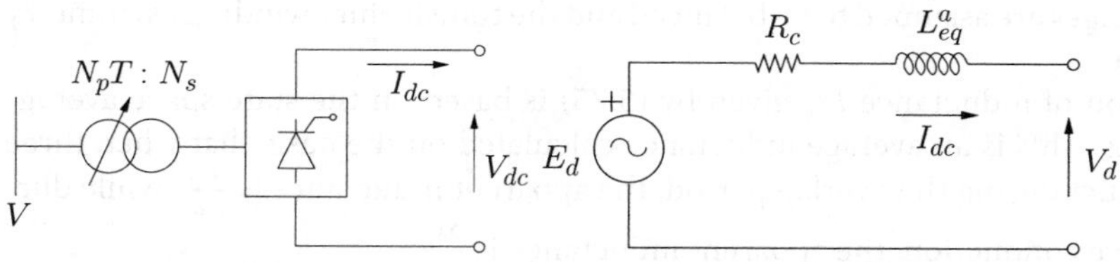

Figure 5.34 Continuous time model of a converter (a) A converter bridge (b) Equivalent circuit.

and the harmonics of the order other than $6n \pm 1$ (where n is an integer) are zero and the magnitude of harmonic of order h is equal to $\frac{I_1}{h}$. In general, for a p pulse converter, the characteristic harmonics are of the order $np \pm 1$.

Modeling of the LCC-HVDC system

Here, we consider the continuous time, average DC voltage model.

Simplified continuous time model

Consider a converter bridge fed from a transformer, which may be connected in star/star or star/delta (see Figure 5.34a). The equivalent circuit of the converter bridge is shown in Figure 5.34b, where

$$E_d = kaV \cos\theta$$

where V is the converter bus AC voltage expressed per unit on its own base (rated) voltage. θ is the delay angle α (for the rectifier) and the angle of advance is β (for the inverter). $a = \frac{1}{T}$ is the off-nominal tap ratio. k is defined as

$$k = \frac{3\sqrt{2}}{\pi}\frac{N_S}{N_P}\frac{V_{\alpha B}}{V_{dB}}$$

where $V_{\alpha B}$ is the base AC voltage and V_{dB} is the base (rated) voltage. By selecting rated DC current (I_{dn}) as the base DC current we can express $\overline{R}_c$ (per unit) as

$$\overline{R}_c = \frac{\overline{X}_c}{2}k \tag{5.74}$$

Note that $\overline{R}_c$ is the commutation resistance. The equivalent inductance L^a_{eq} is the average inductance defined as

$$L^a_{eq} = \frac{X_c}{\omega_o}\left[2(1-d) + \frac{3}{2}d\right] \tag{5.75}$$

where $d = \frac{3\mu}{\pi}$ and μ = overlap angle in radians.

The transformer winding resistance and valve voltage drop can also be taken into account in a manner similar to the average inductance L^a_{eq}. The equivalent circuit of Figure 5.34b is based on the following assumptions:

1) The harmonics in the DC voltage are neglected.

2) The AC voltages are assumed to be balanced and the transformer winding asymmetry is neglected.
3) The inclusion of inductance L^a_{eq} given by (5.75) is based on the state space averaging concepts. This is an average inductance calculated on the basis that when three valves conduct during the overlap period, the apparent inductance is $\frac{3}{2}\frac{X_c}{\omega_o}$ while during two-valve conduction, the apparent inductance is $\frac{2X_c}{\omega_o}$.
4) The converter control is assumed to be continuous.

The last assumption can be relaxed and a discrete-time model can also be derived [17].

Model of Converter Controller

For sake of convenience, the overall control can be divided into three categories:

1) Power control, auxiliary control, and voltage-dependent current order limiter (VDCOL). The output of this block is the current order. The limits on the current order are modified by the VDCOL. The objective of the VDCOL is to prevent individual thyristors from carrying full current for long periods during commutation failures. However, sufficient time delay is given to prevent the action of VDCOL during common AC system faults, which otherwise would drastically reduce DC power.
2) CC and constant extinction angle (CEA) controls. These are usually of feedback type. However, the extinction angle control can also be of predictive (open loop) type. The output of these controllers is a control voltage that determines the instant of gate pulse generation. The input is taken as the current order (generated locally or at the remote station) or the extinction angle reference (generated locally). The communication delay in transmitting the current order may have to be represented.
3) The gate pulse generator has input from the CC or CEA controller and determines the instant of gate pulse generation for each valve. There are basically two types of firing control schemes: (i) individual phase control (IPC) and (ii) equidistant pulse control (EPC). The latter can be (i) pulse frequency control (PFC) or (ii) pulse phase control (PPC).

The effect of IPC or EPC can be included in the controller model. The basic difference in the two schemes is that the change in the phase of the converter bus AC voltage will affect the delay angle in the case of EPC, while it does not in the case of IPC.

The converter control is usually represented by block diagrams and specifying the transfer function of each block. A typical controller block diagram is shown in Figures 5.35 and 5.36. This does not give all the details but is adequate in system modeling. In Figure 5.36, switch S_1 is closed in the case of EPC and switch S_2 is closed when the synchronizing circuit is provided. $\Delta\delta_v$ is the change in the phase of the converter bus voltage and the transfer function of the synchronizing circuit is typically

$$\mathrm{SYN}(s) = 1/(1 + sT_{ss}),\ T_{ss} = 1/(2\pi B_w) \tag{5.76}$$

where B_w is the bandwidth of the synchronizing circuit in Hz. As the bandwidth increases, the characteristics of EPC tend towards those of IPC. The time delay T_d is

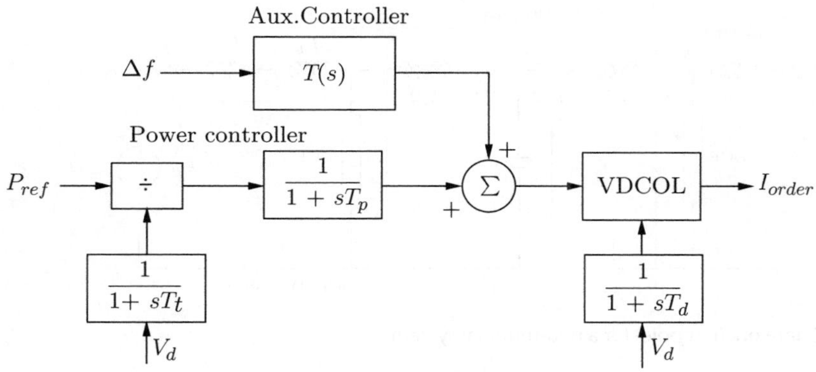

Figure 5.35 Power and auxiliary controller.

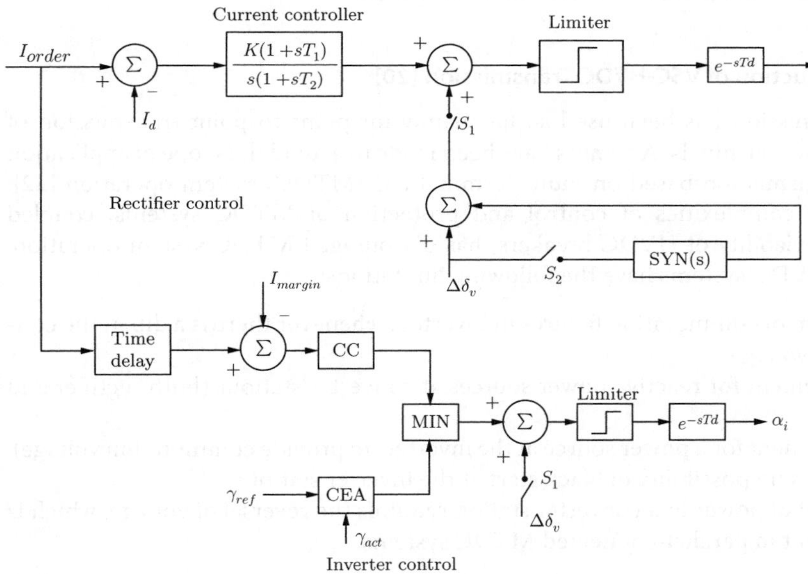

Figure 5.36 Rectifier and inverter controller.

introduced if a continuous time model of the converter is used and is given by

$$T_d = 1/(2pf_o) \tag{5.77}$$

where p is the pulse number and $f_o = \frac{\omega_o}{2\pi}$.

Modeling of the DC Network

The DC network is assumed to consist of a smoothing reactor, DC filters, and the transmission line (see Figure 5.37). The smoothing reactor and DC filters can be represented by lumped parameter linear elements. The DC line can also be modeled as a T or π equivalent if the higher frequency behavior is not of interest. L_{dr} and L_{di} are the

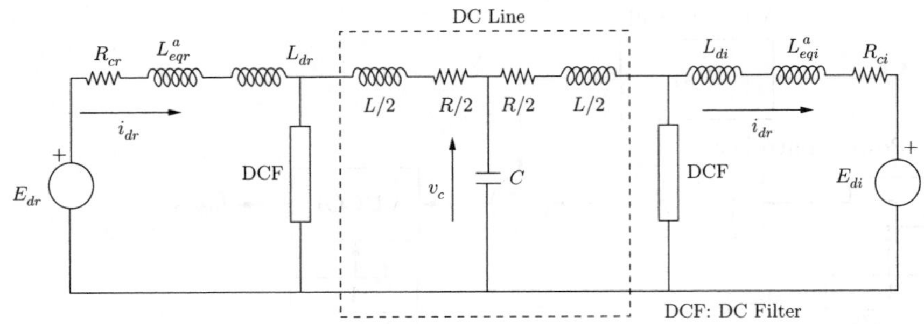

Figure 5.37 DC network (per pole) for a two-terminal system.

inductances of the smoothing reactor at the rectifier and inverter stations, respectively. In a BTB HVDC link, there are no DC filters or DC line.

5.6.4 Introduction of VSC-HVDC Transmission [20]

HVDC transmission has been used so far mainly for point-to-point transmission of power using two terminals. Attempts have been made to extend the scope of application of HVDC transmission based on multi-terminal DC (MTDC) system operation [22]. However, the complexities of control and protection of MTDC systems, coupled with the non-viability of HVDC breakers, has discouraged MTDC system operation. LCC-based HVDC systems have the following limitations:

1) the problem of commutation failures in inverters whenever there is a dip in the converter bus voltage
2) the requirement for reactive power sources at converter stations (both rectifier and inverter)
3) the requirement for a power source at the inverter (to provide commutation voltage), thus there is no possibility of black start at the inverter stations
4) the reversal of power in a converter station requires the reversal of voltage, which is inconvenient in parallel-connected MTDC systems.

In general, satisfactory operation of HVDC converter stations is based on the SCR or ESCR [23]. This also affects the operation of multi-infeed DC (MIDC) systems where more than one converter station is in proximity, feeding the same area [24, 25].

Many of the problems affecting LCC-based HVDC systems are eliminated by the introduction of VSC-based HVDC transmission. For example, a VSC does not require reactive power, it can actually supply reactive power to the AC system if required. There are no commutation failure problems which affect LCC converters. There is no need to reverse the voltage for power reversal. This implies that VSC-based HVDC links are equipped with extruded polymer insulated (XLPE) HVDC cables, which are easy to install, instead of oil-filled, impregnated paper insulated cables.

A VSC-based HVDC link can be viewed as two STATCOMs at two terminals that are interconnected by a DC link. A BTB VSC-based HVDC link is shown in Figure 5.38. By closing the switches S_1 and S_2 and opening the switch S, the configuration changes from two parallel connected STATCOMs to a BTB HVDC link. It is interesting to note

Figure 5.38 A BTB VSC HVDC link.

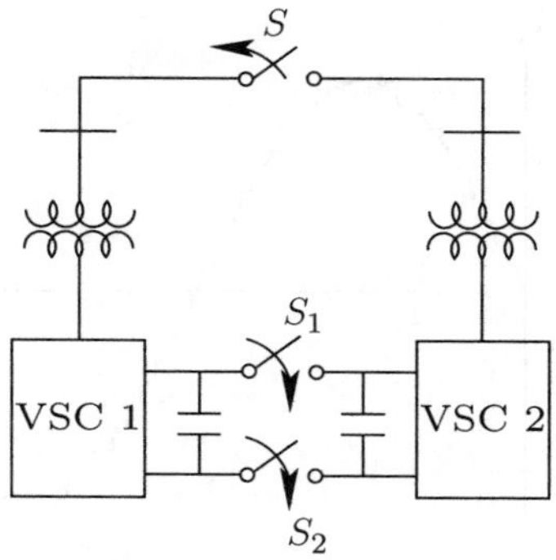

that while SVC and TCSC were offshoots of the conventional LCC-based HVDC technology based on thyristor valves, VSC-HVDC is an extension of the emerging FACTS controllers based on VSC.

The advantages of VSC-HVDC over LCC-HVDC links can be summarized as follows:

1) A VSC converter station permits independent control over both active and reactive power outputs. It can supply or absorb reactive power from the AC network and help in regulating the AC bus voltage. In a two-terminal VSC-HVDC, one terminal is used to control the power flow in the link, while the other terminal regulates the DC-link voltage.
2) Since VSC does not require a voltage source for satisfactory operation, the inverter station can also supply passive loads, which implies black start of the link is feasible when connected to offshore wind power plants.
3) The harmonic filtering required in VSC-HVDC is much simpler as PWM is used to shift the frequency spectrum of the converter output.
4) Fast communication is not required for control purposes.
5) Since power reversal does not require voltage reversal, parallel connected multi-terminal operation is feasible.

A major drawback in VSC-HVDC links is the problem of clearing DC line faults as the freewheeling diodes in the VSC feed the fault from the AC system. The only solution is the use of DC or AC breakers. This limitation has prompted the application of VSC-HVDC to cable transmission in urban areas or when sea-crossing VSC-based HVDC links have been mainly applied for power transfer from offshore wind power plants and supply to off-shore platforms. ABB has commissioned several projects using two- or three-level VSC, using PWM based on selective harmonic elimination. However, the ratings did not exceed 350 MW. The use of PWM at modulation frequencies up to 2 kHz implies higher switching losses (2–3% per terminal) compared to 0.7% in thyristor valve-based LCC.

In 2010, Siemens introduced the modular multilevel converter (MMC) topology for VSC which resulted in reduction of losses to about 1%.

MMC Topology [26–28]

Instead of two or three levels, several levels of injected AC voltage by the VSC result in reducing the size of voltage steps and related voltage gradients. With a high number

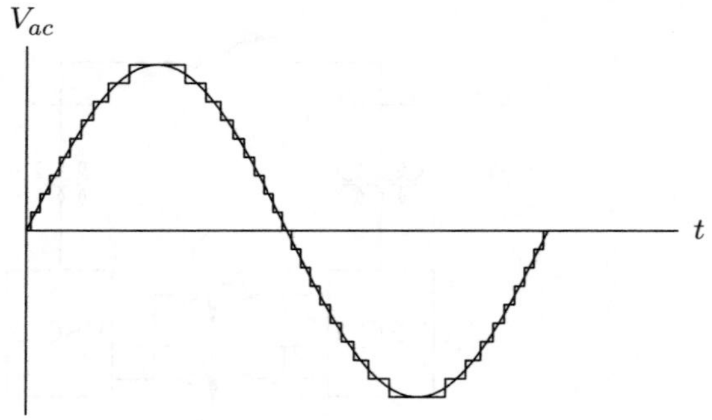

Figure 5.39 Waveform of injected AC voltage by MMC.

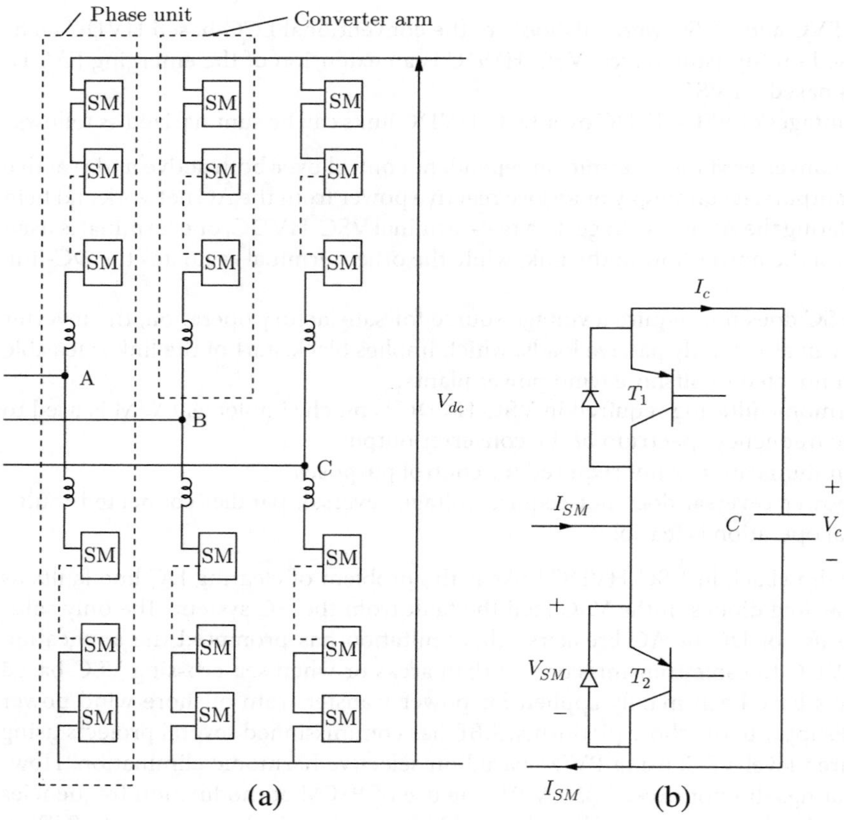

Figure 5.40 MMC representation (a) MMC topology (b) Submodule.

of levels, the switching frequency of individual IGBT devices is reduced. The basic idea in a MMC is to generate an approximate sine wave using several steps, as shown in Figure 5.39, which shows the waveform injected AC voltage by the MMC. An MMC consists of six converter arms with two each in one phase unit, as shown in Figure 5.40a. Each arm comprises a high number of submodules (also called power modules) connected in series with a reactor. Each submodule contains (i) an IGBT half bridge as switching element and (ii) a DC capacitor unit for energy storage. The submodule is shown in Figure 5.40b. The half bridge consists of two switches (T_1 and T_2) each consisting of an IGBT and a free wheeling diode connected in antiparallel. In normal operation, only one switch (T_1 or T_2) is on and the other is off. Depending on the current direction, the capacitor can charge or discharge.

The advantages of MMCs are listed below:

1) The converter design is flexible and has excellent scalability.
2) There is no need for AC filters.
3) Small voltage steps cause very little radiated or conducted high frequency noise.
4) Low switching frequencies of individual devices result in low switching losses.
5) Standard AC transformers can be used.
6) It is easy to replace existing components by state-of-the-art ones, since the switching characteristics of each individual submodule are independently determined. In contrast, in a two-level converter all the devices should have identical switching characteristics.

Since the introduction of MMC topology, the ratings of VSC-HVDC links have excluded 1000 MW.

Remarks

1) The concept of a cascaded multilevel VSC was proposed by Lai and Feng [29]. The first application of this concept was made by GEC Alsthom [30]. They termed it a chain-link converter made up of a series connection of full bridge (FB) modules in each phase.
2) It is also possible to apply FB modules in VSC-HVDC. Siemens apply half-bridge (HB) modules in contrast (see Figure 5.41). From HB MMC to FB MMC, the required

Figure 5.41 Representation of switched capacitors (a) Half-bridge module (b) Full-bridge module.

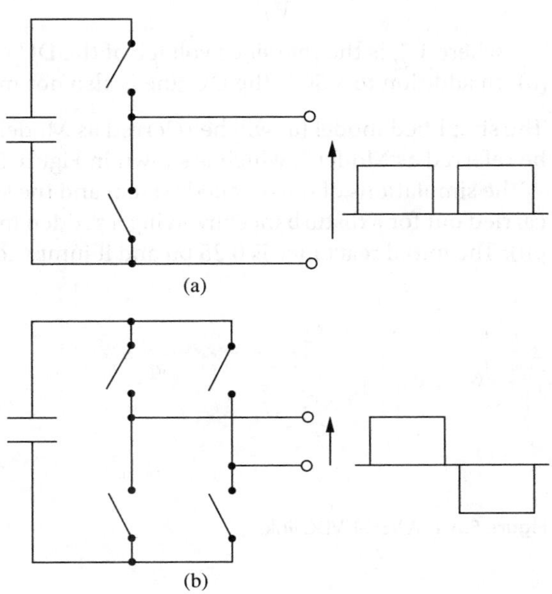

number of semiconductors is increased by a factor of approximately two. (It should be noted that for the two-level converter to HB module, the number of semiconductors is increased by a factor of two for the same rated DC voltage). However, the FB module can produce a bipolar DC voltage output (including a zero state) which implies that during a DC line fault, the fault current can be blocked if the capacitor voltage is sufficiently high.

3) The modeling of MMC HVDC converters for simulation of electromagnetic transients is described in reference [28].

5.A Case Study of a VSC-HVDC Link [31]

We investigate the performance of a VSC-HVDC link transmitting power from a generating station (modeled by a voltage source in series with an impedance) to a load centre represented by an infinite bus (see Figure 5.A.1). VSC1 operates as a rectifier and VSC2 as the inverter station. The DC voltage control can be applied at VSC1 or VSC2 while the other converter station operates on power control. Normally the rectifier station will control the power while the inverter regulates the DC voltage. Both stations can either control reactive power or regulate the AC bus voltage. Normally, both converter stations have to be modeled in detail along with the DC line (typically a cable). In this case study, the impact of a sudden increase in the AC line impedance (due to the tripping of a line) at the inverter side is investigated. The performance of detailed model is compared with the two simplified models as given below:

(a) While VSC2 with the DC line is represented in detail, VSC1 and the associated AC system is represented by a current source (I) defined by

$$I = \frac{P_{ref1} - K(V_{d2}^* - V_{d2})}{V_{d1}} \tag{5.A.1}$$

where V_{d2}^* is the reference voltage of the DC voltage controller at VSC2.

(b) In addition to VSC1, the DC line is also not modeled.

The simplified model (a) will be referred as Model 1, while the simplified model (b) will be referred as Model 2, which is shown in Figure 5.A.2.

The simulations of the detailed system and the simplified system (Models 1 and 2) are carried out for a disturbance involving a sudden increase in the line reactance ($\Delta X = 0.4$ pu). The initial reactance is 0.25 pu and it jumps to 0.65 pu at $t = 0.5$ s. The variations of

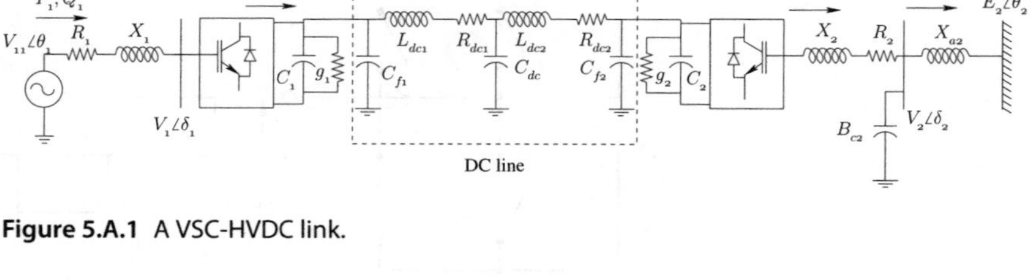

Figure 5.A.1 A VSC-HVDC link.

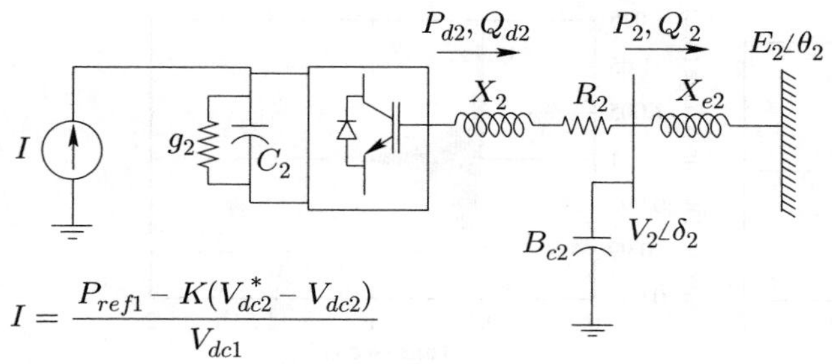

Figure 5.A.2 Model 2 of the VSC-HVDC link.

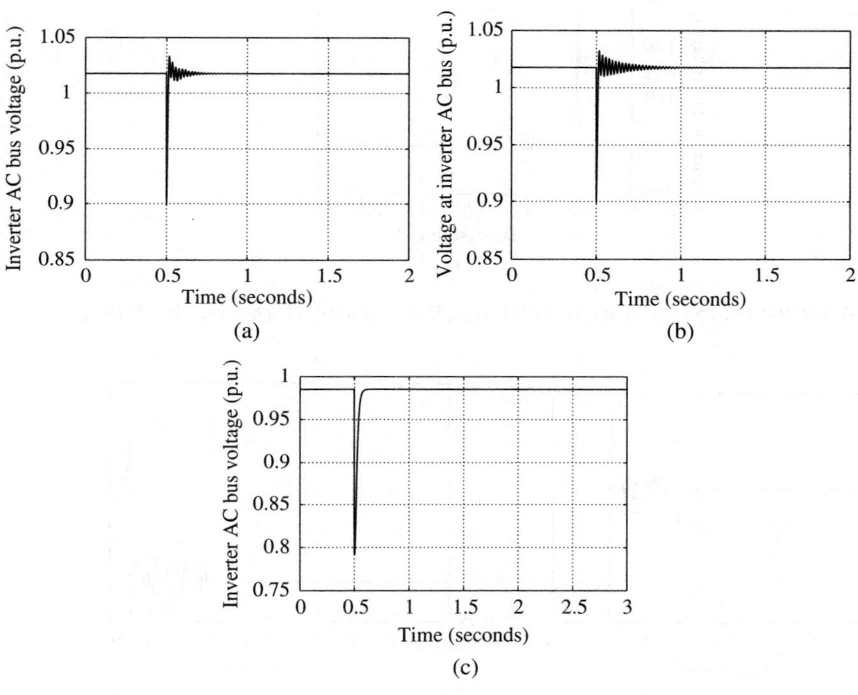

Figure 5.A.3 Variation of voltage at inverter AC bus prior to voltage collapse (a) Detailed model (b) Model 1 (c) Model 2.

the inverters AC and DC bus voltage magnitudes for the three cases (i) detailed system model, (ii) Model 1, and (iii) Model 2 are shown in Figures 5.A.3 and 5.A.4.

While Figure 5.A.3 shows the variations of the VSC2 (inverter) AC bus voltage for the three cases considered, Figure 5.A.4 shows the variations of the inverter DC bus voltage for the same three cases. It is interesting to note that in the study of disturbances on the inverter side it is feasible to simplify the model of the rectifier side (Model 1) where the VSC1 along with the associated AC system models are neglected. However, Model 2 is not as accurate as Model 1 although the trends are similar.

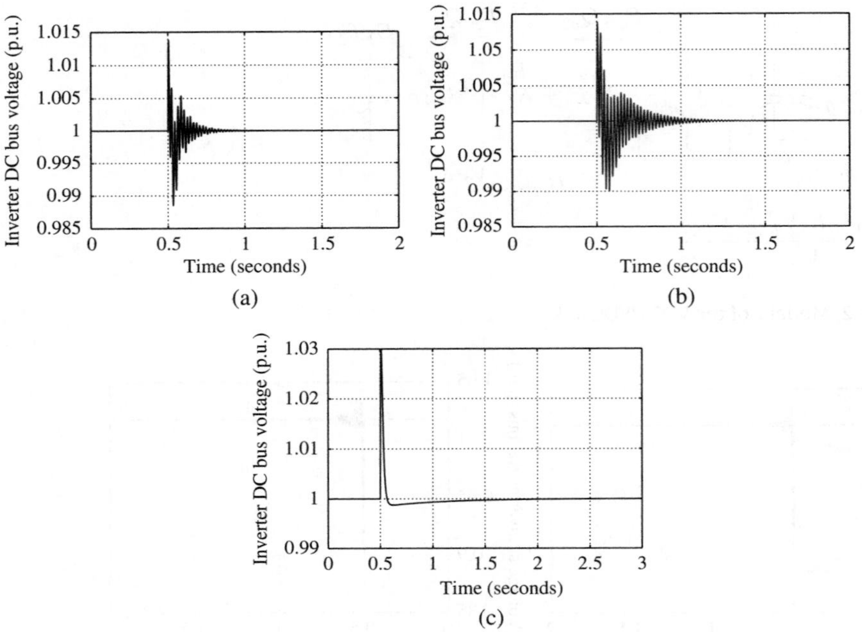

Figure 5.A.4 Variation of voltage at inverter DC bus (a) Detailed model (b) Model 1 (c) Model 2.

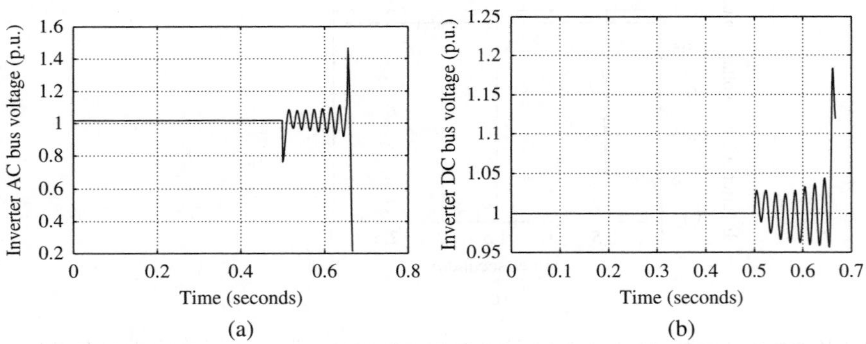

Figure 5.A.5 Variation of voltage at inverter bus (reduced SCR) (a) Inverter AC bus (b) Inverter DC bus.

Voltage Stability

If the apparent impedance of the AC system viewed from the inverter bus increases beyond a limit, there is a voltage collapse at the inverter bus. For a sudden increase in the line reactance by 0.51 pu, the variation of the inverter AC bus voltage and the DC bus voltage are shown in Figures 5.A.5a and 5.A.5b, respectively. Since there is little difference between the results of the detailed model and Model 1, only the results of Model 1 are shown here.

Remarks

1) The DC voltage increases as it is determined by the power injected by the rectifier station. The DC voltage starts falling as the DC line current falls to zero.
2) It is essential to add a term involving change in the inverter DC voltage (see (5.A.1)) to limit the rise in the DC bus voltage.
3) The critical SCR value which results in the voltage collapse at the inverter AC bus is 1.315. This is much smaller than the critical SCR with the LCC-HVDC link (1.644).

References

1 Vithayathil, J., Taylor, C., Klinger, M., and Mittelstadt, W. (1988) Case studies of conventional and novel methods of reactive power control on an ac transmission system, in *CIGRE, SC 38-02*, Paris.

2 Hingorani, N.G. (1993) Flexible AC Transmission. *IEEE Spectrum*, **30** (4), 40–45.

3 Hingorani, N.G. and Gyugyi, L. (2000) *Understanding FACTS*, IEEE Press, New York.

4 Mathur, R.M. and Varma, R.K. (2002) *Thyristor-based FACTS Controllers for Electrical Transmission Systems*, IEEE Press, New York.

5 Padiyar, K.R. (2016) *FACTS Controllers in Power Transmission and Distribution*, New Age International, New Delhi.

6 Dugan, R.C., McGranaghan, M.F., and Beaty, H.W. (1996) *Electrical Power Systems Quality*, McGraw-Hill, New York.

7 Hingorani, N.G. (1995) Introducing Custom Power. *IEEE Spectrum*, **32** (6), 41–48.

8 Miller, T.J.E. (1982) *Reactive Power Control in Electric Systems*, John Wiley & Sons, New York.

9 Christi, N., Hedin, R., Johnson, R. *et al.* (1991) Power system studies and modelling for the Kayenta 230 kV substation advanced series compensation, in *International Conference on AC and DC Power Transmission*, IET, London, pp. 33–37.

10 Cristl, N., Hedin, R., Krause, P.E., and McKenna, S.M. (1992) Advanced series compensation (ASC) with thyristor controlled impedance, in *CIGRE 14/37/38-05*, Paris.

11 Padiyar, K.R., Geetha, M.K., and Rao, K.U. (1996) A novel power flow controller for controlled series compensation, in *Sixth International Conference on AC and DC Power Transmission*, London, pp. 329–334.

12 Kulkarni, A.M. and Padiyar, K.R. (1999) Damping of power swings using series facts controllers. *International Journal of Electrical Power & Energy Systems*, **21** (7), 475–495.

13 Padiyar, K.R. and Rao, K.U. (1997) Discrete control of series compensation for stability improvement in power systems. *International Journal of Electrical Power & Energy Systems*, **19** (5), 311–319.

14 Ängquist, L., Ingeström, G., and Jönsson, H.Å. (1996) Dynamical performance of TCSC schemes, in *CIGRE Paper 14-302*.

15 Piwko, R.J., Wegner, C.A., Damsky, B.L. *et al.* (1994) The Slatt thyristor-controlled series capacitor project – Design, installation, commissioning and system testing, in *CIGRE Paper 14-104*, Paris.

16 Gole, A., Sood, V.K., and Mootoosamy, L. (1989) Validation and analysis of a grid control system using d-q-z transformation for static compensator systems, in *Canadian Conference on Electrical and Computer Engineering*, Montreal, Canada, pp. 745–758.

17 Padiyar, K.R. (2015) *HVDC Power Transmission Systems*, New Age International, New Delhi, 3rd edn.

18 Arrillaga, J. (1998) *High Voltage Direct Current Transmission*, IEE, London, 2nd edn.

19 Hingorani, N.G. (1996) High-voltage DC transmission: A power electronics workhorse. *IEEE Spectrum*, **33** (4), 63–72.

20 Willis, L. and Nilsson, S. (2007) HVDC transmission: Yesterday and today. *IEEE Power and Energy Magazine*, **5** (2), 22–31.

21 Flourentzou, N., Agelidis, V.G., and Demetriades, G.D. (2009) VSC-based HVDC power transmission systems: An overview. *IEEE Transactions on Power Electronics*, **24** (3), 592–602.

22 Kangiesser, K.W., Bowles, J.P., Ekstrom, A. *et al.* (1974) HVDC multi-terminal systems, in *CIGRE 14-08*, Paris.

23 Thallam, R.S. (1992) Review of the design and performance features of HVDC systems connected to low short circuit ratio AC systems. *IEEE Transactions on Power Delivery*, 7 (4), 2065–2073.

24 Szechtman, M., Pilotto, L.A.S., Ping, W.W., and Salgado, E. (1992) The behaviour of several HVDC links terminating in the same load area, in *CIGRE 14-201*, Paris.

25 Bui, L.X., Sood, V.K., and Laurin, S. (1991) Dynamic interactions between HVDC systems connected to ac buses in close proximity. *IEEE Transactions on Power Delivery*, **6** (1), 223–230.

26 Marquardt, R. and Lesnicar, A. (2004) New concept for high voltage-modular multilevel converter, in *IEEE Power Electronics Specialists Conference*, Aachen, pp. 1–5.

27 Friedrich, K. (2010) Modern HVDC PLUS application of VSC in Modular Multi-level Converter topology, in *IEEE International Symposium on Industrial Electronics (ISIE)*, Bari, Italy, pp. 3807–3810.

28 Gnanarathna, U.N., Gole, A.M., and Jayasinghe, R.P. (2011) Efficient modeling of modular multilevel HVDC converters (MMC) on electromagnetic transient simulation programs. *IEEE Transactions on Power Delivery*, **26** (1), 316–324.

29 Lai, J. and Peng, F.Z. (1996) Multilevel converters – a new breed of power converters. *IEEE Transactions on Industry Applications*, **32** (3), 509–517.

30 Ainsworth, J.D., Davies, M., Fitz, P.J. *et al.* (1998) Static var compensator (STATCOM) based on single-phase chain circuit converters. *IEE Proceedings – Generation, Transmission and Distribution*, **145** (4), 381–386.

31 Saichand, K. (2011) *Study of voltage stability in multi-infeed HVDC systems*. M.E. Project Report, Indian Institute of Science, Bangalore.

6

Damping of Power Swings

6.1 Introduction

The stability of the relative motion between the rotors of synchronous machines which are interconnected by AC transmission lines (a synchronous grid) is an important concern for power system engineers. This is also known as the angular stability problem. The relative motion is oscillatory when the system is subjected to small disturbances. The frequencies of these oscillations typically lie between 0.2 to 2 Hz. Usually these oscillations are stable, that is, they die out and the system settles down. Under certain operating conditions, the oscillations may become unstable, that is, they grow in amplitude instead of dying out. The growing oscillations may eventually become large and sustained, or may increase to such an extent that synchronism between the generators is irrevocably lost. In either case, this leads to large fluctuations in voltage, power flows, generator torques and speeds, which cannot be tolerated for a long duration; manual or automatic control/protective actions have to be initiated under these circumstances.

Real-life measurements of both stable and unstable oscillations, as seen in the frequency at various locations in the Indian grid, are shown in Figure 6.1. These were measured using a wide-area frequency measurement system [1]. The unstable oscillations in Figure 6.1b were initiated not by some large disturbance but by a gradual change in system loading. The unstable oscillations then grew into sustained oscillations of near-constant amplitude which lasted for several minutes before dying out when system loading changed further. Unstable oscillations were also observed during the well-known grid disturbance that occurred on August 10, 1996 in the Western Systems Coordinating Council (WSCC) system in North America [2]. In this case, the oscillations grew to large magnitudes and eventually caused the system to separate into islands.

This chapter investigates the behavior of these relative rotor angle oscillations, which are also known as swings. The damping of these oscillations is a complex function of the operating conditions and system parameters. A major cause of the instability of these oscillations is the action of automatic voltage regulators (AVRs). Since modern generators generally have large synchronous reactances, the field excitation has to be changed according to the loading. This is ensured by an AVR, which regulates the terminal voltage of the generator by controlling its field voltage. AVRs also boost the field voltage transiently during large disturbances like faults, which helps in maintaining synchronism between generators. To ensure the stability of the swings while retaining these benefits of AVRs, auxiliary feedback loops are added not only to the AVRs themselves, but also to the controllers of power electronics systems like high-voltage direct current

Dynamics and Control of Electric Transmission and Microgrids, First Edition.
K. R. Padiyar and Anil M. Kulkarni.

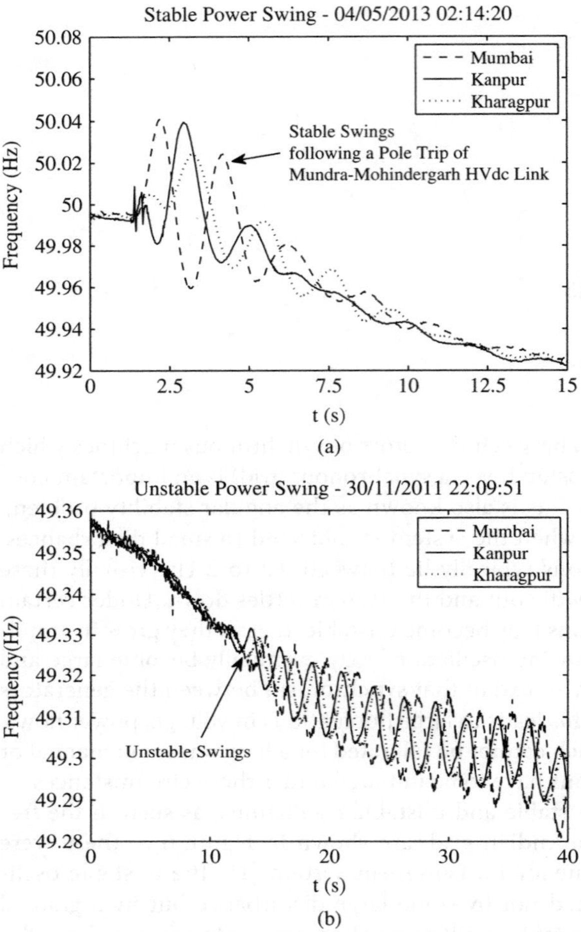

Figure 6.1 Stable and unstable swings observed in the Indian grid. (a) Stable swings (b) Unstable swings

(HVDC) and flexible AC transmission systems (FACTS) devices. These controllers are called power swing damping controllers (PSDCs). The principles of PSDC design are also presented in this chapter.

6.2 Origin of Power Swings

The low-frequency oscillations described in the introduction are observed in the *relative* angles and speeds of synchronous machines interconnected by AC transmission lines. Therefore, the simplest system which can exhibit this phenomenon consists of two synchronous machines connected by an AC transmission line. If one of the machines is assumed to be of a very large rating, then the speed and terminal voltage magnitude of that machine can be assumed to be constant (an infinite bus). Such a system is called a single-machine infinite bus (SMIB) system.

Consider the SMIB system shown in Figure 6.2. The generator is modeled as a voltage source with a constant magnitude E'_q behind its transient reactance x'_d. This model is

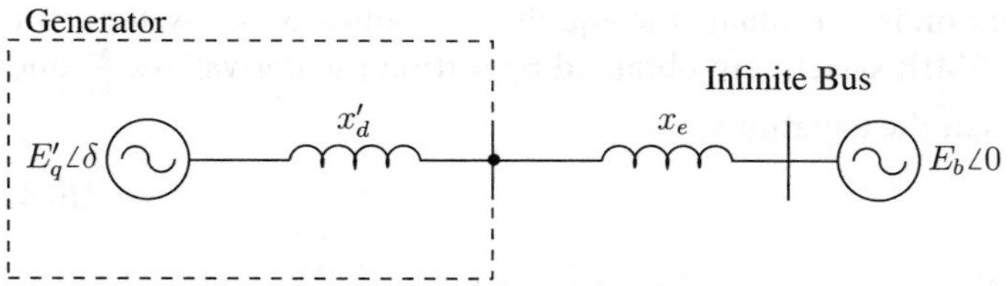

Figure 6.2 SMIB system.

known as the classical model or Model (0.0) (see Section 3.6). Without loss of generality, the infinite bus voltage phase angle is assumed to be zero. The angle δ shown in the figure is related to the electrical angular position of the machine rotor θ, by $\theta = \omega_b t + \delta$, where ω_b is the frequency of the infinite bus in rad/s. The total reactance of the external network (transformer and transmission line) is denoted by x_e. In this study, the speed of the machine is assumed to be near the base speed ω_B even during transients. Hence, the per-unit torque and per-unit power are approximately equal.

The differential equations describing this system are given by

$$\frac{d\delta}{dt} = \omega - \omega_b \tag{6.1}$$

$$M\frac{d(\omega - \omega_b)}{dt} = T_m - T_e - D(\omega - \omega_b) \approx P_m - P_e - D(\omega - \omega_b) \tag{6.2}$$

where

$\omega = \frac{d\theta}{dt}$: electrical speed in rad/s

ω_B: base electrical speed in rad/s

T_m: per-unit mechanical torque

T_e: per-unit electrical torque

P_m: per-unit mechanical power

P_e: per-unit electrical power

$M = \frac{2H}{\omega_B}$; H: inertia constant in MJ/MVA

$H = \frac{\frac{1}{2}J\omega_{Bm}^2}{\text{base VA}}$

J: moment of inertia of turbine and generator in kg-m^2

ω_{Bm}: base mechanical speed in rad/s

D: damping coefficient in per-unit torque/(rad/s)

ω_b: infinite bus frequency in rad/s

The mechanical torque T_m is assumed to be constant. The term $D(\omega - \omega_b)$ is used to model frequency-dependent torque variations, which are not captured in T_m and T_e. The electrical network is assumed to be in quasi-sinusoidal steady state. Hence, the electrical torque T_e is given by the expression

$$T_e \approx P_e = \frac{E'_q E_b}{(x'_d + x_e)} \sin\delta \tag{6.3}$$

In general, numerical integration techniques are required to evaluate the response of this system following disturbances. Real-life observations show that power oscillations may be excited even by small disturbances. This indicates that their behavior may be studied using a *small-signal* model. A small-signal model is derived by assuming that the deviations around an equilibrium point are small. It is obtained from (6.1), (6.2), and (6.3) by the process of linearization, as given in Section 1.2.2. Since a small-signal model is linear, the response can be obtained analytically.

The first step of linearization is to evaluate the equilibrium points of the system. The equilibrium points of the SMIB system are obtained by setting the derivatives $\frac{d\delta}{dt}$ and $\frac{d(\omega-\omega_b)}{dt}$ to zero. This results in the equations

$$\omega_o = \omega_b \tag{6.4}$$

$$\delta_o = \sin^{-1}\left(\frac{T_{eo} \times (x_d' + x_e)}{E_q' E_b}\right) = \sin^{-1}\left(\frac{T_m \times (x_d' + x_e)}{E_q' E_b}\right) \tag{6.5}$$

Note that the subscript o denotes the equilibrium value of the variable.

From the torque-angle curve shown in Figure 6.3, it can be seen that there are two values of δ_o (δ_s and δ_u) corresponding to a specified value of T_m ($T_m < T_{max}$).

Let $\Delta\delta = \delta - \delta_o$ and $\Delta\omega = \omega - \omega_o$ be the deviations from the equilibrium point (δ_o, ω_o). For small deviations the differential equations can be written as

$$\frac{d}{dt}\begin{bmatrix}\Delta\delta \\ \Delta\omega\end{bmatrix} = \begin{bmatrix} 0 & 1 \\ -\frac{K}{M} & -\frac{D}{M}\end{bmatrix}\begin{bmatrix}\Delta\delta \\ \Delta\omega\end{bmatrix} = A\begin{bmatrix}\Delta\delta \\ \Delta\omega\end{bmatrix} \tag{6.6}$$

where

$$K = \left.\frac{\partial T_e}{\partial \delta}\right|_{\delta=\delta_o} = \frac{E_q' E_b}{(x_d' + x_e)}\cos\delta_o \tag{6.7}$$

Equation (6.6) represents the small-signal model of this system.

The small-signal stability of the system can be determined from the eigenvalues of A, which are given by:

$$\lambda = -\frac{D}{2M} \pm \sqrt{\frac{D^2}{4M^2} - \frac{K}{M}} \tag{6.8}$$

If K and D are positive then both the eigenvalues have negative real parts, and therefore the system is stable (see Section 1.2.2).

D is generally quite small. Therefore if K is negative, then one of the eigenvalues is real and positive. Note that $K < 0$ for $\delta_o = \delta_u$. Hence, the system is unstable at this equilibrium point.

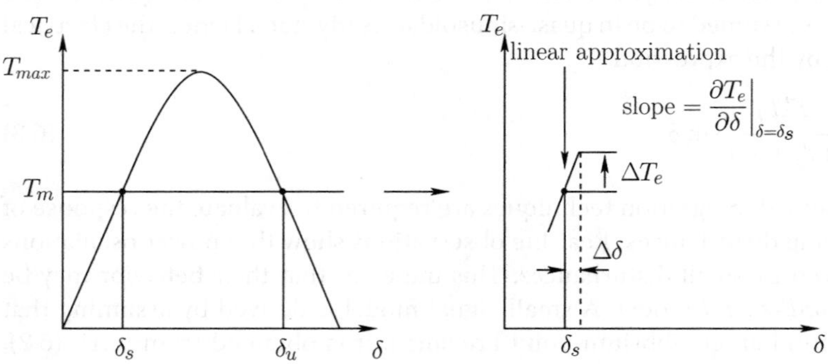

Figure 6.3 Linearization around an equilibrium point.

On the other hand, if $\delta_o = \delta_s$, then $K > 0$. Since D is small, the eigenvalues are complex and are given by

$$\lambda = -\frac{D}{2M} \pm j\sqrt{\frac{K}{M} - \frac{D^2}{4M^2}} \tag{6.9}$$

The system is stable at $\delta_o = \delta_s$ since D is generally greater than zero. The response to small disturbances is a decaying oscillation since the eigenvalues are a complex conjugate pair with a negative real part.

In this section, we have considered a simplified representation of a synchronous machine. The electrical torque T_e is dependent only on the state δ, and the damping effects are captured in a gross fashion by the frequency-dependent term $D(\omega - \omega_b)$ in the swing equation. When detailed models of a synchronous machine and its excitation system are considered, the electrical torque is a complex function of the states of the system. This introduces damping effects that are of electrical origin.

6.3 SMIB Model with Field Flux Dynamics and AVR

The classical model of a synchronous machine assumes that the field flux ψ_f remains constant during transient conditions. In reality, the field flux does not change instantaneously when a disturbance is applied, but it does change thereafter. The variation of flux affects the electrical torque and consequently the frequency and damping of the power swings. The field flux dynamics is also affected by the voltage applied across the field winding. The field voltage is obtained from an excitation system, which is controlled by an AVR. The function of an AVR is to regulate the terminal voltage.

A simplified model of a static excitation system with an AVR is shown in Figure 6.4. V is the generator terminal voltage magnitude. E_{fd} is the per-unit field voltage. Note that E_{fd} is equal to the stator voltage in per-unit quantities if the generator is open-circuited and is running at the rated speed (see the derivation of synchronous machine models in Chapter 3).

The aim of this section is to analyze the effect of field flux dynamics and the excitation system on power swings. For simplicity, we make the following approximations:

1) Model (1.0) of the synchronous machine is used, that is, the field flux dynamics is modeled but the effects of the damper windings are neglected. This model can be obtained from Model (1.1) of the synchronous machine, which is described in Section 3.6, by setting $x'_q = x_q$.

Figure 6.4 Simplified model of a static excitation system with an AVR.

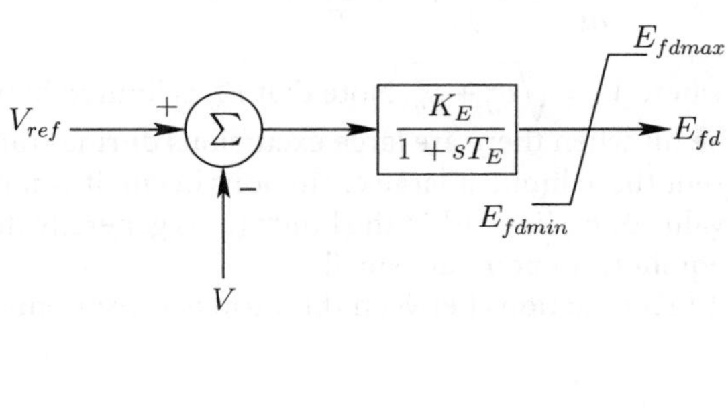

2) The differential equations of the stator flux are approximated by algebraic equations by neglecting the transformer electromotive force (emf) terms, $\frac{1}{\omega_B}\frac{d\psi_d}{dt}$ and $\frac{1}{\omega_B}\frac{d\psi_q}{dt}$. Similarly, the differential equations of the current flowing through the external reactance (see Section 2.4) are approximated by neglecting the time derivative terms of currents, $\frac{1}{\omega_B}\frac{di_d}{dt}$ and $\frac{1}{\omega_B}\frac{di_q}{dt}$. These approximations remove the fast transient components from the response and are therefore valid for the study of the relatively slow power swings.
3) The deviations of the rotor speed ω from the base speed ω_B are assumed to be small.
4) The resistances of the stator winding and the network are neglected.

The following algebraic equations are obtained based on these assumptions:

$$E'_q + x'_d i_d - v_q = 0 \tag{6.10}$$

$$-x_q i_q - v_d = 0 \tag{6.11}$$

$$x_e i_d + (v_q - E_b \cos\delta) = 0 \tag{6.12}$$

$$-x_e i_q + (v_d + E_b \sin\delta) = 0 \tag{6.13}$$

Note that $-E_b \sin\delta$ and $E_b \cos\delta$ are the d and q components of the infinite bus voltage, respectively. This is because we have taken the phase angle of the infinite bus to be zero, and $\theta = \omega_b t + \delta$. The last two equations may also be written compactly as a single complex equation: $(v_q + jv_d) - E_b e^{-j\delta} - jx_e(i_q + ji_d) = 0$.
The differential equation describing the field flux dynamics is given by:

$$T'_{do}\frac{dE'_q}{dt} = E_{fd} - E'_q + (x_d - x'_d)i_d \tag{6.14}$$

Note that E'_q is proportional to the field flux and $T'_{do} = \frac{x_f}{R_f\omega_B}$.
As before, the differential equations describing the dynamics of δ and ω are given by:

$$\frac{d\delta}{dt} = \omega - \omega_b \tag{6.15}$$

$$M\frac{d(\omega - \omega_b)}{dt} = T_m - T_e = T_m - [E'_q i_q + i_d i_q(x'_d - x_q)] \tag{6.16}$$

Note that $T_e = E'_q i_q + i_d i_q(x'_d - x_q)$. The term $D(\omega - \omega_b)$ is neglected here as the aim of this section is to highlight the damping caused by the electrical system.

The equations describing the dynamics of the excitation system can be written as follows (see Figure 6.4):

$$\frac{dE_{fd}}{dt} = -\frac{1}{T_E}E_{fd} + \frac{K_E}{T_E}(V_{ref} - V) \tag{6.17}$$

where $V = \sqrt{v_d^2 + v_q^2}$. Note that E_{fd} is limited between E_{fdmax} and E_{fdmin}. The limits may be hit when there are large excursions during transients. While it is necessary to represent these limits if large excursions occur, it is not necessary to do so if the equilibrium value of E_{fd} lies within the limits (as is generally the case) and the excursions around the equilibrium point are small.
The interactions between the various subsystems are depicted pictorially in Figure 6.5.

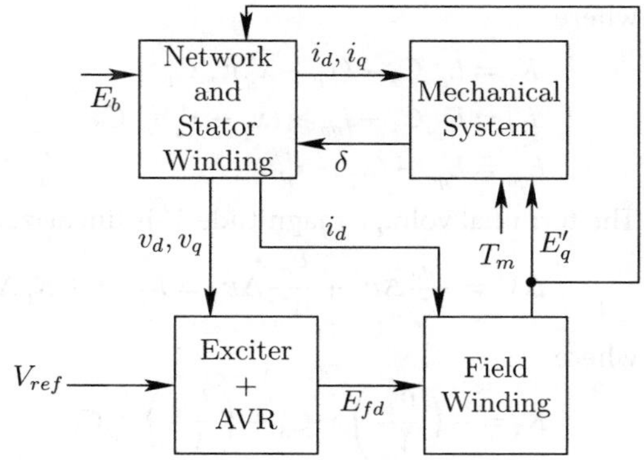

Figure 6.5 Block diagram of the system, including the excitation system and AVR.

6.3.1 Small-Signal Model and Eigenvalue Analysis

The small-signal model of a system is obtained by linearizing the nonlinear differential and algebraic equations of the system at the equilibrium point of interest. This involves evaluating the partial derivatives of the functions with respect to the variables at that equilibrium point, as given in Section 1.2.2. The model derived earlier in this section is a combination of differential and algebraic equations. The algebraic variables (Δv_d, Δv_q, Δi_d, and Δi_q) can be expressed in terms of the state variables ($\Delta\delta$, $\Delta\omega$, $\Delta E'_q$, and ΔE_{fd}) as described below.

By eliminating Δv_d and Δv_q using the equations (6.10) to (6.13), we obtain:

$$i_d = \frac{E_b \cos\delta - E'_q}{(x'_d + x_e)}, \quad i_q = \frac{E_b \sin\delta}{(x_q + x_e)} \tag{6.18}$$

By linearizing these equations, we obtain the equations

$$\Delta i_d = C_1 \Delta\delta + C_2 \Delta E'_q \tag{6.19}$$

$$\Delta i_q = C_3 \Delta\delta + C_4 \Delta E'_q \tag{6.20}$$

where

$$C_1 = -\frac{1}{(x'_d + x_e)} E_b \sin\delta_o, \quad C_2 = -\frac{1}{(x'_d + x_e)}$$

$$C_3 = \frac{1}{(x_q + x_e)} E_b \cos\delta_o, \quad C_4 = 0$$

As before, the subscript o affixed to a variable indicates its equilibrium value. Similarly, we obtain the following equations for Δv_d and Δv_q:

$$\Delta v_q = x'_d C_1 \Delta\delta + (1 + x'_d C_2) \Delta E'_q \tag{6.21}$$

$$\Delta v_d = -x_q C_3 \Delta\delta - x_q C_4 \Delta E'_q \tag{6.22}$$

The torque expression can be linearized as follows:

$$\Delta T_e = [E'_{qo} - (x_q - x'_d) i_{do}] \Delta i_q + i_{qo} \Delta E'_q - (x_q - x'_d) i_{qo} \Delta i_d \tag{6.23}$$

Therefore

$$\Delta T_e = K_1 \Delta\delta + K_2 \Delta E'_q \tag{6.24}$$

where

$$K_1 = E_{qo}C_3 - (x_q - x'_d)i_{qo}C_1$$
$$K_2 = E_{qo}C_4 + i_{qo} - (x_q - x'_d)i_{qo}C_2$$
$$E_{qo} = E'_{qo} - (x_q - x'_d)i_{do}$$

The terminal voltage magnitude V is linearized as follows:

$$\Delta V = \frac{v_{do}}{V_o}\Delta v_d + \frac{v_{qo}}{V_o}\Delta v_q = K_5\Delta\delta + K_6\Delta E'_q \quad (6.25)$$

where

$$K_5 = -\left(\frac{v_{do}}{V_o}\right)x_qC_3 + \left(\frac{v_{qo}}{V_o}\right)x'_dC_1$$
$$K_6 = -\left(\frac{v_{do}}{V_o}\right)x_qC_4 + \left(\frac{v_{qo}}{V_o}\right)(1 + x'_dC_2)$$
$$V_o = \sqrt{v_{do}^2 + v_{qo}^2}$$

The linearized differential equations of the system are given by

$$\frac{d\Delta\delta}{dt} = \Delta\omega \quad (6.26)$$

$$\frac{d\Delta\omega}{dt} = \frac{\omega_B}{2H}[\Delta T_m - \Delta T_e] = \frac{1}{M}[\Delta T_m - \Delta T_e] \quad (6.27)$$

$$T'_{do}\frac{d\Delta E'_q}{dt} = \Delta E_{fd} - \Delta E'_q + (x_d - x'_d)(C_1\Delta\delta + C_2\Delta E'_q) \quad (6.28)$$

$$\frac{d\Delta E_{fd}}{dt} = -\frac{1}{T_E}\Delta E_{fd} + \frac{K_E}{T_E}(\Delta V_{ref} - \Delta V) \quad (6.29)$$

These may be compactly written as follows (with the assumption that $\Delta T_m = 0$):

$$\dot{x} = [A]x + [B]\Delta V_{ref} \quad (6.30)$$

where

$$x^t = [\Delta\delta \;\; \Delta\omega \;\; \Delta E'_q \;\; \Delta E_{fd}],$$

$$[A] = \begin{bmatrix} 0 & 1 & 0 & 0 \\ -\frac{K_1}{M} & 0 & -\frac{K_2}{M} & 0 \\ -\frac{K_4}{T'_{do}} & 0 & -\frac{1}{T'_{do}K_3} & \frac{1}{T'_{do}} \\ -\frac{K_EK_5}{T_E} & 0 & -\frac{K_EK_6}{T_E} & -\frac{1}{T_E} \end{bmatrix}, \quad [B]^t = \begin{bmatrix} 0 & 0 & 0 & \frac{K_E}{T_E} \end{bmatrix}$$

$$K_3 = \frac{1}{1 - (x_d - x'_d)C_2} \quad (6.31)$$

$$K_4 = -(x_d - x'_d)C_1 \quad (6.32)$$

The matrix A is dependent on the operating conditions. K_1 to K_6 are known as the Heffron–Phillip constants [3]. The effect of the operating conditions and the various parameters (such as K_E and T_E) on the stability of the system can be determined from the eigenvalues of A.

Example 6.1 ***Eigenvalues of a Single-Machine System*** Consider the SMIB system shown in Figure 6.6.

The data for the system is given below.

Generator

The base frequency ω_B is $2\pi \times 60$ rad/s.

$x_d = 1.6$ pu, $x'_d = 0.32$ pu, $x_q = 1.55$ pu,

$T'_{do} = 6.0$ s, $H = 5$ MJ/MVA, $D = 0$. Generator resistance is assumed to be negligible.

Exciter and AVR

$K_E = 250.0$ pu/pu, $T_E = 0.025$ s, $E_{fdmax} = 6.0$ pu, and $E_{fdmin} = -6.0$ pu (refer to Figure 6.4).

Network

$x_e = 0.2$ pu ($x_T = 0.1$ pu and $x_{line} = 0.1$ pu)

The generator and network reactances are expressed in per unit on a common base.

The infinite bus voltage is $1.0\angle 0$ and $\omega_b = 2\pi \times 60$ rad/s.

1) Compute the eigenvalues of this system for the following cases:
   (a) The generator is represented by the classical model.
   (b) The field flux dynamics is considered but the AVR is not included. This implies that E_{fd} is constant, that is, $\Delta E_{fd} = 0$.
   (c) The AVR and field flux dynamics are considered.
2) Comment on the stability of the system.
3) Simulate the system response to a small pulse in P_m by numerically integrating the nonlinear differential-algebraic equations.

Two operating conditions are to be considered: $P_m = 0.5$ pu and $P_m = 1.0$ pu. The quiescent terminal voltage magnitude in both cases is $V_o = 1.0$ pu.

Solution

The equilibrium values of variables for the two operating conditions are given in Table 6.1. At the equilibrium points, $\frac{d\delta}{dt} = 0$. Therefore, $\omega_o = \omega_b$. Note that with increased loading, E_{fd} has to be larger to maintain the terminal voltage at 1.0 pu. The eigenvalues of the system for the two operating points are shown in Table 6.2.

Remarks

1) The classical model highlights the swing mode, which is undamped since $D = 0$.
2) The participation factor [4] of the ith state in a mode is defined as the product of the ith component of the right eigenvector and the ith component of the normalized left eigenvector corresponding to the eigenvalue associated with that mode. The participation factor of a state is a measure of the sensitivity of the eigenvalue to local feedback around that state [4]. The participation factors corresponding to different

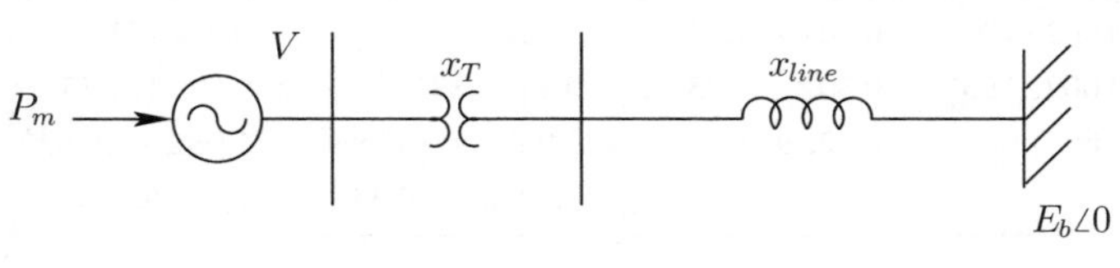

Figure 6.6 Single-machine infinite bus system.

Table 6.1 Equilibrium values (in pu) for $P_m = 0.5$ pu and $P_m = 1.0$ pu, $V_o = 1.0$ pu.

P_m	T_{mo}	P_{eo}	T_{eo}	δ_o	E'_{qo}	E_{fdo}	V_{refo}	i_{do}	i_{qo}
0.5	0.5	0.5	0.5	42.46°	0.904	1.312	1.005	−0.319	0.386
1.0	1.0	1.0	1.0	64.81°	0.874	1.977	1.008	−0.866	0.517

Table 6.2 Eigenvalues of the SMIB system.

$P_m = 0.5$ pu			
Classical model	**Model (1.0) (no AVR)**	**Model (1.0) + AVR**	**Comments**
$0.0 \pm j8.46$	$-0.15 \pm j6.61$	$-0.35 \pm j5.93$	Swing mode
–	-0.27	$-19.94 \pm j10.97$	Field flux/exciter mode
$P_m = 1.0$ pu			
Classical model	**Model (1.0) (no AVR)**	**Model (1.0) + AVR**	**Comments**
$0.0 \pm j8.29$	$-0.21 \pm j7.71$	$\mathbf{0.23 \pm j7.93}$	Swing mode
–	-0.17	$-20.52 \pm j0.94$	Field flux/exciter mode

eigenvalues are shown in Table 6.3 for the operating condition $P_m = 0.5$ pu. The swing mode (eigenvalue shown in bold) is identified by the fact that the rotor angle and rotor speed states have higher participation in it relative to the other states.

3) When the field flux dynamics is considered without an AVR, the swing mode has positive damping. A real eigenvalue is introduced due to the field flux dynamics. When the field voltage is constant, the system is stable for both loading conditions that have been considered here, as all the eigenvalues have negative real parts. It should be noted that the real eigenvalue moves towards the origin with increased loading. We shall see in the next section that this can be correlated to a reduction in the synchronizing torque with increased loading.
4) With the AVR considered, the system is stable for $P_m = 0.5$ pu. For $P_m = 1.0$ pu, the presence of AVR destabilizes the swing mode. The change in eigenvalues with the operating point is not surprising as the linearized model is a function of the equilibrium values of the states.

Table 6.3 Participation factors of states in the various modes ($P_m = 0.5$ pu).

	Classical model	Model (1.0) (no AVR)		Model (1.0) + AVR	
	$\mathbf{j8.46}$	-0.27	$\mathbf{-0.15+j6.61}$	$\mathbf{-0.35+j5.93}$	$-19.94+j10.97$
$\Delta\delta$	$0.5\angle 0°$	$0.002\angle 180°$	$0.50\angle -1.33°$	$0.52\angle 2.83°$	$0.02\angle 171.75°$
$\Delta\omega$	$0.5\angle 0°$	$0.002\angle 180°$	$0.50\angle -1.33°$	$0.52\angle 2.83°$	$0.02\angle 171.75°$
$\Delta E'_q$	–	$1.004\angle 0$	$0.02\angle 94.64°$	$0.063\angle -132.85°$	$1.08\angle -59.87°$
ΔE_{fd}	–	–	–	$0.009\angle -40.35°$	$1.05\angle 62.03°$

Figure 6.7 Response of δ for a pulse change in P_m. (a) With $P_m = 0.5$ pu, $V = 1.0$ pu (b) With $P_m = 1.0$ pu, $V = 1.0$ pu.

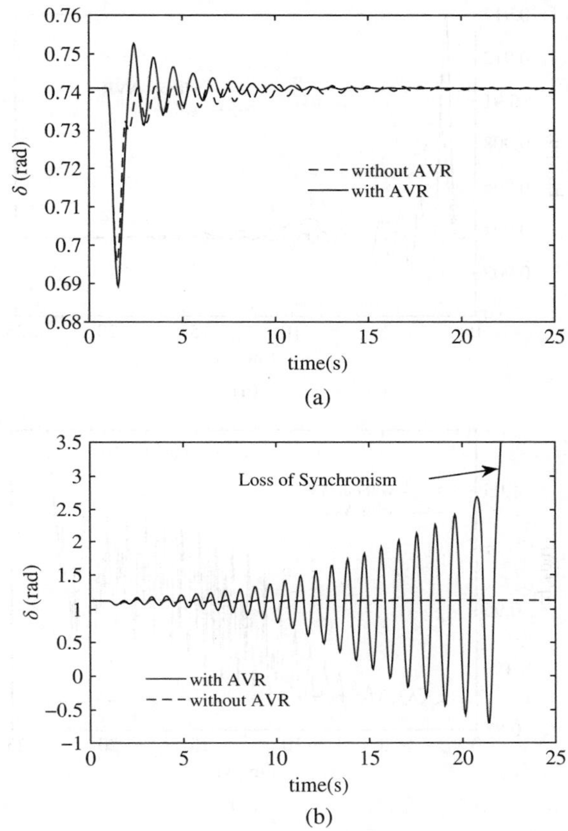

5) The simulated response of δ, E'_q, and E_{fd} for a 1.0 s pulse in P_m of −5% is shown in Figures 6.7–6.9. The simulations are carried out by numerically integrating the nonlinear system equations using Runge–Kutta fourth-order explicit method with a time step of 5 ms. The system is stable for $P_m = 0.5$ pu, but is unstable for $P_m = 1.0$ pu as predicted by eigenvalue analysis.
6) In the unstable case, initially there are growing oscillations. After a point, the rotor angle starts increasing monotonically and the generator irrevocably loses synchronism with the infinite bus. It should be noted that this qualitative change of behavior (from a growing oscillation to a monotonic growth) occurs because the system differential equations are nonlinear. This behavioral change cannot be predicted by the small-signal model. A detailed analysis of the behavior under large deviations from the equilibrium will be carried out separately in Chapter 7.

6.4 Damping and Synchronizing Torque Analysis

Power swings are primarily associated with the rotor angle and speed of a synchronous machine, that is, the participation of δ and ω in the swing mode is generally much higher than that of the other states (see Table 6.3). With detailed models, swing mode damping

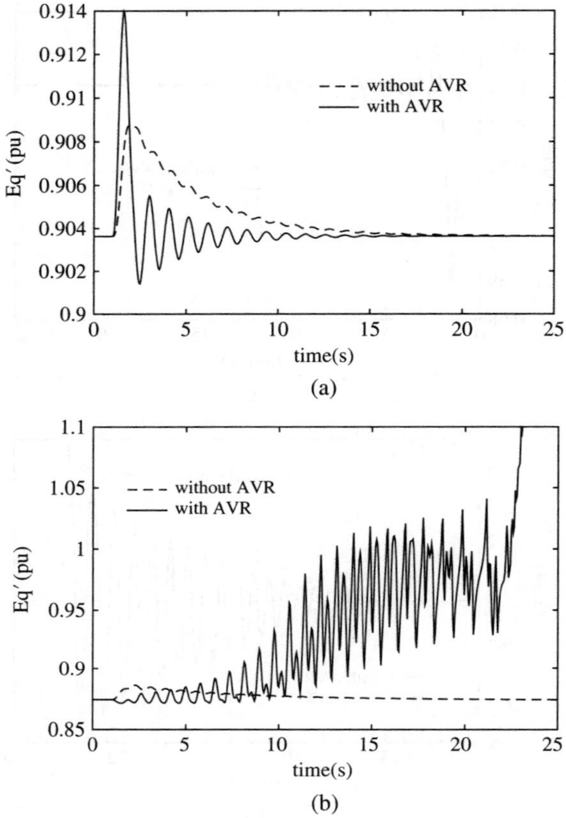

Figure 6.8 Response of E_q' for a pulse change in P_m. (a) With $P_m = 0.5$ pu, $V = 1.0$ pu (b) With $P_m = 1.0$ pu, $V = 1.0$ pu.

and frequency are altered due to the coupling of the rotor mechanical dynamics with dynamics of other states (like the field flux). Although it is possible to analyse the system by evaluating the eigenvalues and eigenvectors of the matrix $[A]$, as shown in the previous section, an *approximate* method of analysis known as damping and synchronizing torque analysis is often used to understand the effect of the electrical system on the swing mode.

Damping and synchronizing torque analysis is based on an approximate second-order model of the system, as given below:

$$\frac{d\Delta\delta}{dt} = \Delta\omega \tag{6.33}$$

$$\frac{2H}{\omega_B}\frac{d(\Delta\omega)}{dt} = -\Delta T_e \approx -T_{De}(\Omega)\Delta\omega - T_{Se}(\Omega)\Delta\delta \tag{6.34}$$

T_{De} and T_{Se} are known as the damping and synchronizing torque coefficients, respectively. T_{De} and T_{Se} are obtained at the swing frequency Ω, from the frequency response of the rest of the system, $\frac{\Delta T_e(j\Omega)}{\Delta\delta(j\Omega)}$, as shown in Figure 6.10. T_{De} and T_{Se} are obtained

Figure 6.9 Response of E_{fd} for a pulse change in P_m. (a) With $P_m = 0.5$ pu, $V = 1.0$ pu (b) With $P_m = 1.0$ pu, $V = 1.0$ pu.

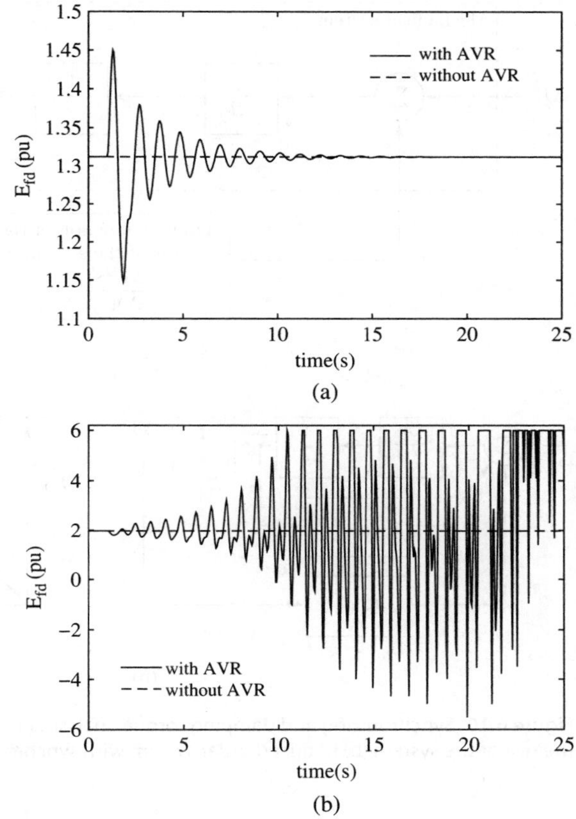

as follows:

$$T_{De}(\Omega) = \Re\left(\frac{\Delta T_e(j\Omega)}{\Delta\omega(j\Omega)}\right) = \frac{1}{\Omega}\Im\left(\frac{\Delta T_e(j\Omega)}{\Delta\delta(j\Omega)}\right)$$
$$T_{Se}(\Omega) = \Re\left(\frac{\Delta T_e(j\Omega)}{\Delta\delta(j\Omega)}\right)$$

where $\Re$ and $\Im$ denote the real and imaginary parts, respectively. Note that both T_{De} and T_{Se} are real numbers.

Remarks

1) The characteristic equation of the system described by (6.34) is given by $\frac{2H}{\omega_B}s^2 + T_{De}(\Omega)s + T_{Se}(\Omega) = 0$. The roots of this equation approximate the eigenvalues corresponding to the swing mode.
2) Both T_{De} and T_{Se} should be positive to ensure the stability of the swing mode, as the roots have negative real parts only under this condition. If $T_{De}(\Omega) < 0$ and $T_{Se}(\Omega) > 0$ then the system will exhibit unstable oscillations. A larger positive value of $T_{De}(\Omega)$ indicates better damping of oscillations. If $T_{Se}(\Omega) < 0$ then one root of the characteristic equation will be real and positive, indicating monotonic instability.

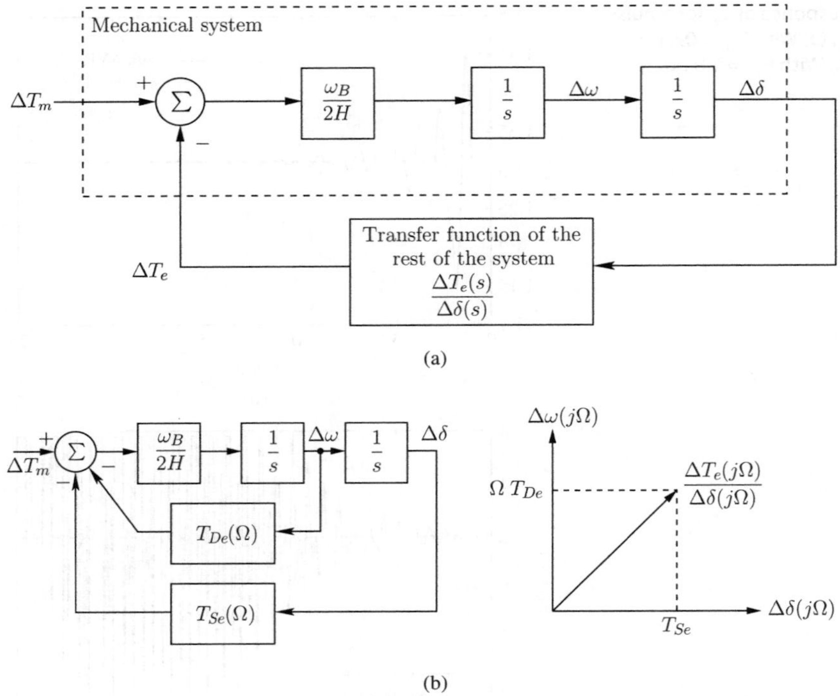

Figure 6.10 Synchronizing and damping torque analysis. (a) Block diagram of mechanical system with the rest of the system (b) Reduced order system with synchronizing and damping torque coefficients.

3) The frequency of the swing mode Ω may not be known *a priori*, and depends on the parameters of the rest of the system. Hence synchronizing and damping torque coefficients should be evaluated over a range of frequencies, say 0–3 Hz.
4) The advantage of this approximate analysis is that it encapsulates the effect of the rest of the system in two frequency-dependent coefficients, T_{De} and T_{Se}. The sensitivity of swing mode damping to the parameters of the rest of the system can thus be determined conveniently.

For illustration, consider the block diagram of the system shown in Figure 6.11. This block diagram is obtained from the linearized model given by (6.30). Note that $\Delta T_e = \Delta T_{e1} + \Delta T_{e_2} = K_1\Delta\delta + \Delta T_{e_2}$. If K_E is large and T_E is very small, as in a static exciter with a high gain AVR, then

$$\Delta T_{e2}(s) \approx -\frac{K_2K_5K_E}{(T'_{do}s + K_6K_E)}\Delta\delta(s) = -\left(\frac{\frac{K_2K_5}{K_6}}{s\frac{T'_{do}}{K_6K_E} + 1}\right)\Delta\delta(s) \tag{6.35}$$

The constants K_1, K_2, K_3, K_4, and K_6 are positive. This is because x_e is positive, δ_o is generally less than 90°, and i_{qo} is positive. Therefore, if K_5 is negative, then $T_{De}(\Omega) = \frac{1}{\Omega}\Im(\frac{\Delta T_{e2}(j\Omega)}{\Delta\delta(j\Omega)})$ is negative, implying oscillatory instability. In Example 6.1, the value of K_5 is 0.0577 for the operating condition $P_m = 0.5$ pu, while it is −0.0309 for $P_m =$

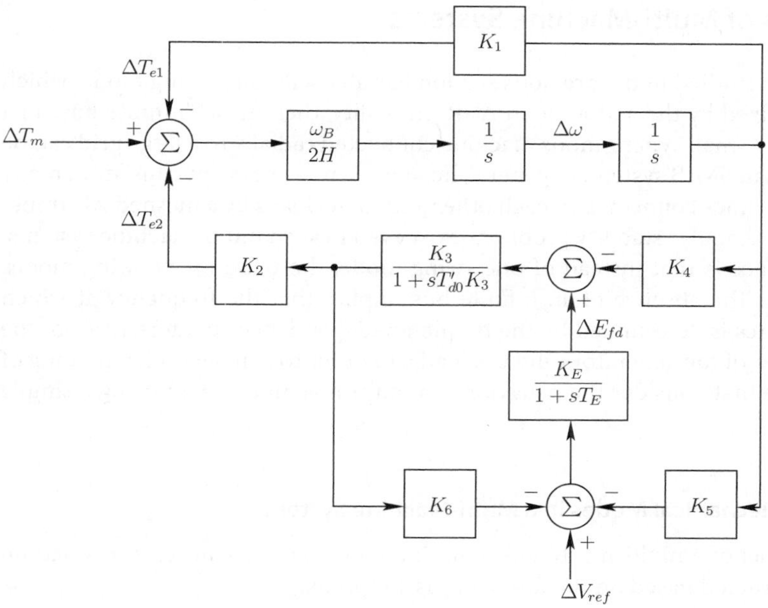

Figure 6.11 Transfer function block diagram, including the excitation system.

1.0 pu. Therefore at $P_m = 1.0$ pu, the AVR causes negative damping torque, leading to oscillatory instability.

5) The synchronizing and damping coefficients in steady state ($\Omega = 0$), when an AVR is not present (E_{fd} is constant), are given by:

$$T_{De} = K_2 K_3^2 K_4 \, T'_{do} \quad \text{and} \quad T_{Se} = K_1 - K_2 K_3 K_4 = \left. \frac{\partial P_{ess}}{\partial \delta} \right|_{\delta=\delta_o} \tag{6.36}$$

where P_{ess} denotes the steady-state power, which is given by:

$$P_{ess}(\delta) = \frac{E_b E_{fd}}{x_d + x_e} \sin\delta + \frac{E_b^2 (x_d - x_q)}{2\,(x_d + x_e)(x_q + x_e)} \sin 2\delta$$

As the power output (loading) is increased or the external reactance is increased, the steady-state value of T_{Se} reduces and may even become negative [5]. Even if T_{Se} is positive, a smaller value makes the system more susceptible to monotonic instability following disturbances that weaken the external network.

In Example 6.1, for the case without the AVR, the steady-state value of T_{Se} is 0.5391 pu/rad for $P_{eo} = 0.5$ pu and 0.4574 pu/rad for $P_{eo} = 1.0$ pu. The real eigenvalue moves from −0.27 to −0.17 when the loading is increased from 0.5 pu to 1.0 pu.

The steady-state value of T_{Se} drops to 0.0156 pu/rad if x_e is increased to 0.4 pu, with $P_{eo} = 1.0$ pu and $E_{fdo} = 2.0$ pu (without an AVR). The real eigenvalue in this case comes very close to the origin (−0.008). The other eigenvalues in this case are $-0.23 \pm j5.83$.

6.5 Analysis of Multi-Machine Systems

The SMIB system studied in the previous section has an oscillatory swing mode, which could be destabilized by the action of an AVR. In reality, there is no "infinite bus" in a system, although a small synchronous machine connected radially to a large grid can be approximated by an SMIB system. In general, however, a power system consists of many synchronous machines connected to each other and to the loads by a meshed AC transmission network. Angular stability problems exist even in such multi-machine systems. The main difference is that instead of one swing mode, there are many swing modes in such a system. The absence of an infinite bus implies that the frequency at which the system operates is determined by the frequency-dependence characteristics of the mechanical power of the generators and the loads. In order to gain an understanding of these aspects, we first consider the behavior of a multi-machine system using a simple model.

6.5.1 Electro-Mechanical Modes in a Multi-Machine System

A simplified model of a multi-machine system interconnected by an AC transmission network is constructed based on the following assumptions.

1) The synchronous generators are described by the classical model, that is, as voltage sources $E'_{qi}\angle\delta_i$ behind their transient reactances x'_{di} (for the ith generator). The rotor position of the ith generator is defined as $\theta_i = \omega_o t + \delta_i$ where ω_o is the equilibrium frequency. The bus to which $E'_{qi}\angle\delta_i$ is connected is called the internal bus of the generator.
2) Although the loads are actually distributed across the buses of the network, it is assumed that their effect can be approximated by connecting them to the internal buses of the nearest generators.
3) The AC transmission network is reduced to the internal buses of the generators, as illustrated in Figure 6.12 for a three-machine system.
4) Losses in the network are neglected.
5) Mechanical damping is neglected.

Given the gross nature of the assumptions, this model cannot be expected to capture the dynamic behavior accurately in practical studies. Nonetheless, it can be used to qualitatively understand the nature of the swing modes in a multi-machine system. The

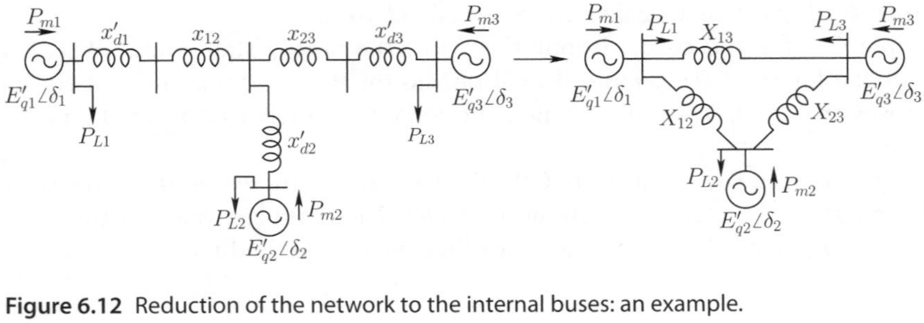

Figure 6.12 Reduction of the network to the internal buses: an example.

linearized system equations corresponding to this model can be written as

$$\begin{bmatrix} I & 0 \\ 0 & M \end{bmatrix} \begin{bmatrix} \dfrac{d\Delta\delta}{dt} \\ \dfrac{d\Delta\omega}{dt} \end{bmatrix} = \begin{bmatrix} 0 & I \\ -K & 0 \end{bmatrix} \begin{bmatrix} \Delta\delta \\ \Delta\omega \end{bmatrix} + \begin{bmatrix} 0 \\ \Delta P_m - \Delta P_L \end{bmatrix} \tag{6.37}$$

where M is a $m \times m$ diagonal matrix with $M_{jj} = \frac{2H_j}{\omega_B}$ (H_j is the inertia constant of the jth synchronous machine). m is the number of machines. ΔP_m and ΔP_L are the vectors of the deviations in the mechanical power input to the generators and the loads connected to the generator internal buses, respectively. 0 and I denote zero and identity matrices of appropriate dimensions. K is a $m \times m$ "stiffness" matrix, whose (i, j)th term is given by:

$$K_{ij} = \left.\frac{\partial P_{ei}}{\partial \delta_j}\right|_{\delta=\delta_o} = -\frac{E'_{qi}E'_{qj}}{X_{ij}} \cos(\delta_{io} - \delta_{jo}) \qquad \text{for} \quad i \neq j \tag{6.38}$$

$$K_{ii} = \left.\frac{\partial P_{ei}}{\partial \delta_i}\right|_{\delta=\delta_o} = \sum_{j=1, j\neq i}^{m} \frac{E'_{qi}E'_{qj}}{X_{ij}} \cos(\delta_{io} - \delta_{jo}) = -\sum_{j=1, j\neq i}^{m} K_{ij} \tag{6.39}$$

where P_{ei} is the power output of the ith machine and δ_i is the rotor angle of the ith machine referred to a reference frame rotating at a fixed speed ω_o. X_{ij} is the equivalent reactance of the branch connecting the generator buses i and j. E'_{qi} and E'_{qj} are the generator internal voltage magnitudes, which are assumed to be constant (classical model). K_{ij} is evaluated at the equilibrium point; the subscript o denotes the equilibrium value of the corresponding variable. The time-response of this system can be obtained from the eigenvalues and eigenvectors of the state matrix $A = \begin{bmatrix} I & 0 \\ 0 & M \end{bmatrix}^{-1} \begin{bmatrix} 0 & I \\ -K & 0 \end{bmatrix} = \begin{bmatrix} 0 & I \\ -M^{-1}K & 0 \end{bmatrix}$ and the initial conditions of the states (see Section 1.2.2).

Properties of the Eigenvalues of *A*

The nature of the eigenvalues of A can be inferred from the properties of the related matrices $A_r = -M^{-1}K$ and $A_s = -M^{-\frac{1}{2}}KM^{-\frac{1}{2}}$.

1) The matrix K is symmetric because $K_{ij} = K_{ji}$, while the matrix M is non-singular, diagonal, and positive definite.
2) Since M is diagonal and K is symmetric, A_s is also symmetric. Hence its eigenvalues are real.
3) A_r is related to A_s by a similarity transformation, that is, $A_r = M^{-\frac{1}{2}}A_sM^{\frac{1}{2}}$. Hence the eigenvalues of both matrices are equal. It follows that the eigenvalues of A_r are also real.
4) K is singular because $\sum_{i=1}^{m} K_{ij} = 0$ for all j. Therefore A_r is singular (at least one eigenvalue is zero). The singularity of K is not surprising since power flows are functions of phase angular *differences* only.
5) For normal operating conditions, the magnitude of angular differences $|\delta_i - \delta_j|$ are lesser than 90°. Therefore $K_{ii} > 0$ for all i. In addition, since $\sum_{i=1}^{m} K_{ij} = 0$, it follows from Gerschgorin theorem (see Appendix 6.A) that all eigenvalues of K will be non-negative. Therefore K is a positive semi-definite matrix. If the system does

not have disconnected parts (islands), then only one eigenvalue of K is zero (see Appendix 6.A).

6) From these properties of K, and the fact that M is positive definite, we can infer that A_s is negative semi-definite with one zero eigenvalue. In other words, A_s has one zero eigenvalue and the rest of the $(m-1)$ eigenvalues of A_s are negative and real. The same holds true for A_r (see point 3).
7) The eigenvalues of A are equal to the square roots of the eigenvalues of A_r. This is because the characteristic equation of A can be shown to be $\det(s^2I - A_r) = 0$, while the characteristic equation of A_r is $\det(sI - A_r) = 0$. Therefore, if the eigenvalues of A_r are $-\mu_i$, then the eigenvalues of A are $\pm j\sqrt{\mu_i}$.

It follows that the matrix A has $(m-1)$ conjugate pairs of purely imaginary eigenvalues $(\pm j\sqrt{\mu_i} = \pm j\Omega_i, i = 1 \cdots m-1)$ and one pair of zero eigenvalues.

Properties of the Eigenvectors of A

1) Since A_s is symmetric, it has m linearly independent eigenvectors. Consequently A_r also has m linearly independent eigenvectors.
2) Let the right eigenvector corresponding to the ith eigenvalue of A_r be denoted by v_i. It satisfies the relationship $A_r v_i = -\mu_i v_i$.
3) The eigenvectors of A_s can be shown to be $M^{\frac{1}{2}} v_i$. The eigenvectors corresponding to the distinct eigenvalues of A_s are orthogonal due to the symmetry of A_s. Therefore $(v_i^t M^{\frac{1}{2}})(M^{\frac{1}{2}} v_j) = v_i^t M v_j = 0$ if $i \neq j$ and if the corresponding eigenvalues are distinct.
4) Since the eigenvalues of A_r are real, v_i can be chosen to be real.
5) Let us denote the right eigenvector corresponding to the zero eigenvalue of A_r as v_m. Therefore, $A_r v_m = 0 \times v_m$ which implies that $-Kv_m = 0 \times v_m$. Since $\sum_{i=1}^{m} K_{ij} = 0$ for all j, it follows that all the components of the eigenvector v_m are equal.
6) It can be easily verified that since $A_r v_i = -\mu_i v_i$, the eigenvectors of A have the form $\begin{bmatrix} v_i \\ \pm j\sqrt{\mu_i} \times v_i \end{bmatrix} = \begin{bmatrix} v_i \\ \pm j\Omega_i \times v_i \end{bmatrix}$.
7) The eigenvectors that correspond to the pair of zero eigenvalues are $\begin{bmatrix} v_m \\ \pm 0 \times v_m \end{bmatrix} = \begin{bmatrix} v_m \\ 0 \end{bmatrix}$. Therefore, we do not have two linearly independent eigenvectors for the pair of zero eigenvalues.

Natural Response

If ΔP_m and ΔP_L are zero, then the natural response of (6.37) can be expressed as follows (see Section 1.2.2):

$$\begin{aligned} \Delta\delta(t) &= \alpha v_m\, e^{0\times t} + \beta v_m\, t\, e^{0\times t} + \sum_{i=1}^{m-1} v_i(c_i \cos\Omega_i t + d_i \sin\Omega_i t) \\ &= \alpha v_m + \beta v_m\, t + \sum_{i=1}^{m-1} v_i(c_i \cos\Omega_i t + d_i \sin\Omega_i t) \end{aligned} \tag{6.40}$$

The speed deviation $\Delta\omega(t)$ is obtained by taking the derivative of this expression:

$$\Delta\omega(t) = \beta v_m + \sum_{i=1}^{m-1} \Omega_i v_i(-c_i \sin\Omega_j t + d_i \cos\Omega_i t) \tag{6.41}$$

where $\alpha, \beta, c_1, ..., c_{m-1}, d_1, d_2, ..., d_{m-1}$ are scalars which depend on the initial conditions. Note that the $te^{0\times t}$ term appears due to the fact that the eigenvectors corresponding to the pair of zero eigenvalues are not linearly independent [6]. The oscillatory responses are due to the complex conjugate pairs of eigenvalues.

Centre-of-Inertia Mode

The first two terms of the response in (6.40) are associated with the pair of zero eigenvalues of the matrix A. These correspond to the "common" motion of the system because all the components of v_m are equal, as discussed earlier. In other words, this motion is not visible in the relative motion between the rotors of different machines. The common motion is also known as the centre-of-inertia (COI) mode. The COI mode can also be isolated from the other modes by summing up the differential equations corresponding to the rotor angle and speed of each machine as follows:

$$\Sigma M_i \frac{d\delta_i}{dt} = \Sigma M_i(\omega_i - \omega_o) \tag{6.42}$$

$$\Sigma M_i \frac{d\omega_i}{dt} = \Sigma P_{mi} - \Sigma P_{ei} = \Sigma P_{mi} - \Sigma P_{Li} \tag{6.43}$$

Note that $\Sigma P_{ei} = \Sigma P_{Li}$ at every instant, since we assume a lossless network. These equations may now be rewritten as

$$\frac{d\delta_{COI}}{dt} = \omega_{COI} - \omega_o \tag{6.44}$$

$$M_{COI}\frac{d\omega_{COI}}{dt} = \Sigma P_{mi} - \Sigma P_{Li} \tag{6.45}$$

where $M_{COI} = \Sigma M_i$, $\omega_{COI} = \frac{\Sigma M_i \omega_i}{\Sigma M_i}$, and $\delta_{COI} = \frac{\Sigma M_i \delta_i}{\Sigma M_i}$.

The deviations from the equilibrium can therefore be written as follows:

$$\begin{bmatrix} \frac{d\Delta\delta_{COI}}{dt} \\ \frac{d\Delta\omega_{COI}}{dt} \end{bmatrix} = \begin{bmatrix} 0 & 1 \\ 0 & 0 \end{bmatrix} \begin{bmatrix} \Delta\delta_{COI} \\ \Delta\omega_{COI} \end{bmatrix} + \frac{1}{M_{COI}} \begin{bmatrix} 0 \\ \Sigma\Delta P_{mi} - \Sigma\Delta P_{Li} \end{bmatrix} \tag{6.46}$$

Note that the COI motion is affected by the *overall* load-generation unbalance. For a given amount of unbalance, the rate of change of COI speed is inversely proportional to the cumulative inertia of the system M_{COI}.

It is clear from the matrix $\begin{bmatrix} 0 & 1 \\ 0 & 0 \end{bmatrix}$ in (6.46) that the COI is associated with a pair of zero eigenvalues and these do not have a pair of linearly independent eigenvectors. This is consistent with the generalized analysis given earlier in this section.

We have assumed here that ΔP_{mi} and ΔP_{Li} are not dependent on $\Delta\omega_{COI}$, that is, they are external inputs. If the total load is less (more) than the total generation, then the speed will increase (decrease) at a rate which is proportional to the imbalance and inversely proportional to M_{COI}. There is no equilibrium condition in this situation. However, if ΔP_{mi} and ΔP_{Li} are dependent on the frequency in the following manner

$$\Delta P_{mi} = \Delta P_{mi_o} - k_{mi}\Delta\omega_{COI} \text{ and } \Delta P_{Li} = \Delta P_{li_o} + k_{li}\Delta\omega_{COI},$$

then one eigenvalue is real and negative, while the other eigenvalue is zero. This is because the augmented equation is given by

$$\begin{bmatrix} \frac{d\Delta\delta_{COI}}{dt} \\ \frac{d\Delta\omega_{COI}}{dt} \end{bmatrix} = \begin{bmatrix} 0 & 1 \\ 0 & \frac{-D_{COI}}{M_{COI}} \end{bmatrix} \begin{bmatrix} \Delta\delta_{COI} \\ \Delta\omega_{COI} \end{bmatrix} + \begin{bmatrix} 0 \\ \frac{\Sigma\Delta P_{mi_o} - \Sigma\Delta P_{Li_o}}{M_{COI}} \end{bmatrix} \tag{6.47}$$

where $D_{COI} = \Sigma k_{mi} + \Sigma k_{li}$. The non-zero eigenvalue is given by $\frac{-D_{COI}}{M_{COI}}$. If there is load-generation imbalance, then the COI speed deviates but settles to a new equilibrium which depends on the value of D_{COI}. Generators with speed-governors and loads like fans and pumps bring about this frequency dependence in P_{mi} and P_{Li}, respectively. Strictly speaking, the dependence is usually on the *local* frequency and not on ω_{COI}. The COI motion in such a situation is not completely decoupled from the other states.

The COI mode is associated with the *frequency stability problem* in power systems. An SMIB system may be viewed as a two-machine system, with one machine having infinite inertia. There is no COI motion in an SMIB system since $M_{COI} = \infty$.

Swing Modes

The oscillatory terms in the natural response in (6.40) represent the swing modes. There are $(m-1)$ such modes in a m machine system, each associated with a characteristic frequency Ω_j and mode-shape v_j. The components of v_j determine the relative observability of various machines in the swing mode.

Unlike the COI mode, the swing modes involve *relative* motion between the generators. It can be shown that the sign of all the components of v_j cannot be the same. This is because (i) $v_m^t M v_j = 0$, due to the orthogonality of the eigenvectors as discussed earlier, (ii) v_j can be chosen to be real, and (iii) v_m has all its components equal. Consequently, the angular deviations of all machines when a swing mode is excited cannot be in the same direction, that is, the machines "swing" against each other. The swing modes are associated with the *angular stability problem* in multi-machine systems.

In a realistic power system there may be a large number of machines and, correspondingly, a large number of swing modes. The oscillatory modes can be grouped into three broad categories:

A. Intra-plant modes in which only the generators within a power plant participate. The oscillation frequencies are generally high in the range of 1.5 to 2.0 Hz.
B. Local modes or inter-plant modes in which several generators in an area participate. The frequencies of oscillations are in the range of 0.8 to 1.8 Hz.
C. Interarea modes in which groups of generators over an extensive area participate. The oscillation frequencies are low and in the range of 0.2 to 0.5 Hz.

The above categorization can be illustrated with the help of a system consisting of two areas connected by a weak (large reactance) AC tie line (see Figure 6.13). Area 2 is represented by a single generator G_4. Area 1 contains three generators G_1, G_2, and G_3. The generators G_1 and G_2 are connected in parallel and participate in the intra-plant

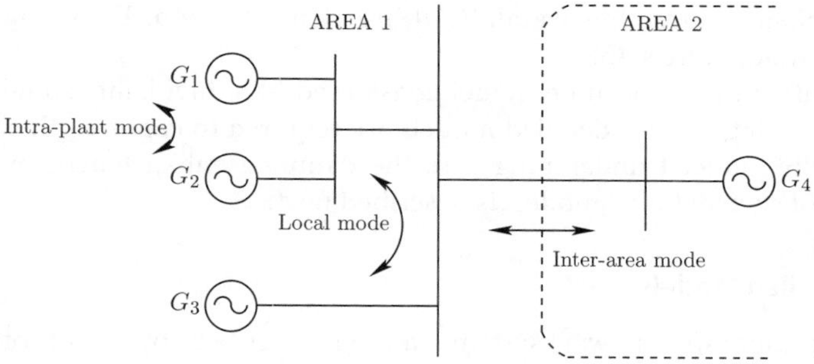

Figure 6.13 Broad categories of swing modes.

oscillations which have higher frequency due to the lower effective reactance between the two machines and also smaller inertias. The local mode oscillation is associated with G_1 and G_2 swinging together against G_3.

The distinction between local modes and inter-area modes applies mainly for those systems which can be divided into distinct areas which are separated by long distances. For systems in which the generating stations are distributed uniformly over a geographic area, it is difficult to distinguish between local and inter-area modes from physical considerations. However, a common observation is that the inter-area modes have the lowest frequency and are observable over a wide geographic area. The COI mode is observed in the rotor angles and speeds throughout the system, and to the same extent everywhere.

Most disturbances result in one or more modes being excited. For example, the response of bus frequencies to a generation tripping event in the Indian grid is shown in Figure 6.14. The tripping excited more than one swing mode as well as the COI mode.

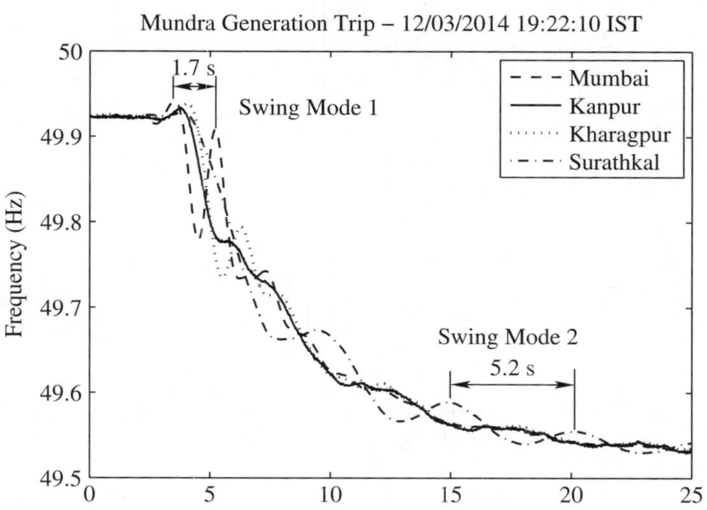

Figure 6.14 Excitation of multiple swing modes and the COI mode after a generator tripping event in the Indian grid.

The load-generation imbalance caused the overall (COI) frequency to drop. The swing modes as well as the COI mode were stable.

While the highly simplified multi-machine model considered here highlights some qualitative features of the swing modes, detailed models are required to capture other aspects like the electrically induced undamping, and the damping enhancement by PSDCs. The formulation of such detailed models is described next.

6.5.2 Analysis with Detailed Models

For the study of swing modes, a power system can be described by a set of differential-algebraic equations. The algebraic equations arise from the neglect of the derivative terms in the *d*-*q* equations describing the fast-acting subsystems, like the stator windings of generators, the transmission network, and power-electronic converters.

The power system model can be constructed in a modular fashion; the equations of various subsystems like generators and loads are interfaced to the equations of the network as depicted in Figure 6.15. For shunt-connected subsystems the interface variables are injected currents and bus voltages at the connection points, while for series-connected subsystems the interface variables are injected voltages and currents flowing through the series-connected branches. The network currents and voltages are expressed in terms of the *D*-*Q* variables by applying a *common* transformation (Kron's transformation) to all the phase (*abc*) currents and voltages. Recall from Section 2.4 that Kron's transformation is given by:

$$\begin{bmatrix} f_a \\ f_b \\ f_c \end{bmatrix} = \sqrt{\frac{2}{3}} \begin{bmatrix} \cos \omega_o t & \sin \omega_o t & \frac{1}{\sqrt{2}} \\ \cos\left(\omega_o t - \frac{2\pi}{3}\right) & \sin\left(\omega_o t - \frac{2\pi}{3}\right) & \frac{1}{\sqrt{2}} \\ \cos\left(\omega_o t + \frac{2\pi}{3}\right) & \sin\left(\omega_o t + \frac{2\pi}{3}\right) & \frac{1}{\sqrt{2}} \end{bmatrix} \begin{bmatrix} f_D \\ f_Q \\ f_o \end{bmatrix} \tag{6.48}$$

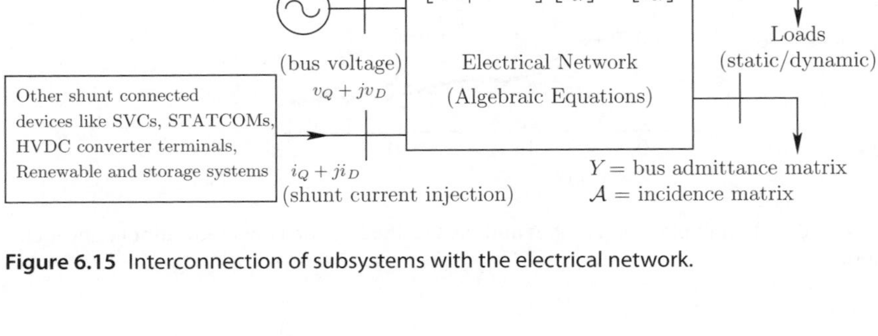

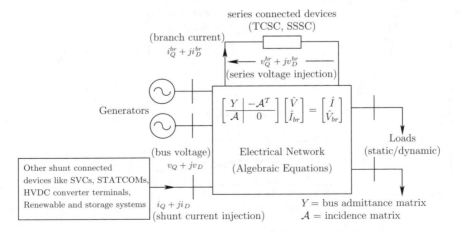

Figure 6.15 Interconnection of subsystems with the electrical network.

The frequency ω_o used in the transformation is constant and is usually taken to be the pre-disturbance equilibrium frequency of the system.

The currents and voltages injected by the shunt and series devices are also expressed in this common reference frame in order to facilitate interfacing with the network equations. The modeling and interfacing of the equations in order to obtain the complete system model is discussed in detail in the following subsections.

Representation of Shunt-Connected Subsystems

Examples of shunt-connected subsystems are generators, loads, SVCs, STATCOMS, HVDC converters, and the converters of renewable energy systems and energy storage systems. The modeling of a few of these subsystems is described below. The models generally consist of two sets of equations: (i) the equations which describe the device and its controllers, and (ii) the algebraic equations which relate the current injected by the device to the states of the device and the voltage of the bus to which the device is connected.

1) Synchronous generators: Model (1.0) of a synchronous generator along with a static excitation system and AVR has been presented in Section 6.3. Although turbine-governor models have not been considered in the study so far (i.e., T_m has been assumed to be constant), these can also be included in the overall model. Since turbine-governors are relatively slow-acting, they are unlikely to affect the relatively higher frequency intra-plant and local swing modes, but they may affect the lower frequency inter-area swing modes [7].

   A few changes have to be made in the model considered in Section 6.3 to make it suitable for multi-machine analysis.

   (a) The frequency of the infinite bus ω_b in the swing equation (6.16) is replaced by ω_o, the frequency used in Kron's transformation. The rotor angle δ is thus measured with respect to a frame rotating at the frequency ω_o.

   (b) The differential equations of the generators may be retained in the *d-q* variables. However, the generator currents and bus voltages which are expressed in Park's reference frame (*d-q* variables) have to be converted into the common Kron's reference frame (*D-Q* variables) to facilitate interfacing with the network equations (which are expressed in the *D-Q* variables).

   Since the rotor position θ which is used in Park's transformation is expressed as $\theta = \omega_o t + \delta$ it follows that (refer Section 2.4):

$$(i_Q + ji_D) = (i_q + ji_d)e^{j\delta} \qquad (v_q + jv_d) = (v_Q + jv_D)e^{-j\delta} \tag{6.49}$$

   As a result, the algebraic equations for the (1.0) model, (6.10) and (6.11),

$$\begin{bmatrix} E'_q \\ 0 \end{bmatrix} + \begin{bmatrix} 0 & x'_d \\ -x_q & 0 \end{bmatrix} \begin{bmatrix} i_q \\ i_d \end{bmatrix} = \begin{bmatrix} v_q \\ v_d \end{bmatrix} \tag{6.50}$$

   can be expressed in terms of *D-Q* variables as follows:

$$P^{-1} \begin{bmatrix} E'_q \\ 0 \end{bmatrix} + P^{-1} \begin{bmatrix} 0 & x'_d \\ -x_q & 0 \end{bmatrix} P \begin{bmatrix} i_Q \\ i_D \end{bmatrix} = \begin{bmatrix} v_Q \\ v_D \end{bmatrix} \tag{6.51}$$

   where

$$P = \begin{bmatrix} \cos\delta & \sin\delta \\ -\sin\delta & \cos\delta \end{bmatrix} \tag{6.52}$$

(c) The equations of the external network and the infinite bus ((6.12) and (6.13)) are replaced by a generalized network representation in *D*-*Q* variables, which will be considered shortly.

The differential-algebraic equations for higher-order generator models are presented in Section 3.6. These may be used instead of Model (1.0) for greater accuracy if the corresponding data required for these models are available. The modifications given above are applicable to these higher-order models as well.

2) Static VAr Compensator (SVC):
The block diagram of a SVC voltage regulator is shown in Figure 6.16. The equations of the voltage regulator are

$$\dot{B}_i = K_i(V_{ref} - V) = K_i\left(V_{ref} - \sqrt{v_Q^2 + v_D^2}\right) \tag{6.53}$$

$$B_{ref} = K_p(V_{ref} - V) + B_i \tag{6.54}$$

The dynamics of the SVC susceptance may be represented by the first-order transfer function

$$\dot{B}_{svc} = \frac{1}{T_p}[-B_{svc} + B_{ref}] \tag{6.55}$$

T_p depends on the response of the SVC susceptance controller and is typically around one to two cycles. This is a gross fundamental frequency representation of the SVC which is suitable for slow-transient studies. This representation neglects the detailed switching behavior of the underlying power electronic devices and the fast-transient behavior of the associated firing and synchronizing controllers. Similar simplifications have been done earlier (implicitly) in the modeling of the excitation system presented in Section 6.3.

The algebraic equations which relate the SVC susceptance to the injected current can be written as

$$i_Q + ji_D = -jB_{svc}(v_Q + jv_D) \tag{6.56}$$

3) STATCOM/shunt-connected voltage source converter:
A STATCOM is a shunt-connected voltage source converter (VSC) based device which is used for reactive power control. It has a fast-acting closed-loop reactive current controller (see Chapter 5). Therefore, the STATCOM can be modelled as

$$\frac{di_r}{dt} = \frac{1}{T_{ps}}[-i_r + i_r^*] \tag{6.57}$$

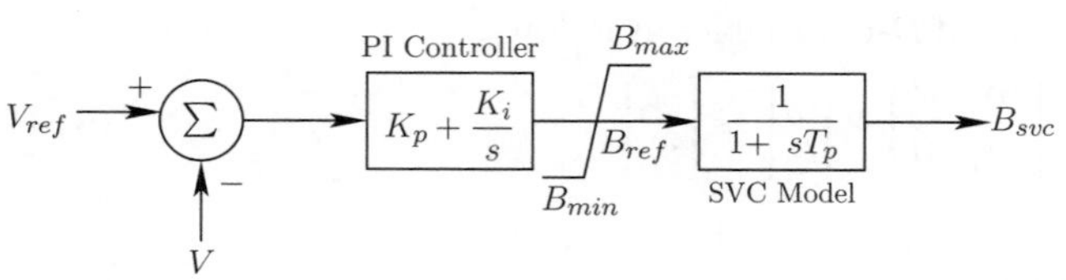

Figure 6.16 Block diagram of an SVC with a voltage regulator.

The voltage regulator of a STATCOM sets the reactive current reference i_r^*. The algebraic equation relating the injected current to the reactive current is given by

$$i_Q + j i_D = -j i_r \frac{(v_Q + j v_D)}{\sqrt{v_D^2 + v_Q^2}} \tag{6.58}$$

It is assumed here that the real power exchange with the system is zero (losses are neglected). As per the convention used here, a positive value of i_r implies that the STATCOM is injecting reactive power (behaving like a capacitor). In general, shunt-connected VSC devices may also be used for real power injection if there is an energy source at the DC link of the converter. The source of energy could be a wind or solar energy extraction system or a storage system like a battery. In such a case, the real-current i_p also has to be considered as follows:

$$i_Q + j i_D = -(i_p + j i_r) \frac{(v_Q + j v_D)}{\sqrt{v_D^2 + v_Q^2}} \tag{6.59}$$

i_p is obtained from the real current reference, which may be set by power scheduling or modulation controllers.

In the case of multiple VSC converters that are connected to a grid and have a common DC link (but without an energy source at the DC link), the real current references of all the converters cannot be set independent of each other. This is because the overall power drawn by all the converters, minus the losses, has to be zero in steady state. In practice, this is ensured by regulating the voltage across the capacitor which is connected across the DC link of the VSCs. Since this control is generally quite fast, an algebraic power-balance equation may be used to represent this constraint (instead of representing the dynamics of the capacitor and its voltage regulator in detail). An example of such converter systems are VSC-based HVDC-links.

In two terminal line-commutated converter-based HVDC transmission systems the DC-link current is generally regulated. The real and reactive power injections by the converters into the AC buses are not independent, and can be related to the DC-link current by algebraic equations [8, 9]. As before, the overall power drawn by the converters (in a two-terminal or multi-terminal link), minus the losses, has to be zero in steady state.

4) Loads:

The current injected by a load is given by the equation

$$(i_Q + j i_D) = -\frac{(P_L - jQ_L)}{(v_Q - j v_D)} \tag{6.60}$$

where P_L and Q_L denote the real and reactive power drawn by the load. The relationship between the real and reactive power and the voltage magnitude and frequency for a "static" load can be represented as follows:

$$P_L = P_{Lo}\left(p_1 + p_2\left(\frac{V}{V_o}\right) + p_3\left(\frac{V}{V_o}\right)^2\right)\left(1 + k_{pf}\left(\frac{\Delta f}{f_o}\right)\right) \tag{6.61}$$

$$Q_L = Q_{Lo}\left(q_1 + q_2\left(\frac{V}{V_o}\right) + q_3\left(\frac{V}{V_o}\right)^2\right)\left(1 - k_{qf}\left(\frac{\Delta f}{f_o}\right)\right) \tag{6.62}$$

where V and f denote the bus voltage magnitude and frequency, respectively, while the subscript o denotes the quiescent or nominal value. p_1, p_2, and p_3 denote the fraction of the constant power, constant current, and constant impedance component of the real power, respectively, while q_1, q_2, and q_3 denote the same for reactive power. Note that $p_1 + p_2 + p_3 = q_1 + q_2 + q_3 = 1$.

Frequency deviation Δf is the derivative of the bus phase angle ϕ, where $\phi = \tan^{-1}\frac{v_D}{v_Q}$. This relationship may be approximately modeled as follows:

$$\dot{x}_f = \frac{1}{T_L}(\phi - x_f) \tag{6.63}$$

$$\Delta f = \frac{1}{T_L}(\phi - x_f) \tag{6.64}$$

Note that in the Laplace domain, $\Delta f(s) = \frac{s}{sT_L+1}\phi(s)$, which approximates the derivative relationship between f and ϕ for slow transients, for small values of T_L.

A static load model captures the steady-state voltage and frequency dependence of loads, but it cannot capture the transient behavior accurately. Therefore, some loads like large induction motors may be represented by their dynamic models. The induction machine model that was given in Chapter 4, which follows the generator convention in the current and torque directions, may be adapted to represent a motor load. The transformer emfs associated with the stator fluxes may be neglected (i.e., $\frac{1}{\omega_B}\frac{d\psi_D}{dt} \approx 0$ and $\frac{1}{\omega_B}\frac{d\psi_Q}{dt} \approx 0$) in the study of low-frequency power swings.

It is usually not feasible to model each and every load separately because of the large number of loads. Therefore, similar types of loads may be aggregated into a single equivalent load [10]. If diverse loads are present, then one may categorize the loads at a given bus into, say, residential, commercial, industrial, agricultural, and power plant loads. The loads belonging to the same category are aggregated and modeled using typical parameters corresponding to that category [11]. Diurnal and seasonal changes in load composition may be accounted for while modeling the loads in practical studies.

Representation of Series-Connected Subsystems

FACTS devices like thyristor-controlled series compensators (TCSCs) and static synchronous series compensators (SSSCs) inject a voltage in series with a transmission line. See Chapter 5 for a detailed description of these devices. While the TCSC is a variable reactance device, an SSSC injects a variable reactive voltage, that is, a voltage in quadrature to the line current.

The dynamics of the TCSC reactance may be approximated as a first-order transfer function, as in the case of an SVC. Thus, a TCSC can be represented by the following equations:

$$\dot{X}_{TCSC} = \frac{1}{T_p}[-X_{TCSC} + X_{ref}] \tag{6.65}$$

$$v_Q^{br} + jv_D^{br} = -jX_{TCSC}(i_Q^{br} + ji_D^{br}) \tag{6.66}$$

The superscript br is used for the branch voltage and branch current (to distinguish them from the shunt current injection and bus voltages in the case of shunt devices). The set-point X_{ref} is determined by power scheduling controllers.

Similar equations may be obtained for an SSSC for the reactive voltage v_R. The SSSC output equation is given by:

$$v_Q^{br} + jv_D^{br} = -jv_R \frac{(i_Q^{br} + ji_D^{br})}{\sqrt{i_D^2 + i_Q^2}} \tag{6.67}$$

Remarks

1) Devices like unified power flow controllers (UPFCs) involve both series voltage injection and shunt current injection (see Section 5.2.2). The series and shunt converter may be modeled as an SSSC and a STATCOM, respectively, with the possibility of real power injection via either or both converters. However, an additional algebraic constraint equation, representing the fact that the total power exchange is zero (if there is no energy source at the DC link of the device), has to be included in the model.
2) Voltage regulators have been modeled for an SVC and a generator excitation system so far. In general, all power electronic actuators (generator excitation systems, HVDC systems, FACTS, and renewable-grid interfaces) not only have such current or voltage regulators, but also have auxiliary controllers which are used to improve the stability of the system. The controllers which can affect the transients in the frequency range (0.2–2 Hz) have to be considered in the model for studying power swings.
3) A representation of device limits is needed to accurately capture the behavior following disturbances, especially if the disturbances result in large excursions. For example, see [12] for a discussion on representation of TCSC limits in stability studies. The limiting behavior of an SVC and a STATCOM has been discussed in Chapter 5. However, for small-signal studies around an equilibrium point we may assume that the devices are within the limits during the transients provided that the equilibrium values themselves are within the limits.

Representation of the Transmission Network

A network model suitable for the study of power swings is obtained by neglecting the derivative terms in the D-Q equations of the transmission network elements. This effectively removes the faster transients associated with the network voltages and currents. The resulting algebraic equations in the D-Q variables can be written compactly as complex equations (see equation (2.24) in Section 2.4). If only shunt devices are present in the system then the transmission network equations are of the following form:

$$\hat{I} = [Y]\hat{V} \tag{6.68}$$

where $\hat{I}$ is the complex vector of currents injected by the shunt devices $(i_Q + ji_D)$ and $\hat{V}$ is the complex vector of bus voltages $(v_Q + jv_D)$. $[Y]$ is the bus admittance matrix of the network. If series-connected devices are also present, then the transmission network equations are modified as follows:

$$\hat{I} = [Y]\hat{V} - \mathcal{A}^t\hat{I}^{br} \tag{6.69}$$

$$\hat{V}^{br} = \mathcal{A}\hat{V} \tag{6.70}$$

where $\hat{I}^{br}$ and $\hat{V}^{br}$ are the complex vectors of the branch currents $(i_Q^{br} + ji_D^{br})$ and voltages injected $(v_Q^{br} + jv_D^{br})$ by the series devices, respectively. $\mathcal{A}$ is a $s \times n$ incidence matrix which relates the s injected voltages to the n node voltages. For example, if only one series device, say a TCSC, is present and is connected between buses k and l, then $\mathcal{A}$ is a $1 \times n$ vector with its kth and lth components being $+1$ and -1, respectively. These are the only non-zero entries of $\mathcal{A}$ in this case. Note that $\hat{V}^{br}$ and $\hat{I}$ are obtained from the series and shunt device output equations (e.g., from equations (6.56) and (6.66) for a SVC and a TCSC, respectively).

Formulation of the Overall System Model

The overall model is obtained by gathering all the differential and algebraic equations which describe the various components of the system. The model can be written as follows:

$$\dot{x} = f(x, y, u) \tag{6.71}$$

$$0 = g(x, y) \tag{6.72}$$

where x are the state variables, corresponding to various dynamical subsystems (like generators, exciters, loads, FACTS, HVDC etc.), u are the set-points of the controllers, and y are the algebraic variables, that is, the currents injected by the shunt-connected devices, bus voltages, voltages injected by the series-connected devices, and the corresponding branch currents.

Note that the transmission network equations (6.69) and (6.70), which are linear, are a subset of the algebraic equations $g(x, y) = 0$. The other algebraic equations are the device injection equations (e.g., (6.51), (6.56), and (6.66) for a generator, SVC, and TCSC, respectively).

A small-signal model can be obtained by linearizing the differential-algebraic equations (6.71) and (6.72) around an equilibrium point as follows:

$$\Delta\dot{x} = \left[\frac{\partial f}{\partial x}\right]\Delta x + \left[\frac{\partial f}{\partial y}\right]\Delta y + \left[\frac{\partial f}{\partial u}\right]\Delta u \tag{6.73}$$

$$0 = \left[\frac{\partial g}{\partial x}\right]\Delta x + \left[\frac{\partial g}{\partial y}\right]\Delta y \tag{6.74}$$

$\left[\frac{\partial f}{\partial x}\right]$ denotes the matrix with the partial derivative $\frac{\partial f_i}{\partial x_j}$ as its (i, j)th element. The other matrices are similarly defined. The partial derivatives are evaluated at the operating point. Eliminating Δy from the above equations yields the state space form as follows

$$\Delta\dot{x} = \left(\left[\frac{\partial f}{\partial x}\right] - \left[\frac{\partial f}{\partial y}\right]\left[\frac{\partial g}{\partial y}\right]^{-1}\left[\frac{\partial g}{\partial x}\right]\right)\Delta x + \left[\frac{\partial f}{\partial u}\right]\Delta u \tag{6.75}$$

$$= [A_r]\Delta x + [E_r]\Delta u \tag{6.76}$$

For large systems, A_r is generally not sparse. Hence, the following unreduced form may be retained in order to preserve sparsity:

$$\begin{bmatrix} I & 0 \\ 0 & 0 \end{bmatrix}\begin{bmatrix} \Delta\dot{x} \\ \Delta\dot{y} \end{bmatrix} = \begin{bmatrix} \left[\frac{\partial f}{\partial x}\right] & \left[\frac{\partial f}{\partial y}\right] \\ \left[\frac{\partial g}{\partial x}\right] & \left[\frac{\partial g}{\partial y}\right] \end{bmatrix}\begin{bmatrix} \Delta x \\ \Delta y \end{bmatrix} + \begin{bmatrix} \left[\frac{\partial f}{\partial u}\right] \\ 0 \end{bmatrix}\Delta u \tag{6.77}$$

The generalized eigenvalues (λ_i) and the corresponding generalized right eigenvectors (v_i) can then be obtained from the following equation [13].

$$\begin{bmatrix} \left[\frac{\partial f}{\partial x}\right] & \left[\frac{\partial f}{\partial y}\right] \\ \left[\frac{\partial g}{\partial x}\right] & \left[\frac{\partial g}{\partial y}\right] \end{bmatrix} v_i = \lambda_i \begin{bmatrix} I & 0 \\ 0 & 0 \end{bmatrix} v_i \tag{6.78}$$

The finite generalized eigenvalues are also the eigenvalues of the matrix $[A_r]$ in (6.76). For large systems which may have thousands of states, selective eigenvalue computation techniques [14] are used to compute only the eigenvalues of interest, which in the present study are those corresponding to the low-frequency swing modes of the system.

Example 6.2 ***Eigenvalues of a Three-Machine System*** Consider the three-machine system shown in Figure 6.17. The data for this system is adapted from [15] and is given in Appendix 6.B.

Three models are considered in this analysis:

1) Base model: Machines 1 and 2 are modeled by Model (1.1) (see Section 3.6) while machine 3 is modeled by Model (1.0). The loads are modeled using the following parameters: $p_1 = 0.0$, $p_2 = 0.0$, $p_3 = 1.0$, $q_1 = 0.0$, $q_2 = 0.0$, $q_3 = 1.0$, $k_{pf} = 0.0$, and $k_{qf} = 0.0$. Note that with these parameters the loads behave like constant impedances. The exciters and AVRs are modeled as shown in Figure 6.4.
2) Base model with frequency-dependent loads: This is the same as the base model, except that $k_{pf} = 1.5$ and $k_{qf} = 2.0$.
3) Classical model with constant real power loads: All generators are modeled by the classical model, while the loads have the following parameters: $p_1 = 1.0$, $p_2 = 0.0$, $p_3 = 0.0$, $q_1 = 0.0$, $q_2 = 0.0$, $q_3 = 1.0$, $k_{pf} = 0.0$, and $k_{qf} = 0.0$.

The mechanical power input to the generators is assumed to be constant.

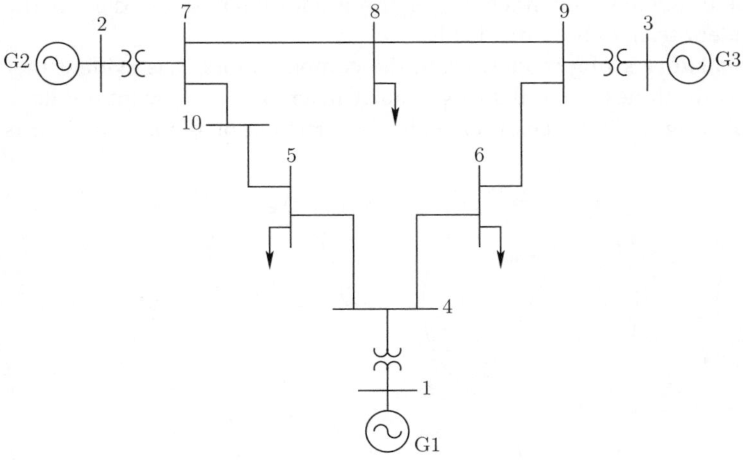

Figure 6.17 Three-machine system.

Table 6.4 Eigenvalues of the three-machine system.

Eigenvalues			
Classical model	**Base model**	**Frequency-dependent loads**	**Comments**
–	–	$-1651.5, -1543, -492.48$	
–	$-9.660 \pm j19.52$	$-9.639 \pm j19.52$	
–	$-10.448 \pm j13.66$	$-10.452 \pm j13.66$	
–	$-11.258 \pm j11.34$	$-11.257 \pm j11.34$	
$\pm j\mathbf{13.37}$	$\mathbf{-0.451 \pm j13.16}$	$\mathbf{-0.460 \pm j13.15}$	Swing mode 2
$\pm j\mathbf{8.55}$	$\mathbf{-0.039 \pm j8.59}$	$\mathbf{-0.069 \pm j8.58}$	Swing mode 1
–	-4.345	-4.342	
–	-3.251	-3.251	
$\pm j0.0073\,(0,0)$	$\pm j0.0086\,(0,0)$	$-0.0005\,(0);-0.149$	COI (common) mode

Compute the eigenvalues of the system. Examine the eigenvectors corresponding to the swing mode and the common mode.

Solution

The eigenvalues of this system for the various system models are given in Table 6.4. The right eigenvector components associated with generator speeds for the swing modes and the common-mode are shown using compass plots in Figs. 6.18, 6.19 and 6.20.

Remarks

1) The base model without frequency dependence has 14 states, and therefore has 14 eigenvalues, while the classical model has six states and six eigenvalues. The base model with frequency dependence has three additional states due to the frequency-dependent term in the three loads.
2) There are two oscillatory swing modes. From the compass plots, it is evident that for swing mode 1, machines 2 and 3 swing against machine 1. For swing mode 2, machines 2 and 3 swing against each other, with low participation of machine 1. This

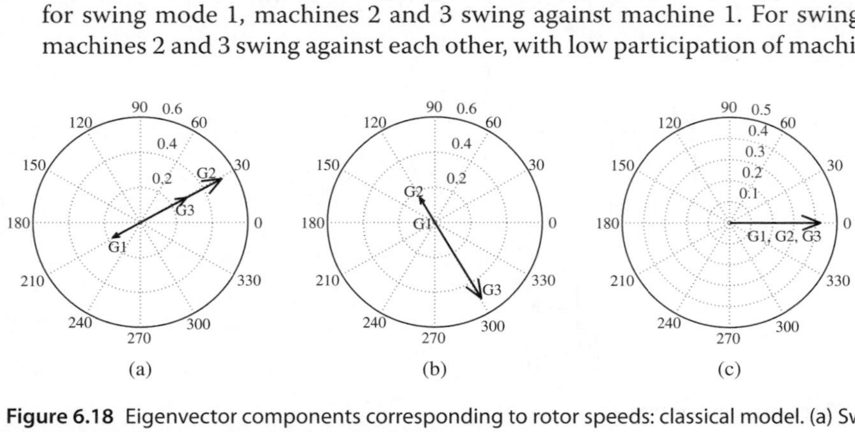

Figure 6.18 Eigenvector components corresponding to rotor speeds: classical model. (a) Swing mode 1 ($j8.545$) (b) Swing mode 2 ($j13.373$) (c) COI mode (0).

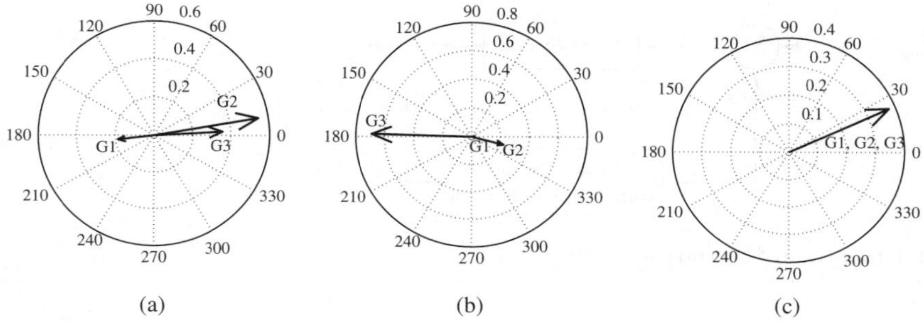

Figure 6.19 Eigenvector components corresponding to rotor speeds: base model. (a) Swing mode 1 ($-0.0394 \pm j8.585$) (b) Swing mode 2 ($-0.4510 \pm j13.157$) (c) COI mode (0).

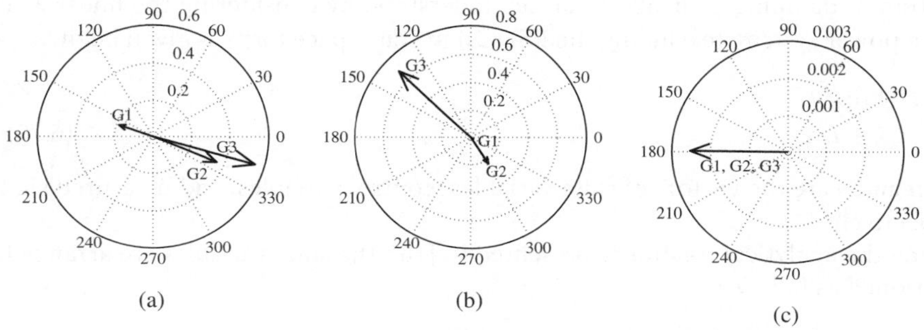

Figure 6.20 Eigenvector components corresponding to rotor speeds: base model with frequency-dependent loads. (a) Swing mode 1 ($-0.0693 \pm j8.583$) (b) Swing mode 2 ($-0.460 \pm j13.15$) (c) COI mode (-0.1489).

is true for all models. However, for the classical model, the machines swinging against each other have a phase difference of exactly 180°. This is not true for the other models, although the phase difference is nearly equal to 180°.

3) There are several other modes which are associated with the AVRs, field and damper windings of the machines for the base model with and without load-frequency dependence.
4) The common or COI mode is associated with two zero eigenvalues when load-frequency dependence is not considered. The computed eigenvalues corresponding to the common mode are not exactly zero due to numerical precision errors (hence the zero values are shown in brackets). When load-frequency dependence is considered, one of the eigenvalues corresponding to the COI mode is real and negative, indicating that the COI speed will settle to an equilibrium if it is disturbed due to load–generation imbalance.
5) For the COI mode, all the eigenvector components corresponding to the rotor speeds have the same value, indicating the common movement associated with this mode.

6.6 Principles of Damping Controller Design [16]

It was shown in Section 6.3 that the action of an AVR may destabilize the swing mode under certain operating conditions. Since feedback controllers like AVRs are necessary

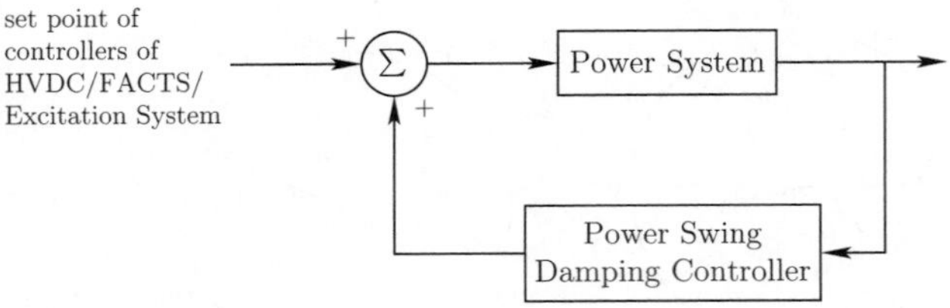

Figure 6.21 Power swing damping controllers.

for important regulation functions, the damping of the swings may be ensured by adding auxiliary feedback loops which modulate the set-points of the existing controllers in the system. Such feedback controllers are called PSDCs (see Figure 6.21).

The action of damping controllers can be understood by considering the linearized model of a power system (excluding the PSDCs) in state space form as given below:

$$\dot{x} = Ax + Bu$$
$$y = Cx + Du \tag{6.79}$$

where x, u and y are the vectors of state variables, control variables, and measured signals, respectively.

Following the analytical treatment presented in [16], the state variables are arranged and partitioned as follows:

$$\begin{aligned} x &= [\Delta\delta_1\ \Delta\delta_2\ \dots\ \Delta\delta_n\ \ \Delta\omega_1\ \Delta\omega_2\ \dots\ \Delta\omega_n\ \ z^t]^t \\ &= [\Delta\delta^t\ \ \Delta\omega^t\ \ z^t]^t \end{aligned} \tag{6.80}$$

where $\Delta\delta_i$ and $\Delta\omega_i$ are the rotor angle and speed deviation (in rad/s) of the ith generator, respectively, and z denotes all the other states. The matrices A, B, and C may be correspondingly partitioned as shown in equation (6.81):

$$A = \begin{bmatrix} 0 & I & 0 \\ A_{21} & A_{22} & A_{23} \\ A_{31} & A_{32} & A_{33} \end{bmatrix} \quad B = \begin{bmatrix} 0 \\ B_2 \\ B_3 \end{bmatrix} \quad C = \begin{bmatrix} C_1 & C_2 & C_3 \end{bmatrix} \tag{6.81}$$

This partitioning separates out the generator angles and speeds (expressed here in rad/s). The generator angles and speeds are transformed using the eigenvectors of A_{21} (modal decomposition). If V is the right eigenvector matrix of A_{21} then $V^{-1}A_{21}V = \Lambda$, where Λ is a diagonal matrix whose diagonal entries are the normalized modal synchronizing coefficients. The transformation to the modal variables (x_m) is given by

$$x_m = U^{-1}x \quad U = \begin{bmatrix} V & 0 & 0 \\ 0 & V & 0 \\ 0 & 0 & I \end{bmatrix} \tag{6.82}$$

The equations in the transformed variables are given by:

$$\dot{x}_m = U^{-1}AUx_m + U^{-1}Bu = A_m x_m + B_m u \tag{6.83}$$

$$y = CUx_m + Du = C_m x_m + Du \tag{6.84}$$

The structure of A_m is as follows:

$$A_m = \begin{bmatrix} 0 & I & 0 \\ \Lambda & A_{m22} & A_{m23} \\ A_{m31} & A_{m32} & A_{m33} \end{bmatrix} \tag{6.85}$$

If the states in (6.84) are rearranged such that modal angle and modal speed corresponding to a dominant swing mode, say the ith mode λ_i, are the first and second variables, then the system can be represented as follows:

$$\begin{bmatrix} \Delta\dot{\delta}_{mi} \\ \Delta\dot{\omega}_{mi} \\ \dot{z}_{mi} \end{bmatrix} = \begin{bmatrix} 0 & 1 & 0 \\ -k_{mi} & -d_{mi} & -A_{d23} \\ A_{d31} & A_{d32} & A_{d33} \end{bmatrix} \begin{bmatrix} \Delta\delta_{mi} \\ \Delta\omega_{mi} \\ z_{mi} \end{bmatrix} + \begin{bmatrix} 0 \\ -B_{d2} \\ B_{d3} \end{bmatrix} u \tag{6.86}$$

$$y = \begin{bmatrix} C_{d1} & C_{d2} & C_{d3} \end{bmatrix} \begin{bmatrix} \Delta\delta_{mi} \\ \Delta\omega_{mi} \\ z_{mi} \end{bmatrix} + Du \tag{6.87}$$

z_{mi} consists of all variables except the modal angle and modal speed corresponding to swing mode λ_i. The interactions of these variables with the ith modal angle and speed are depicted in the block diagram shown in Figure 6.22. Consistent with the convention followed in this chapter, ω_{mi} is in rad/s.

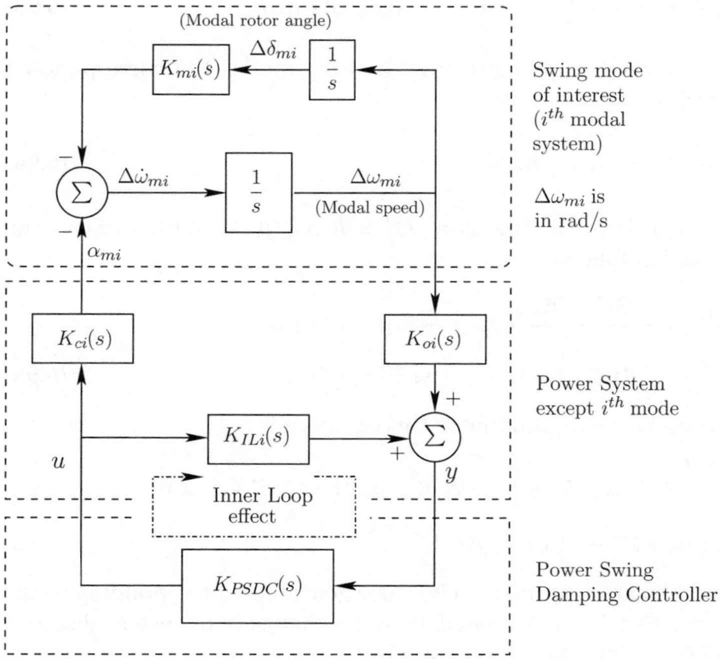

Figure 6.22 Multi-modal decomposition block diagram. Source: [16] Fig. 2.1, p. 949, reproduced with permission ©IEEE 1995.

The transfer functions in the block diagram are obtained by applying Laplace transform to equations (6.86) and (6.87) and eliminating z_{mi} from the equations. The expression for $K_{mi}(s)$ is

$$K_{mi}(s) = k_{mi} + sd_{mi} + A_{d23}(sI - A_{d33})^{-1}(A_{d31} + sA_{d32}) \tag{6.88}$$

$K_{ci}(s)$, $K_{oi}(s)$, and $K_{ILi}(s)$ are the modal controllability, observability, and inner-loop transfer functions, respectively, whose expressions are

$$K_{ci}(s) = A_{d23}(sI - A_{d33})^{-1}B_{d3} + B_{d2} \tag{6.89}$$

$$K_{oi}(s) = C_{d3}(sI - A_{d33})^{-1}\left(\frac{A_{d31}}{s} + A_{d32}\right) + \left(\frac{C_{d1}}{s} + C_{d2}\right) \tag{6.90}$$

$$K_{ILi}(s) = C_{d3}(sI - A_{d33})^{-1}B_{d3} + D \tag{6.91}$$

$K_{PSDC}(s)$ is the PSDC transfer function. The effective feedback due to the PSDC is given by

$$K_{ei}(s) = K_{oi}(s)\frac{K_{PSDC}(s)}{1 - K_{PSDC}(s)K_{ILi}(s)}K_{ci}(s) \tag{6.92}$$

The damping and synchronizing torque concepts presented in Section 6.4 can be applied to the modal system corresponding to each individual mode. Therefore, a system without a PSDC can be approximately described as follows:

$$\frac{d\Delta\omega_{mi}}{dt} = \frac{d^2\Delta\delta_{mi}}{dt^2} = -\frac{\Im[K_{mi}(j\Omega_i)]}{\Omega_i}\Delta\omega_{mi} - \Re[K_{mi}(j\Omega_i)]\Delta\delta_{mi}, \tag{6.93}$$

where Ω_i is the frequency corresponding to the ith swing mode. The corresponding characteristic polynomial is

$$s^2 + \left(\frac{\Im[K_{mi}(j\Omega_i)]}{\Omega_i}\right)s + \Re[K_{mi}(j\Omega_i)] \tag{6.94}$$

We assume that $\Im[K_{mi}(j\Omega_i)]$ is small. Therefore, $\Omega_i^2 \approx \Re[K_{mi}(j\Omega_i)]$. With a PSDC, the equation (6.93) is modified as follows:

$$\begin{aligned}\frac{d\Delta\omega_{mi}}{dt} = \frac{d^2\Delta\delta_{mi}}{dt^2} &= -\frac{\Im[K_{mi}(j\Omega_i)]}{\Omega_i}\Delta\omega_{mi} - \Re[K_{mi}(j\Omega_i)]\Delta\delta_{mi} \\ &\quad -\Re[K_{ei}(j\Omega_i)]\Delta\omega_{mi} + \Omega_i\Im[K_{ei}(j\Omega_i)]\Delta\delta_{mi}\end{aligned} \tag{6.95}$$

The characteristic polynomial corresponding to this equation is

$$\begin{aligned}&s^2 + \left(\frac{\Im[K_{mi}(j\Omega_i)]}{\Omega_i} + \Re[K_{ei}(j\Omega_i)]\right)s + \Re[K_{mi}(j\Omega_i)] - \Omega_i\Im[K_{ei}(j\Omega_i)] \\ &\approx s^2 + \Re[K_{ei}(j\Omega_i)]s + \Omega_i^2 - \Omega_i\Im[K_{ei}(j\Omega_i)]\end{aligned} \tag{6.96}$$

The roots of this polynomial are (approximately) the eigenvalues corresponding to the swing mode. If we assume that $K_{ei}(j\Omega_i)$ is small, then the change in the eigenvalue $\lambda_i = \sigma_i + j\Omega_i$, due to the PSDC, is (approximately) given by

$$\Delta\lambda_i = \Delta\sigma_i + j\Delta\Omega_i \approx -\frac{1}{2}\Re[K_{ei}(j\Omega_i)] - \frac{j}{2}\Im[K_{ei}(j\Omega_i)] = -\frac{1}{2}K_{ei}(j\Omega_i) \tag{6.97}$$

With reference to (6.92), it is clear that the PSDC transfer function $K_{PSDC}(s)$ should compensate for $\angle K_{ci}(j\Omega_i) + \angle K_{oi}(j\Omega_i)$, so that the movement of the eigenvalue λ_i is towards the left half of the complex plane. In other words, $K_{ei}(j\Omega_i)$ should be positive and real to obtain a pure damping effect. The negative of $\angle K_{ci}(j\Omega_i) + \angle K_{oi}(j\Omega_i)$ is the phase lead that has to be provided by $K_{PSDC}(j\Omega_i)$ to achieve this. This is called the *controller phase index (CPI)* [16]. A small negative imaginary component in $K_{ei}(j\Omega_i)$ is not altogether undesirable, as it improves the synchronizing effect.

6.6.1 Actuator Location and Choice of Feedback Signals

The actuators for PSDCs are generally power electronic converters (e.g., HVDC, FACTS, and generator excitation systems) as they have sufficient bandwidth to affect the swing modes, in addition to having the vernier control capability necessary for modulation. The actuator location has a significant impact on the *controllability* of swing modes. A lower modal controllability implies that a larger control effort (or controller range) is needed to achieve a given damping performance.

The modal *observability* is also an important consideration for the choice of a feedback signal. A high modal observability of the targeted swing mode(s) and low observability of the untargeted modes is desirable so that the design does not get unduly constrained by the movement of untargeted modes. There is some flexibility in the choice of a feedback signal. While a locally available signal or a signal synthesized from local measurements is preferable (since communication glitches or failures are not a concern), the advent of wide-area measurement systems (WAMS) [17] opens up the possibility of synthesizing highly customized signals using system-wide measurements.

Although modal observability is the main consideration for the choice of a feedback signal, there are other important considerations, as described below.

1) The CPIs corresponding to the swing modes, which are controllable and observable, should all preferably be about the same value. This makes the design of a PSDC easier because the phase of K_{PSDC} does not have to be significantly different for each swing mode; note that swing modes lie in a small frequency range of about 0.2–2 Hz. It is also desirable that the CPI remains nearly constant over different operating points so that a controller designed at one operating point works satisfactorily at other operating points.
2) Given two candidate feedback signals with similar modal observability $|K_{oi}(j\Omega_i)|$, the signal with lesser inner-loop gain $|K_{ILi}(j\Omega_i)|$ is preferred over the other. A high inner-loop gain limits the achievable damping influence on the mode due to the denominator term in (6.92). To capture this, an index called the *maximum damping influence (MDI)* has been proposed [16], which is defined as follows: **MDI(i)** $= \frac{|K_{ci}(j\Omega_i)||K_{oi}(j\Omega_i)|}{2\sqrt{10}|K_{ILi}(j\Omega_i)|}$. This index, which is defined for each swing mode, is a measure of the maximum eigenvalue shift that can be achieved before the inner-loop dominates the damping control. In the definition of this index, a 10 dB gain margin is assumed in the inner-loop feedback system.
3) The content of noise and high-frequency oscillatory modes in the feedback signal should be low. This will obviate the need for complicated or heavy filtering.

In addition to these considerations, there may be specific constraints associated with certain feedback signals depending on the type of actuator. For example, in generator excitation systems, the use of the electrical power signal as a PSDC feedback signal can cause undesirable changes in terminal voltage when mechanical power ramping is done [7].

6.6.2 Components of a PSDC

The typical structure of a single-input single-output PSDC transfer function is shown in Figure 6.23. The overall PSDC transfer function is

$$K_{PSDC}(s) = KG_f(s)G_w(s)G_c(s)$$

The various blocks in this controller are described below.

1) *Filter and transducer block*
The filter block includes the low-pass filtering required to exclude the high frequency components in the feedback signal. It also models the delays in the synthesis and conditioning of the feedback signals. Generally the phase delays associated with the filter transfer function at the swing mode frequencies are small.
2) *Washout block*
The washout block is provided to eliminate steady-state bias in the output of a PSDC which will modify the set-point of the device. The PSDC is expected to respond only to transient variations in the input signal and not to the DC offsets in the signal. This is achieved by subtracting the low-frequency components of the signal from the signal itself. Thus, the transfer function of a washout block is as follows:

$$G_w(s) = 1 - \frac{1}{1+sT_w} = \frac{sT_w}{1+sT_w} \tag{6.98}$$

Since swing mode frequencies can be between 0.2 and 2 Hz, the time constant T_w may typically be chosen to be 10–15 seconds. The washout block may add some phase lead at low frequencies corresponding to the inter-area swing modes. This should be accounted for in the overall PSDC design.
3) *Dynamic compensator*
The compensator ensures that the desired phase of the overall transfer function $K_{ei}(j\Omega_i)$ is achieved. A typical dynamic compensator is made up of one or more lead-lag stages as given below:

$$G_c(s) = \frac{(1+sT_1)(1+sT_3)}{(1+sT_2)(1+sT_4)} \tag{6.99}$$

Note that $\frac{(1+sT_1)}{(1+sT_2)}$ introduces a phase lead, $\phi_{lead} = \tan^{-1}\sqrt{\frac{T_1}{T_2}} - \tan^{-1}\sqrt{\frac{T_2}{T_1}}$ at the centre frequency $f_c = \frac{1}{2\pi\sqrt{T_1T_2}}$.

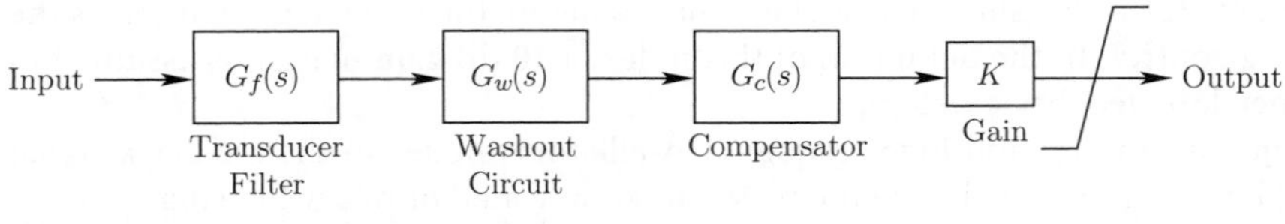

Figure 6.23 Component blocks of a PSDC.

The ratio $\frac{T_1T_3}{T_2T_4}$ is generally kept less than 10 in order to avoid amplification of high-frequency noise in the signal. For the initial design, the PSDC transfer function is approximated to $G_c(s)$, the transfer function of the dynamic compensator. The effects of the washout circuit, transducer, and filter are expected to be incidental and may be neglected in the initial design. They must, however, be considered in the final evaluation of the design.

4) *Gain*

In a large system, an individual PSDC will not be able to affect all the swing modes to the same extent. The choice of feedback signals and the design of a PSDC should be done so that improvements in damping of all controllable swing modes are achieved, while specifically targeting the ones which are highly controllable and poorly damped. A damping ratio of at least 5% for the targeted swing modes (cumulatively achieved through all PSDCs) can be considered as satisfactory, as this will ensure that there is adequate margin to account for changes in damping due to changes in operating conditions. The 5% target for the damping ratio is, however, not a hard and fast rule.

The movement of all affected eigenvalues with increasing gain should be tracked. With increasing gain, it is possible that some eigenvalues may move towards the imaginary axis of the complex plane. Therefore, the gain is limited to provide adequate margins to avoid destabilization. Since the $K_{oi}(s)$, $K_{ci}(s)$, and $K_{ILi}(s)$ are functions of the operating point, a compromise value of gain may be chosen which gives reasonable damping performance for a wide variety of operating points.

5) *Limiter*

A PSDC modulates the set-point of an actuator (e.g., the voltage reference of an AVR in an excitation system) to improve the small-signal stability of a system. Limits are generally provided to avoid excessive modulation which is not commensurate with the rating of the actuator, and to avoid large deviations in quantities which we seek to regulate (like generator terminal voltage). Moreover, large deviations in the set-point may endanger the large disturbance stability of the system.

6.6.3 PSDCs based on Generator Excitation Systems: Power System Stabilizers

PSDCs which use the generator excitation system as an actuator are called power system stabilizers (PSSs). Typically, a PSS output modulates the set-point V_{ref} of the AVR. This means that the PSS output, denoted by V_s, is added at the summing junction at the AVR input as shown in Figure 6.24. The following example demonstrates the damping introduced by a PSS in an SMIB system.

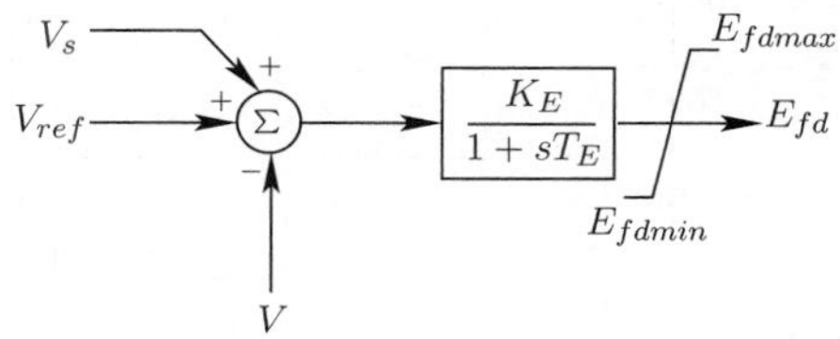

Figure 6.24 AVR with PSS output added at the AVR summing junction.

Example 6.3 ***Power System Stabilizer in an SMIB system*** Consider the SMIB system of Example 6.1.

1) Evaluate $K_c(s)$ for various values of the mechanical power input P_m and external reactance x_e.
2) Design a PSS using the slip signal $S = \frac{\omega - \omega_B}{\omega_B}$.

Solution

There is only one swing mode in this system. $\Delta\delta$ and $\Delta\omega$ are themselves the modal variables and therefore we drop the subscripts i and mi. The controller input is $u = \Delta V_s$ and the feedback signal is $y = \Delta S$. Note that the angular acceleration $\Delta\dot{\omega}$ is given by (see Figure 6.22)

$$\Delta\dot{\omega}(s) = -\frac{\omega_B}{2H}\Delta T_e(s) = -K_m(s)\Delta\delta(s) - K_c(s)\Delta V_s(s)$$

Therefore

$$K_c(s) = \left.\frac{\omega_B}{2H}\frac{\Delta T_e(s)}{\Delta V_s(s)}\right|_{\Delta\delta(s)=0}$$

With reference to Figure 6.22, note that $\alpha(s) = K_c(s)\Delta V_s(s)$. This transfer function can be computed from the linearized differential-algebraic equations of the system. The computed frequency response of $K_c(s)$ is shown in Figure 6.25 for various values of P_m and x_e. The phase angle $\angle K_c(j\Omega) + \angle K_o(j\Omega)$ is equal to $\angle K_c(j\Omega)$ for a slip signal since $K_o(s) = \frac{1}{\omega_B}$, which is a real number. This phase lag has to be compensated by the PSS at the swing mode frequency, which is approximately 7.9 rad/s, in order to achieve damping. From the figure it can be noted that the maximum phase lag and maximum gain of $K_c(s)$ occurs when $P_m = 1.0$ pu and $x_e = 0.2$ pu. A speed or slip input PSS is generally tuned at this

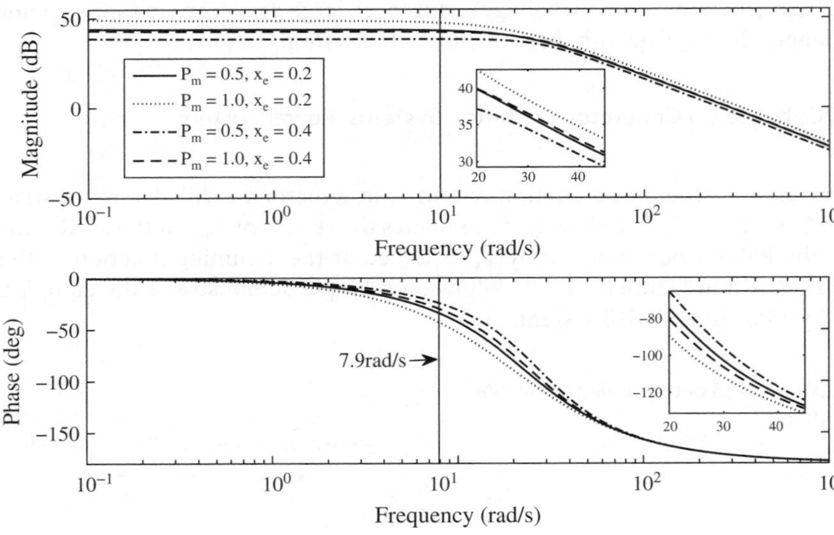

Figure 6.25 Frequency response of $K_c(s) = \left.\frac{\omega_B}{2H}\frac{\Delta T_e(s)}{\Delta V_s(s)}\right|_{\Delta\delta(s)=0}$.

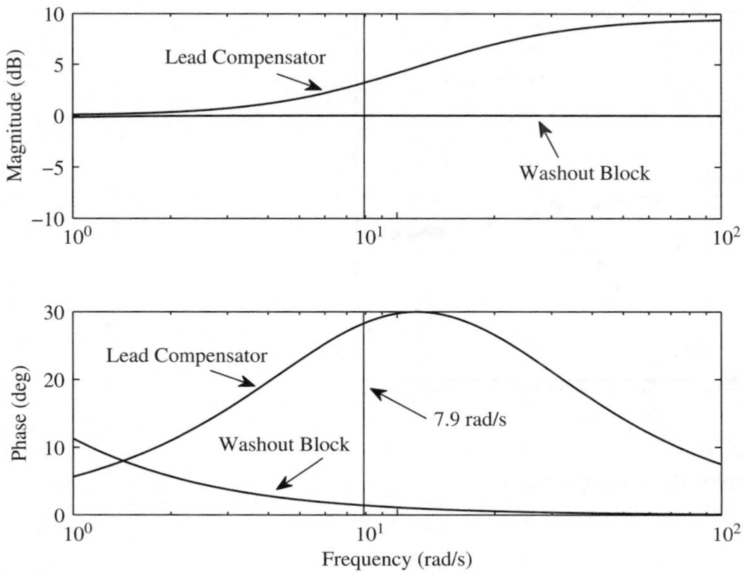

Figure 6.26 Frequency response of PSS transfer functions.

operating condition [18]. Note that since the slip is not dependent on any other state or input other than the speed (which in this case is also the modal speed), $K_{IL}(s) = 0$.

Consider a PSS consisting of a washout and phase lead block as given below:

$$K_{PSS}(s) = K \frac{T_w s}{1 + T_w s} \frac{1 + T_1 s}{1 + T_2 s}$$

where $T_1 = 0.15$ s, $T_2 = 0.05$ s, and $T_w = 5$ s. The Bode plots of the lead compensator and washout transfer functions are shown in Figure 6.26. The PSS slightly under-compensates the phase lag of $K_c(s)$ at the swing mode frequency (which is about 42° for $P_m = 1.0$ pu and $x_e = 0.2$ pu), but this is acceptable and will contribute to synchronizing torque.

The root locus plot is shown in Figure 6.27, where the PSS gain K is varied from 0 to 5 in steps of 0.5. The improved damping due to the PSS is evident from the root locus plot. Since $K_{IL} = 0$ for the slip signal, a significant damping influence can be obtained with increasing gain. The system, which was unstable without a PSS, is stabilized. The eigenvalues of the system with $K = 2$ pu/pu are:

$$-31.17, -14.03 \pm j7.7442, \mathbf{-0.675 \pm j8.10}, -0.201$$

The swing mode is highlighted in bold. The simulated response of the system to a pulse disturbance in mechanical power input (−5% lasting for 1 s) is shown in Figure 6.28. The stabilizing effect of the PSS is evident from the simulated response. The limits on the output of the PSS are set to +0.1 pu and −0.05 pu, but these are not hit for this disturbance.

An SMIB system is a good approximation for the study of the local mode of a small generator radially connected to a large power system. However, most generating plants consist of several generator units. Therefore, intra-plant swing modes are present in

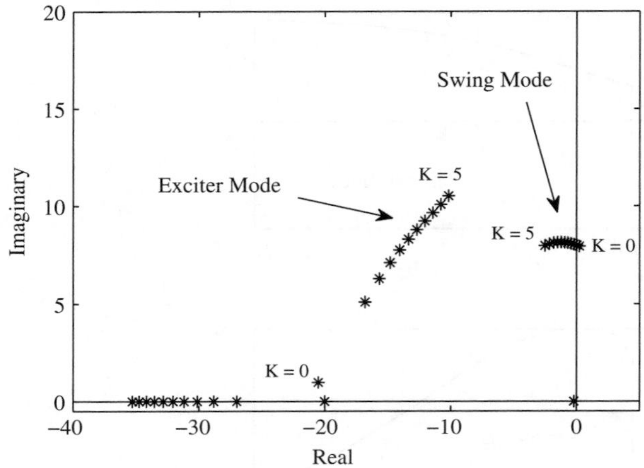

Figure 6.27 Root locus plot ($P_m = 1.0, x_e = 0.2$ pu).

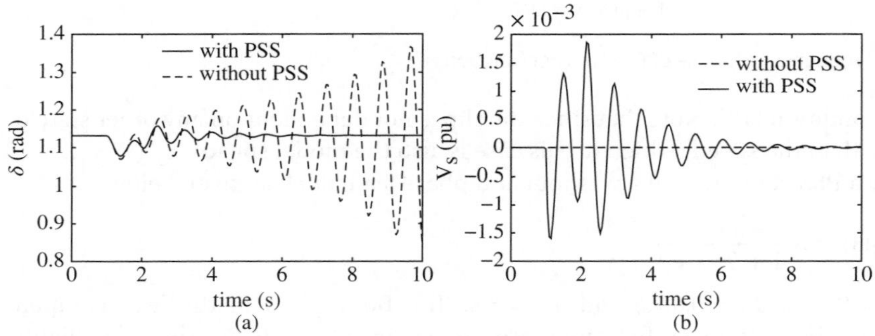

Figure 6.28 Rotor angle and PSS output with and without a PSS. PSS gain $K = 2.0$ pu/pu. (a) Rotor angle (b) PSS output.

addition to the inter-plant (local) modes (see Figure 6.13). In such a situation, if the PSS has been designed based on the compensation requirements for damping the local mode then its effect on the intra-plant mode must be carefully examined. This effect of the PSS on the intra-plant mode is illustrated in the following example.

Example 6.4 ***PSS performance in the presence of intra-plant modes*** Consider the SMIB system of Example 6.1. Let us split the single generator into two identical units of half the rating. On the system base, H is halved and the generator reactances are doubled. This system is shown in Figure 6.29. The reactance values are $x_{T1} = x_{T2} = 0.2$ pu and $x_{line} = 0.1$ pu. The mechanical power input to each machine is 0.5 pu and $V_1 = V_2 = 1.0$ pu.

1) Compute the eigenvalues of this two-generator infinite bus system and compare them with the SMIB system.

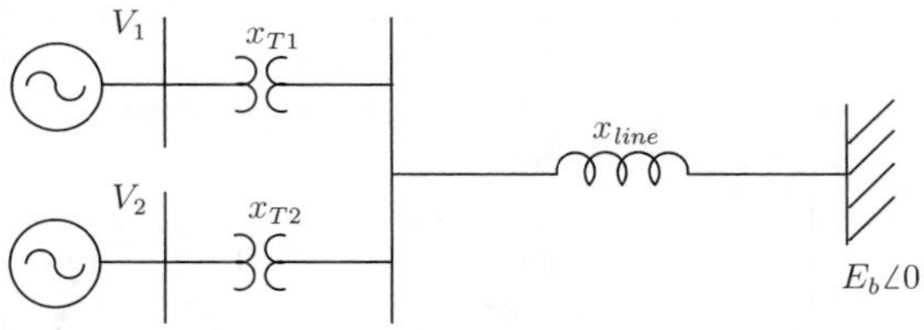

Figure 6.29 Power plant with two units connected to an infinite bus.

Table 6.5 Eigenvalues of the system with two identical generator units (without PSS).

Classical model	Model (1.0) (E_{fd} constant)	Model (1.0) + AVR	Comments
$0.0 \pm j\,8.29$	$-0.205 \pm j\,7.70$	$0.232 \pm j\,7.93$	Local mode
$0.0 \pm j\,9.43$	$-0.233 \pm j\,8.44$	$-0.099 \pm j\,8.34$	Intra-plant mode
–	-0.168	$-20.52 \pm j\,0.99$	
–	-0.210	$-7.927, -32.55$	

2) If the PSS given in the previous example is present on both machines (modulating the respective AVR and using the respective slip input), obtain the root loci for increasing PSS gain. Compare this with the root loci obtained in the previous example.

Solution

The eigenvalues of the two-generator infinite bus system (without PSSs) are given in Table 6.5. The eigenvalues corresponding to the local swing mode and exciter mode are the same as those obtained in Example 6.1 (for $P_m = 1.0$ pu and $x_e = 0.2$ pu). There are additional modes, namely the intra-plant swing mode and intra-plant exciter/field flux mode. The system will behave exactly as in Examples 6.1 and 6.3 for disturbances external to the plant, for which the intra-plant modes are not excited.

If the slip-input PSS of the previous example is applied to the two units, then the root locus plot for increasing gain (same gain in both units) is shown in Figure 6.30 (gain varied from 0 to 5 in steps of 0.5). The movement of the local mode is the same as before. The intra-plant mode exhibits an upward movement (increased modal frequency) in addition to an increase in damping.

For the analysis of this system, modal decomposition has to be performed to obtain the local and intra-plant mode using the eigenvectors of the A_{21} matrix of the system. Since both generators have identical parameters and operating conditions, and are connected in a similar fashion, it is easy to infer that the right eigenvector matrix $V = \begin{bmatrix} 1 & 1 \\ 1 & -1 \end{bmatrix}$. Note that $V^{-1} = \frac{1}{2}V^t$. In other words, $\omega_{m1} = \frac{\omega_1+\omega_2}{2}$, $\omega_{m2} = \frac{\omega_1-\omega_2}{2}$, $\omega_1 = \omega_{m1} + \omega_{m2}$, and $\omega_2 = \omega_{m1} - \omega_{m2}$.

The block diagram of Figure 6.22 can be generalized to consider two modes at a time. In such a case the gains shown in the figure will be 2×2 matrices while u, y, and α_m are 2×1 vectors. If both PSSs are identical and both are simultaneously active, then the effective feedback due to the PSSs is given by

$$\begin{bmatrix} \alpha_{m1}(s) \\ \alpha_{m2}(s) \end{bmatrix} = V \begin{bmatrix} \beta(s) & \gamma(s) \\ \gamma(s) & \beta(s) \end{bmatrix} \begin{bmatrix} K_{PSS}(s) & 0 \\ 0 & K_{PSS}(s) \end{bmatrix} \begin{bmatrix} \Delta S_1 \\ \Delta S_2 \end{bmatrix}$$

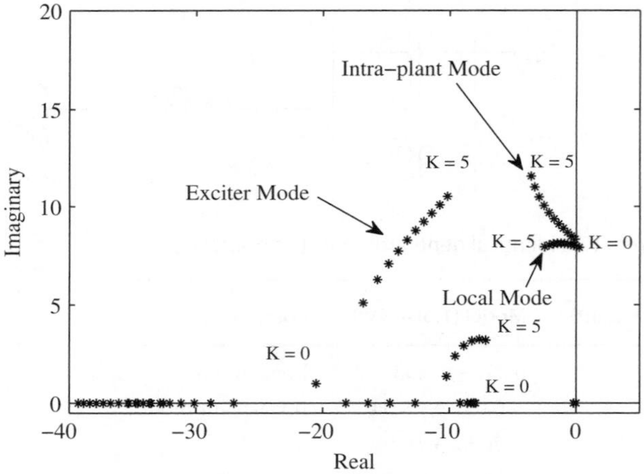

Figure 6.30 Root locus plot for the two-unit system with PSS in both units.

$$= V \begin{bmatrix} \beta(s) & \gamma(s) \\ \gamma(s) & \beta(s) \end{bmatrix} \begin{bmatrix} \dfrac{K_{PSS}(s)}{\omega_B} & 0 \\ 0 & \dfrac{K_{PSS}(s)}{\omega_B} \end{bmatrix} V^{-1} \begin{bmatrix} \Delta\omega_{m1} \\ \Delta\omega_{m2} \end{bmatrix}$$

$$= \frac{K_{PSS}(s)}{\omega_B} \begin{bmatrix} K_{c1}(s) & 0 \\ 0 & K_{c2}(s) \end{bmatrix} \begin{bmatrix} \Delta\omega_{m1} \\ \Delta\omega_{m2} \end{bmatrix} \tag{6.100}$$

where

$$\beta(s) = \frac{\omega_B}{2H_1} \frac{\Delta T_{e1}(s)}{\Delta V_{s1}(s)}\bigg|_{\substack{\Delta V_{s2}(s)=0 \\ \Delta\delta_1(s)=0 \\ \Delta\delta_2(s)=0}} = \frac{\omega_B}{2H_2} \frac{\Delta T_{e2}(s)}{\Delta V_{s2}(s)}\bigg|_{\substack{\Delta V_{s1}(s)=0 \\ \Delta\delta_1(s)=0 \\ \Delta\delta_2(s)=0}}$$

$$\gamma(s) = \frac{\omega_B}{2H_1} \frac{\Delta T_{e1}(s)}{\Delta V_{s2}(s)}\bigg|_{\substack{\Delta V_{s1}(s)=0 \\ \Delta\delta_1(s)=0 \\ \Delta\delta_2(s)=0}} = \frac{\omega_B}{2H_2} \frac{\Delta T_{e2}(s)}{\Delta V_{s1}(s)}\bigg|_{\substack{\Delta V_{s2}(s)=0 \\ \Delta\delta_1(s)=0 \\ \Delta\delta_2(s)=0}}$$

$K_{c1}(s) = \beta(s) + \gamma(s)$ and $K_{c2}(s) = \beta(s) - \gamma(s)$, which correspond to the local and intra-plant modes, respectively. Note that $H_1 = H_2 = 0.5H$. Therefore, $K_{c1}(s)$ is identical to the transfer function $K_c(s)$ considered in the previous example.

The phase responses of $K_{c1}(s)$ and $K_{c2}(s)$ are shown in Figure 6.31. The PSS which was considered in the previous example will significantly under-compensate for the intra-plant mode. This is because the phase lag of $K_{c2}(s)$ at the intra-plant mode frequency is higher than $K_{c1}(s)$ at the local-mode frequency by about 17°, but the phase lead provided by $K_{PSS}(s)$ is practically the same at both modal frequencies. Since the PSS under-compensates for the intra-plant mode, a significant increase in the modal *frequency* occurs as the gain is increased.

To avoid affecting the intra-plant modes adversely, appropriate phase compensation should be provided in a PSS at the modal frequencies, and the effect of the PSS on these

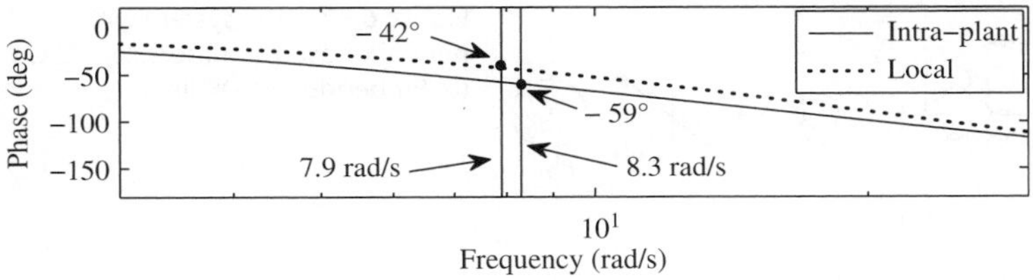

Figure 6.31 Phase responses of $K_{c1}(s)$ (local mode) and $K_{c2}(s)$ (intra-plant mode).

modes should be tracked during PSS design. Alternatively, feedback signals with low observability of the intra-plant modes like the frequency at the common high-voltage bus or the average speed of the units may be used instead of the slip or speed signals of the individual units. This ensures that the intra-plant modes, which are generally well-damped to begin with, are not affected by the PSS.

6.6.4 Adverse Torsional Interactions with the Speed/Slip Signal

The speed or slip signal contains torsional oscillations of the generator-turbine shaft (this aspect is discussed in detail in Chapter 10). A slip or speed input PSS may inadvertently reduce the damping of the torsional modes. This is not evident in our present study since we have not modeled shaft torsional dynamics. In practice, speed or slip signals are not used for PSSs unless the signals are carefully filtered to exclude the torsional modes. Alternatively, signals which are closely related to the speed signals, but have lower observability of the torsional oscillations may be used. Examples of such feedback signals are electrical power and bus frequency. The impact of these signals on other aspects of the dynamic response should be carefully considered. For example, an electrical power based PSS may affect the terminal voltage during power ramping [7] while the frequency signal may be susceptible to noise due to nearby loads like arc furnaces [19]. A modified speed signal synthesized using both the electrical power and the speed signal of a generator, called the delta-P-omega signal has been proposed [20]. This has low observability of the torsional modes, but is practically identical to the speed signal in the low-frequency range. The synthesis of this signal and its impact on torsional modes is presented and contrasted with a conventional speed-input PSS in Section 10.4.2.

6.6.5 Damping of Swings using Grid-Connected Power Electronic Systems

Although PSSs can influence inter-area swing mode damping in addition to damping the local modes, actuators like FACTS and HVDC systems which are deployed in the transmission network may have greater controllability over the inter-area modes. While the detailed model of the system is required to design and evaluate a damping controller, an intuitive understanding of the damping provided by a HVDC and a TCSC can be obtained from an SMIB system. The classical model of the generator is considered in this analysis, as shown in Figure 6.32. The TCSC is modeled as a controllable capacitive reactance (x_{TCSC}). Since the electrical power output of the generator is given by

$$P_e = \frac{E'_q E_b \sin\delta}{x'_d + x_T + x_{line} - x_{TCSC}} \tag{6.101}$$

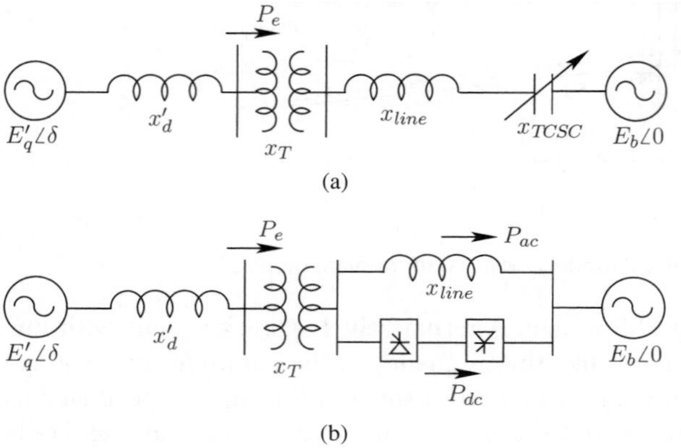

Figure 6.32 SMIB system with network-based actuators. (a) TCSC (b) Embedded HVDC link.

it follows that

$$\Delta P_e = \frac{E'_q E_b \cos\delta_o \Delta\delta}{x'_d + x_T + x_{line} - x_{TCSCo}} + \frac{E'_q E_b \sin\delta_o \Delta x_{TCSC}}{(x'_d + x_T + x_{line} - x_{TCSCo})^2} \tag{6.102}$$

If Δx_{TCSC} is varied in proportion to the speed deviation $\Delta\omega$ then

$$\begin{aligned}\Delta P_e &= \frac{E'_q E_b \cos\delta_o\ \Delta\delta}{x'_d + x_T + x_{line} - x_{TCSCo}} + \frac{E'_q E_b \sin\delta_o\ k_{TCSC}\ \Delta\omega}{(x'_d + x_T + x_{line} - x_{TCSCo})^2}\\ &= K\Delta\delta + D_{TCSC}\Delta\omega\end{aligned} \tag{6.103}$$

It is assumed here for simplicity that the TCSC response to the modulation signal $\Delta\omega$ is instantaneous, that is, the time-constant of the TCSC plant, T_p in (6.65), is negligible. D_{TCSC} represents the damping introduced by the TCSC. Note that when the expression for power deviation is put in the swing equations (see Section 6.2) it causes damping in a manner similar to the mechanical damping term D.

In the case of an HVDC link connected in parallel to the AC line, as shown in Figure 6.32,

$$\Delta P_e = \Delta P_{ac} + \Delta P_{dc} \tag{6.104}$$

If the power in the DC link is modulated in proportion to the generator speed deviation, then for a small proportional gain k_{dc}

$$\Delta P_e = \frac{E'_q E_b \cos\delta_o\ \Delta\delta}{x'_d + x_T + x_{line} - x_{TCSCo}} + k_{dc}\Delta\omega = K\Delta\delta + D_{dc}\Delta\omega \tag{6.105}$$

As in the case of a TCSC, damping is introduced due to the term $D_{dc}\Delta\omega$. Note that the modulating signals Δx_{TCSC} and ΔP_{dc} are added to the set-point values of the TCSC reactance x_{TCSCo} and HVDC link power P_{dco}, respectively.

In a large meshed network with many machines, local signals at the FACTS or HVDC locations like bus frequency, voltage angular/frequency differences across transmission lines, AC transmission line current magnitudes, AC line power flows, or a combination of these signals may be used. Signals synthesized from measurements taken at one or more remote locations, or system-wide signals may also be considered.

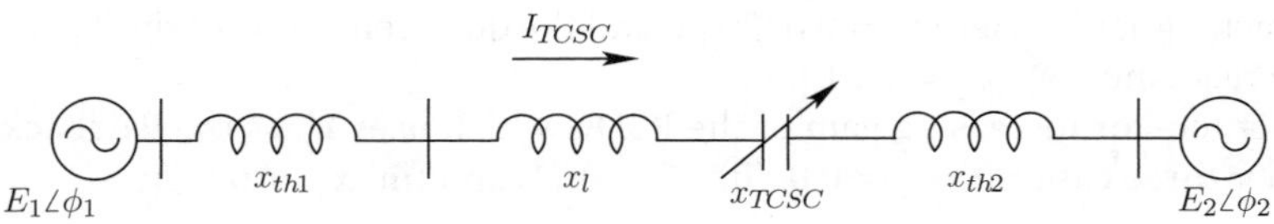

Figure 6.33 Equivalent system for computing the voltage across the transmission path.

In the following example, damping introduced by a TCSC in a three-machine system is presented. The signal V_p, which is mathematically defined as (see Figure 6.33)

$$V_p = I_{TCSC}(x_{th1} + x_{th2} + x_l - x_{TCSC}) = I_{TCSC}(x_{th} + x_l - x_{TCSC}) \tag{6.106}$$

is used as the feedback signal. x_l is the reactance of the transmission line, x_{TCSC} is the capacitive reactance of the TCSC and I_{TCSC} is the line current magnitude. x_{th} is a tunable parameter which may be tuned to attain the best damping influence. Note that V_p represents the voltage difference across the transmission path. This signal is similar to the angular difference of synthesized remote voltages on each side of the TCSC, which is examined in [16]. V_p can be obtained from the local measurements of current and the voltage magnitude across the TCSC. Note that for small deviations,

$$\Delta V_p = (x_{th} + x_l - x_{TCSCo})\Delta I_{TCSC} - I_{TCSCo}\Delta x_{TCSC} \tag{6.107}$$

Δx_{TCSC} is the variation in the TCSC reactance due to the PSDC.

Example 6.5 ***TCSC-based PSDC in a Three-Machine System*** Consider the three-machine system of Example 6.2. Note that there are two swing modes in this system. A TCSC with a quiescent capacitive reactance of 0.01 pu is introduced in lines 4–6 as shown in Figure 6.34. A PSDC is to be used, which modulates this reactance.

1) Evaluate $K_o(j\Omega)$, $K_c(j\Omega)$, $K_{IL}(j\Omega)$, MDI, and CPI corresponding to both swing modes for the following candidate feedback signals:
   (a) TCSC current magnitude I_{TCSC}

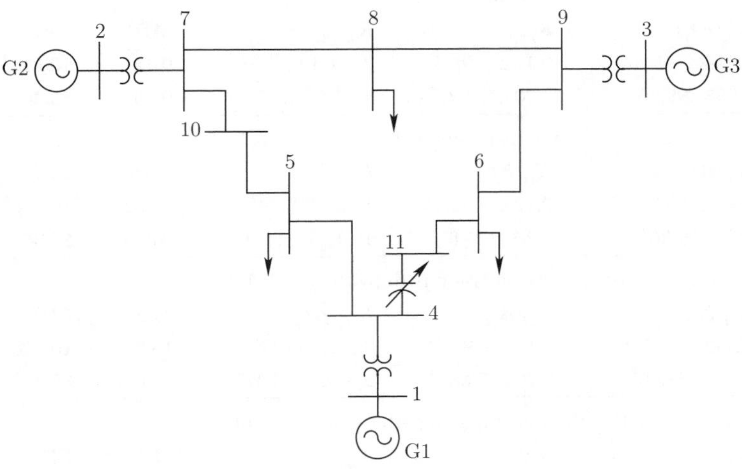

Figure 6.34 Three-machine system with a TCSC introduced in line 4-6.

(b) The feedback signal V_p. Here $x_l = 0.0920$ pu and the quiescent value of the TCSC capacitive reactance, $x_{TCSCo} = 0.01$ pu.

2) Obtain the root loci for increasing gain of the PSDC which uses V_p as the feedback signal. Consider three cases: (i) $x_{th} = 0.0$, (ii) $x_{th} = 0.41$, and (iii) $x_{th} = 0.6$ pu.

Solution

The quiescent value of the TCSC reactance is 0.01 pu. The design indices K_{oi}, K_{ci}, K_{ILi}, MDI, and CPI obtained for the I_{TCSC} and V_p as feedback signals are given in Table 6.6. These are shown for the two swing modes of this system.

Note that the CPI is near +90° for both swing modes and for all the candidate signals. Therefore the compensator design for these signals is straightforward: a compensator which gives a phase shift of 90° in the frequency range corresponding to the swing modes may be used to achieve damping.

Interestingly, it is shown analytically [21] that the CPI is exactly +90° for all swing modes and operating conditions, for certain dual pairs of actuators and feedback signals (this holds true for a simplified model of a power system with the synchronous machines represented by the classical model). The line current magnitude and the TCSC reactance form such a dual pair [21]. Since ΔV_p is also practically proportional to ΔI_{TCSC} if the PSDC gain (and therefore Δx_{TCSC}) is small, the result holds true for this signal as well. As we have considered a detailed model of the system here, the computed CPIs in the table are not exactly +90°.

The MDI for the voltage across the transmission path, with a tuned value of $x_{th} = 0.41$ pu, is the highest among the three signals. In this case, the increase in damping with increasing PSDC gain is expected to be quite significant as the inner-loop effect is relatively small.

Consider the PSDC shown in Figure 6.35. The parameters of this controller are $T_w = 5.0$ s and $T = 0.01$ s, so that an approximate lead of 90° is achieved at the swing mode

Table 6.6 Computed design indices.

		With current signal			
$\lambda_i = \sigma_i + j\Omega_i$	$K_{oi}(j\Omega_i)$	$K_{ci(j\Omega_i)}$	$K_{ILi(j\Omega_i)}$	MDI	CPI
$-0.035 \pm j8.61$	$48.14\angle -91.21°$	$0.12\angle -9.25°$	$2.99\angle 1.05°$	0.31	100.46°
$-0.45 \pm j13.18$	$24.98\angle 99.86°$	$0.20\angle 174.87°$	$4.03\angle -2.65°$	0.20	85.26°
		Voltage across the transmission path (with $x_{th} = 0.0$)			
$\lambda_i = \sigma_i + j\Omega_i$	$K_{oi}(j\Omega_i)$	$K_{ci}(j\Omega_i)$	$K_{ILi}(j\Omega_i)$	MDI	CPI
$-0.035 \pm j8.61$	$3.95\angle -91.21°$	$0.12\angle -9.25°$	$1.23\angle 179.79°$	0.062	100.46°
$-0.45 \pm j13.18$	$2.05\angle 99.86°$	$0.20\angle 174.87°$	$1.14\angle -179.24°$	0.056	85.26°
		Voltage across the transmission path (with $x_{th} = 0.41$)			
$\lambda_i = \sigma_i + j\Omega_i$	$K_{oi}(j\Omega_i)$	$K_{ci}(j\Omega_i)$	$K_{ILi}(j\Omega_i)$	MDI	CPI
$-0.035 \pm j8.61$	$23.59\angle -91.21°$	$0.12\angle -9.25°$	$0.03\angle 109.47°$	15.94	100.46°
$-0.45 \pm j13.18$	$12.24\angle 99.86°$	$0.20\angle 174.87°$	$0.51\angle -10.37°$	0.76	85.26°
		Voltage across the transmission path (with $x_{th} = 0.6$)			
$\lambda_i = \sigma_i + j\Omega_i$	K_{oi}	K_{ci}	K_{ILi}	MDI	CPI
$-0.035 \pm j8.61$	$32.83\angle -91.21°$	$0.12\angle -9.25°$	$0.57\angle 3.78°$	1.11	100.46°
$-0.45 \pm j13.18$	$17.04\angle 99.86°$	$0.20\angle 174.87°$	$1.28\angle -5.70°$	0.41	85.26°

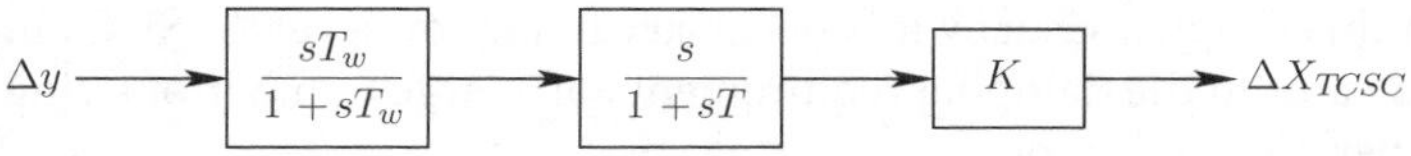

Figure 6.35 TCSC damping controller.

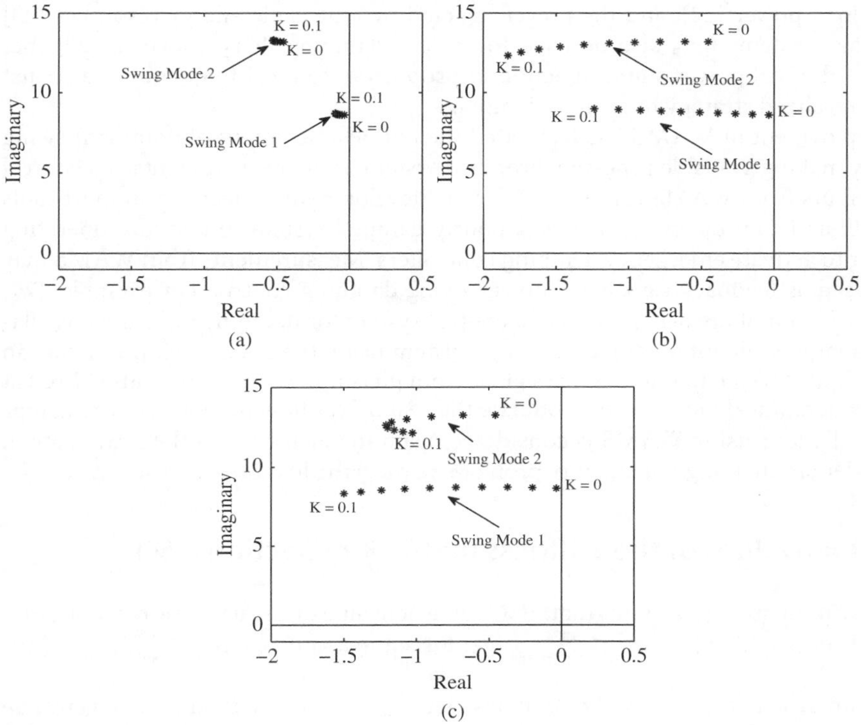

Figure 6.36 Movement of the swing modes with V_p as the PSDC input. (a) With x_{th} = 0.0 pu (b) With x_{th} = 0.41 pu (c) With x_{th} = 0.60 pu.

frequencies. Additional filtering is not considered here, although it would be necessary in practice to attenuate the gain at high frequencies. The root loci for a PSDC using V_p as an input signal, with $x_{th} = 0.0$, $x_{th} = 0.41$, and $x_{th} = 0.6$ pu, are shown in Figure 6.36. It is clear that the movement of both swing modes is towards the left half of the complex plane for lower gains. For higher gains, the movement in the case of $x_{th} = 0.0$ and $x_{th} = 0.6$ is inhibited due to the inner-loop effect discussed earlier; after an initial increase in damping there is a "turning back" of the eigenvalue. This is evident for $x_{th} = 0.0$ and $x_{th} = 0.6$ pu. The movement of the swing modes is best achieved with $x_{th} = 0.41$ pu, and this may be taken as the tuned value for this operating condition.

6.7 Concluding Remarks

The damping of power swings is affected by the operating conditions. Since a power system experiences wide variations in the operating conditions, there is a need to continually assess the stability of power swings. Instability can be avoided by deployment of

PSDCs, as discussed in this chapter. Usually it is necessary to deploy several PSDCs in a large power system to cater to the damping requirements of a large number of swing modes under various operating conditions.

While generator excitation systems, FACTS, and HVDC systems are the primary actuators for implementing damping controllers, decentralized modulation control of end-use load power [22], and the power injected by renewable energy resources [23] and storage systems may also be used to achieve damping. It is shown in [22] that control of the real power shunt injection in proportion to local frequency is a robust and decentralized strategy to achieve damping.

The deployment of WAMS has facilitated the online assessment of dominant swing modes by making available time-synchronized system-wide measurements at a control centre. Inputs from WAMS can be used for the development of decision support tools which will assist an operator to steer a poorly damped system to a secure operating region, or to activate emergency damping controllers. Measurements from WAMS may also be used as feedback signals for power swing damping control. For example, [24] describes the initial open-loop tests of a control system for damping inter-area oscillations in a large-scale interconnected power system using the Pacific DC Intertie as an actuator. It uses frequency-difference signals from phasor measurement units (PMU) at various geographical locations to modulate the real power flow in this link. The damping of oscillations using WAMS is considered again in Chapter 9. In the next chapter, we consider another angular stability problem, namely the loss of synchronism.

6.A Eigenvalues of the Stiffness matrix K of Section 6.5.1

Gerschgorin theorem [13] states that: If λ is an eigenvalue of a square matrix A of order m, then λ must satisfy $|\lambda - a_{ii}| \leq \sum_{j=1 j\neq i}^{m} |a_{ij}|$ for some i, and $|\lambda - a_{ii}| \leq \sum_{j=1 j\neq i}^{m} |a_{ji}|$ for some i.

For a symmetric matrix this implies that if we let $\sum_{j=1 j\neq i}^{m} |a_{ij}| = R_i$ then every eigenvalue must lie inside at least one of the regions $|\lambda - a_{ii}| = R_i$. As discussed in Section 6.5.1, K is a real symmetric $m \times m$ matrix such that $K_{ij} \leq 0$ for $i \neq j$ and $\sum_{j=1}^{m} K_{ij} = 0$ for all i. Therefore, it can be inferred from Gerschgorin Theorem that all eigenvalues of K are non-negative.

Note that K is singular and we are also interested in finding the number of zero eigenvalues of K. Since K is a $m \times m$ symmetric matrix, it has m linearly independent eigenvectors. Therefore finding the number of zero eigenvalues is equivalent to finding the number of linearly independent eigenvectors of K, denoted here by x, for which $Kx = 0$. Since $\sum_{j=1}^{m} K_{ij} = 0$, one such eigenvector is $v_m^t = [1\ 1\ ....\ 1]$.

K can be written as follows: $K = QK_dQ^t$, where Q is the node-branch incidence matrix with entries 0, +1, and −1, and K_d is a $n_{br} \times n_{br}$ diagonal matrix where n_{br} is the number of branches. All the entries of K_d are positive if we assume that $|\delta_i - \delta_j| < 90°$ (see Section 6.5.1). Hence, K_d is a symmetric positive definite matrix and is non-singular. Q is a $m \times n_{br}$ matrix as given below:

$$Q_{ij} = \begin{cases} +1 & \text{if the reference direction of power in branch } j \text{ is out of bus } i \\ -1 & \text{if the reference direction of power in branch } j \text{ is into bus } i \\ 0 & \text{if branch } j \text{ is not connected to bus } i \end{cases} \tag{6.108}$$

Table 6.7 Transmission line and transformer data.

Bus numbers		Resistance	Reactance	Shunt susceptance (total)
4	6	0.0170	0.0920	0.1580
6	9	0.0390	0.1700	0.3580
9	8	0.0119	0.1008	0.2090
8	7	0.0085	0.0720	0.1490
10	5	0.0160	0.0805	0.1530
7	10	0.0160	0.0805	0.1530
5	4	0.0100	0.0850	0.1760
1	4	0.0000	0.0576	0.0000
9	3	0.0000	0.0586	0.0000
7	2	0.0000	0.0625	0.0000

Table 6.8 Generator data.

Bus	R_a	x_d	x'_d	T'_{d0}	x_q	x'_q	T'_{q0}	H	D	General model
1	0.0	0.1460	0.0608	8.9600	0.0969	—	–	23.64	0.0	(1.0)
2	0.0	0.8958	0.1198	6.0000	0.8645	0.1969	0.5350	6.400	0.0	(1.1)
3	0.0	1.3125	0.1813	5.8900	1.2578	0.2500	0.6000	3.010	0.0	(1.1)

Table 6.9 Power flow data.

Bus	Voltage	Angle (degrees)	Generated real power	Generated reactive power	Load real power	Load reactive power
1	1.0400	0.0000	2.9329	0.9793	0.00	0.00
2	1.0253	−1.8243	2.2109	0.5318	0.00	0.00
3	1.0253	−8.4713	1.1232	0.3589	0.00	0.00
4	0.9991	−9.3573	0.0000	0.0000	0.00	0.00
5	0.9702	−16.8045	0.0000	0.0000	2.25	0.75
6	0.9482	−16.9167	0.0000	0.0000	1.90	0.60
7	1.0020	−9.5543	0.0000	0.0000	0.00	0.00
8	0.9691	−15.4418	0.0000	0.0000	2.00	0.65
9	1.0068	−12.1270	0.0000	0.0000	0.00	0.00
10	0.9902	−13.1914	0.0000	0.0000	0.00	0.00

Every column of Q adds up to 0 and every column of Q has exactly one +1 and one −1 entry. The rest of the entries are 0. For example, K of the network given in Figure 6.12 may be written as follows:

$$K = \begin{bmatrix} 1 & 1 & 0 \\ -1 & 0 & 1 \\ 0 & -1 & -1 \end{bmatrix} \begin{bmatrix} K_{13} & 0 & 0 \\ 0 & K_{12} & 0 \\ 0 & 0 & K_{23} \end{bmatrix} \begin{bmatrix} 1 & -1 & 0 \\ 1 & 0 & -1 \\ 0 & 1 & -1 \end{bmatrix}$$

If x is an eigenvector of K corresponding to zero eigenvalue, then $x^t K x = 0$. Hence, $x^t Q K_d Q^t x = 0$, which can be written as $(Q^t x)^t K_d (Q^t x) = 0$. Since K_d is a symmetric postive definite matrix, the solution of $Q^t x = 0$ are the eigenvectors of K corresponding to its zero eigenvalues. Hence, we need to find the number of linearly independent vectors x such that $Q^t x = 0$. It is known that for a connected network (i.e., no islands are present in the network), the rank of its incidence matrix Q is $m - 1$, where m is the number of nodes (see Lemma 2.2 in [25]). Hence for such a network, K has only one linearly independent eigenvector corresponding to a zero eigenvalue. Therefore for a connected network, K has exactly one zero eigenvalue.

6.B Three-Machine System Data

The data for this system is given in Tables 6.7, 6.8 and 6.9.

1) The per-unit data is given on a 100 MVA base. The base frequency $\omega_B = 2 \times \pi \times 60$ rad/s. The time constants are in seconds.
2) A fixed capacitive shunt with a susceptance value $B = 0.5$ pu is present at bus 5. This is in addition to the line capacitance given in Table 6.7.
3) Each machine is equipped with an AVR and a static exciter represented as shown in Figure 6.4. The parameters of the transfer function are $K_E = 200$ pu/pu and $T_E = 0.05$ s.

References

1 Salunkhe, K.A., Gajjar, G., Soman, S.A., and Kulkarni, A.M. (2014) Implementation and applications of a wide area frequency measurement system synchronized using Network Time Protocol, in *2014 IEEE PES General Meeting*, National Harbor, Maryland, pp. 1–5.

2 Hauer, J., Trudnowski, D., Rogers, G. *et al.* (1997) Keeping an eye on power system dynamics. *IEEE Computer Applications in Power*, **10** (4), 50–54.

3 Heffron, W.G. and Phillips, R.A. (1952) Effect of a modern amplidyne voltage regulator on underexcited operation of large turbine generators. *Transactions of the American Institute of Electrical Engineers. Part III: Power Apparatus and Systems*, **71** (1), 692–697.

4 Pagola, F.L., Perez-Arriaga, I.J., and Verghese, G.C. (1989) On sensitivities, residues and participations: Applications to oscillatory stability analysis and control. *IEEE Transactions on Power Systems*, **4** (1), 278–285.

5 Padiyar, K.R. (2002) *Power System Dynamics: Stability and Control*, BS Publications, Hyderabad, 2nd edn.

6 Chen, C.T. (1995) *Linear System Theory and Design*, Oxford University Press, New York, 2nd edn.

7 Rogers, G. (2000) *Power System Oscillations*, Kluwer, Boston.

8 Padiyar, K.R. (2015) *HVDC Power Transmission Systems*, New Age International Publishers, New Delhi, 3rd edn.

9 Kundur, P. (1994) *Power System Stability and Control*, McGraw-Hill, New York.

10 Rogers G. J, Manno, J.D. and Alden, R.T.H. (1984) An aggregate induction motor model for industrial plants. *IEEE Transactions on Power Apparatus and Systems*, **PAS-103** (4), 683–690.

11 Vaahedi, E., El-Kady, M.A., Libaque-Esaine, J.A., and Carvalho, V.F. (1987) Load models for large-scale stability studies from end-user consumption. *IEEE Power Engineering Review*, **PER-7** (11), 29–29.

12 Paserba, J.J., Miller, N.W., Larsen, E.V., and Piwko, R.J. (1995) A thyristor controlled series compensation model for power system stability analysis. *IEEE Transactions on Power Delivery*, **10** (3), 1471–1478.

13 Golub, G.H. and Van Loan, C.F. (2013) *Matrix Computations*, John Hopkins University Press, Baltimore, 4th edn.

14 Pai, M.A., Sengupta, D.P., and Padiyar, K.R. (2004) *Small Signal Analysis of Power Systems*, Alpha Science International Ltd, Harrow, Middlesex.

15 Anderson, P.M. and Fouad, A.A. (2003) *Power system control and stability*, IEEE Press Power Engineering Series, IEEE Press, New York, 2nd edn.

16 Larsen, E.V., Sanchez-Gasca, J.J., and Chow, J.H. (1995) Concepts for design of FACTS controllers to damp power swings. *IEEE Transactions on Power Systems*, **10** (2), 948–956.

17 Phadke, A.G. and Thorp, J.S. (2008) *Synchronized Phasor Measurements and Their Applications*, Springer,USA.

18 Larsen, E.V. and Swann, D.A. (1981) Applying power system stabilizers Part I: General concepts, Part II: Performance objectives and tuning concepts, and Part III: Practical considerations. *IEEE Transactions on Power Apparatus and Systems*, **PAS-100** (6), 3017–3046.

19 Busby, E.L., Hurley, J.D., Keay, F.W., and Raczkowski, C. (1979) Dynamic stability improvement at Monticello station-analytical study and field tests. *IEEE Transactions on Power Apparatus and Systems*, **PAS-98** (3), 889–901.

20 Lee, D.C., Beaulieu, R.E., and Service, J.R.R. (1981) A power system stabilizer using speed and electrical power inputs-design and field experience. *IEEE Transactions on Power Apparatus and Systems*, **PAS-100** (9), 4151–4157.

21 Mhaskar, U.P. and Kulkarni, A.M. (2006) Power oscillation damping using FACTS devices: modal controllability, observability in local signals, and location of transfer function zeros. *IEEE Transactions on Power Systems*, **21** (1), 285–294.

22 Samuelsson, O. (2001) Load modulation for damping of electro-mechanical oscillations, in *IEEE Power Engineering Society Winter Meeting*, Columbus, pp. 241–246, Vol. 1.

23 Singh, M., Allen, A., Muljadi, E., and Gevorgian, V. (2014) Oscillation damping: A comparison of wind and photovoltaic power plant capabilities, in *IEEE Symposium*

on Power Electronics and Machines for Wind and Water Applications, Milwaukee, pp. 1–7.

24 Pierre, B.J., Wilches-Bernal, F., Schoenwald, D.A. *et al.* (2017) Open-loop testing results for the Pacific DC Intertie wide area damping controller, in *IEEE Manchester PowerTech*, Manchester, pp. 1–6.

25 Bapat, R.B. (2014) *Graphs and Matrices*, Hindustan Book Agency (India), New Delhi, 2nd edn.

7

Analysis and Control of Loss of Synchronism

7.1 Introduction

In the previous chapter we saw that a synchronous grid – a grid having synchronous machines connected by AC transmission lines – may occasionally experience small-signal angular instability, manifested as negatively damped power swings. Auxiliary controllers like power system stabilizers (PSSs) in generator excitation systems or power swing damping controllers in flexible AC transmission systems (FACTS) and high-voltage direct current (HVDC) systems can add damping to these power swings. While small-signal stability is essential for system operation, it does not guarantee that the system will be stable for any arbitrarily large deviation of the states from the equilibrium point. It is seen in synchronous grids that if the rotor angle and speed deviations (due to disturbances) lie outside the *region of attraction* of the equilibrium point, then it may lead to an unbounded increase in the relative rotor angles and speeds. The system then cannot settle back to a synchronous operating condition. This phenomenon is known as loss of synchronism (LoS).

A system is vulnerable to LoS when the region of attraction is small or when disturbances like faults cause large deviations from the equilibrium. If the system is small-signal unstable or does not have an equilibrium, then the region of attraction is non-existent. Under these circumstances too, the relative rotor angles and speeds may increase and eventually cause LoS. The terms pole-slipping, out-of-step operation and LoS refer to the same phenomenon. When LoS occurs due to large disturbances like faults, then it is also known as transient instability or large disturbance angular instability.

LoS arises because of the nonlinear relationship between the electrical power flows and phase angular differences in a synchronous grid. Like the undamped power swings studied in the previous chapter, LoS is an instability of the relative angular motion between the rotors of synchronous machines. Therefore, LoS does not occur in a system consisting of a single synchronous generator connected to a passive load, and in a system with synchronous machines that are connected to each other by asynchronous (DC) transmission links.

7.2 Effect of LoS

LoS is accompanied by large deviations in rotor angles and speeds, and fluctuating voltages, currents, and generator torques. This cannot be allowed to continue for long as it

Dynamics and Control of Electric Transmission and Microgrids, First Edition.
K. R. Padiyar and Anil M. Kulkarni.

may result in equipment damage. The large variations in voltages, currents, and speeds generally trigger protective relay actions, causing the tripping of equipment. The tripping of transmission lines due to such protective relay actions may lead to the separation of a large synchronous grid into smaller islands. While the integrated system prior to the disturbance may have had a balance between load and generation, this may not be the case in the individual islands that are formed due to an uncontrolled separation. Large load-generation imbalances in the separated islands cause a sharp rise or decline in the voltages and frequencies. If this causes further equipment trippings, then it may be unviable to operate the system (a "blackout").

If a LoS situation does occur, only a very small window of time is available for taking emergency control actions to stabilize the system and retain its integrity. If separation cannot be avoided, it may be possible to prevent an uncontrolled separation by blocking uncoordinated trippings caused by the equipment-protection relays. The disconnection points may then be chosen so that the islands that are formed have a better load-generation balance and have generators within them remaining in synchronism. Load and generation tripping may also be done in order to improve the load-generation balance in the islands. The islands thus formed are more likely to survive, and can then be resynchronized to each other expeditiously once the transients have died down.

The blackouts of 30th and 31st July 2012 in the Indian grid were caused by LoS, triggered by the tripping of heavily loaded transmission lines [1]. The frequency trajectories at two locations in the grid during these disturbances are shown in Figure 7.1. The frequencies were measured using a wide-area frequency measurement system [2]. The frequencies, which were equal to each other before the initiating disturbance, eventually diverged from each other after the disturbance, indicating a LoS situation. On both days, the system separated into two islands due to the tripping of additional transmission lines a few seconds after the initial line tripping. The additional lines tripped because the distance relays on these lines mistook the large voltage and current variations following the LoS to be due to a fault. On both days, one of the islands had a severe generation deficit and it blacked out due to under-frequency.

Whether or not a system will lose synchronism following a disturbance depends on the nature and location of the disturbance and the operating conditions. The analysis and control of LoS is more complicated than the analysis and control of small-signal angular instability, since it involves the consideration of the nonlinear characteristics of the

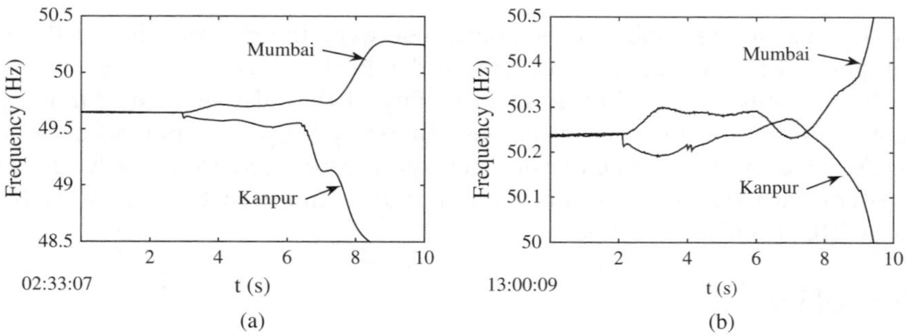

Figure 7.1 Frequencies measured at two locations in the Indian grid during the blackouts. (a) 30th July 2012; (b) 31st July 2012.

system. The techniques for the study of LoS generally involve the numerical simulation of a large number of credible disturbances. A system should normally be operated so as to have an adequate stability margin to withstand credible disturbances without LoS. Stability margins can often be improved by operating the system at low loading levels, that is, by keeping the quiescent voltage angular differences across transmission lines at a low value. Unfortunately, this means that generation may have to be rescheduled or loads may need to be tripped to curtail power flows, which has significant economic implications. Stability can be improved by strengthening the transmission network through the installation of new transmission lines and series/shunt compensation equipment. In addition, controllers of generator excitation systems, FACTS, and HVDC systems may be designed to improve stability by exploiting the transient capabilities of these devices.

7.3 Understanding the LoS Phenomenon

Since LoS is a phenomenon caused by the instability of relative angular motion, it can be understood using the example of a single-machine infinite bus (SMIB) system, as shown in Figure 7.2. The dynamics of the rotor angle δ and the rotor electrical speed ω are described by the following equations:

$$\frac{d\delta}{dt} = \omega - \omega_b \tag{7.1}$$

$$M\frac{d(\omega - \omega_b)}{dt} = T_m - T_e \approx P_m - P_e \tag{7.2}$$

where $M = \frac{2H}{\omega_B}$. H is the inertia constant of the machine in MJ/MVA, ω_B is the base electrical frequency in rad/s and ω_b is the frequency of the infinite bus. Damping is neglected in the model. Note that the per-unit mechanical and electrical torques (T_m and T_e) and per-unit powers (P_m and P_e) are approximately equal since $\frac{\omega}{\omega_B} \approx 1$ in practical angular stability studies.

If the generator is represented using the classical model (constant E'_q), as shown in the figure, and the network is assumed to be in quasi-sinusoidal steady state, then P_e is given

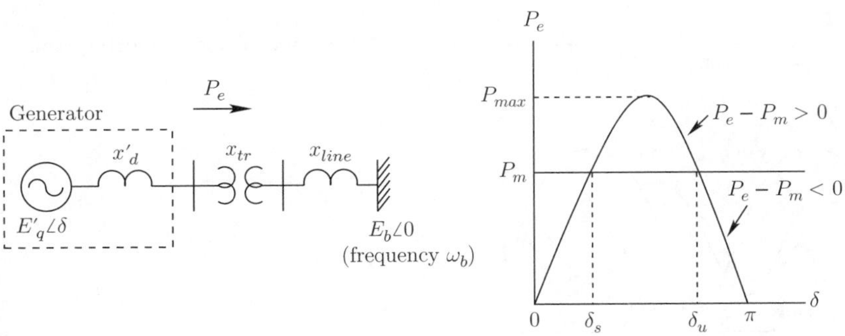

Figure 7.2 Single-machine infinite bus system.

by the equation

$$P_e = \frac{E'_q E_b}{(x'_d + x_{tr} + x_{line})} \sin\delta = P_{max}\sin\delta \quad (7.3)$$

Note that P_e is a nonlinear function of δ. For δ greater than 90°, P_e decreases with an increase in δ. $P_m - P_e$ is positive for $\delta > \delta_u$ (see Figure 7.2) where $\delta_u = 180° - \delta_s$. Therefore, during the transients following a disturbance, if δ becomes greater than δ_u while $\omega - \omega_b > 0$, then $P_m - P_e$ is no longer restorative and the system is pushed further away from the equilibrium point $\delta = \delta_s, \omega = \omega_b$, resulting in LoS.

To illustrate this, let us assume that the synchronous generator is initially at equilibrium ($\omega = \omega_b, \delta = \delta_s$). A three-phase fault occurs in the external system, which causes E_b to dip to zero during the fault. Since E_b is zero during the fault, P_e is also zero in that duration, and therefore the synchronous machine accelerates from its pre-fault equilibrium. When the fault is cleared, E_b returns to its pre-fault value, but δ and ω are no longer at the post-fault equilibrium, and hence a transient ensues. Note that in this example the pre-fault and post-fault equilibria are the same since the pre-fault and post-fault networks are the same (which is not true in general).

Figure 7.3 shows the response of this system obtained by numerically integrating the swing equations (7.1) and (7.2). The system parameters are $x'_d = 0.32$ pu, $x_{tr} = 0.1$ pu, $x_{line} = 0.5$ pu, $H = 5$ s, $P_m = T_m = 1.0$ pu, and $\omega_B = 2\pi \times 60$ rad/s. $E_b = 1.0$ pu prior to and after the fault. Note that $\delta_s = 53°$ and $E'_q = 1.152$ pu.

As the clearing time of the fault T_{clear} is increased, δ and ω move further away from the equilibrium. Consequently, the maximum amplitude of δ during the transient following the clearing of the fault also increases. After a point (critical clearing time, CCT), the system loses synchronism. Incidentally, for smaller fault clearing times the deviation in δ is almost like a sinusoidal oscillation (swing), as predicted by the small-signal analysis carried out in Chapter 6. However, as the disturbance magnitude is increased (larger fault clearing times) the response progressively appears less sinusoidal. Eventually δ increases monotonically when the fault clearing time is above the critical value. Note that this behavior cannot be predicted by small-signal analysis.

The rotor speed, generator current, generator terminal voltage, and electrical torque are shown in Figure 7.4. LoS cannot be allowed to continue for long because the large fluctuations may cause equipment damage, therefore the system has to be disconnected from the infinite bus if LoS occurs. In a LoS situation, the voltage magnitude dips very

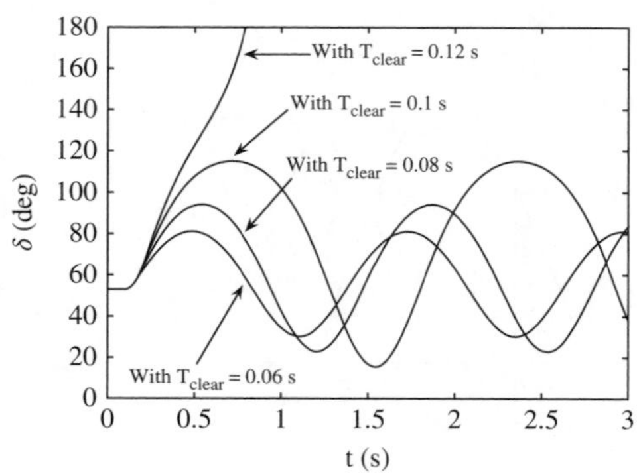

Figure 7.3 Illustration of loss of synchronism.

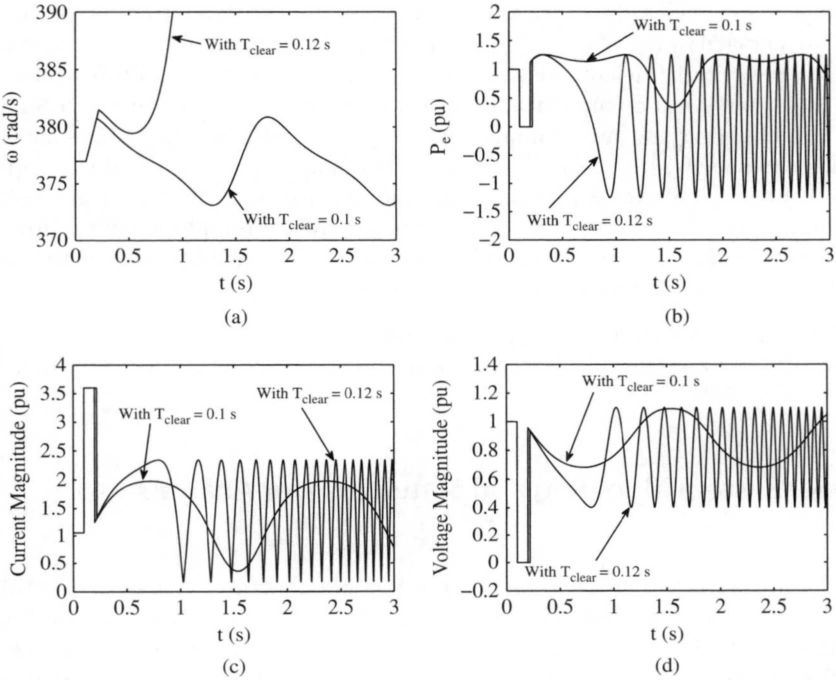

Figure 7.4 Response under stable and unstable (LoS) conditions. (a) Speed (b) Electric power (c) Current (d) Terminal voltage.

significantly while simultaneously the current rises (after the fault clearing) due to the large angular separation. A distance relay in the line will therefore see a low apparent impedance and will trigger the tripping of the line if this impedance is in its tripping zone. This leads to a "natural" but uncontrolled separation of the system. Alternatively, specialized out-of-step relays may be used to detect loss of synchronism conditions.

The synchronous machine loses synchronism in this case due to the large deviation from the equilibrium caused by a large disturbance. Large disturbance instability is the most common mechanism by which LoS occurs, and is also called *transient instability*. LoS may also occur if the system is small-signal unstable (see Figure 6.7 in Example 6.1, Chapter 6) or in the trivial situation where an equilibrium point itself does not exist (e.g., if $P_m > P_{max}$).

7.4 Criteria for Assessment of Stability

The evaluation of a power system's ability to withstand large disturbances and transition to an acceptable operating condition without LoS is commonly known as *transient stability analysis*. Real-life disturbances may consist of a complex sequence of events, which may involve not only faults (short circuits) and tripping of the faulted element(s), but also unexpected trippings due to incorrect relay settings. From practical considerations, the assessment of stability is generally done by analyzing the system response to a limited set of pre-decided credible contingencies.

The contingencies which a power system should be able to withstand (transient stability criteria) are decided by the grid regulators; different grids may have different criteria. In most cases, the criteria include the ability of the system to withstand a three-phase bus fault of a certain duration at critical locations such as terminals of heavily loaded generators and lines carrying large amounts of power. Although three-phase bus faults are convenient to specify, statistical data indicate that single-line-to-ground faults on HV and EHV lines are the most prevalent [3]. The misoperation of relays can result in double contingency faults which are more common than three-phase faults. However, three-phase faults are more severe and therefore transient stability criteria usually include the ability to withstand the following disturbances:

1) three-phase faults cleared by primary protection
2) single-line-to-ground faults cleared by back-up protection (due to a stuck primary breaker).

7.5 Power System Modeling and Simulation for Analysis of LoS

In Section 7.3 we were able to demonstrate the LoS phenomenon in a simple SMIB system. The estimate of the disturbance magnitude that will cause a LoS (CCT in the case of faults) is dependent on the model used for the analysis. The model of a power system for the study of LoS is similar to the one given in Section 6.5.2. The fast transients associated with the network, stator fluxes, and power electronic actuators are neglected in LoS studies in the same way as in power-swing damping studies. The major difference is that linearization of the nonlinear equations cannot be used in the study of LoS, since the deviations from equilibrium which lead to LoS are not small. Limits of actuators like excitation systems, FACTS, and HVDC systems, and the limiters present in their controllers also need to be represented in detail.

The equations are formulated as nonlinear differential-algebraic equations, as given below (refer Section 6.5.2):

$$\dot{x} = f(x, y, u) \tag{7.4}$$

$$0 = g(x, y) \tag{7.5}$$

x represents the states (generator angles, speeds, rotor fluxes, and the states corresponding to the plant and controllers of excitation systems, FACTS, HVDC, renewable energy systems, and dynamic loads), y represents the bus voltages and injected currents (for shunt-connected devices) and the branch currents and series injected voltages (for series-connected devices), and u represents inputs like controller setpoints.

To predict the behavior for large deviations from the equilibrium, numerical integration (simulation) of the differential-algebraic equations is necessary. The simulation of the equations can be carried out using the partitioned-explicit method or the simultaneous-implicit method, which are explained in Appendix 7.A. In a simulation program, events like faults and line-trippings are simulated by making appropriate changes to the algebraic equations (7.5).

For planning or off-line studies, the initial conditions for the simulation are obtained from a realistic power-flow scenario. Several such scenarios are considered for transient stability assessment. For online assessment during actual operation, the initial conditions are obtained from the measurements using state-estimation tools. Various disturbances are then simulated as per the transient stability criteria, and stability is assessed by examination of the trajectory of the relative rotor speeds and angles.

Example 7.1 ***Some factors that affect LoS (SMIB system revisited)*** In this example we examine the effect of factors like the generator model, excitation controller, and changes in the operating conditions by numerical simulation studies of the SMIB system shown in Figure 7.5.

The generator is connected to an infinite bus via a transformer and a pair of identical transmission lines. The system is initially at equilibrium. The transformer and transmission lines are modeled as reactances. A three-phase fault occurs at $t = 2$ s near the transformer end of one of the transmission lines. The fault is cleared by tripping the transmission line. The post-fault system has only one transmission line and is, therefore, weaker than the pre-fault system. $x_T = 0.1$ pu and $x_{line} = 0.1$ pu for the base case. The pre-fault total external reactance is $x_e = x_T + \frac{x_{line}}{2} = 0.15$ pu, while the post-fault external reactance is $x_e = x_T + x_{line} = 0.2$ pu. The equilibrium electrical power output for the base case is $P_{eo} = 1.0$ pu, which is equal to P_m, while the pre-fault terminal voltage V_o is 1.0 pu. The base frequency is $\omega_B = 2\pi \times 60$ rad/s. The infinite bus voltage E_b is 1.0 pu and its frequency $\omega_b = \omega_B$. Note that the post-fault system is the same as the system considered in Example 6.1.

Compute the system response for different fault clearing times. Note that a larger clearing time implies a greater deviation of the rotor angle and speeds from the equilibrium point. Therefore the CCT can be considered as a measure of stability. Find the CCT and examine/compare the responses for the following cases:

1) synchronous generator modeled by the classical model (see Section 7.3)
2) synchronous generator modeled by Model (1.0) (the model equations are given in Section 6.3):
   (a) with AVR inactive
   (b) with AVR and PSS active.
3) In addition, examine (one at a time) the effect of:
   (a) doubling the transmission line reactance, x_{line}
   (b) decreasing the quiescent power output to $P_{eo} = 0.5$ pu
   (c) making the generator absorb reactive power.

The generator and AVR data are the same as in Example 6.1, and are given here again for convenience. $x_d = 1.6$ pu, $x'_d = 0.32$ pu, $x_q = 1.55$ pu, $T'_{do} = 6.0$ s, $H = 5$ MJ/MVA, $K_E = 250.0$ pu/pu, $T_E = 0.025$ s, $E_{fdmax} = 6.0$ pu, and $E_{fdmin} = -6.0$ pu (refer to Figure 6.4).

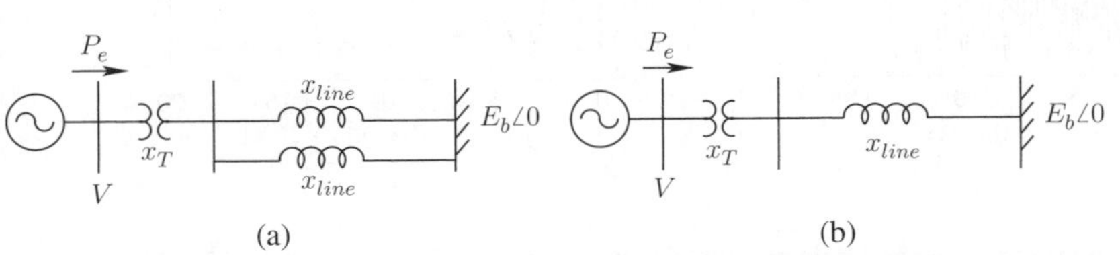

Figure 7.5 The pre-fault and post-fault systems considered in Example 7.1. (a) Pre-fault system (b) Post-fault system.

The PSS block diagram and parameters are the same as those considered in Example 6.3 ($T_1 = 0.15$ s, $T_2 = 0.05$ s, $T_w = 5$ s, and $K = 2$ pu/pu). The output of the PSS is limited between +0.1 and −0.05 pu. Note that a PSS is necessary as the post-fault operating point is small-signal unstable with only an AVR (refer to Examples 6.1 and 6.3).

Solution

In this example, Runge–Kutta fourth order explicit method is used for numerical integration, with a time step of 1 ms.

7.5.1 Effect of System Model

Figure 7.6 shows the effect of the model on the response of the system. It is seen that the CCT is the least for Model (1.0) of the generator without an AVR. While E'_q is constant for the classical model, it dips and recovers slowly for Model (1.0) without an AVR. The AVR, when present, transiently boosts the field voltage E_{fd} after fault clearing and contributes to quicker recovery of E'_q. This results in an improvement in stability. Note that E_{fd} hits its upper limit (6.0 pu) during the transients as shown in Figure 7.7. The PSS simultaneously modulates the AVR voltage set-point. Since the deviations are large, the PSS is also constrained by its own limits (+0.1, −0.05) pu. The tighter lower limit ensures that the PSS does not lower the field voltage (via the AVR) excessively after the initial acceleration of the rotor. It is seen from Figure 7.6c that LoS need not always

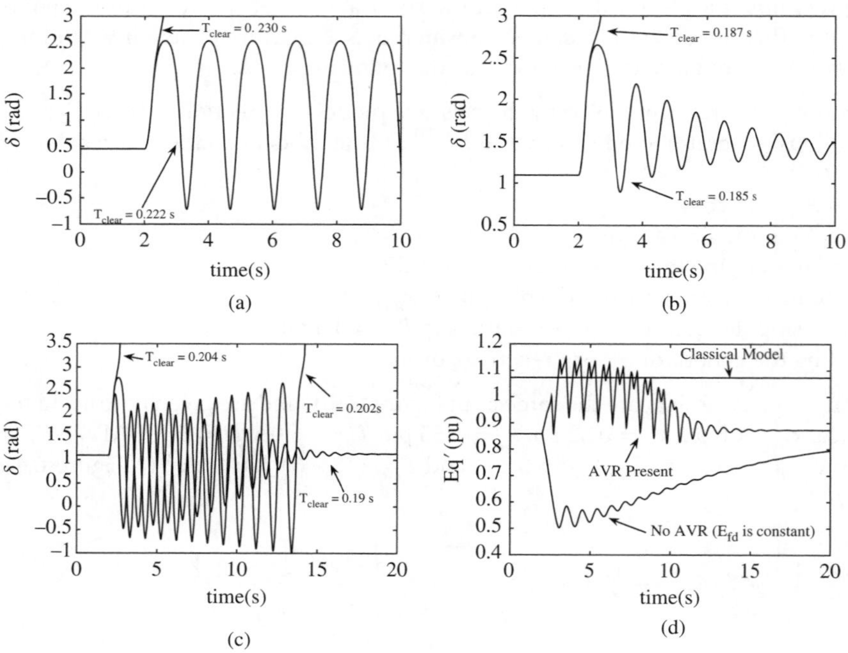

Figure 7.6 Simulated responses: effect of generator modeling. (a) Classical model, CCT = 0.222 s (b) Model (1.0), no AVR, CCT = 0.185 s (c) Model (1.0) with AVR and PSS, CCT = 0.198 s (d) Response of E'_q, T_{clear} = 0.185 s.

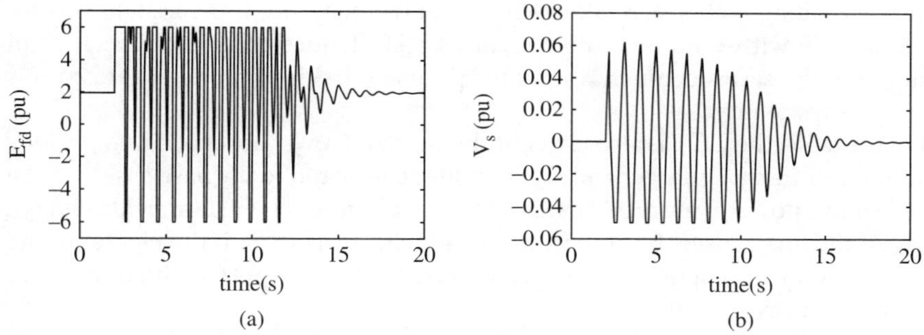

Figure 7.7 Field voltage and PSS output (with AVR and PSS). (a) Field voltage, $T_{clear} = 0.190$ s (b) PSS output, $T_{clear} = 0.190$ s.

occur at the "first swing". In the case where AVR and PSS are considered, the damping of power swings significantly improves when the amplitude of rotor oscillations becomes smaller. This is because the initial damping action is inhibited due to the AVR and PSS limiters.

7.5.2 Effect of Changing Operating Conditions

The effects of (i) doubling the transmission line reactance, x_{line}, (ii) decreasing the quiescent power output to $P_{eo} = 0.5$ pu, and (iii) making the generator absorb reactive power on the CCT are shown in Table 7.1. Note that the generator is made to absorb reactive power by increasing the infinite bus voltage slightly and reducing the field voltage.

The CCT is lower for the case with a larger external reactance, indicating that stability has worsened. On the other hand, a reduction in active power loading significantly improves stability. This can be understood from the power-angle curve shown in Figure 7.2. For larger loadings and larger external reactance:

(a) the pre-fault equilibrium rotor angle is larger
(b) the restorative (decelerating) effect $P_e - P_m$ is lower (Note that for the larger external reactance P_{max} is lower, and for a larger loading P_m is higher.)

Table 7.1 CCT for different operating conditions (with AVR and PSS), where Q_{eo} is the reactive power injected by the machine.

Pre-fault conditions	CCT (s)	Remarks
$P_{eo} = 1.0$ pu, $Q_{eo} = 0.0754$ pu, $V_o = 1.0$ pu, $E_{fdo} = 1.9533$ pu, $\delta_o = 62.851°$, $E_b = 1.0$ pu, $x_{line} = 0.1$ pu	0.198	Base case
$P_{eo} = 1.0$ pu, $Q_{eo} = 0.1010$ pu, $V_o = 1.0$ pu, $E_{fdo} = 1.977$ pu, $\delta_o = 64.807°$, $E_b = 1.0$ pu, $x_{line} = 0.2$ pu	0.142	Increased line reactance
$P_{eo} = 0.5$ pu, $Q_{eo} = 0.0188$ pu, $V_o = 1.0$ pu, $E_{fdo} = 1.3041$ pu, $\delta_o = 41.284°$, $E_b = 1.0$ pu, $x_{line} = 0.1$ pu	0.313	Decreased loading
$P_{eo} = 1.0$ pu, $Q_{eo} = -0.2189$ pu, $V_o = 0.98$ pu, $E_{fdo} = 1.7471$ pu, $\delta_o = 76.752°$, $E_b = 1.025$ pu, $x_{line} = 0.1$ pu	0.173	Reactive power absorption

Reduction in loading is therefore a preventive control measure if the system is insecure (i.e., unable to withstand a credible disturbance). Reduction of the reactance of transmission lines by series compensation (using capacitors) is also one of the options to improve system stability.

As seen in the table, the CCT is also lower under reactive power absorption conditions since the initial rotor angle is larger and E_{fd} is smaller compared to the base case. For the case where reactive power is absorbed and an AVR is *not* present (constant field voltage), the tripping of the line causes the system to lose synchronism even if the clearing time is very small, as shown in Figure 7.8. This is because there is no post-fault equilibrium for this condition, as explained below.

E_{fd} is relatively smaller under reactive power absorption conditions (under-excitation) and is not boosted up after the line tripping (no AVR). Therefore, the $P_m = 1.0$ pu becomes greater than the maximum steady-state power that can be transferred by the *post-fault* network. The maximum steady-state power can be computed from the expression for steady-state power P_{ess}:

$$P_{ess}(\delta) = \frac{E_b E_{fd}}{x_d + x_e} \sin\delta + \frac{E_b^2 (x_d - x_q)}{2\,(x_d + x_e)(x_q + x_e)} \sin 2\delta$$

This case study demonstrates that it is advisable to limit under-excitation during high loading conditions, if an AVR is not present.

7.6 Loss of Synchronism in Multi-Machine Systems

LoS may also occur in a multi-machine system, but with some differences compared to an SMIB system. These are discussed below.

1) Since there is no infinite bus in a real system, the rotor angles and speeds of individual machines in a multi-machine system have a component corresponding to the centre-of-inertia (COI) motion, which is affected by the load-generation balance. As discussed in Section 6.5.1, COI motion is common to the rotor angles and speeds of all the machines in a system. Since relative angular and speed dynamics are the primary focus in LoS studies and not the common motion, it is convenient to view

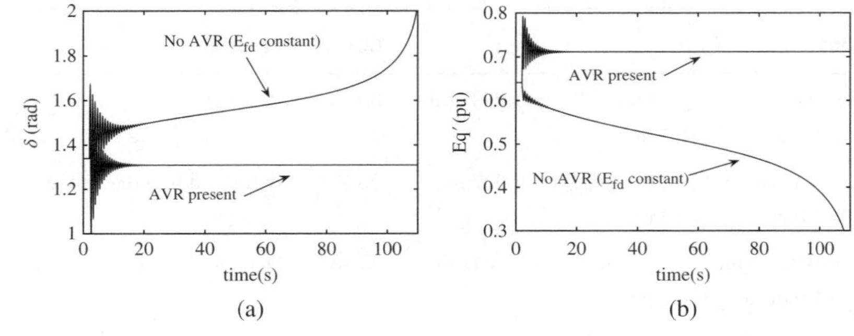

Figure 7.8 Response for the reactive power absorption case, $T_{clear} = 0.05$ s. (a) Rotor angle δ (b) Field flux E_q'.

the individual generator rotor angles and speeds *relative* to the COI angle and speed $\left(\delta_{COI} = \frac{\sum H_i \delta_i}{\sum H_i}, \omega_{COI} = \frac{\sum H_i \omega_i}{\sum H_i}\right)$ as this will remove the COI component from the rotor angles and speeds.

2) LoS in a multi-machine system is manifested as a separation of two or more groups of synchronous machines from one another (each group having one or more machines). The angular separation between the machines within such groups is often quite small compared to the angular separation between two machines belonging to different groups. In other words, the machines within a group are "coherent". The grouping of machines is determined by the nature and location of the disturbance and the initial operating condition. This is in contrast to LoS in an SMIB system wherein there is only one mode of separation, namely, the separation of the rotor-angle and speed of the machine from the infinite bus angle and frequency. A two-machine system also has only one mode of separation, but unlike an SMIB system the rotor angles and speeds of a two-machine system may also additionally exhibit COI motion.
3) The initial deceleration/acceleration of machines after fault clearing may not be uniform, and the time instant of the maxima/minima of all the rotor angles with respect to the COI angle may not coincide, since there are many modes of relative motion. Therefore, one may not be able to identify a common "first swing" in the response.
4) The largest fluctuations in currents and voltages following LoS occur on the branches of the network which form the cut-set of the separation. The location of the cut-set is dependent on the grouping of machines, which depends on the disturbance and initial operating conditions. The large fluctuations in currents and voltages may cause the tripping of distance relays of the transmission lines belonging to the cut-set, causing a natural but uncontrolled separation of the system into islands.
5) If a system is vulnerable to LoS for a certain disturbance, then the corrective actions needed to improve stability are often specific to the disturbance and the initial operating conditions.

Example 7.2 ***LoS in a Ten-Machine System*** The single line diagram of the ten-machine, New England power system is shown in Figure 7.9. The system data are given in Appendix 7.B and are adapted from [4]. Unless otherwise specified, the loads are modeled as constant impedances and are independent of frequency. The mechanical power inputs to the machines are assumed to be constant during the transients. Examine the effect of the following:

1) disturbance location
2) load model change
3) series compensation of a line
4) change of the pre-fault power schedule of the generators.

Solution

For these simulation studies, the partitioned-explicit method is used. The Runge–Kutta fourth order method is used for numerical integration with a time step of 5 ms. The dummy-differential equation concept discussed in Section 7.A.2 is used to simplify the solution of the differential-algebraic equations in the cases where the load model is of constant-power type and generator dynamic saliency is considered. A time constant $T = 10$ ms is used in these dummy-differential equations.

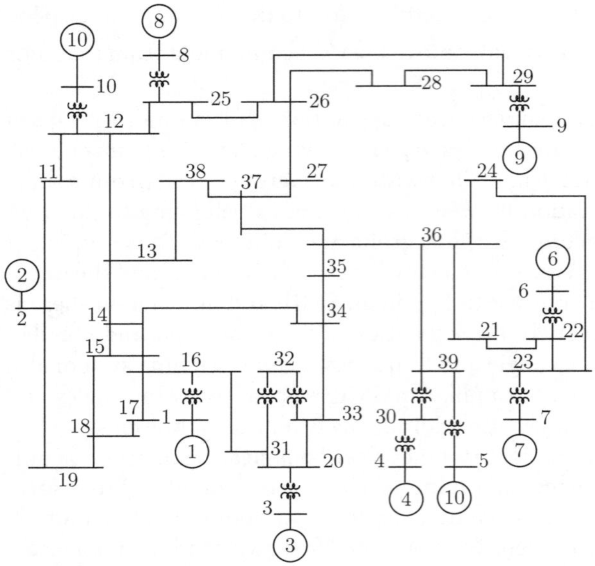

Figure 7.9 Ten-machine, 39-bus system.

7.6.1 Effect of Disturbance Location on Mode of Separation:

The disturbance nature and location has a bearing on how the machines separate in case of LoS. Consider the following disturbances (occurring at 0.5 s):

1) three-phase fault on line 34-35 at bus 34, which is cleared by tripping the line after 0.34 s
2) three-phase fault on line 37-38 at bus 37, which is cleared by tripping the line after 0.376 s
3) three-phase fault on line 26-28 at bus 26, which is cleared by tripping the line after 0.173 s.

The trajectories in each case are unstable because the clearing time in each case is greater than the corresponding CCT. The interesting aspect is that the machines/groups of machines which separate are different in each case (see Figure 7.10).

Note that all rotor angles are expressed with reference to the COI angle. Therefore they do not exhibit any COI component. The COI frequency during the fault-on period and the post-fault period increases because the overall load (constant-impedance model) decreases due to the low voltage. This causes an increase in COI angle as well, as shown in Figure 7.11.

7.6.2 Effect of the Load Model

Consider two frequency-independent static load models (see Section 6.5.2 for the mathematical description of a static load model):

(a) a constant-impedance model for real and reactive power loads ($k_{pv} = k_{qv} = 2.0$)
(b) a constant-power model for a real power and constant susceptance model for reactive power ($k_{pv} = 0.0$, $k_{qv} = 2.0$).

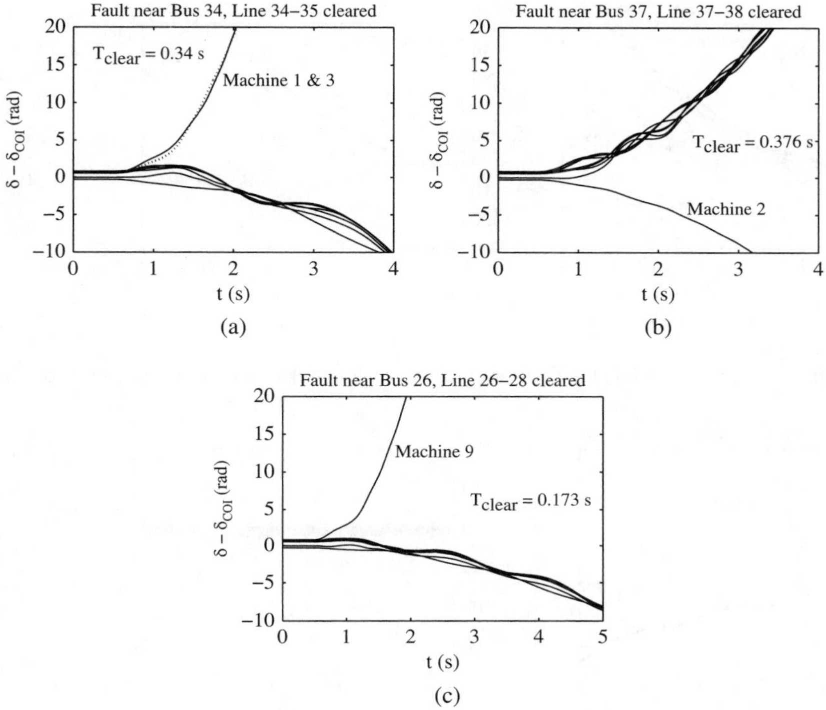

Figure 7.10 Effect of disturbance location on mode of separation. (a) Rotor angle (Case 1) (b) Rotor angle (Case 2) (c) Rotor angle (Case 3).

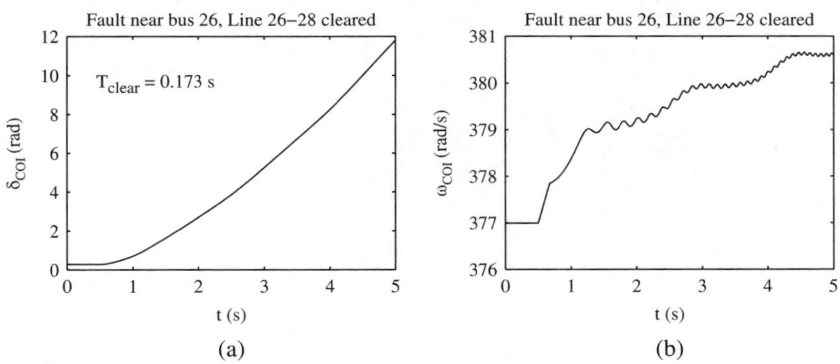

Figure 7.11 COI measurements. (a) COI angle (b) COI speed.

For a three-phase fault on line 37-38 at bus 37, which is cleared by tripping the line, the CCT for the constant-power load model (0.291 s) is significantly smaller than for the constant-impedance load model (0.375 s). Therefore, for a clearing time of 0.376 s, LoS occurs in both cases. However, the way the machines separate is different in each case, as shown in Figure 7.12. For the constant-power model there is a four-way splitting, while in the case of the constant-impedance model, only machine 2 splits from the rest of the system.

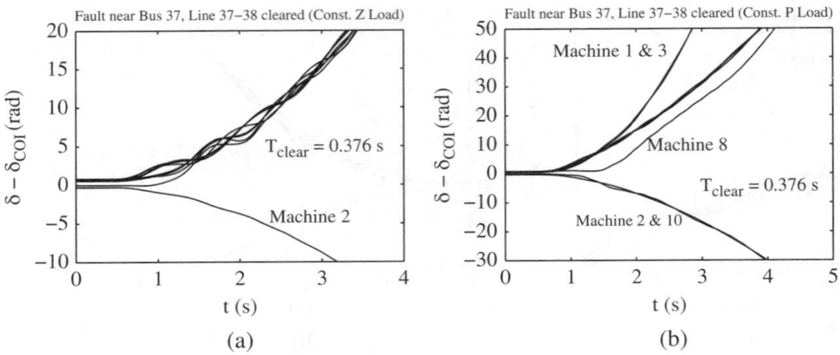

Figure 7.12 Effect of load model. (a) With constant-impedance load (b) With constant-power load.

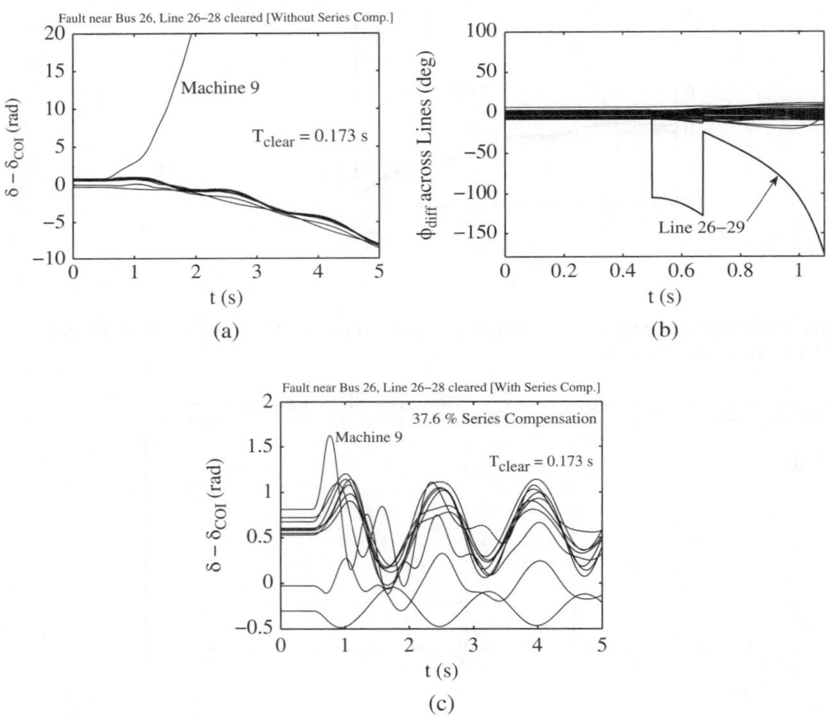

Figure 7.13 Effect of series compensation in a critical line. (a) Rotor angle (without compensation) (b) Phase difference (without compensation) (c) Rotor angle (with compensation).

7.6.3 Effect of Series Compensation in a Critical Line

Consider a three-phase fault on line 26-28 at bus 26, which is cleared after 0.173 s by tripping the line. The system is unstable for this situation (see Figure 7.13). The transmission line 26-29 is the critical line for this disturbance. The phase angular difference between the voltages at both ends of the line, plotted in Figure 7.13, shows that the angular difference increases to very large values. Insertion of 37.6% capacitive series compensation in line 26-29 just after the fault is cleared stabilizes the system. Clearly, the use of series compensation in the critical line improves stability for this disturbance.

7.6.4 Effect of a Change in the Pre-fault Generation Schedule

A three-phase fault on line 34-35 at bus 34, which is cleared by tripping the line after 0.34 s, leads to the group of machines 1 and 3 separating out from the rest of the system. Therefore, reducing the loading on the transmission lines connecting these two machines to the rest of the system is a possible method of improving stability.

Consider a decrease in the quiescent power output of both Generator 1 and Generator 3 by 0.5 pu. The power output of Generator 2 is increased by 1.0 pu in order to maintain the load-generation balance. The change in transmission losses due to this rescheduling is very marginal. This rescheduling is able to stabilize the system for the given disturbance (see Figure 7.14).

It should be noted that actions like generation rescheduling and insertion of series capacitors in a line have significant economic implications. In the case of generation rescheduling, the power may have to be rescheduled to a costlier generator from a cheaper one, while series capacitors entail a high capital cost. The choice of the location of series capacitors and their size, the candidate generators for rescheduling, and the quantum of rescheduling are specific to the disturbances considered.

A comprehensive study involving the simulation of many disturbances is generally required to determine the rescheduling /compensation requirements. An approximate technique for assessment of stability and quantification of the stability margin, which is called the energy function method, may be used in conjunction with simulation tools to facilitate this. This method will be introduced in Section 7.8.

7.6.5 Voltage Phase Angular Differences across Critical Lines/Apparent Impedance seen by Relays

Consider a three-phase fault on line 37-38 at bus 37, which is cleared by tripping the line. Two clearing times, 0.2 s and 0.376 s, are considered, the system being stable for 0.2 s and unstable for 0.376 s.

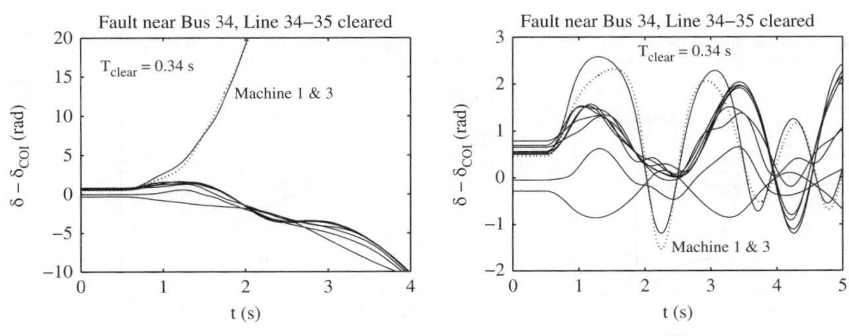

Figure 7.14 Effect of change in the pre-fault generation schedule. (a) Without rescheduling (b) With rescheduling.

The rotor angles and the voltage phase angular differences across the branches of the network are plotted in Figures 7.15 and 7.16, respectively.

It is seen that in the unstable case, the voltage phase angular differences across the transmission lines 18-19 and 11-12 increase to very large values, but the phase angles across other lines do not increase as much. In Figures 7.17 and 7.18, quadrilateral

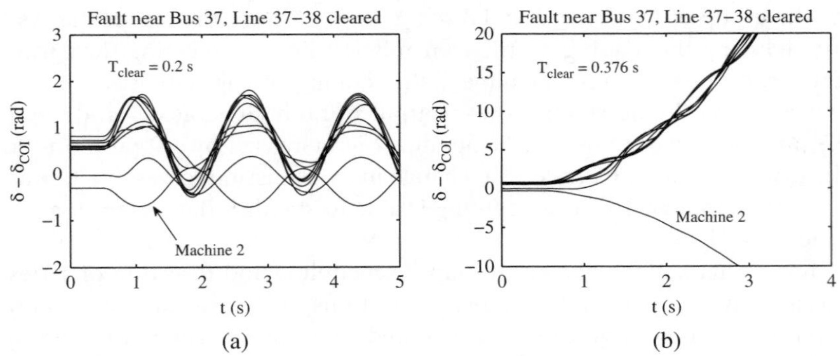

Figure 7.15 Rotor angles for a stable and unstable case. (a) Stable case (b) Unstable case.

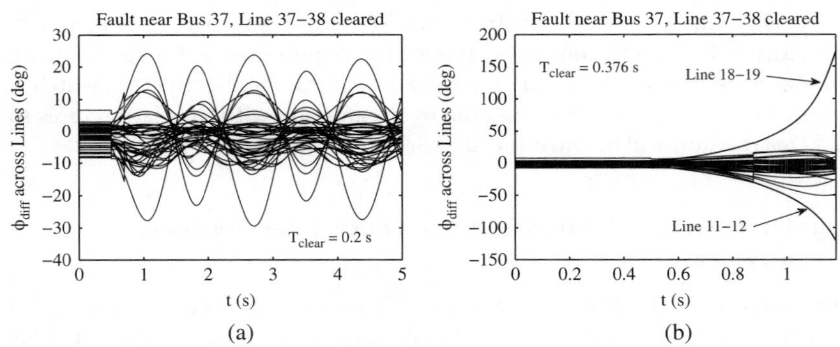

Figure 7.16 Phase angle difference across lines and transformers. (a) Stable case (b) Unstable case.

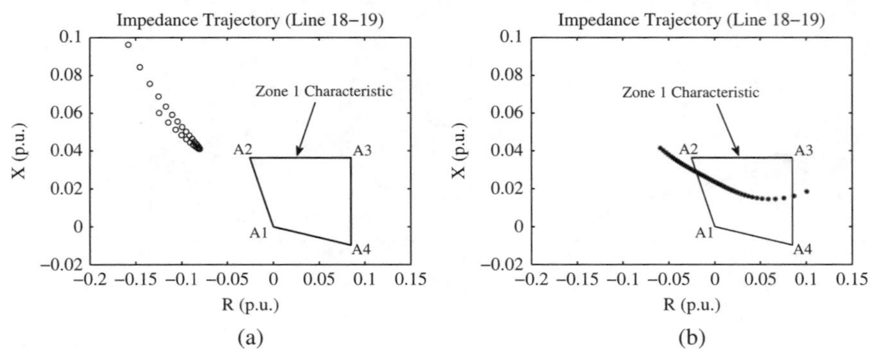

Figure 7.17 Impedance trajectory for line 18-19. (a) Stable case (b) Unstable case.

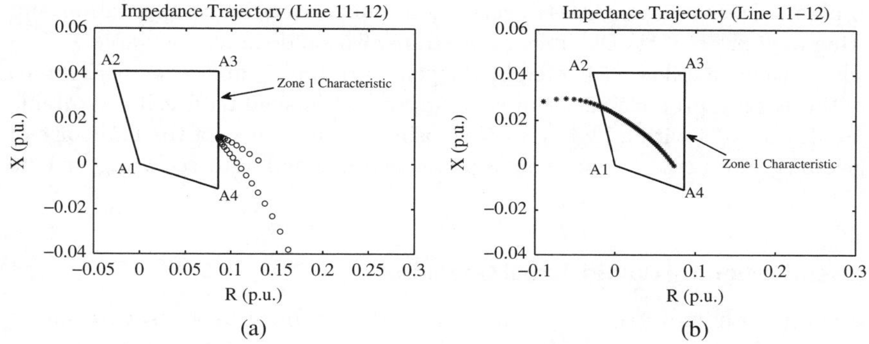

Figure 7.18 Impedance trajectory for line 11-12. (a) Stable case (b) Unstable case.

characteristics are drawn in the R–X plane, which represent plausible primary tripping zones (Zone 1) of the distance relays on lines 18-19 and 11-12. The impedance trajectories seen at one end of lines 18-19 and 11-12 are shown. In the stable case, the impedance trajectory stays away from the distance relay Zone 1 characteristic. In the unstable case, the apparent impedances of the lines enter their respective quadrilaterals, which will trigger the distance relays (not simulated here). This is not surprising because when the phase angular differences across these lines rise, voltage magnitudes at various points on these transmission lines tend to drop while current magnitudes rise, thereby decreasing the apparent impedance seen by the relay. Since the lines 18-19 and 11-12 form a cut-set which isolates Generator 2, tripping of these lines will result in islanding of this generator.

Note that the cut-set of separation may sometimes include transformer branches. Hence the distance relays on the adjacent transmission lines may not get triggered. Therefore, special relays (called out-of-step relays) may be deployed to specifically detect LoS conditions and trigger controlled islanding of the system. If natural separation or controlled separation does not occur, then generator speed deviations may become quite large and result in trippings due to over/under-speed protection.

7.7 Measures to Avoid LoS

Series compensation of transmission lines as well as operational measures like generation rescheduling, which were discussed in Example 7.2, are among several possible measures to improve stability. The strategy for avoiding transient instability is three-pronged, as discussed below.

7.7.1 System Planning and Design

As a part of system planning studies, stability is assessed by applying the transient stability criteria (see Section 7.4) to plausible future operating scenarios. Augmentation of the transmission system may be suggested if the criteria are not satisfied. These augmentations include (i) building of additional AC transmission circuits, (ii) strengthening of existing or planned transmission lines by series and shunt capacitive compensation, and (iii) exploring the option of HVDC links where transmission distances are large.

Although transient stability is affected by the generator inertia and reactances, modification of these is not practicable. As most faults are of transient nature, it is desirable to have circuit breakers with a high-speed reclosing feature. Most of the faults are of single-phase to ground type. Hence single-phase tripping and auto-reclosing may be explored.

7.7.2 Preventive Control During Actual Operation

Preventive control is initiated by an operator if the system is found to be insecure during actual operation. The assessment of the system as secure or insecure – also known as dynamic security assessment (DSA) – is based on a study involving computer simulation tools. The initial conditions for this study are obtained from online measurements (state-estimation). These tools assess whether the system can withstand disturbances as per the transient stability criteria. A DSA tool should also provide a stability margin, which should be relatable to the additional loading that a stable system can take without violating the transient stability criteria. For unstable systems, the stability margin is negative. The margin helps an operator to determine the quantum of preventive control actions, such as generation rescheduling, that are needed to make the system secure.

7.7.3 Emergency Control

System design and preventive control are based on an assessment of stability for potential disturbances. If an actual disturbance occurs, then automatic controllers react to it and affect the evolving trajectories. While manual control ("heroic action") is also possible, the available time window is often too small for a system operator to react. For the LoS event in the Indian grid (see Figure 7.1), an uncontrolled separation of the system occurred within 5 s of the initiating disturbance.

Emergency controls are specifically designed for improving stability and are inactive during normal operating conditions. They are usually triggered after large disturbances when the triggering criteria are met, and have a high impact. These controllers exploit the transient capability of actuators like generator excitation systems, HVDC systems, and FACTS. For example, the power flow in AC transmission lines can be diverted to a parallel-connected HVDC link by boosting the power set-point of the HVDC controller. A HVDC transmission system can use its transient capability for a few seconds in order to facilitate this (e.g., the Chandrapur–Padghe link in India has a continuous power rating of 1500 MW, but can allow 2000 MW for 5 s during transient conditions [5]). Some actuators may be used only for the purpose of improving transient stability. An example of this is the shunt-connected 1400 MW dynamic braking resistor installed at Bonneville Power Administration's Chief Joseph Substation [6].

Control of the turbine power may also be exploited to improve transient stability. Modern steam turbines have the capability for fast actuation of the control valve and the intercept valve (within a fraction of a second). These features were primarily meant to control turbine overspeed under load rejection conditions, but they can also be triggered by a special logic upon detection of an abrupt loss of power output during faults.

Generator or load tripping are the other options for improving transient stability. If the LoS is irrevocable despite the control actions, then controlled separation into stable islands with minimal load-generation imbalance is the next best option, so that the islands can be resynchronized in the earliest possible time. Controlled separation has to be designed in conjunction with load and generation tripping schemes because the islands may have a poor load-generation balance in them.

Emergency control actions need to be made adaptive to avoid false alarms and incorrect actions. This would require supervision of such schemes using system-wide information. Chapter 9 describes an enabling technology (wide-area measurement systems, WAMS) which can potentially improve the reliability of emergency control schemes.

7.8 Assessment and Control of LoS Using Energy Functions [7, 8]

Transient stability evaluation using simulation is done by numerically integrating the nonlinear differential-algebraic equations over a time interval of 10–15 s. This is computationally burdensome, particularly in view of the fact that transient stability is not only dependent on the prevailing network topology and operating condition but is also a function of the disturbance. Therefore, the simulations have to be carried out for a large number of credible disturbances. A fast method for stability evaluation which does not require the explicit computation of the post-disturbance response of the system is desirable. A stability evaluation method should also yield a quantitative measure of the stability margin and the sensitivities of this margin to parameter changes and control actions. This has motivated the development of the energy function method.

The basic idea is as follows: $W(x)$, which is a function of the system states x, is evaluated at the instant at which the disturbance is cleared. If this value is less than a certain threshold W_{cr}, which depends on the the post-disturbance system, then the system is classified as stable. Ideally, the function $W(x)$ should have the following properties:

1) $W(x)$ is constant (or nearly constant) as the states change along the post-disturbance trajectory.
2) $W(x)$ at any instant is a function of the states x at that instant only, and is not dependent on the path taken by the states up to that instant. In other words, W is path-independent.

Therefore, this function is called an "energy" function, and W_{cr} is called the critical energy. The energy margin W_m is then defined as follows:

$$W_m = W_{cr} - W(x) \tag{7.6}$$

If $W_m > 0$ then the system is classified as stable, otherwise it is classified as unstable. It is assumed here that a suitable energy function and its critical value can be determined so that an accurate classification can be done. For the simple SMIB system considered in Section 7.3, with the generator modeled by the classical model, it is indeed possible to obtain such an energy function and its critical value, and the stability classification is exact. When multi-machine systems with detailed models of various power system components and their controllers are considered, this approach can only give approximate answers.

7.8.1 Energy Function Method Applied to an SMIB System

For the system considered in Section 7.3, with the generator modeled by the classical model, an energy function $W(\delta, \omega)$ can be constructed as given below:

$$W_{KE}(\omega) = \frac{1}{2}M(\omega - \omega_b)^2 \tag{7.7}$$

$$W_{PE}(\delta) = -\frac{E'_q E_b}{(x'_d + x_e)}\cos\delta - P_m\delta \tag{7.8}$$

$$W(\delta, \omega) = W_{KE} + W_{PE} \tag{7.9}$$

Note

1) The energy function W is defined for the *post-fault* system. Hence the post-fault value of the external reactance x_e has to be used in the energy calculations.
2) It is easy to verify from the swing equations that $\frac{dW}{dt} = 0$ along the post-fault trajectory. Therefore W is constant along the post-fault trajectory.
3) The critical value of the energy W_{cr} is the value of W evaluated at the *unstable equilibrium point* ($\omega = \omega_b, \delta = \delta_u$; see Figure 7.2). Therefore, $W_{cr} = W(\delta_u, \omega_b)$.
4) The energy at the fault clearing point is determined by the integration of the faulted system equations. For a three-phase fault at the generator terminals, $P_e = 0$ during the fault. If the system is at equilibrium before the fault ($\delta = \delta_s$ and $\omega = \omega_b$), then at fault clearing,

$$\delta_{cl} = \delta(T_{clear}) = \delta_s + \frac{1}{2}\frac{\omega_B P_m}{2H}T_{clear}^2$$

$$\omega_{cl} = \omega(T_{clear}) = \omega_b + \frac{\omega_B P_m}{2H}T_{clear}$$

5) The criterion for stability is therefore $W(\delta_{cl}, \omega_{cl}) - W(\delta_u, \omega_b) < 0$, which yields

$$\frac{1}{2}M(\omega_{cl} - \omega_b)^2 - \frac{E'_q E_b \cos\delta_{cl}}{(x'_d + x_e)} - P_m\delta_{cl} < -\frac{E'_q E_b \cos\delta_u}{(x'_d + x_e)} - P_m\delta_u$$

It is not difficult to see that this method will accurately determine stability. W is constant along the post-disturbance trajectory and therefore if $W(\delta_{cl}, \omega_{cl}) > W_{cr}$ then $(\omega - \omega_b) > 0$ when the rotor angle reaches the value $\delta = \delta_u$. Consequently, the rotor angle will increase beyond δ_u and cannot be pulled back towards the stable equilibrium point of the post-fault system. This is because the restoring power is negative for $\delta > \delta_u$.

The energy function method applied to this system yields exact results, and is equivalent to the equal area criterion (EAC) given in Appendix A. The energy function method may be used not only to evaluate the stability of an SMIB system, but also to quantify the corrective measures in case the system is found to be unstable. This is illustrated in the following example.

Example 7.3 ***Assessment of Stability and Control in an SMIB system*** A single machine is connected to an infinite bus as shown in Figure 7.19. The generator is initially operating in steady state with $P_e = 1.0$ pu and $V = 1.0$ pu. $E_b = 1.0\angle 0$ pu and $\omega_b = \omega_B = 2\pi \times 60$ rad/s.

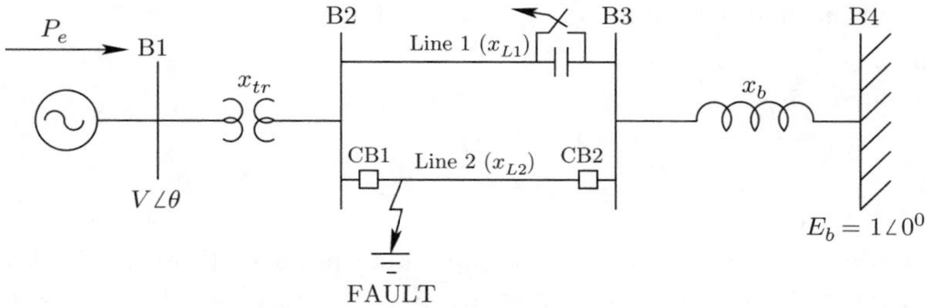

Figure 7.19 Single-machine infinite bus system.

The parameters of the system are given below:

Generator: $H = 5$ MJ/MVA, $x'_d = 0.32$ pu. The generator is modeled by the classical model as in Section 7.3.

Network: $x_{L1} = x_{L2} = 0.4$ pu, $x_{tr} = x_b = 0.1$ pu.

A three-phase fault occurs on line 2 near B2 and is cleared by tripping line 2 (opening CB1 and CB2).

1) If the clearing time of the fault is 0.11 s, and the capacitor shown in the figure is kept bypassed, determine whether the system is stable or not by computing the energy margin.
2) If the system is unstable, compute the minimum value of capacitor reactance which, if inserted immediately after fault clearing, will stabilize the system.

Solution

The pre-fault equilibrium values of the system variables are $E'_q = 1.114$ pu, $\delta_o = 40.276° = 0.703$ rad, and $\omega_o = \omega_B$.

Step I: To check whether the system is unstable for $T_{clear} = 0.11$ s. The values of δ and ω at the clearing time are computed by integrating the swing equations of the SMIB system. This yields

$$\omega_{cl} = \frac{\omega_B P_m}{2H} T_{clear} + \omega_o = 381.138 \text{ rad/s}$$

$$\delta_{cl} = \frac{\omega_B P_m}{4H} T_{clear}^2 + \delta_o = 0.931 \text{ rad.}$$

The value of the energy function (W_{cl1}) is

$$\begin{aligned} W_{cl1} &= W(\omega = \omega_{cl},\ \delta = \delta_{cl}) \\ &= \frac{1}{2}M(\omega_{cl} - \omega_b)^2 - \frac{E'_q E_b}{(x'_d + x_{L1} + x_{tr} + x_b)} \cos\delta_{cl} - P_m \delta_{cl} \\ &= -1.426 \end{aligned}$$

The critical value of the energy function (W_{cr1}) is evaluated as follows:

$$\begin{aligned} W_{cr1} &= W(\omega = \omega_b, \delta = \delta_u) \\ &= -\frac{E_q' E_b}{(x_d' + x_{L1} + x_{tr} + x_b)} \cos\delta_u - P_m \delta_u \\ &= -1.487 \end{aligned}$$

δ_u is the rotor angle at the post-fault unstable operating point (without the series capacitor in the circuit). Note that $\delta_u = \pi - \delta_s = 2.169$ rad, where $\delta_s = 0.972$ rad is the post-disturbance equilibrium value of the rotor angle. It can be seen that $W_{cl1} > W_{cr1}$. Hence it can be concluded that the system is unstable for $T_{clear} = 0.11$ s.

Step II: To find the minimum value of series capacitor reactance.

To stabilize the system for $T_{clear} = 0.11$ s, a capacitor is inserted in series with line 1 immediately after the fault is cleared. For this case, the value of the energy function at fault clearing (W_{cl2}) is

$$\begin{aligned} W_{cl2} &= W(\omega = \omega_{cl},\ \delta = \delta_{cl}) \\ &= \frac{1}{2} M(\omega_{cl} - \omega_b)^2 - P_C \cos\delta_{cl} - P_m \delta_{cl} \\ &= 0.2281 - 0.5970 P_C - 0.9310 \end{aligned}$$

Note that

$$P_C = \frac{E_q' E_b}{(x_d' + x_{tr} + x_{L1} - x_c + x_b)} = \frac{P_m}{\sin\delta_s'}$$

where δ_s' is the post-disturbance equilibrium value of the rotor angle (with the series capacitor in the circuit). Therefore,

$$W_{cl2} = 0.2281 - \frac{0.5970}{\sin\delta_s'} - 0.9310$$

The value of the rotor angle at the post-disturbance unstable equilibrium point is $\delta_u' = \pi - \delta_s'$. The critical value ($W_{cr2}$) of the energy function is then evaluated as:

$$\begin{aligned} W_{cr2} &= W(\omega = \omega_b, \delta = \delta_u') \\ &= -P_C \cos\delta_u' - P_m \delta_u' \\ &= \frac{\cos\delta_s'}{\sin\delta_s'} - (\pi - \delta_s') \end{aligned}$$

If the system is stabilized by the series capacitor (inserted after fault clearing) then $W_{cl2} \le W_{cr2}$. The limiting condition ($W_{cl2} = W_{cr2}$) is a nonlinear equation in terms of δ_s'. This equation can be solved using a nonlinear equation solver like the Gauss–Seidel method, which yields $\delta_s' = 0.914$ rad, $P_C = 1.263$ pu, and $x_c = 0.038$ pu. Note that since we have taken the limiting condition, the computed value of x_c is the minimum reactance required for stabilizing the system.

To verify that the system is indeed stabilized by the insertion of the series reactance, we numerically simulate the behavior of the system for this disturbance. The plots of the rotor angle δ for $T_{clear} = 0.11$ s without and with the series capacitor in the post-fault circuit are shown in Figure 7.20. It is observed that the system is stabilized by the insertion of the series capacitor.

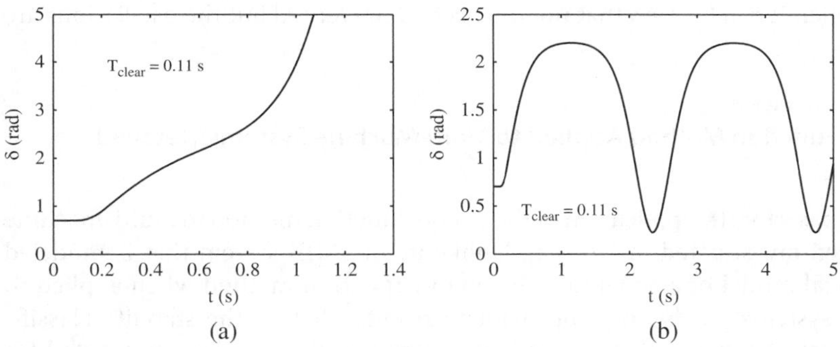

Figure 7.20 Plot of δ vs t. (a) Without capacitor (b) With series capacitor ($x_c = 0.038$ pu).

Discrete Control of Series Compensation

The SMIB system of Example 7.3 is marginally stable when the series capacitor (with a reactance $x_c = 0.038$ pu) is inserted in the post-fault system. The oscillations do not die out since there is no mechanism for damping them in this system. It is possible to damp the oscillations if the series capacitance is not merely inserted but also controlled. Vernier control of the inserted capacitance can be achieved by using a thyristor-controlled series capacitor (TCSC). It is also possible to use "bang-bang" control wherein the capacitor is inserted and bypassed at discrete instants of time after the fault is cleared.

If the following strategy is used to determine the switching instants [9], then the oscillations can be damped out.

1) The capacitor is inserted for the first time as soon as the fault is cleared.
2) The capacitor is switched off when $dW_{PE}/dt = 0$ and $d\delta/dt < 0$.
3) The capacitor is reinserted when $d\delta/dt$ is at a maximum, provided $(d\delta/dt) \geq \epsilon$.

ϵ is a small positive value in order to prevent chattering once the system has come close enough to the equilibrium point. Figure 7.21 shows the response of the generator rotor angle, W_{KE}, and W_{PE} when this strategy is used. The clearing time is 0.11 s as before and $x_c = 0.04$ pu, which is only slightly higher than the critical value of the capacitor

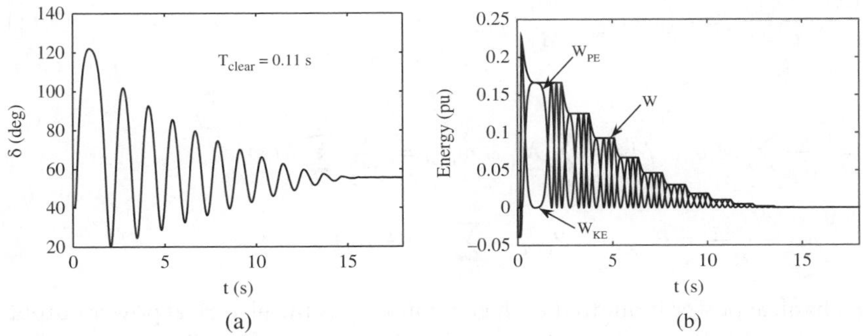

Figure 7.21 "Bang-bang" control. (a) Rotor angle (b) Kinetic energy, potential energy, and total energy.

determined earlier. It can be seen that not only is LoS prevented but the oscillations are also damped.

7.8.2 Energy Function Method Applied to Multi-Machine Systems/Detailed Models

The theoretical basis for the application of the energy function method to multi-machine systems has been investigated in [10–12]. Unlike in an SMIB system that is modeled using the classical model of a generator, the energy function method when applied to multi-machine systems yields only approximate results, that is, the stability classification is not exact. While a path-independent energy function can be constructed for a simplified model of a multi-machine system (see the special case in the following subsection), the computation of the critical energy is not straightforward. This is because unlike in an SMIB or two-machine system, the grouping of machines when LoS occurs in a multi-machine system is not known *a priori*. The grouping is dependent on the location and nature of the disturbance, and the operating conditions. This aspect is also reflected in the fact that there are several unstable equilibrium points (UEPs) in a multi-machine system. Therefore, if the critical energy is to be computed based on the value of the energy function at a UEP, as in an SMIB system, then we are also faced with the problem of finding the *appropriate* UEP corresponding to the disturbance.

When detailed component models are considered, it may not be possible to find an energy function that is constant or nearly constant and is also path-independent. The use of *energy functions* which do not satisfy both these criteria makes the method heuristic. Despite these difficulties, several expedients have been devised to make the energy function method practically useful for stability evaluation in multi-machine systems [8].

Energy Functions for Multi-Machine Systems

An energy function of a multi-machine system can be constructed from the kinetic and potential energy components in a manner similar to an SMIB system. The individual swing equations of the synchronous machines are

$$\frac{d(\delta_i - \delta_{COI})}{dt} = \omega_i - \omega_{COI} \tag{7.10}$$

$$M_i \frac{d(\omega_i - \omega_{COI})}{dt} = P_{mi} - P_{ei} - \frac{M_i}{M_T} P_{COI} \tag{7.11}$$

where

$$\delta_{COI} = \frac{1}{M_T} \sum_{i=1}^{n_g} M_i \delta_i\,,\quad M_T = \sum_{i=1}^{n_g} M_i\,,\quad \omega_{COI} = \frac{1}{M_T} \sum_{i=1}^{n_g} M_i \omega_i,$$

$$M_T \frac{d\omega_{COI}}{dt} = P_{COI}, \qquad P_{COI} = \sum_{i=1}^{n_g} (P_{mi} - P_{ei})$$

P_{mi} is the mechanical power input to the ith generator, P_{ei} is the electrical power output of the ith generator, H_i is the inertia constant of the ith generator, $M_i = 2H_i/\omega_B$, and n_g is the number of generators. The rotor angles and speeds in (7.11) are expressed relative to the center of inertia (COI). This ensures that the COI component, which is not relevant for LoS studies, does not affect the results.

An *energy function* W which satisfies $\frac{dW}{dt} = 0$ is then given by

$$W = W_{KE} + W_{PE} \tag{7.12}$$

where

$$W_{KE} = \sum_{i=1}^{n_g} \frac{1}{2} M_i (\omega_i - \omega_{COI})^2 \tag{7.13}$$

$$W_{PE} = -\sum_{i=1}^{n_g} P_{mi}\theta_i + \sum_{i=1}^{n_g} \int_{\theta_i^0}^{\theta_i} \left(P_{ei} + \frac{M_i}{M_T} P_{COI}\right) d\theta_i \tag{7.14}$$

where $\theta_i = \delta_i - \delta_{COI}$. The initial value θ_i^0 may be chosen to be the value of θ_i at the instant at which the fault is applied. The mechanical power inputs to the generators P_{mi} are assumed to be constant. While W_{KE} is not path-dependent, the second term in the right-hand side of (7.14) may be path-dependent if detailed models of devices and their controllers are considered.

A Special Case

If we assume that:

1) generators are represented by the classical model (constant-magnitude voltage source behind the transient reactance)
2) real power loads are of constant-power type and reactive power loads are of constant reactance type
3) mechanical input power P_m is constant
4) transmission losses are neglected
5) the cumulative load equals the cumulative generation ($P_{COI} = 0$),

then it can be shown that the energy function is path-independent, and is given by the expression [13]

$$W = W_{KE} + W_{PE} \tag{7.15}$$

$$W_{KE} = \sum_{i=1}^{n_g} \frac{1}{2} M_i (\omega_i - \omega_{COI})^2 \tag{7.16}$$

$$W_{PE} = -\sum_{i=1}^{n_g} P_{mi}\theta_i + \sum_{i=1}^{n_L} P_{Li}\phi_i - \sum_{i=1}^{n_b} \frac{1}{2} V_i^2 (b_i - B_{Li}) + \sum_{k=1}^{n_l} \frac{1}{2} Q_k^{br} \tag{7.17}$$

where

$$Q_k^{br} = \frac{V_{Fk}^2}{x_k^{br}} + \frac{V_{Tk}^2}{x_k^{br}} - 2\frac{V_{Fk} V_{Tk} \cos \phi_k^{br}}{x_k^{br}} \tag{7.18}$$

and

P_{Li}	real power load at the ith bus
V_i	voltage magnitude of the ith bus
ϕ_i	phase angle of voltage of the ith bus with respect to the COI angle
b_i	shunt capacitive susceptance at the ith bus (including the line susceptance)

B_{Li}	shunt inductive susceptance of constant reactance load at the ith bus
Q_k^{br}	reactive power loss in the kth transmission branch
V_{Fk}, V_{Tk}	voltage magnitudes at the two nodes of the kth transmission branch
ϕ_k^{br}	voltage phase difference between the two nodes of the kth transmission branch
x_k^{br}	reactance of the kth transmission branch
n_b	number of buses excluding the generator internal bus behind the transient reactance
n_L	number of load buses
n_l	number of transmission branches

Note that the "transmission branches" considered here include transmission lines, transformers, and the generator transient reactances.

The analytical expression of the energy function for this special case is not only path-independent but is also structure-preserving, as it is expressed in terms of the parameters of the individual components of the system. This allows us to conveniently derive the sensitivity of the energy margins to various parameters like branch reactances, shunt susceptances, and mechanical powers [14, 15].

Two-Machine Equivalent of a Multi-Machine System [16]

Under LoS conditions, it is usually observed that the unstable system initially splits into two groups of machines. Let us denote these as Group I (the accelerating group) and Group II (the decelerating group), as depicted in Figure 7.22. The relative motion between the machines within a group is relatively small. Each group can then be represented by its COI angle and speed, and the energy function is constructed using these variables. Effectively, the system is reduced to a two-machine system. The relative angular dynamics of such a system is like an SMIB system. Therefore the two-machine equivalent facilitates the straightforward application of the energy function method.

The equations of motion of the COI angles and speeds of each group of machines are described by the following equations:

$$\frac{d\delta_I}{dt} = \omega_I - \omega_o, \quad M_I \frac{d(\omega_I - \omega_o)}{dt} = P^I_{COI} \tag{7.19}$$

$$\frac{d\delta_{II}}{dt} = \omega_{II} - \omega_o, \quad M_{II} \frac{d(\omega_{II} - \omega_o)}{dt} = P^{II}_{COI} \tag{7.20}$$

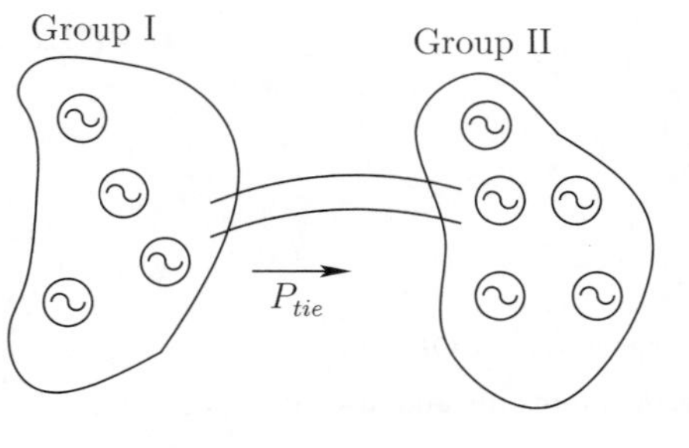

Figure 7.22 Grouping of machines for the given disturbance.

where ω_o is the frequency of the synchronously rotating reference frame and

$$\delta_I = \frac{1}{M_I}\sum_{i\epsilon I} M_i\delta_i, \quad \omega_I = \frac{1}{M_I}\sum_{i\epsilon I} M_i\omega_i, \quad M_I = \sum_{i\epsilon I} M_i$$

$$\delta_{II} = \frac{1}{M_{II}}\sum_{i\epsilon II} M_i\delta_i, \quad \omega_{II} = \frac{1}{M_{II}}\sum_{i\epsilon II} M_i\omega_i, \quad M_{II} = \sum_{i\epsilon II} M_i$$

$$P^I_{COI} = \sum_{i\epsilon I}(P_{mi} - P_{ei}), \quad P^{II}_{COI} = \sum_{i\epsilon II}(P_{mi} - P_{ei}).$$

Note that $i \in I$ and $i \in II$ denote the set of generators belonging to Group I and Group II, respectively. The equations describing the relative motion between the groups of machines is then given by

$$\frac{d(\delta_I - \delta_{II})}{dt} = (\omega_I - \omega_{II}) \tag{7.21}$$

$$\frac{d(\omega_I - \omega_{II})}{dt} = \frac{1}{M_I}P^I_{COI} - \frac{1}{M_{II}}P^{II}_{COI} \tag{7.22}$$

For the equivalent system, the energy function can be obtained as follows:

$$W = W_{KE} + W_{PE} \tag{7.23}$$

where

$$W_{KE} = \frac{1}{2}M_{eq}(\omega_I - \omega_{II})^2 \tag{7.24}$$

$$\begin{aligned} W_{PE} &= \int_{(\delta_I^0-\delta_{II}^0)}^{(\delta_I-\delta_{II})} \left(-\frac{M_{eq}}{M_I}P^I_{COI} + \frac{M_{eq}}{M_{II}}P^{II}_{COI}\right) d(\delta_I - \delta_{II}) \\ &= \int_{(\delta_I^0-\delta_{II}^0)}^{(\delta_I-\delta_{II})} \left(-\frac{M_{eq}}{M_I}P^I_{GL} + \frac{M_{eq}}{M_{II}}P^{II}_{GL} + P_{tie}\right) d(\delta_I - \delta_{II}) \end{aligned} \tag{7.25}$$

where $M_{eq} = \frac{M_I M_{II}}{M_I + M_{II}}$,

$$P^I_{COI} = \sum_{i\epsilon I}(P_{mi} - P_{ei}) = \sum_{i\epsilon I} P_{mi} - \sum_{j\epsilon I} P_{Lj} - P_{tie} = P^I_{GL} - P_{tie},$$

$$P^{II}_{COI} = \sum_{i\epsilon II}(P_{mi} - P_{ei}) = \sum_{i\epsilon II} P_{mi} - \sum_{j\epsilon II} P_{Lj} + P_{tie} = P^{II}_{GL} + P_{tie}$$

P_{tie} denotes the power flow in the set of tie-lines which interconnect the areas containing the two groups of machines as shown in Figure 7.22.

$\sum_{j\epsilon I}P_{Lj}$ and $\sum_{j\epsilon II}P_{Lj}$ denote the sum of load powers in each area while P^I_{GL} and P^{II}_{GL} denote the generation-load balance in each area. The transmission system is assumed to be lossless. The energy function is path-independent if the integrand on the right-hand side of (7.25) can be expressed approximately as a function of $(\delta_I - \delta_{II})$. If this cannot be done then the path-dependent integral in the expression for W_{PE} may be evaluated using numerical integration.

7.8.3 Evaluation of Critical Energy in a Multi-Machine System

The critical energy can be approximately determined by the following methods:

a) computing the energy at the appropriate unstable equilibrium point (UEP) for the given disturbance, also known as the controlling UEP method [12, 17].
b) computing the local maximum of the potential energy under sustained fault conditions, also known as potential energy boundary surface (PEBS) method [18].

The critical energy is easier to compute with the PEBS method. Moreover, the PEBS method can be used with path-dependent energy functions as well. We briefly describe this method here. A more detailed treatment can be found in [7, 8].

For the SMIB system shown in Figure 7.2, the variation of potential energy with the rotor angle δ is shown in Figure 7.23a. The potential energy reaches a local maximum at $\delta = \delta_u$. This is also the stability boundary of the system; $(\omega_i - \omega_b)$ must reduce to zero before δ reaches δ_u for the system to be stable.

It is assumed that even for multi-machine systems, W_{PE} reaches a local maximum at the stability boundary in a similar manner. The critical value of energy W_{cr} is approximated to be the value of W_{PE} when it reaches a local maximum along the *sustained-fault* trajectory [18]. This is depicted in Figure 7.23b. The computation of the sustained fault trajectory requires the numerical integration of the system equations with the fault applied (without clearing it). Note that the critical energy is dependent on the fault location; it has to be evaluated for each fault separately.

7.9 Generation Rescheduling Using Energy Margin Sensitivities

As seen in Example 7.2, it is possible to prevent machines from losing synchronism for a disturbance if generation rescheduling is done, that is, the mechanical power input to the generators is changed. If the system is unstable for a certain disturbance then the energy margin, which is negative, along with the energy margin sensitivities to generator mechanical powers can be used to determine the rescheduling required to stabilize

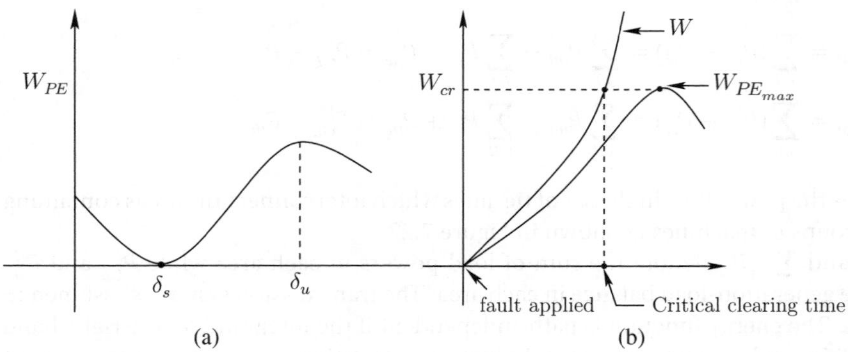

Figure 7.23 Variation of potential energy and the determination of critical energy. (a) Variation in W_{PE} in an SMIB system (b) Determination of W_{cr}: PEBS approach.

the system for that disturbance. The use of energy margin sensitivities requires lesser computational effort as compared to an unguided simulation-based search.

The analytical expression for the sensitivity of the energy margin to generation change (rescheduling) has been derived in [15] using a simplified model of the system. In deriving the expression, it is assumed that generation rescheduling will not alter the grouping of the machines following the disturbance. If the energy margin W_m is known then the amount of generation to be rescheduled, that is, shifted from the accelerating group of machines to the decelerating group, can be obtained from the following expression:

$$\Delta P_m = \frac{W_m}{\hat{S} + C_1 + C_2} = \frac{W_{cr} - W_{cl}}{\hat{S} + C_1 + C_2} \tag{7.26}$$

where

$$\hat{S} = (\delta_I - \delta_{II})_{cl} - (\delta_I - \delta_{II})_{cr}, \quad C_1 = -(\omega_I - \omega_{II})_{cl}\, T_{clear},$$

$$C_2 = \left(\frac{M_{eq}}{M_I} P^I_{COI} - \frac{M_{eq}}{M_{II}} P^{II}_{COI}\right)_{cl} \frac{1}{2M_{eq}}\, T^2_{clear}$$

Generation rescheduling alters the acceleration of the machines during the fault. While $\hat{S}$ is the energy sensitivity expression for the post-fault system due to the generation change, the terms C_1 and C_2 account for the change in the energy at fault clearing due to the change in the acceleration of the generators during the fault. The subscripts *cl* and *cr* denote that the variables are evaluated at the instant of fault clearing and the instant at which W_{PE} reaches its maximum value, respectively. This maximum may be evaluated along the unstable trajectory (with fault clearing) instead of the sustained fault trajectory (as in the PEBS method), since the rescheduling has to be done for a system which is unstable.

The procedure to obtain the generation shift is given below.

1) For the given disturbance, determine the instant when the potential energy component (7.25) of the energy function evaluated along the post-fault unstable trajectory reaches its maximum. $(\delta_I - \delta_{II})_{cr}$ is evaluated using δ_I and δ_{II} at this instant.
2) Evaluate $(\delta_I - \delta_{II})_{cl}$, $(\omega_I - \omega_{II})_{cl}$, and P_{COI} for Groups I and II immediately after the fault is cleared. This requires the numerical integration of the system for the duration of the fault.
3) Compute the energy margin W_m from maximum value of potential energy W_{cr} and the total energy W_{cl} (7.23) at the fault clearing time.
4) Determine the generation change ΔP_m using (7.26). The proportion in which the rescheduling has to be done within a group may be primarily determined from economic criteria. Another criterion is the inertia of the machines; to avoid a change in the grouping of generators after the rescheduling, which will invalidate an assumption in the derivation of (7.26), the generation change should preferably be done on the larger inertia machines in the group.

Note that the generation rescheduling is specific to the disturbance and can be expected to alleviate only the instability due to that disturbance. In practice, the energy margin and the corresponding sensitivities should be obtained for the set of critical disturbances as per the transient stability criteria. The energy margins and the sensitivities evaluated for these disturbances can be used to guide the rescheduling

Table 7.2 Generation rescheduling ($T_{clear} = 0.34$ s and constant-impedance type loads).

Step	W_m	$\hat{S}$	C_1	C_2	Generation shift	Amount	Error
1	−3.9482	−0.7905	−2.1386	−1.1442	ΔP_m	0.97	−3%
	ΔP_m (actual value obtained from an unguided simulation search)					1.0	–

process so that the final generation schedule ensures system stability for all these disturbances.

7.9.1 Case Study: Generation Rescheduling

Consider the ten-machine system of Example 7.2. A three-phase fault on line 34-35 at bus 34 is cleared by tripping the line. The system is unstable for a clearing time of 0.34 s, as shown in Figure 7.14a. Assuming that the loads are of constant-impedance type, determine the amount of rescheduling to be done to stabilize the system for this disturbance.

Solution
For this disturbance, machines 1 and 3 are in the accelerating group while the rest are in the decelerating group. Using the procedure explained earlier in this section, the amount of generation shift required to stabilize the system is obtained. The energy margin, $\hat{S}$, C_1, and C_2 are given in Table 7.2. Machines 1 and 3 are the only generators in the accelerating group. The mechanical powers of both machines are reduced equally, that is, by $\Delta P_m/2$. In the decelerating group, the mechanical power of machine 2, which has the largest inertia, is increased by ΔP_m. The actual value of the generation shift required to stabilize the system is also found using an unguided simulation search. In this case study, there is a good match between the values obtained via the sensitivity calculations and the simulation search. The rotor angle plots for the rescheduled generation $\Delta P_m = 1.0$ pu are shown in Figure 7.14b. These demonstrate the stabilizing effect of this action.

Several other case studies carried out for the same system are reported in [15]. It is seen that carrying out only a partial generation shift and then repeating the process again gives more accurate results than the single-step procedure given here.

7.A Simulation Methods for Transient Stability Studies

Transient stability simulations involve the numerical solution of a set of nonlinear differential-algebraic equations (given the initial conditions), which are of the form:

$$\dot{x} = f(x, y, u) \tag{7.A.1}$$

$$0 = g(x, y) \tag{7.A.2}$$

There are two numerical approaches for solving these equations, which are described in the following subsections.

7.A.1 Simultaneous Implicit Method

In the simultaneous implicit method, the algebraic equations and differential equations are solved together. The differential equations are discretized using an implicit integration method. The discretized equations when the trapezoidal rule is used are as follows:

$$x_{k+1} = x_k + \frac{h}{2}(f_i(x_k, y_k, u_k) + f(x_{k+1}, y_{k+1}, u_{k+1})) \tag{7.A.3}$$

$$0 = g(x_{k+1}, y_{k+1}) \tag{7.A.4}$$

where h is the integration step size.

The discretized equations (7.A.3) and (7.A.4) are solved simultaneously to obtain x_{k+1} *and* y_{k+1} from x_k, y_k, u_k, and u_{k+1}. Since these algebraic equations are nonlinear they may be solved by the Newton–Raphson (NR) method, which involves an iterative solution at every time step. To reduce the computational burden, the system Jacobian (which is required by the NR method) may be calculated only once in a time step and the same Jacobian is then used for the iterations within the time step. The Jacobian may also be kept constant over a few time steps to further reduce the computational burden.

Implicit methods like the trapezoidal rule and the backward Euler method have good numerical stability and thereby allow for larger step sizes, although the overall computation time depends on the convergence of iterations at each time step. These methods may face convergence problems when the network topology changes suddenly due to a fault or equipment tripping.

Since fast transients associated with the network, stator fluxes, and fast-acting actuators are generally not modeled in transient stability simulations, use of variable integration step sizes are unlikely to yield major benefits. However, if a long-term simulation involving a study of the slower prime-mover controllers (say, after a system separation) is to be carried out, then the use of variable step sizes with an implicit integration method is likely to be beneficial.

7.A.2 Partitioned Explicit Method

In this method, the algebraic equations and differential equations are solved alternately. The differential equations are numerically integrated using an explicit method. For example, if the forward Euler (FE) method is used, x_{k+1} is obtained from x_k, y_k, and u_k as follows:

$$x_{k+1} = x_k + hf(x_k, y_k, u_k) \tag{7.A.5}$$

Note: FE is used here only for the purpose of illustration; it is not preferred in practical studies because of its poor numerical performance. Alternatives like the Runge–Kutta fourth-order method may be used instead.

The following algebraic equations are solved thereafter to yield the value of y_{k+1}:

$$0 = g(x_{k+1}, y_{k+1}) \tag{7.A.6}$$

The updated value of y is then used for solving the differential equation in the next time step.

A large subset of the algebraic equations are those pertaining to the linear network (see Section 6.5.2). There are, however, other algebraic equations corresponding to the shunt- and series-connected devices that are nonlinear. Therefore, the overall solution of the algebraic equations will require the use of an iterative solver at each time step. Considerable computational efficiency could be achieved if (i) *all* algebraic equations are linear in the variable y, that is, the algebraic equations are of the form $\mathcal{B}\, y_{k+1} = s(x_{k+1})$, and (ii) the matrix $\mathcal{B}$ does not vary with the time step for large contiguous blocks of time, that is, changes occur only when faults and equipment trippings occur. The solution of the algebraic equations can then be efficiently implemented using sparsity-oriented matrix triangularization of $\mathcal{B}$ and the backward-forward substitution methodology [19]. We now briefly describe an approximate method to achieve this end.

Dummy Differential Equation Approach

The nonlinear algebraic equation corresponding to a device i can be represented as follows:

$$0 = h(x_i, y_i) \tag{7.A.7}$$

This equation can be approximated by a "dummy" differential equation and a *linear* algebraic equation as follows:

$$T\dot{z}_i = -z_i + [h(x_i, y_i) - A_i y_i - r(x_i)] \tag{7.A.8}$$

$$0 = A_i y_i + z_i + r(x_i) \tag{7.A.9}$$

Note that the differential equation is nonlinear, but the algebraic equation is linear in the variable y. If $T = 0$, the differential-algebraic equations are equivalent to (7.A.7). If T is small but non-zero, then we expect the equations to be a good approximation. For most transient stability studies, a reasonable value of T is in the range of 5–10 ms. The use of the dummy differential equations is an extension of the approach used specifically for handling generator saliency [13].

The dummy differential equations are clubbed and solved along with the other differential equations of the system. The linear algebraic equations are clubbed with the other linear network equations and are solved using a linear system solver. The overall solution is non-iterative, but this is at the cost of adding the low time-constant dummy differential equations to our original set of equations, which forces the use of smaller step sizes. This is often an acceptable trade-off, given the overall simplicity which results due to the approximation.

For example, the current injected by a constant real and reactive power load (P_L, Q_L),

$$(i_Q + ji_D) = -\frac{(P_L - jQ_L)}{(v_Q - jv_D)} \tag{7.A.10}$$

can be approximated by:

$$T\frac{d}{dt}(i_Q^d + ji_D^d) = -(i_Q^d + ji_D^d) - \frac{(P_L - jQ_L)}{(v_Q - jv_D)} + (G_L - jB_L)(v_Q + jv_D) \tag{7.A.11}$$

$$(i_Q + ji_D) = -(G_L - jB_L)(v_Q + jv_D) + (i_Q^d + ji_D^d) \tag{7.A.12}$$

In this case, the algebraic variables are i_D, i_Q, v_D, and v_Q, and the dummy state variables are i_D^d and i_Q^d. $G_L - jB_L$ may be chosen to be the equivalent admittance of the load at nominal voltage. The approximation is depicted by the equivalent circuit shown in Figure 7.A.1.

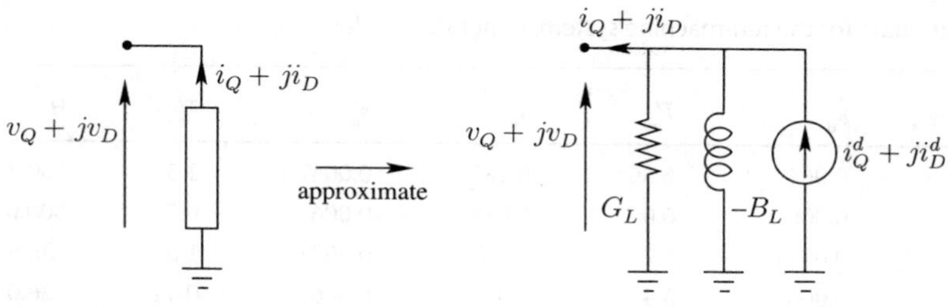

Figure 7.A.1 Approximate equivalent of a nonlinear load.

This approach may also be used to handle dynamic saliency, as discussed below [13]. The output equations of Model (1.0) of a generator (damper windings neglected) (see Section 6.5.2) are given by:

$$P^{-1}\begin{bmatrix} E'_q \\ 0 \end{bmatrix} + P^{-1}XP\begin{bmatrix} i_Q \\ i_D \end{bmatrix} = \begin{bmatrix} v_Q \\ v_D \end{bmatrix} \tag{7.A.13}$$

where

$$P = \begin{bmatrix} \cos\delta & \sin\delta \\ -\sin\delta & \cos\delta \end{bmatrix} \quad \text{and} \quad X = \begin{bmatrix} 0 & x'_d \\ -x_q & 0 \end{bmatrix} \tag{7.A.14}$$

For Model (1.0), dynamic saliency is present if $x'_d \neq x_q$ (this is generally the case). Therefore, the matrix $P^{-1}XP$ is a function of δ, which changes at every time step. To avoid this, we use the approximate equations given below:

$$P^{-1}\begin{bmatrix} E'_q \\ E' \end{bmatrix} + P^{-1}\begin{bmatrix} 0 & x'_d \\ -x'_d & 0 \end{bmatrix} P\begin{bmatrix} i_Q \\ i_D \end{bmatrix} = P^{-1}\begin{bmatrix} E'_q \\ E' \end{bmatrix} + \begin{bmatrix} 0 & x'_d \\ -x'_d & 0 \end{bmatrix}\begin{bmatrix} i_Q \\ i_D \end{bmatrix} = \begin{bmatrix} v_Q \\ v_D \end{bmatrix} \tag{7.A.15}$$

and

$$T\frac{dE'}{dt} = -E' + (x_q - x'_d)i_q \tag{7.A.16}$$

The equation 7.A.15 may be written compactly as follows:

$$(E'_q + jE')e^{j\delta} - jx'_d(i_Q + ji_D) = (v_Q + jv_D) \tag{7.A.17}$$

Equations 7.A.16 and 7.A.17 are used in the simulation. Note that the coefficient of $(i_Q + ji_D)$ is a constant. Dynamic saliency in higher-order generator models can be handled in a similar fashion.

7.B Ten-Machine System Data

The data for this system is adapted from [4] and is given in Tables 7.B.1–7.B.4. All values are in per unit on a 100 MVA base.

The excitation system parameters (see Figure 6.4) on all machines are as follows: K_E = 25 pu, T_E = 0.025 s, E_{fdmax} = 6.0 pu, and E_{fdmin} = −6.0 pu. Machine 9 is provided with a PSS as shown in Figure 7.B.1.

Table 7.B.1 Generator data for the ten-machine system, generator Model (1.1).

Bus	x_d	x'_d	T'_{do}	x_q	x'_q	T'_{qo}	H
1	0.295	0.0647	6.56	0.282	0.0647	1.5	30.3
2	0.02	0.006	6.0	0.019	0.006	0.7	500.0
3	0.2495	0.0531	5.7	0.237	0.0531	1.5	35.8
4	0.33	0.066	5.4	0.31	0.066	0.44	26.0
5	0.262	0.0436	5.69	0.258	0.0436	1.5	28.6
6	0.254	0.05	7.3	0.241	0.05	0.4	34.8
7	0.295	0.049	5.66	0.292	0.049	1.5	26.4
8	0.290	0.057	6.7	0.280	0.057	0.41	24.3
9	0.2106	0.057	4.79	0.205	0.057	1.96	34.5
10	0.2	0.004	5.7	0.196	0.004	0.5	42.0

Table 7.B.2 Transformer data for the ten-machine system.

From	To	R	X	Tap
39	30	0.0007	0.0138	1.0
39	5	0.0007	0.0142	1.0
32	33	0.0016	0.0435	1.0
32	31	0.0016	0.0435	1.0
30	4	0.0009	0.0180	1.0
29	9	0.0008	0.0156	1.0

From	To	R	X	Tap
25	8	0.0006	0.0232	1.0
23	7	0.0005	0.0272	1.0
22	6	0.0000	0.0143	1.0
20	3	0.0000	0.0200	1.0
16	1	0.0000	0.0250	1.0
12	10	0.0000	0.0181	1.0

Table 7.B.3 Transmission line data for the ten-machine system.

From	To	R	X	B_c
37	27	0.0013	0.0173	0.3216
37	38	0.0007	0.0082	0.1319
36	24	0.0003	0.0059	0.0680
36	21	0.0008	0.0135	0.2548
36	39	0.0016	0.0195	0.3040
36	37	0.0007	0.0089	0.1342
35	36	0.0009	0.0094	0.1710
34	35	0.0018	0.0217	0.3660
33	34	0.0009	0.0101	0.1723
28	29	0.0014	0.0151	0.2490
26	29	0.0057	0.0625	1.0290
26	28	0.0043	0.0474	0.7802
26	27	0.0014	0.0147	0.2396
25	26	0.0032	0.0323	0.5130
23	24	0.0022	0.0350	0.3610
22	23	0.0006	0.0096	0.1846
21	22	0.0008	0.0135	0.2548

From	To	R	X	B_c
20	33	0.0004	0.0043	0.0729
20	31	0.0004	0.0043	0.0729
19	2	0.0010	0.0250	1.2000
18	19	0.0023	0.0363	0.3804
17	18	0.0004	0.0046	0.0780
16	31	0.0007	0.0082	0.1389
16	17	0.0006	0.0092	0.1130
15	18	0.0008	0.0112	0.1476
15	16	0.0002	0.0026	0.0434
14	34	0.0008	0.0129	0.1382
14	15	0.0008	0.0128	0.1342
13	38	0.0011	0.0133	0.2138
13	14	0.0013	0.0213	0.2214
12	25	0.0070	0.0086	0.1460
12	13	0.0013	0.0151	0.2572
11	12	0.0035	0.0411	0.6987
11	2	0.0010	0.0250	0.7500

Table 7.B.4 Load flow data for the ten-machine system, negative P and Q indicate loads.

Bus	V	$\phi°$	P	Q
1	0.982	0.00	5.52	1.620
			−0.092	−0.046
2	1.030	−10.97	10.00	2.262
			−11.04	−2.500
3	0.983	2.34	6.500	1.660
4	1.012	3.17	5.080	1.551
5	0.997	4.19	6.320	0.838
6	1.049	5.20	6.500	2.810
7	1.064	7.991	5.600	2.297
8	1.028	1.84	5.400	0.276
9	1.027	7.55	8.300	0.597
10	1.048	−4.01	2.500	1.839
11	1.035	−9.32	0.00	0.00
12	1.017	−6.44	0.00	0.00
13	0.986	−9.44	−3.220	−0.024
14	0.950	−10.37	−5.000	−1.840
15	0.951	−9.12	0.00	0.00
16	0.952	−8.35	0.00	0.00
17	0.944	−10.80	−2.338	−0.840
18	0.945	−11.36	−5.220	−1.760
19	1.007	−11.18	0.00	0.00
20	0.958	−5.59	0.00	0.00
21	0.985	−4.34	−2.740	−1.150
22	1.015	0.19	0.00	0.00
23	1.012	−0.08	−2.745	−0.847
24	0.973	−6.80	−3.086	−0.922
25	1.026	−4.97	−2.240	−0.472
26	1.012	−6.21	−1.390	−0.170
27	0.992	−8.33	−2.810	−0.755
28	1.016	−2.47	−2.060	−0.276
29	1.0187	0.46	−2.835	−0.269
30	0.984	−2.02	−6.280	−1.030
31	0.955	−6.53	0.00	0.00
32	0.935	−6.51	−0.075	−0.880
33	0.956	−6.38	0.00	0.00
34	0.955	−8.22	0.00	0.00
35	0.957	−8.53	−3.200	−1.530
36	0.974	−6.89	−3.294	−0.323
37	0.981	−8.10	0.00	0.00
38	0.981	−9.09	−1.580	−0.300
39	0.985	−1.02	0.00	0.00

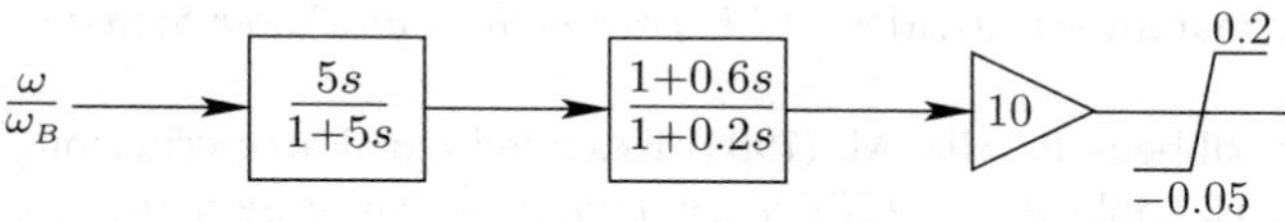

Figure 7.B.1 PSS on machine 9.

References

1 Powergrid Corporation of India (2012), Report on the Grid Disturbance on 30th July 2012 and grid disturbance on 31st July 2012. Submitted to Central Electricity Regulatory Commission, New Delhi.

2 Salunkhe, K.A., Gajjar, G., Soman, S.A., and Kulkarni, A.M. (2014) Implementation and applications of a wide area frequency measurement system synchronized using network time protocol, in *IEEE PES General Meeting*, National Harbor, Maryland.

3 (1992) IEEE committee report on single phase tripping and auto reclosing of transmission lines. *IEEE Transactions on Power Delivery*, 7 (1), 182–192.

4 Behera, A.K. (1988) *Transient stability analysis of multimachine power systems using detailed machine models*, Ph.D. thesis, University of Illinois at Urbana-Champaign.

5 Adhikari, T., Isacsson, G., and Ambekar, V.D. (1999) The Chandrapur–Padghe HVDC bipole transmission, in *CIGRE Symposium of Power System Issues in Rapidly Industrializing Countries*, Kuala Lumpur, pp. 20–23.

6 Shelton, M.L., Winkelman, P.F., Mittelstadt, W.A., and Bellerby, W.J. (1975) Bonneville Power Administration 1400-MW braking resistor. *IEEE Transactions on Power Apparatus and Systems*, **94** (2), 602–611.

7 Pai, M.A. (1989) *Energy Function Analysis for Power System Stability*, Kluwer, Boston.

8 Padiyar, K.R. (2013) *Structure Preserving Energy Functions in Power Systems: Theory and Applications*, CRC Press, Boca Raton.

9 Padiyar, K.R. and Rao, K.U. (1997) Discrete control of series compensation for stability improvement in power systems. *International Journal of Electrical Power & Energy Systems*, **19** (5), 311–319.

10 Chiang, H.D., Wu, F., and Varaiya, P. (1987) Foundations of direct methods for power system transient stability analysis. *IEEE Transactions on Circuits and Systems*, **34** (2), 160–173.

11 Chiang, H.D., Wu, F.F., and Varaiya, P.P. (1988) Foundations of the potential energy boundary surface method for power system transient stability analysis. *IEEE Transactions on Circuits and Systems*, **35** (6), 712–728.

12 Chiang, H.D., Wu, F.F., and Varaiya, P.P. (1994) A BCU method for direct analysis of power system transient stability. *IEEE Transactions on Power Systems*, **9** (3), 1194–1208.

13 Padiyar, K.R. (2008) *Power System Dynamics: Stability and Control*, BS publications, Hyderabad, 2nd edn.

14 Shubhanga, K.N. and Kulkarni, A.M. (2002) Application of structure preserving energy margin sensitivity to determine the effectiveness of shunt and series FACTS devices. *IEEE Transactions on Power Systems*, **17** (3), 730–738.

15 Shubhanga, K.N. and Kulkarni, A.M. (2004) Stability-constrained generation rescheduling using energy margin sensitivities. *IEEE Transactions on Power Systems*, **19** (3), 1402–1413.

16 Xue, Y., Custem, T.V., and Ribbens-Pavella, M. (1989) Extended equal area criterion justifications, generalizations, applications. *IEEE Transactions on Power Systems*, **4** (1), 44–52.

17 Fouad, A.A. and Vittal, V. (1988) The transient energy function method. *International Journal of Electrical Power & Energy Systems*, **10** (4), 233–246.

18 Kakimoto, N., Ohsawa, Y., and Hayashi, M. (1978) Transient stability analysis of electric power system via Luré type Lyapunov functions, Part I : New critical value for transient stability; and Part II : Modification of Luré type Lyapunov with effect of transconductance. *Trans. IEE of Japan*, **98** (5/6), 62–71; 72–79.

19 Crow, M. (2003) *Computational Methods for Electric Power Systems*, CRC Press, Boca Raton.

8

Analysis of Voltage Stability and Control

8.1 Introduction

Power system stability has been primarily associated with rotor angle (synchronous) stability in the past. The transient (angle) stability relates to the stability under large disturbances such as line or bus faults followed by clearing of the fault. The advent of fast-acting static exciters with high gain automatic voltage regulators (AVRs) in the 1960s introduced the new problem of small, low-frequency oscillations (0.2–2.0 Hz) that can grow and lead to loss of synchronism. In the 1970s, the problem of torsional oscillations (10–50 Hz), which can be sustained or negatively damped, led to the investigation of subsynchronous resonance (SSR) and its various facets.

In the 1980s, the problem of voltage instability and collapse resulted in several major system blackouts [1–3]. For example, on July 23, 1987 in Tokyo, Japan, loads increased at 400 MW/minute after the noon hour. Despite connection of all available shunt capacitors, the voltage decayed, with voltages on the 500 kV system at 460 kV at 13.15 hours and at 370 kV at 13.19 hours. Collapse began at 13.19 hours; 8168 MW was interrupted. Similar events occurred in France on December 18, 1978 and 29 GW of load was interrupted with energy outage of 100 GWh. Similar incidents have also been experienced in other countries.

Voltage collapse incidents have also been observed in HVDC systems. In the Nelson River HVDC system, Winnipeg, Canada, partial voltage collapse occurred on April 13, 1986 during energization of a converter transformer. Inrush current depressed the AC voltage, resulting in commutation failures. To overcome this, the converter firing angle (α) was reduced and this led to the collapse of the inverter AC voltage to 57%, which recovered after temporary DC blocking. A second voltage collapse led to the shut down of three of four DC poles, and undervoltage load shedding.

In south-east Brazil, at the inverter station of the Itaipu HVDC system, the AC voltage dropped to 0.85 pu on November 30, 1986, after several outages in the AC system. Repetitive commutation failures occurred and the DC power control increased DC current, which led to increased converter reactive power consumption. A complete DC system shut down and an AC system break up resulted. Over 1200 MW of load was shed. This problem finally led to the modification in the converter control.

In this chapter, we will examine the nature of voltage stability issues, including mechanisms of collapse. We will also look into the comparison of voltage and angle stability and factors affecting voltage instability. The voltage stability can be improved by the

Dynamics and Control of Electric Transmission and Microgrids, First Edition.
K. R. Padiyar and Anil M. Kulkarni.

provision of FACTS controllers, in particular STATCOM. The analysis will be illustrated with the presentation of examples.

8.2 Definitions of Voltage Stability

The CIGRE Study Committee 38 and IEEE Power System Dynamic Performance Committee set up a Joint Task Force, which published its report in May 2004 [4], on the topic of definitions and classification of power system stability. The classification of power system stability is shown in Figure 8.1.

"Voltage stability refers to the ability of a power system to maintain steady voltages at all buses in the system after being subjected to a disturbance from a given initial operating condition. It depends on the ability to maintain/restore equilibrium between load demand and supply from the power system. Instability that may result occurs in the form of a progressive fall or rise of voltages of some buses. A possible outcome of voltage instability is loss of load in an area, or tripping of transmission lines by their protective systems leading to cascading outages". Loss of synchronism of some generators may result from these outages or from operating conditions that violate field current limits [3]. The term "voltage collapse" refers to the process by which the sequence of events accompanying voltage instability leads to a blackout or abnormally low voltages in a significant part of the power system [1].

One form of voltage stability problem that results in uncontrolled overvoltages is the self-excitation of synchronous machines due to excessive capacitive loads. The examples are (i) open-ended high-voltage lines and (ii) shunt capacitors and filter banks from HVDC stations [5]. Negative field current capability of the exciter of the synchronous machine has a beneficial influence on the limits of self-excitation.

Large Disturbance Voltage Stability

This refers to the system's ability to maintain steady voltages following large disturbances such as system faults, loss of generation or circuit contingencies. This ability

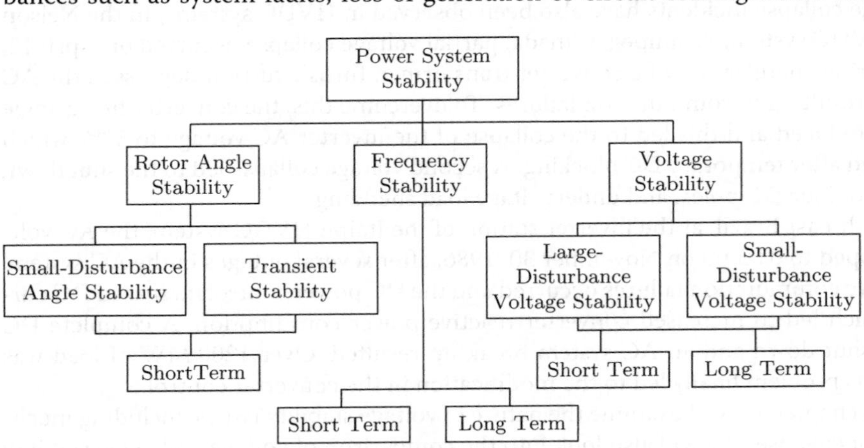

Figure 8.1 Classification of power system stability. Source: [4] Fig. 1, p. 1390, reproduced with permission ©IEEE 2004.

is determined by the system and load characteristics, and interactions of both continuous and discrete controls and protections. The study period of interest may extend from a few seconds to tens of minutes. The performances of motor drives, on-load tap changers (OLTCs), and generator field current limiters play a role in the response of the system.

Small Disturbance Voltage Stability

This refers to the system's ability to maintain steady voltage when subjected to small perturbations such as incremental changes in the load. This form of stability is influenced by the characteristics of the loads, continuous controls, and discrete controls at a given time.

Short-Term Voltage Stability

This involves dynamics of fast-acting load components such as induction motors, electronically controlled loads, and HVDC converters. The study period of interest is of the order of several seconds.

Long-Term Voltage Stability

This involves slower acting equipment such as tap-changing transformers, thermostatically controlled loads, and generator current limiters. The study period of interest may extend to several minutes.

Remarks

1. The power uncontrollability is a steady-state problem accompanied by low voltages, when switching in more load results in reduced power. A related concept is maximum power transfer capability or maximum loadability, which is computed from the static characteristics of the system. For example, consider an AC voltage source (E) supplying a load through a transmission line (modeled by a series reactance $X = x_e + x_g$). B_c is the susceptance of the shunt capacitor (see Figure 8.2).
   It can be shown that

$$P = \frac{E_g^2 \sin 2\delta}{2X(1 - B_c X)} = P_L \tag{8.1}$$

$$V = \frac{E_g \cos \delta}{(1 - B_c X)} \tag{8.2}$$

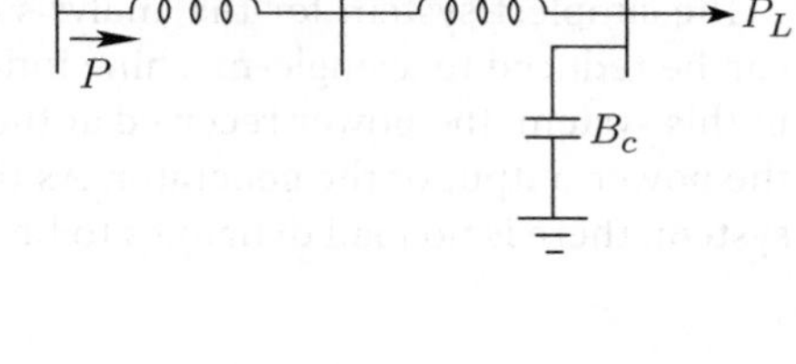

Figure 8.2 A simplified SMLB system.

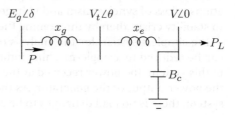

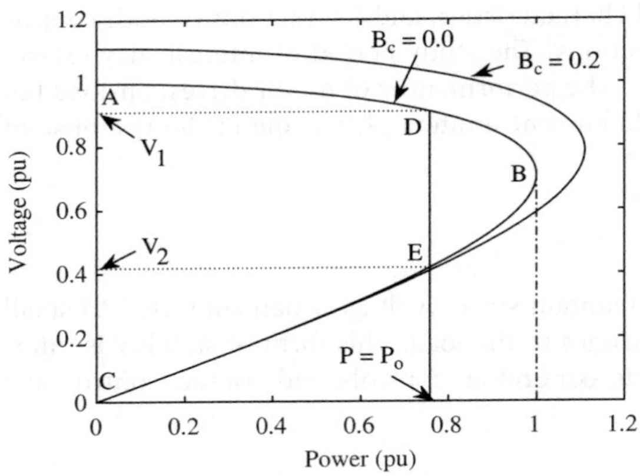

Figure 8.3 *P–V* diagram at load bus.

The plots of voltage V as a function of P for a unity power factor load are shown in Figure 8.3 for two values of B_c (0.0 and 0.2). It is assumed that $E_g = 1.0$ pu and $X = 0.5$ pu.

The curves shown in Figure 8.3 are termed "nose" curves (for obvious reasons). Similar curves are obtained even if the power factor of the load is less than unity. If the load (say a lamp load) is switched on incrementally, the observed load voltage traces the curve ADBEC (when $B_c = 0.0$). The point B corresponds to the maximum power that can be supplied. If additional loads (say lamps) are switched on, the voltage keeps dropping until P reaches zero.

From Figure 8.3, it is observed that for a constant power load of $P = P_o$, there are two voltage solutions, V_1 and V_2 (for $B_c = 0.0$). In the literature, it is often mentioned that the high-voltage solution V_1 is stable while the low-voltage solution V_2 is unstable. This statement requires an explanation based on system dynamics. It is incorrect to say that the lower portion of the nose curve (BC) is unstable. In reality, the region BC corresponds to "power uncontrollability" and maximum loadability is determined by point B.

8.3 Comparison of Angle and Voltage Stability

The analysis of power system stability in the past pertained to the power angle dynamics expressed in terms of swing equations of synchronous generators. In the context of voltage stability and the dynamics related to it, the stability of the system based on power angle dynamics can be termed angle stability, where the major concern is the loss of synchronism. On the other hand, in the analysis of voltage stability, the major concern is the voltage collapse, although the generators may remain in synchronism. Both situations – loss of synchronism and voltage collapse – can affect system security and lead to stability crisis, thereby threatening the integrity of the system.

The simplest system for the analysis of angle stability is a two-machine system that can be reduced to a single-machine infinite bus (SMIB) system. It should be noted that in this system, the power received at the infinite bus in steady state is determined from the power output of the generator. As there is no load bus in the simple two-bus SMIB system, there is no load dynamics to be considered. Only the generator dynamics affect

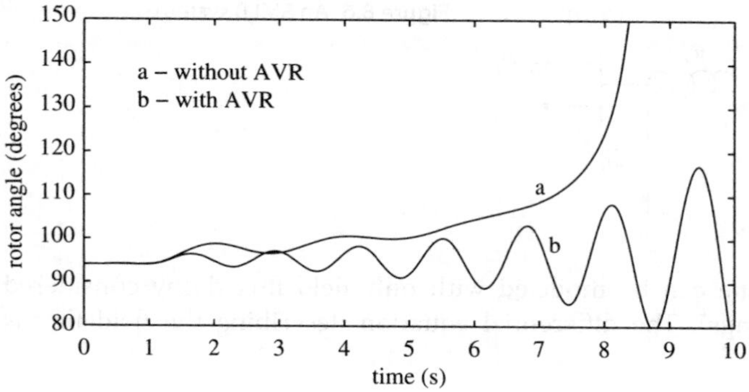

Figure 8.4 Swing curves.

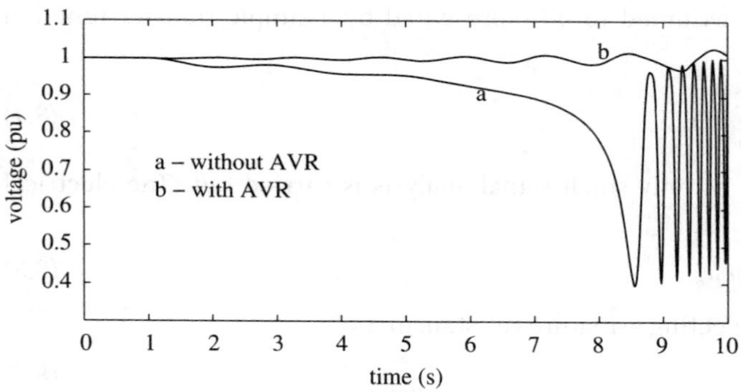

Figure 8.5 Variation of terminal voltage.

(angle) stability. However, it should be noted that loss of synchronism in a SMIB system is accompanied by voltage instability at the generator terminal bus (see Figures 8.4 and 8.5). The disturbance considered is an increase in T_m by 0.01 pu at $t = 1$ s. If AVR is not considered, there is aperiodic or monotonic decrease in terminal voltage. On the other hand, with a fast-acting excitation system and high gain AVR there is oscillatory instability, which can lead to subsequent loss of synchronism. It should be noted that negatively damped oscillations in the rotor angle are also accompanied by voltage oscillations that increase in amplitude. When the generator loses synchronism, the terminal voltage has large oscillations with increasing frequency. However, voltage stability is normally associated with load buses. Thus, the simplest system that can be considered for the study of voltage stability is the single-machine load bus (SMLB) system shown in Figure 8.6.

8.3.1 Analysis of the SMLB System

The system shown in Figure 8.6 contains an equivalent machine of several coherent generators (swinging together). The load is initially considered to be of static type, with voltage-dependent active and reactive power characteristic. A shunt capacitor is connected across the load for reactive power compensation.

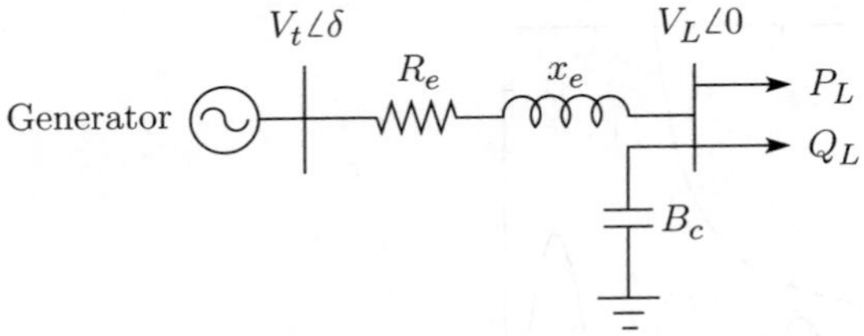

Figure 8.6 An SMLB system.

The synchronous machine can be modeled with only field flux decay considered (neglecting damper windings). The differential equation describing the field flux is given by

$$T'_{do}\frac{dE'_q}{dt} = E_{fd} - E'_q + (x_d - x'_d)i_d \tag{8.3}$$

The excitation system is assumed to be represented by a simple transfer function given by

$$E_{fd} = \frac{K_E}{1+sT_E}(V_{ref} - V_t) \tag{8.4}$$

The limiters are ignored as only small signal analysis is carried out. The electrical torque T_e is given by

$$T_e = E'_q i_q - (x_q - x'_d)i_d i_q \tag{8.5}$$

The stator equations (neglecting armature resistance) are

$$E'_q + x'_d i_d = v_q \tag{8.6}$$

$$-x_q i_q = v_d \tag{8.7}$$

The network and load equations are

$$v_q = -x_e i_d + R_e i_q + v_{qL} \tag{8.8}$$

$$v_d = R_e i_d + x_e i_q + v_{dL} \tag{8.9}$$

$$v_{qL} + jv_{dL} = \frac{P_L + j(Q_L - V_L^2 B_c)}{(i_q - ji_d)},\ V_L = \sqrt{v_{dL}^2 + v_{qL}^2} \tag{8.10}$$

$$P_L = P_{Lo}\left(\frac{V_L}{V_{Lo}}\right)^{m_p} \tag{8.11}$$

$$Q_L = Q_{Lo}\left(\frac{V_L}{V_{Lo}}\right)^{m_q} \tag{8.12}$$

Linearizing (8.5) to (8.12) and simplifying, it can be shown that

$$\Delta i_d = Y_d \Delta E'_q \tag{8.13}$$

$$\Delta i_q = Y_q \Delta E'_q \tag{8.14}$$

$$\Delta T_e = K'_2 \Delta E'_q \tag{8.15}$$

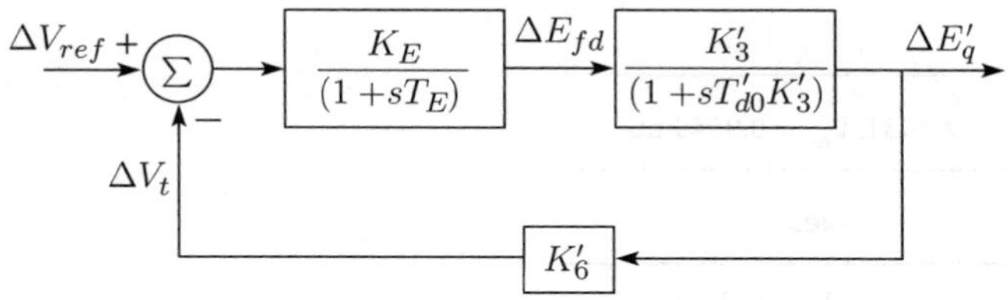

Figure 8.7 Block diagram of the voltage control loop.

It is interesting to note that the torque deviation ΔT_e depends only on $\Delta E_q'$ in this case, in contrast to a SMIB system where

$$\Delta T_e = K_1 \Delta\delta + K_2 \Delta E_q' \tag{8.16}$$

where K_1 and K_2 are a part of Heffron–Philips constants.

Thus, an SMLB system cannot exhibit angle instability. The power output of the generator (as well as the electrical torque per unit) is determined by the load power. In the absence of angle dynamics, the only relevant dynamics is that corresponding to the field flux decay and excitation system (see block diagram in Figure 8.7). Here K_3' and K_6' are analogous to Heffron–Phillips constants. The characteristic equation for the system shown in Figure 8.7 is given by

$$s^2 + \left(\frac{1}{T_{do}'K_3'} + \frac{1}{T_E}\right)s + \frac{1 + K_3'K_6'K_E}{T_{do}'K_3'T_E} = 0 \tag{8.17}$$

If $K_E K_6' K_3' \gg 1$, then the criteria for stability are

$$\frac{1}{T_{do}'K_3'} + \frac{1}{T_E} > 0 \quad \text{and} \quad K_6' > 0 \tag{8.18}$$

If the generator is represented by the classical model of a constant voltage source E_g behind a constant reactance x_g, there is no possibility of voltage instability and collapse as there is no voltage or load dynamics.

Example 8.1 Consider an SMLB system with the following data: $R_e = 0, x_e = 0.4$, $x_d = 1.6, x_q = 1.55, x_d' = 0.32$ (all in pu), $K_E = 250$, $T_E = 0.25$ s, $T_{do}' = 6.0$ s. The speed of the machine is assumed to be constant at the rated speed. Evaluate the poles of the system shown in Figure 8.7.

Solution

The poles are evaluated for four operating conditions and are shown in Table 8.1.

Remarks

1) It is interesting to note that the SMLB system is unstable if K_6' is negative and a real eigenvalue crosses the imaginary axis. This is unlike the unstable rotor oscillations in a SMIB system where complex eigenvalues cross the imaginary axis. Note that in a SMIB system $\Delta V_t = K_5\, \Delta\delta + K_6\, \Delta E_q'$. If K_5 is negative, the oscillations are negatively damped and the system becomes unstable.

Table 8.1 Poles of the system.

m_p	K'_3	K'_6	Poles
$P_o = 0.9, Q_o = 0.0, V_{Lo} = 0.9, E_{fdo} = 2.1931, V_{to} = 0.9849$ pu			
2	0.4701	0.9554	$-2.18 \pm j12.49$
1	1.5528	1.4346	$-2.05 \pm j15.34$
0	−0.0795	5.7062	$-0.95 \pm j30.69$
$P_o = 0.9, Q_o = 0.0, V_{Lo} = 0.95, E_{fdo} = 2.1194, V_{to} = 1.0228$ pu			
2	0.4912	0.9825	$-2.17 \pm j12.66$
1	1.5299	1.3551	$-2.05 \pm j14.90$
0	−0.2061	2.8395	$-1.60 \pm j21.62$
$P_o = 1.0, Q_o = 0.0, V_{Lo} = 0.9, E_{fdo} = 2.3975, V_{to} = 1.0038$ pu			
2	0.4522	0.9259	$-2.18 \pm j12.29$
1	1.5723	1.5198	$-2.05 \pm j15.80$
0	0.0128	**−27.6067**	−76.47, **59.49**
$P_o = 1.0, Q_o = 0.0, V_{Lo} = 0.95, E_{fdo} = 2.3096, V_{to} = 1.0391$ pu			
2	0.4706	0.9561	$-2.18 \pm j12.49$
1	1.5522	1.4325	$-2.05 \pm j15.33$
0	−0.0823	5.5484	$-0.99 \pm j30.26$

2) It should be noted that E'_q is the voltage generated by the flux linked with the field winding that increases with the field current (as the field excitation is increased under normal conditions). However, under voltage instability conditions, the terminal voltage falls as the field excitation is increased (since K'_6 is negative).

8.4 Mathematical Preliminaries [6, 7]

The dynamics of a power system are described by differential-algebraic equations (DAE) given by

$$\dot{x} = f(x, y, p) \tag{8.19}$$

$$0 = g(x, y, p) \tag{8.20}$$

where x is the vector of n state variables, y is the vector of m algebraic variables, and p is a vector of k parameters. Assuming that the Jacobian $[\frac{\partial g}{\partial y}]$ is a non-singular matrix in the domain of the variables considered, we can express y as a function of x, using implicit function theorem, and obtain

$$\dot{x} = F(x, p) \tag{8.21}$$

where F is a locally unique and smooth function.

For a specified set of parameters, the equilibrium points of the DAE system are obtained as the solutions of the equations

$$f(x, y, p) = 0, \quad g(x, y, p) = 0 \tag{8.22}$$

The stability of the equilibrium points (EP) can be determined by linearizing the above equations to obtain

$$\begin{bmatrix} \Delta\dot{x} \\ 0 \end{bmatrix} = [J_s] \begin{bmatrix} \Delta x \\ \Delta y \end{bmatrix} \tag{8.23}$$

where

$$[J_s] = \begin{bmatrix} f_x & f_y \\ g_x & g_y \end{bmatrix} \tag{8.24}$$

The subscript denotes the variable with respect to which the differential is obtained. For example,

$$f_x = \frac{\partial f}{\partial x} \quad \text{and} \quad g_y = \frac{\partial g}{\partial y}$$

Assuming g_y is non-singular, we can eliminate Δy and obtain finally

$$\Delta\dot{x} = [A]\Delta x, \quad [A] = f_x - f_y g_y^{-1} g_x \tag{8.25}$$

The stability of the equilibrium points of the system is obtained from the eigenvalues of the matrix $[A]$ evaluated at the equilibrium point. If all the eigenvalues are in the left half of the complex plane (LHP), the equilibrium point is stable.

It should be noted that the matrix $[A]$ and hence the stability properties of an equilibrium point depend on the parameters p. As the parameters change, one or more of the eigenvalues of the matrix can cross the imaginary axis (from left to right, assuming the equilibrium point was initially stable). If a real eigenvalue crosses the imaginary axis and moves to the right half of the complex plane (RHP), then we say it is a saddle-node bifurcation (SNB). At this point, the determinant of $[A]$ is zero as one of the eigenvalues becomes zero. From Schur's formula, we have

$$\det(J_s) = \det(g_y) \det A \tag{8.26}$$

Thus, the determinant of $[J_s]$ is also zero when the matrix $[A]$ is singular.

In DAE systems, it is also possible that $\det(J_s) = 0$ even when $\det(A) \neq 0$. This occurs when $\det(g_y) = 0$. In such cases, it is called a singularity induced bifurcation (SIB). Obviously, this happens when the implicit function theorem is not applicable.

When two complex eigenvalues cross the imaginary axis, we say that Hopf bifurcation (HB) occurs and the equilibrium point becomes unstable. While SNB is characterized by monotonic increase or decrease (depending on the sign of the variable), the HB is characterized by growing oscillations.

If the voltage collapse is monotonic, it is caused by SNB and the stability analysis can be performed based on the singularity properties of the system Jacobian (J_s). Several papers essentially consider only the load flow Jacobian (g_y) by assuming that the singularity properties of the system Jacobian can be approximated by the singularity properties of the load flow Jacobian. This is not always applicable.

If the voltage collapse is accompanied by undamped oscillations (caused by Hopf bifurcation) then dynamic analysis is required.

8.5 Factors Affecting Instability and Collapse

Voltage collapse occurs invariably following a large disturbance or large load increase in a heavily stressed system (the transmission lines operating with power flows close to their limits). The increase in the load or tripping of a line results in an increased reactive power consumption in the transmission network and consequent voltage reduction, which triggers control mechanisms for load restoration. It is the dynamics of these controls that often lead to voltage instability and collapse. In induction motor and heating loads, the load restoration mechanism is inherent. However, in other (static) loads, the dynamics is introduced by the OLTC dynamics. It has been observed in practice that blocking the operation of the OLTC at distribution transformers helps to prevent voltage collapse.

The voltage collapse can also occur when the voltage support provided by the synchronous generators feeding the load is lost or reduced. This can happen due to the field current limitation caused by the overexcitation limiter (OEL) of the generators [8]. Normally, the OEL acts as a slow device allowing transient overexcitation of the generator for about 20 s before enforcing the field current limit. In this case, other control mechanisms such as OLTCs have time to act, and the problem becomes a long-term voltage stability issue. However, when transient overexcitation is not allowed above an instantaneous limit ($I_{\max}$) that must be enforced in the short term (a fraction of a second), it becomes a short-term voltage stability problem.

In what follows, the models of the induction motor, HVDC converter [9], OLTC, and generic dynamic load models will be presented [10].

8.5.1 Induction Motor Loads [11]

The simplest induction motor model considers only the slip (S) dynamics defined by

$$2H_m \frac{dS}{dt} = (T_m - T_e) = \left[\frac{P_m}{1-S} - P_e\right] \tag{8.27}$$

where H_m is the inertia constant, P_m is the mechanical load on the induction motor, and P_e is the electrical power drawn by the motor neglecting the stator losses. The steady-state equivalent circuit of an induction motor is shown in Figure 8.8. Here the magnetic core losses are neglected. If X_m is assumed to be large, then it can be shown that

$$P_e = \frac{V_t^2}{\left(\frac{R_r}{S}\right)^2 + (X_r + X_s)^2} \frac{R_r}{S} \tag{8.28}$$

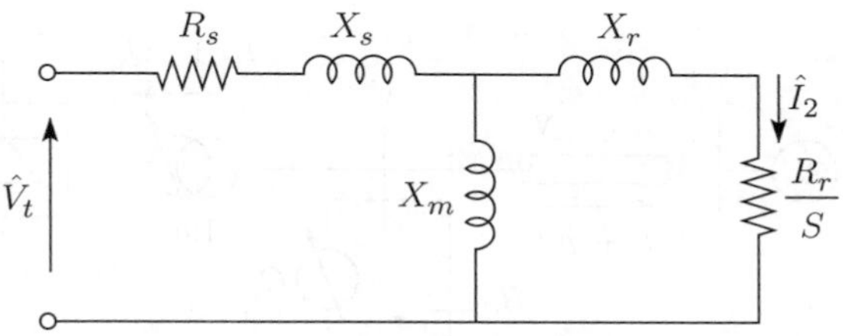

Figure 8.8 Steady-state equivalent circuit of an induction motor.

It should be noted that when the load torque T_m is constant, the motor draws constant power in steady state, irrespective of the voltage applied, as long as the motor operates in the stable region of the torque slip characteristics. Figure 8.9 shows the typical torque slip characteristics of an induction motor. The peak torque (T_p) occurs at slip S_p and the motor operates stably with $S = S_1$ in the region of slip ($0 < S < S_p$), where the mechanical torque applied is less than the peak torque. However, during faults or periods of low voltage, the motor decelerates as $T_m > T_e$ and the slip continues to increase until the fault is cleared and the voltage is restored. If the slip exceeds S_2 ($S_p < S_2 < 1$), then the motor enters the unstable region and the motor continues to decelerate and stalls eventually. A stalled motor draws large currents, which can depress the voltages in the vicinity and lead to the stalling of other motors. Note that S_2 is an unstable equilibrium point.

There is a critical clearing time for a given fault to avoid the stalling of a motor. This can be obtained from solving equation (8.27).

8.5.2 HVDC Converter [9]

Short-term voltage instability is also associated with HVDC converters, particularly at the inverter terminal as voltage reduction can cause commutation failures at the inverter. Figure 8.10 shows a converter terminal connected to an AC system. Q_c is a controllable reactive power source.

The converter equations are

$$V_d = kaV\cos\theta - R_c I_d \tag{8.29}$$

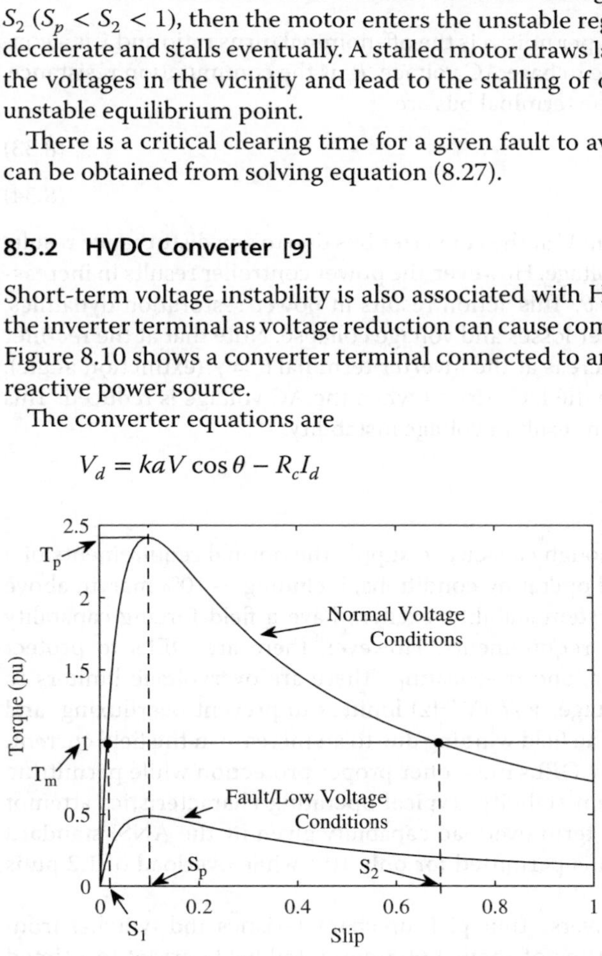

Figure 8.9 Torque-slip characteristics of an induction motor.

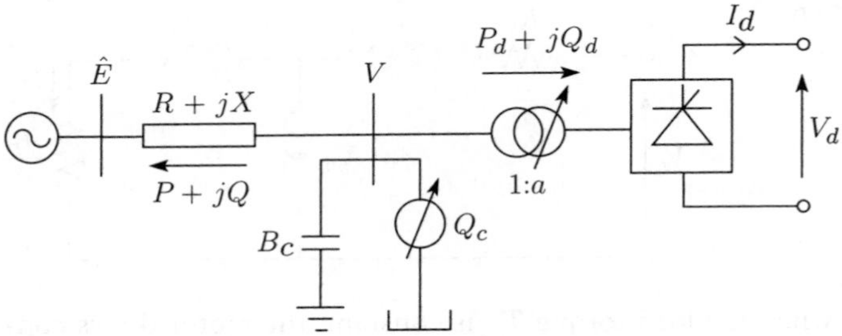

Figure 8.10 An HVDC converter terminal.

$$P_d = V_d I_d \tag{8.30}$$

$$Q_d = |P_d| \tan \phi \tag{8.31}$$

$$\cos \phi = \frac{V_d}{kaV} \tag{8.32}$$

All the variables are expressed in per unit. a is the off-nominal turns ratio and k is a constant that relates base DC voltage to base AC voltage. R_c is the commutation resistance.

The power flow equations at the terminal bus are

$$P + P_d = 0 \tag{8.33}$$

$$Q + Q_d - B_c V^2 = Q_c \tag{8.34}$$

The reduction in the AC voltage V at the converter bus due to any disturbance results in the reduction of the DC bus voltage. However, the power controller results in increasing the DC current by reducing θ. This action results in power restoration dynamics, which can lead to increased power losses and voltage collapse. Note that at the rectifier terminal $\theta = \alpha$ (delay angle), whereas at the inverter terminal $\theta = \gamma$ (extinction angle). γ has to be increased to increase the DC current when the AC voltage is reduced. This worsens the power factor and can result in voltage instability.

8.5.3 Overexcitation Limiters

Most excitation systems have enough capacity to supply the normal requirements of a synchronous generator under all operating conditions, including 5–10% margin above normal rated conditions. For system stability, exciters have a field-forcing capability which exceeds the steady-state requirements. However, there are OELs to protect against overvoltages, overfluxing, and overheating. There are overvoltage limiters to limit the generator terminal voltage, V/f (V/Hz) limiters to prevent overfluxing, and OELs to prevent overheating of the field winding due to an increase in the field current.

The limiting action provided by OELs must offer proper protection while permitting maximum field forcing for system stability. Typical operating characteristics attempt to mimic the field current short-term overload capability given by the ANSI standard in 1977. The overload of 2.08 pu is permitted for only 10 s while overload of 1.2 pu is permitted for 2 min.

A common type of OEL has inverse-time pick-up characteristics and switches from an instantaneous limiter with setting of about 1.60 times rated field current to a timed

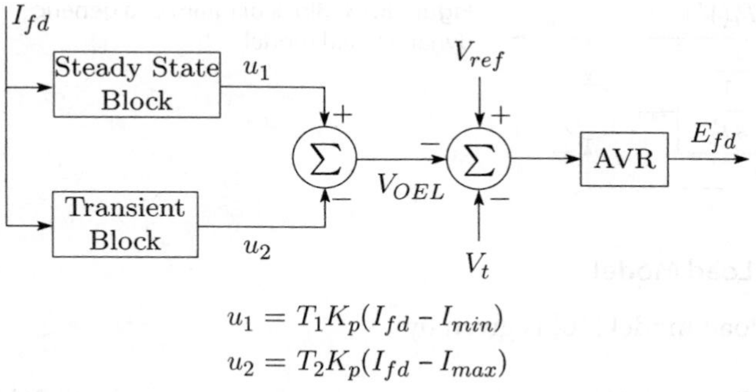

Figure 8.11 Block diagram of a summed type OEL. Source: [12] Fig. 2, p. 792, reproduced with permission ©IEEE 2006.

limiter with a setting of 1.05 times the rated field current. In some types of OEL the current limit is ramped down from the instantaneous value to the timed limiter setting. The ramp rate may be constant or proportional to the level of overexcitation.

The exciter limiter operation can be divided into two categories: (i) takeover and (ii) summed type. The former are also called "hard" limiters and control the limited quantity (say, field current) to the limiter set point within tolerances set by the limiter loop gain. While takeover limiters are active, the normal voltage regulator loop's output may saturate to a large value. This could cause a delay in the response of the excitation system when recovering from the limiter operation.

The "summed" limiters (also called "soft" limiters) can be influenced by other inputs to the voltage regulator summing junction (such as the reference adjuster, PSS). The performance of this limiter type will depend on its gain and bandwidth. Summed limiters do not require rate feedback or transient gain reduction (TGR). A block diagram of a summed type limiter is shown in Figure 8.11. T_1 and T_2 are logic variables taking values of one or zero. Both can take the value of zero, but both cannot be one at the same time.

8.5.4 OLTC Transformers [10]

Consider a OLTC transformer feeding a static load (see Figure 8.12). This is described by the model defined in (8.35). Here the transformer tap 'a' is assumed to be continuous. Although the taps are discrete, the approximation introduced by treating 'a' as continuous, is convenient to use.

$$T\frac{da}{dt} = V_{ref} - aV, \quad a_{min} \le a \le a_{max} \tag{8.35}$$

The load dynamics of thermostatic controlled resistance devices (such as used for space heating) also has a similar structure [10].

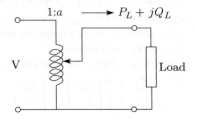

Figure 8.12 An OLTC transformer connected to a load.

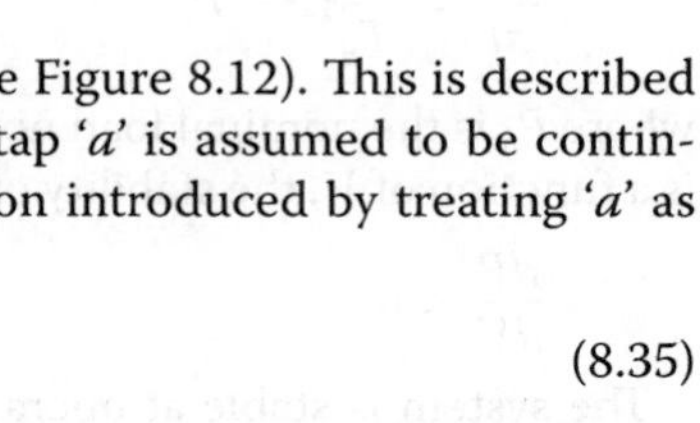

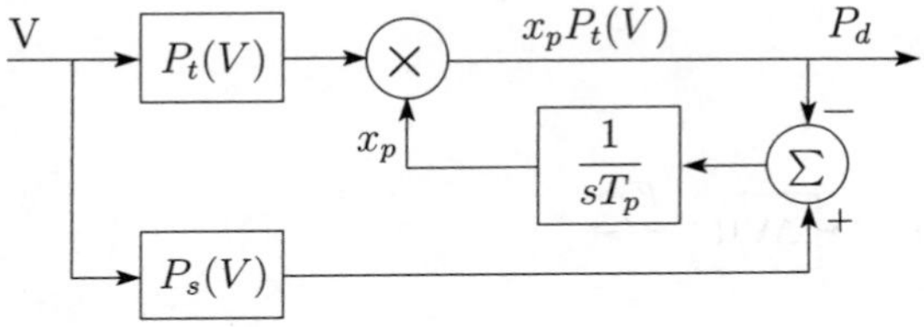

Figure 8.13 Block diagram of a generic dynamic load model.

8.5.5 A Nonlinear Dynamic Load Model

A generic nonlinear dynamic load model [10] is given by

$$T_p \frac{dx_p}{dt} = P_s(V) - P_d, \quad P_d = x_p P_t(V) \tag{8.36}$$

where x_p is the load state variable, and $P_t(V)$ and $P_s(V)$ are the transient and steady-state load characteristics. T_p is the load recovery time constant. The transient and steady-state load characteristics are voltage dependent. The generic dynamic load model can be represented by the block diagram shown in Figure 8.13.

8.6 Dynamics of Load Restoration

As mentioned earlier, an SMIB system cannot exhibit voltage instability if the synchronous machine (generator) is represented by the classical model. A SMLB system, when the load dynamics is modeled, can exhibit voltage instability. In the previous section we considered the dynamics of an induction motor, HVDC converter, and OLTC. It should be noted that in all these cases the inherent dynamics of the load and the controllers results in attempts for restoration of load power which may fail, resulting in voltage collapse.

For example, consider an OLTC or voltage regulator that controls a voltage-sensitive load. The control action can be modeled by the equation

$$\frac{dV}{dt} = \frac{1}{T_p}(P - P_o) \tag{8.37}$$

where P_o is the nominal load power and P is the power received at the load bus. Since P is a function of V, the stability of the equilibrium point is determined from the criterion

$$\frac{dP}{dV} < 0 \tag{8.38}$$

The system is stable at operating point $D\,(V = V_1)$ and unstable at operating point $E\,(V = V_2)$ in Figure 8.3. The region of attraction for the equilibrium point D is $ADBE$. Voltage instability and collapse can result following a disturbance if the transient voltage falls below V_2. The probability of this increases as the operating power P_o is close to the maximum power P_{max}. Alternatively, if $(V - V^*)$ is below a specified margin (where V^* is the voltage corresponding to point B), it can be said that the system is prone to voltage instability and collapse.

At the load bus, the following equations apply, using the Jacobian

$$\begin{bmatrix} J_1 & J_2 \\ J_3 & J_4 \end{bmatrix} \begin{bmatrix} \Delta\delta \\ \Delta V \end{bmatrix} = \begin{bmatrix} \Delta P \\ 0 \end{bmatrix} \tag{8.39}$$

where

$$J_1 = \frac{\partial P}{\partial \delta},\ J_2 = \frac{\partial P}{\partial V},\ J_3 = \frac{\partial Q}{\partial \delta},\ J_4 = \frac{\delta Q}{\partial V}$$

If $J_3 \neq 0$, it can be shown that

$$\frac{dP}{dV} = J_2 - J_1 J_3^{-1} J_4 \tag{8.40}$$

The right-hand side of (8.40) is zero if the determinant of the Jacobian is zero (the Jacobian is singular). Thus, the determinant of the Jacobian can also indicate proximity to voltage collapse.

A better index for the evaluation of the voltage stability margin is obtained if it is recognized that (8.37) can be expressed as

$$T_p \frac{dV}{dt} = -\frac{\partial W}{\partial V} \tag{8.41}$$

where $W = \int_{V_1}^{V} (P_o - P)dV$ is termed potential energy. The energy margin defined by

$$\Delta W = \int_{V_1}^{V_2} (P_o - P)dV \tag{8.42}$$

can be used as an index for voltage security. A little reflection shows that when active power is treated as constant, the reactive power versus voltage $(Q - V)$ curve can be used for evaluating the margin. In this case, (8.37) is replaced by

$$\frac{dV}{dt} = \frac{1}{T_q}[Q(V) - Q_o] \tag{8.43}$$

and (8.42) is replaced by

$$\Delta W' = \int_{V_1}^{V_2} (Q_o - Q)dV \tag{8.44}$$

It should be noted that, in general, the following equation applies at the load bus

$$Q = Q_L - Q_C \tag{8.45}$$

where Q is the received reactive power and Q_C is the reactive power supplied by the reactive compensator at the load bus. It is assumed that

$$Q_o = Q_{Lo} - Q_{Co} \tag{8.46}$$

Instead of (8.43), the reactive power compensator may be described by the equation

$$\frac{1}{T_c}\frac{dQ_C}{dt} = V_o - V \tag{8.47}$$

In this case, the criterion for stability is

$$\frac{dV}{dQ_C} > 0 \tag{8.48}$$

Remarks Based on the discussion given above, the following points are worth noting:

1) In contrast to the SMIB system, in the SMLB system the swing equation has no role to play in stability analysis.

2) While the voltage dynamics in the generator can contribute to voltage instability, the primary factor is the dynamics of load restoration at the load bus. The voltage collapse may take several seconds or minutes based on the time constants T_p or T_q.
3) If load dynamics is ignored, the underside of the nose curve does not indicate voltage instability. It only indicates power uncontrollability.
4) The modeling of load (restoration) dynamics can vary depending upon the physical characteristics of the controller.
5) It has been observed during incidents of voltage collapse that the disabling of OLTC helps to overcome voltage instability. This fact is easily explained from the analysis given above.
6) The proximity to voltage collapse can be measured by several indices. Many of them utilize the fact that the load flow Jacobian is close to being singular (the determinant of the Jacobian is close to zero).
7) Although static analysis (neglecting system dynamics) can be used as an approximation in the study of voltage stability, particularly in long-term voltage stability, dynamic analysis is essential for accurate prediction of the phenomenon.

The analysis of voltage stability will be taken up in the next section.

8.7 Analysis of Voltage Stability and Collapse

8.7.1 Simulation [12, 13]

As mentioned earlier, voltage instability is a dynamic phenomenon and the system is described by nonlinear differential algebraic equations. For the study of voltage stability under large disturbances, it is necessary to simulate the system. Midterm stability programs that can simulate the system up to a few minutes can be utilized for this purpose provided that they can model the load characteristics accurately and also include the dynamics of OLTC and OELs. The representation of OLTCs can significantly increase the simulation time frame as they have an intrinsic time delay of the order of 30 s and an additional 1–5 s is taken for each subsequent tap movement [12]. It is also necessary to correctly model step size, initial tap position, and tap range. If AVR line drop compensation is used to control voltages remote from the generator terminals, it has a major effect on the reactive power outputs of a set of generating units and this needs to be represented as well, in addition to OELs.

Load representation should include not only static loads, which are voltage dependent, but also dynamic loads such as induction motors and thermostatic loads. All the reactive compensation devices – switched shunt reactors, capacitors, SVC (with limiting action) – need to be represented adequately. In addition, special protection schemes such as undervoltage load shedding, OLTC blocking, reactor tripping, and generator runback should be modeled [13].

8.7.2 Small Signal (Linear) Analysis

There are two approaches here: static analysis (considering only algebraic equations) and dynamic analysis (considering system dynamics). It should be understood that static analysis is an approximation that may give reasonably accurate results in identifying critical situations. By defining voltage collapse proximity indicators (VCPIs), it is possible to implement security assessment. The sensitivity information can be used to devise corrective measures to overcome the problem of voltage collapse.

The system equations are

$$\dot{x} = f(x, y) \tag{8.49}$$

$$0 = g(x, y) \tag{8.50}$$

The load flow equations are included in (8.50). Linearizing (8.49) and (8.50) we get

$$\Delta\dot{x} = [J_{SYS}]\Delta x \tag{8.51}$$

where

$$[J_{SYS}] = \left[\frac{\partial f}{\partial x}\right] - \left[\frac{\partial f}{\partial y}\right]\left[\frac{\partial g}{\partial y}\right]^{-1}\left[\frac{\partial g}{\partial x}\right] \tag{8.52}$$

$[J_{SYS}]$ is termed the system Jacobian matrix and is distinct from the load flow Jacobian matrix $[J_{LF}]$. All the eigenvalues of the system Jacobian matrix should lie in the LHP ($\Re(\lambda_i) < 0$) for the equilibrium (operating) point to be stable.

It is assumed that the system Jacobian matrix is dependent on a parameter μ (say, the load at a specified bus) which is varied. For $\mu < \mu_c$, the equilibrium point is stable. At $\mu = \mu_c$, the critical value of the parameter, a bifurcation is said to occur [7] and the equilibrium point becomes unstable for $\mu > \mu_c$. At $\mu = \mu_c$, the system Jacobian matrix can have either

(a) a real eigenvalue which is zero or
(b) a complex pair on an imaginary axis.

The loci of the critical eigenvalues for the two cases are shown in Figure 8.14. In the first case, instability is due to the crossing of a real eigenvalue into the RHP and the bifurcation is said to be of saddle-node type. In the second type, the instability is due to the crossing of a complex pair into the RHP and it is termed a Hopf bifurcation.

If it is assumed that the voltage instability and collapse is due to SNB, then the system Jacobian matrix is singular at the critical value of the parameter $\mu = \mu_c$ as one of its eigenvalues is zero. In this case, static analysis based on the rank properties of the system Jacobian matrix can give accurate results.

Another implicit assumption used in the static analysis is that the rank properties of the system Jacobian matrix are related to that of the load flow Jacobian matrix.

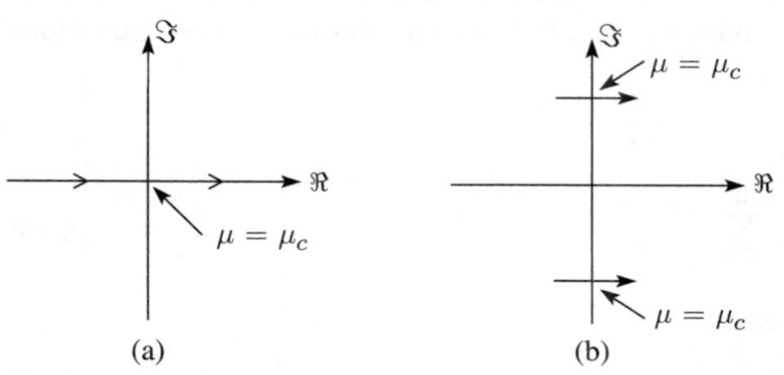

Figure 8.14 Loci of critical eigenvalues. (a) Saddle-node bifurcation (b) Hopf bifurcation.

Venikov *et al.* [14] showed that the rank of the load flow Jacobian matrix is equal to the rank of the system Jacobian provided that:

1. the active power and voltage magnitude are specified at each generator bus
2. loads are constant power (P and Q) type
3. the slack node is an infinite bus.

With these conditions, the determinant of the load flow Jacobian matrix is identical to the product of all the eigenvalues of the system Jacobian matrix. It is shown in [15] that in general the system Jacobian matrix can have a zero eigenvalue even when the load flow Jacobian matrix is nonsingular. The load level which produces a singular load flow Jacobian matrix should be considered as an optimistic upper bound on maximum loadability.

The load flow Jacobian matrix, $[J_{LF}]$, is defined by

$$[J_{LF}]\begin{bmatrix}\Delta\theta\\ \Delta V\end{bmatrix} = \begin{bmatrix}J_{P\theta} & J_{PV}\\ J_{Q\theta} & J_{QV}\end{bmatrix}\begin{bmatrix}\Delta\theta\\ \Delta V\end{bmatrix} = \begin{bmatrix}\Delta P\\ \Delta Q\end{bmatrix} \tag{8.53}$$

The rank of $[J_{LF}]$ is identical to that of the reduced Jacobian matrix $[J_R]$ defined by

$$[J_R] = [J_{QV} - J_{Q\theta}J_{P\theta}^{-1}J_{PV}] \tag{8.54}$$

provided that $[J_{P\theta}]$ is nonsingular. This follows from Schur's formula:

$$\det[J_{LF}] = \det[J_{P\theta}]\cdot\det[J_R] \tag{8.55}$$

Reference [16] suggests a singular value decomposition of the reduced Jacobian matrix $[J_R]$ and the use of the smallest value, σ_n, as a measure of the proximity of the voltage stability limit. An $n\times n$ matrix $[A]$ can be expressed by its singular value decomposition

$$[A] = [U][D][V]^t = \sum_{i=1}^{n}\sigma_i u_i v_i^t \tag{8.56}$$

where $[U]$ and $[V]$ are $n\times n$ orthogonal matrices with columns u_i and v_i, $(i = 1, 2\ldots, n)$, respectively. $[D]$ is a diagonal matrix of singular values σ_i, $i = 1, 2\ldots, n$. Also $\sigma_i \geq 0$ for all i. Without loss of generality, it can be assumed that

$$\sigma_1 \geq \sigma_2 \geq \sigma_3 \geq \ldots\sigma_n \geq 0 \tag{8.57}$$

For a real symmetric matrix, the absolute values of eigenvalues are the singular values. In general, σ_i is the square root of an eigenvalue of the matrix $[AA^t]$ or $[A^tA]$.

The singular value decomposition is well conditioned and the singular values are insensitive to perturbation in the matrix elements. Given the system of linear equations

$$[A]x = b \tag{8.58}$$

the solution can be expressed as

$$x = [A]^{-1}b = [UDV^t]^{-1} = \sum_{i=1}^{n}\frac{u_i^t b}{\sigma_i}v_i \tag{8.59}$$

If $\Delta P = 0$, (8.53) can be expressed as

$$[J_R]\Delta V = \Delta Q \tag{8.60}$$

Applying (8.59) to the solution of (8.60) it is seen that

1. the smallest value σ_n is an indicator of proximity to the voltage stability limit
2. the right singular vector v_n, corresponding to σ_n, is an indicator of affected (sensitive) bus voltages
3. the left singular vector u_n, corresponding to σ_n, is an indicator of the most sensitive direction for changes in the reactive power injections.

Reference [17] uses eigenvalue analysis based on the reduced Jacobian matrix for the study of voltage stability. In [18] it is shown that voltage instability can occur due to Hopf bifurcation, which cannot be predicted from static analysis. The instability is caused by the exciter mode, which can be damped by suitable design of a stabilizer.

The results given in [19] show that the voltage instability at the inverter bus of a HVDC link is due to Hopf bifurcation while the instability at the rectifier bus is normally due to SNB. While the results obtained from static analysis are in agreement with the results obtained from dynamic analysis if the bifurcation is of saddle-node type, the static analysis can give very optimistic and hence misleading results if the bifurcation is Hopf.

8.8 Integrated Analysis of Voltage and Angle Stability

The use of fast-acting excitation systems with high-gain voltage regulators gives rise to the problem of undamped low-frequency oscillations. This problem has been viewed as part of the angle stability problem (see Chapter 6) and the small signal stability analysis of a SMIB system based on Heffron–Phillips constants was presented by De Mello and Concordia [20].

As discussed earlier, a SMIB system cannot exhibit voltage instability and a SMLB system cannot exhibit angle instability. A three-bus system consisting of a generator bus, a load bus, and an infinite bus is the smallest system that can exhibit both voltage and angle instabilities. The work of [20] can be extended to such a system and the results based on a detailed study are reported in [21].

Consider the system shown in Figure 8.15. The load is assumed to be a constant power type (with a constant power factor). The active power received by the infinite bus is kept constant at 1.0 pu. Using Heffron–Phillips constants, it is possible to obtain the region of stability (ROS) for a given load in the K_E–T_E (regulator gain and time constant) plane. It was observed that the ROS shrinks as the load is increased.

Figure 8.15 A three-bus system.

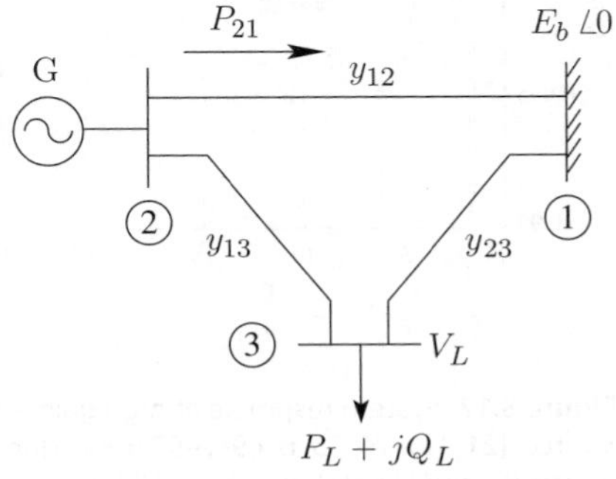

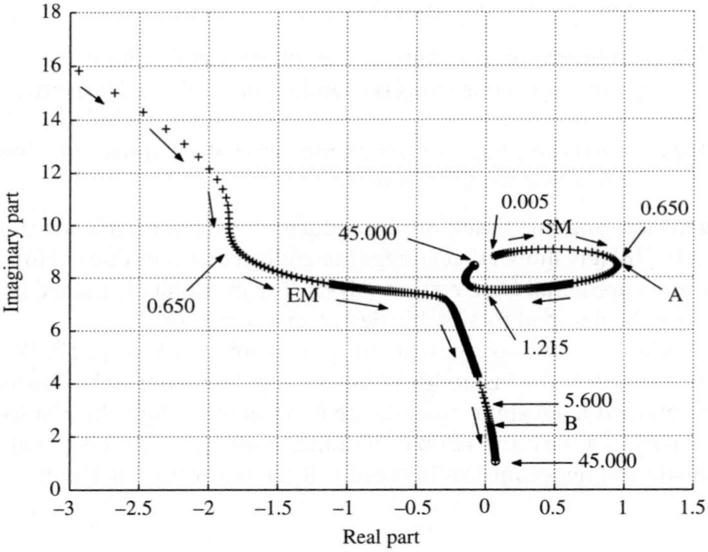

Figure 8.16 Root loci of a three-bus two-line system at $K_E = 350$, $P_L = 3.430$ pu. T_E is changing. Source: [21] Fig. 10, p. 495, reproduced with permission ©Elsevier 2002.

It was observed that when K_E is below a certain low limit, the instability is due to the exciter mode whereas when K_E is above a uniform high limit, the instability is caused by the swing modes.

For values of K_E lying between these two limits, the instability is caused by the swing mode when $T_E < T_1$ and by the exciter mode when $T_E > T_2$. For a specified value of K_E, the system is stable when $T_1 < T_E < T_2$. The stable range of T_E shrinks as K_E is increased. The loci of both swing and exciter modes for $K_E = 350$ as T_E is varied are shown in Figure 8.16. This shows that the system is stable for $1.215 < T_E < 5.050$.

The simulation results for a small disturbance (for $K_E = 350$, $T_E = 7.0$ s) are shown in Figure 8.17. This shows that the exciter mode instability affects the load bus voltage and

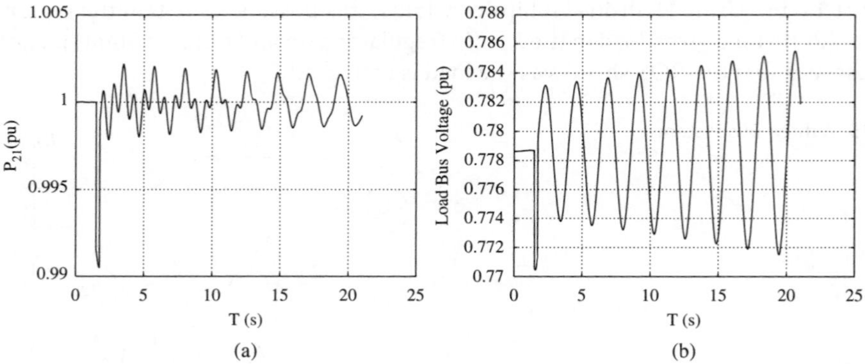

Figure 8.17 System response at high gain and high time constant of the exciter, with $P_L = 3.43$ pu. Source: [21] Fig. 14, 15, p. 496, 497, reproduced with permission ©Elsevier 2002. (a) Oscillations in power P_{21} tend to stabilize, $K_E = 350$, $T_E = 7.0$ s (b) Growing oscillations in load bus voltage for $K_E = 350$ and $T_E = 7.0$ s.

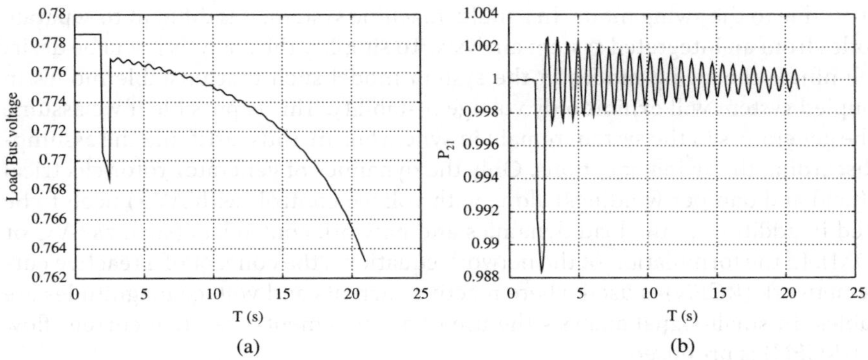

Figure 8.18 System response at low gain and high time constant of the exciter, with $P_L = 3.430$ pu. Source: [21] Fig. 16, 17, p. 497, reproduced with permission ©Elsevier 2002. (a) Monotonic fall in load bus voltage for $K_E = 0.5$ and $T_E = 1.0$ s (b) Oscillations in P_{21} stabilize for $K_E = 0.5$ and $T_E = 1.0$ s.

not the power swing in the line connecting the infinite bus to the generator. The voltage oscillations increase in magnitude while the power swings are damped. The simulation results for $K_E = 0.5$ and $T_E = 1.0$ s are shown in Figure 8.18. This shows similar results with exciter mode instability not affecting power swings. There is monotonic collapse in the load bus voltage due to SNB.

The results for the three-bus system can be extended to multi-machine systems. The following conclusions can be reached on the nature of instability:

1. As the system loading is increased, with normal operation of AVRs, the initial instability is due to the swing mode, which can be stabilized by damping controllers.
2. The voltage instability appears to be the result of the unstable exciter mode at higher loadings, which is affected by the regulator gain and time constant.
3. Low gains and high time constants can result in the instability of the exciter mode with the nature of the instability dependent on the gain. Below a critical gain the instability is due to SNB, while at higher gains the instability occurs at higher time constants.

If it is assumed that an equivalent generator can represent all the generators (which remains in synchronism) in a multi-machine system, the equivalent K_E and T_E are affected by the OEL. Under abnormal conditions, increase in T_E (slowing down of the exciter response) or decrease in the effective K_E can cause voltage instability. Interestingly, these conclusions are not much affected even with dynamic modeling of loads.

Attempts have been made recently [22, 23] to decouple voltage and angle stability if it is assumed that the generators remain in synchronism following a disturbance. Ignoring the rotor swing dynamics has little effect on the prediction of instability of modes leading to voltage instability.

8.9 Analysis of Small Signal Voltage Instability Decoupled from Angle Instability

As shown in the single-machine, three-bus system example, the voltage instability is primarily due to the instability of the exciter mode whereas the synchronous or angle instability is due to the swing mode. In a multi-machine system, it is difficult to separate these modes from an integrated system analysis. To simplify the analysis it will be desirable to eliminate the swing modes in the system model such that unstable modes in the decoupled system will exhibit only voltage instability. This is possible if we assume that all the generators in the system remain in synchronism, thus justifying the assumption of discarding the swing equations. Only the dynamics of generator rotor electrical circuits (field and damper windings) along with voltage control (with AVR) need to be considered in addition to the load dynamics and network controllers (such as SVC or STATCOM). In the formulation of the network equations, the concept of a reactive current flow network (RCFN) is used where reactive currents and voltage magnitudes are the variables. In small-signal analysis the use of an incremental reactive current flow network (IRCFN) is proposed.

8.9.1 Decoupling of Angle and Voltage Variables [22, 23]

To simplify the analysis, it is assumed that:

1. the active power loads are constants
2. the network is lossless and
3. mechanical torque or power at each generator is held constant.

The second assumption can be relaxed if it is assumed that the variation in the network losses can be neglected.

The swing equations (using the centre of inertia (COI)) formulations are given by

$$M_i \frac{d^2\theta_i}{dt^2} = P_{mi} - P_{ei} - \left(\frac{M_i}{M_T}\right) P_{COI} \tag{8.61}$$

where

$$P_{COI} = \sum_{i=1}^{m} (P_{mi} - P_{ei}), \quad M_T = \sum_{i=1}^{m} M_i$$

$$\theta_i = \delta_i - \delta_o, \quad \delta_o = \sum_{i=1}^{m} (M_i \delta_i)/M_T$$

and m is the number of generators.

From energy conservation, $\sum P_{ei} = \sum P_{Lj} + P_{loss}$, where P_{Lj} is the real power at the jth bus. If the generators are assumed to swing together, then the LHS of (8.61) is zero and it can be shown that $P_{mi} = P_{ei}$, if losses are neglected. Thus, $\Delta P_{ei} = 0$ if P_{mi} is constant. This applies even when losses are nonzero, but constant. This fact enables us to consider transmission networks with injection of only reactive currents at various nodes.

Consider a lossless line (k) connecting buses i and j. The reactive current injected into the line at bus i is given by

$$I_{qi}^k = \frac{Q_i}{V_i} = b_k(V_i - V_j \cos\delta_k), \quad \delta_k = \delta_i - \delta_j \tag{8.62}$$

The equivalent RCFN of a lossless transmission line (including the shunt-connected elements representing shunt reactors and line charging) is shown in Figure 8.19. Here b_i^s and b_j^s are shunt susceptances at buses i and j and b_k is the series susceptance $\left(\frac{1}{x_k}\right)$ of

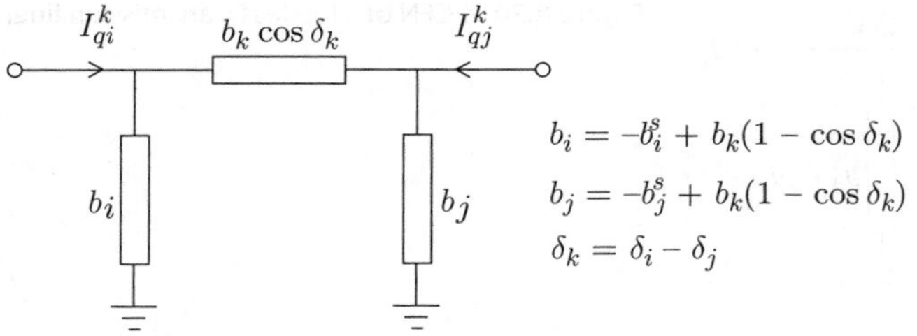

Figure 8.19 RCFN of a lossless transmission line.

the line. It should be noted that δ_k is not a constant but varies with V_i to keep the power flow P_k in the line constant. Note that P_k is given by

$$P_k = b_k V_i V_j \sin \delta_k \tag{8.63}$$

The reactive current flow network (RCFN) is composed of the equivalent circuits of all the elements in the transmission network. While shunt capacitors and reactors are linear elements in RCFN, the transmission lines constitute the nonlinear elements of the network due to dependence of δ_k on bus voltage magnitudes.

8.9.2 Incremental RCFN

When the equations describing RCFN are linearized, we can obtain an incremental reactive current flow network (IRCFN). Linearizing (8.62) we obtain

$$\Delta I_{qi}^k = a_1 \Delta V_i + a_2 \Delta V_j + c_1 \Delta \delta_k \tag{8.64}$$

where $a_1 = b_k$, $a_2 = b_k \cos \delta_k^o$, and $c_1 = b_k V_j^o \sin \delta_k^o$. The superscript o indicates the operating value. Similarly, we can obtain

$$\Delta I_{qj}^k = a_2 \Delta V_i + a_3 \Delta V_j + c_2 \Delta \delta_k \tag{8.65}$$

where $a_3 = b_k = a_1$ and $c_2 = b_k V_i^o \sin \delta_k^o$. Combining (8.64) and (8.64) we obtain

$$\begin{bmatrix} \Delta I_{qi}^k \\ \Delta I_{qj}^k \end{bmatrix} = \begin{bmatrix} a_1 & a_2 \\ a_2 & a_1 \end{bmatrix} \begin{bmatrix} \Delta V_i \\ \Delta V_j \end{bmatrix} + \begin{bmatrix} c_1 \\ c_2 \end{bmatrix} \Delta \delta_k \tag{8.66}$$

Linearizing (8.63), we get

$$\Delta P_k = d_1 \Delta V_i + d_2 \Delta V_j + g \Delta \delta_k \tag{8.67}$$

where $d_1 = b_k V_j^o \sin \delta_k^o$, $d_2 = b_k V_i^o \sin \delta_k^o$, and $g = b_k V_i^o V_j^o \cos \delta_k^o$.

Since power flow P_k is assumed to remain constant, $\Delta P_k = 0$ and hence we obtain

$$\Delta \delta_k = h_1 \Delta V_i + h_2 \Delta V_j \tag{8.68}$$

where $h_1 = -\frac{d_1}{g} = -\frac{\tan \delta_k^o}{V_i^o}$ and $h_2 = -\frac{d_2}{g} = -\frac{\tan \delta_k^o}{V_j^o}$. Substituting (8.68) in (8.66), we obtain

$$\begin{bmatrix} \Delta I_{qi}^k \\ \Delta I_{qj}^k \end{bmatrix} = \begin{bmatrix} y_1 & y_2 \\ y_3 & y_4 \end{bmatrix} \begin{bmatrix} \Delta V_i \\ \Delta V_j \end{bmatrix} \tag{8.69}$$

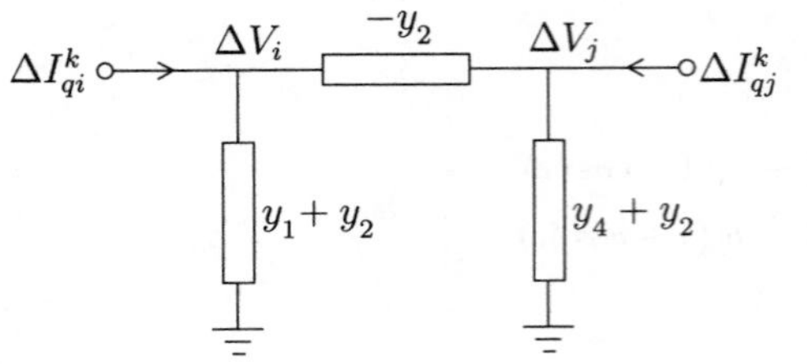

Figure 8.20 IRCFN of a lossless transmission line.

Equation (8.69) represents the IRCFN of the transmission line shown in Figure 8.20. Here the elements y_1, y_2, and y_3 are defined by

$$y_1 = b_k - \left(\frac{V_j^o}{V_i^o}\right) b_k \sin\delta_k^o \tan\delta_k^o$$

$$y_2 = -b_k(\cos\delta_k^o + \sin\delta_k^o \tan\delta_k^o) = y_3$$

$$y_4 = b_k - \left(\frac{V_i^o}{V_j^o}\right) b_k \sin\delta_k^o \tan\delta_k^o$$

The shunt branches of the transmission line (representing charging reactive power) and shunt reactors or capacitors are represented as linear elements in the RCFN and thus will be included in IRCFN without any modification.

8.9.3 Nonlinear Reactive Loads

Loads are assumed to be static with constant real power (P_L) and reactive power given by

$$Q_L = Q_o\left(\frac{V_L}{V_o}\right)^{m_q} \tag{8.70}$$

The reactive current supplied to the load is given by

$$I_{qL} = \frac{Q_L}{V_L} \tag{8.71}$$

Linearizing the above equation, we get

$$\Delta I_{qL} = [(m_q - 1)Q_o/V_o^2]\Delta V_L = b_L \Delta V_L \tag{8.72}$$

Note that b_L is positive for $m_q > 1$. Combining equations for all the transmission lines, shunt elements, and reactive loads, we can obtain the equation describing IRCFN as

$$\Delta I_q = [Y^q]\Delta V \tag{8.73}$$

It should be noted that $[Y^q]$ can be interpreted as the conductances matrix of a resistive network.

8.9.4 Generator Model

The generator is represented by a two-axis model (1.1) with the field winding on the d-axis and a damper winding on the q-axis. The excitation system assumes a static exciter represented by a single time constant transfer function. The linearized state and output equations for a generator are obtained as [21]

$$\dot{x}_g = [A_g]x_g + [B_g]\Delta V_2 \tag{8.74}$$

$$\Delta I_{qg} = [C_g]x_g + [D_g]\Delta V_g \tag{8.75}$$

where

$$x_g^t = \left[\, \Delta E_q' \;\; \Delta E_d' \;\; \Delta E_{fd} \,\right], \Delta I_{qg} = \frac{Q_g}{V_g}$$

It is assumed that $\Delta P_g = 0$.

It is also possible to obtain linearized dynamic models for FACTS controllers such as SVC or STATCOM.

Example 8.2 [23] An SMLB system is considered (see Figure 8.2) whose data are shown in Table 8.2. The power factor of the load is assumed to be unity.

With $R_e = 0.0$ and $x_e = 0.3$, the critical eigenvalues (a complex pair) cross the imaginary axis and move to the RHP as P_L is increased, resulting in Hopf bifurcation. Further increase in P_L results in singularly induced bifurcation (SIB) when the complex pair splits into two real eigenvalues, one of which moves towards $+\infty$, becomes discontinuous and becomes $-\infty$ (for P_L increasing from 1.3287 to 1.3288), and remains in the LHP thereafter. The other real eigenvalue moves towards the origin and then crosses the origin (SNB). The initial instability is due to Hopf bifurcation. The bifurcation diagram is shown in Figure 8.21.

Discussion

1. The behavior of the system with increase in the power level appears to be generic. Similar behavior is observed with (i) variation in the power factor, (ii) change in the reactive power compensation, and (iii) change in the value of m_q. It was observed [23] that the voltage stability limit is reduced for negative values of m_q.
2. In [18], a three-machine, nine-bus system is studied. The authors consider the swing dynamics and trace the unstable mode to the E_q'–R_f pair (E_q' from the generator and R_f from the IEEE Type 1 excitation system). Essentially it can be viewed as an unstable exciter mode which becomes unstable by Hopf bifurcation. The decoupling of

Table 8.2 Generator data.

Generator	$x_d = 1.0, x_q = 0.6, x_d' = 0.4, x_q' = 0.3$ pu, $T_{d0}' = 8.0$ s, $T_{q0}' = 0.07$ s
Exciter	$K_E = 20.0, T_E = 0.1$ s

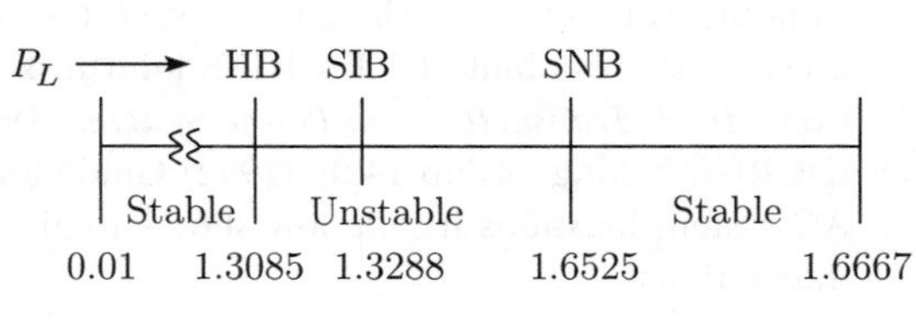

Figure 8.21 Bifurcation diagram for the SMLB.

the swing dynamics simplifies the analysis. Normally, the identification of the state variables leading to unstable modes requires knowledge of participation factors calculated from the left and right eigenvectors.
3. It is interesting to observe that the results given in [23] are close to the results given in [18]. The bifurcation diagram for the three-machine system case given in [23] is similar to that given for the SMLB system.

8.10 Control of Voltage Instability

Voltage instability along with angle instability pose a threat to the system security. Uncontrolled load rejection due to voltage collapse can cause system separation and blackouts. Hence the system must be planned in such a way as to reduce the possibility of voltage instability. The system must also be operated with an adequate margin for voltage stability. In the event of voltage instability due to unforseen contingencies, the system control must prevent widespread voltage collapse and restore the loads as quickly as possible.

The incidence of voltage instability increases as the system is operated close to its maximum loadability limit. Environmental and economic constraints have limited the transmission network expansion while forcing the generators to be sited far away from the load centres. This has resulted in stressing the existing transmission network.

The application of FACTS controllers can help in increasing the transmission capacity without compromising security and raising the power transfer limits closer to the thermal limits. The availability of SVC, STATCOM, and TCSC can be used to augment stability limits in critical transmission lines.

The reactive power compensation close to the load centres as well as critical buses in the network is an important factor in overcoming voltage instability. The location, size, and speed of control have to be selected properly for maximum benefit. The design of suitable protective measures in the event of voltage instability is also necessary. The application of undervoltage load shedding [24], controlled system separation, and adaptive or intelligent control are steps in this direction.

References

1 Taylor, C.W. (1994) *Power System Voltage Stability*, McGraw Hill, New York.
2 Pai, M.A. and Ajjarapu, V. (1993) Voltage stability in power systems – an overview, in *Recent Advances in Control and Management of Energy Systems*, Interline Publishing, Bangalore, pp. 1121–1136.
3 Van Cutsem, T. and Vournas, C. (1998) *Voltage Stability of Electric Power Systems*, Springer.
4 Kundur, P., Paserba, J., Ajjarapu, V. *et al.* (2004) Definition and classification of power system stability IEEE/CIGRE joint task force on stability terms and definitions. *IEEE Transactions on Power Systems*, **19** (3), 1387–1401.
5 CIGRE Working Group 14.05 (1992) Guide for planning DC links terminating at AC system locations having low short-circuit capacities, Part 1:AC/DC system interactions.

6 Ilić, M. and Zaborszky, J. (2000) *Dynamics and Control of Large Electric Power Systems*, John Wiley & Sons, New York.
7 Venkatasubramanian, V., Schattler, H., and Zaborsky, J. (1993) The varied origins of voltage collapse in large power systems. *12th IFAC Congress*, Sydney, Vol. **5**, 451–458.
8 IEEE Task Force on Excitation Limiters (1995) Recommended models for overexcitation limiting devices. *IEEE Transactions on Energy Conversion*, **10** (4), 706–713.
9 Padiyar, K.R. (2013) *HVDC Power Transmission Systems*, New Age International, New Delhi, 3rd edn.
10 Hill, D.J. (1993) Nonlinear dynamic load models with recovery for voltage stability studies. *IEEE Transactions on Power Systems*, **8** (1), 166–176.
11 Balanathan, R., Pahalawaththa, N.C., and Annakkage, U.D. (2002) Modelling induction motor loads for voltage stability analysis. *International Journal of Electrical Power and Energy Systems*, **24** (6), 469–480.
12 Potamianakis, E.G. and Vournas, C.D. (2006) Short-term voltage instability: effects on synchronous and induction machines. *IEEE Transactions on Power Systems*, **21** (2), 791–798.
13 Stubbe, M., Bihain, A., and Deuse, J. (1993) Simulation of voltage collapse. *International Journal of Electrical Power and Energy Systems*, **15** (4), 239–243.
14 Venikov, V.A., Stroev, V.A., Idelchick, V.I., and Tarasov, V.I. (1975) Estimation of electrical power system steady-state stability in load flow calculations. *IEEE Transactions on Power Apparatus and Systems*, **94** (3), 1034–1041.
15 Sauer, P.W. and Pai, M.A. (1990) Power system steady-state stability and the load-flow Jacobian. *IEEE Transactions on Power Systems*, **5** (4), 1374–1383.
16 Lof, P.A., Andersson, G., and Hill, D.J. (1993) Voltage stability indices for stressed power systems. *IEEE Transactions on Power Systems*, **8** (1), 326–335.
17 Morison, G.K., Gao, B., and Kundur, P. (1993) Voltage stability analysis using static and dynamic approaches. *IEEE Transactions on Power Systems*, **8** (3), 1159–1171.
18 Rajagopalan, C., Lesieutre, B.C., Sauer, P.W., and Pai, M.A. (1992) Dynamic aspects of voltage/power characteristics. *IEEE Transactions on Power Systems*, **7** (3), 990–1000.
19 Padiyar, K.R. and Rao, S.S. (1996) Dynamic analysis of voltage instability in AC–DC systems. *International Journal of Electrical Power and Energy Systems*, **18** (1), 11–18.
20 Demello, F.P. and Concordia, C. (1969) Concepts of synchronous machine stability as affected by excitation control. *IEEE Transactions on Power Apparatus and Systems*, **PAS-88** (**4**), 316–329.
21 Padiyar, K.R. and Bhaskar, K. (2002) An integrated analysis of voltage and angle stability of a three-node power system. *International Journal of Electrical Power and Energy Systems*, **24** (6), 489–501.
22 Vournas, C.D., Sauer, P.W., and Pai, M.A. (1996) Relationships between voltage and angle stability of power systems. *International Journal of Electrical Power and Energy Systems*, **18** (8), 493–500.
23 Padiyar, K.R. and Rao, S.S. (1996) Dynamic analysis of small signal voltage instability decoupled from angle instability. *International Journal of Electrical Power and Energy Systems*, **18** (7), 445–452.
24 Taylor, C.W. (1992) Concepts of undervoltage load shedding for voltage stability. *IEEE Transactions on Power Delivery*, **7** (2), 480–488.

9

Wide-Area Measurements and Applications

9.1 Introduction

The technology of wide-area measurement systems (WAMS) [1, 2] aims to make available high-quality system-wide measurements for enhanced monitoring, control, and protection applications. A key feature of this technology is the precise time synchronization of the measurements using a global positioning system (GPS). This feature also allows for the direct measurement of phase angular differences between phasors at different and distant locations in a power grid. The technology overcomes the limitations of conventional supervisory control and data acquisition (SCADA) systems.

SCADA systems bring system-wide measurements at relatively slower rates; typically, data is refreshed once every 1–10 s. These measurements are not accurately time-stamped, and may arrive at the control centres with large communication delays. These factors limit the accuracy and scope of decision-support tools and reduce the situational awareness of a system operator. For the same reasons, measurements from SCADA systems cannot be used for feedback control and fast-acting system protection schemes. WAMS is a superior technology and its advent has spurred the development of novel methods to determine the health of a power grid. Enhanced control and protection applications for improving system stability, which were not feasible earlier, are now being contemplated.

Synchronized phasor measurements were introduced in the mid-1980s [3] and now there are several large-scale implementations across the world. This chapter describes the main features of this technology and also discusses potential applications.

9.2 Technology and Standards

A schematic of a WAMS is shown in Figure 9.1. In this system, measurement units placed at various locations measure, time-stamp, and send the measurements to one or more control centres. Time-stamping is done using timing signals from a reliable and accurate source, such as GPS satellites.

Since the measurements are effectively time-stamped with a common clock, it is possible to determine the relative phase angle between currents and voltages at different locations. It should be noted that any error in time synchronization results in an error

Dynamics and Control of Electric Transmission and Microgrids, First Edition.
K. R. Padiyar and Anil M. Kulkarni.

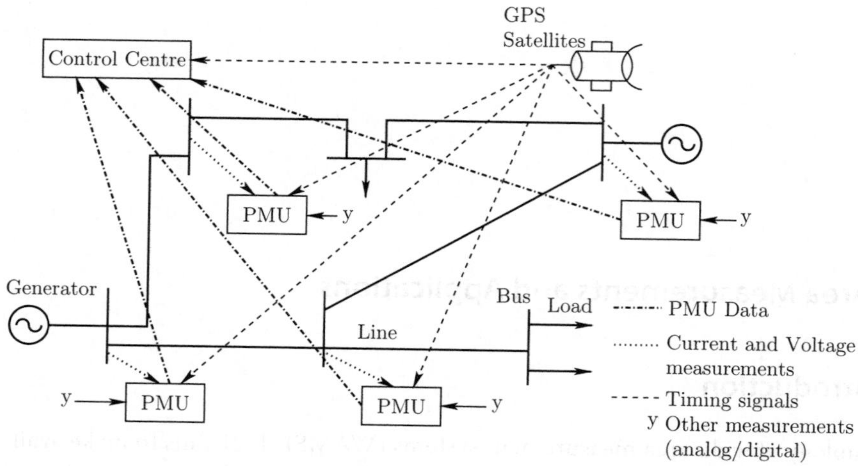

Figure 9.1 A wide-area measurement system.

in the measured phase angle. For example, a typical time sychronization error of 1 μs will cause a 0.018° error in the phase angle of a 50 Hz waveform.

As the measurement units primarily measure current and voltage phasors, they are called phasor measurement units (PMUs). A block diagram of the PMU hardware is shown in Figure 9.2. In addition to voltage and current phasors, PMUs may also be used to transmit other information like bus frequency, rate of change of frequency, circuit breaker status, and generator speed measurements.

The equipment which collates the data from various PMUs is called a phasor data concentrator (PDC). The collated data is available for applications at a control centre,

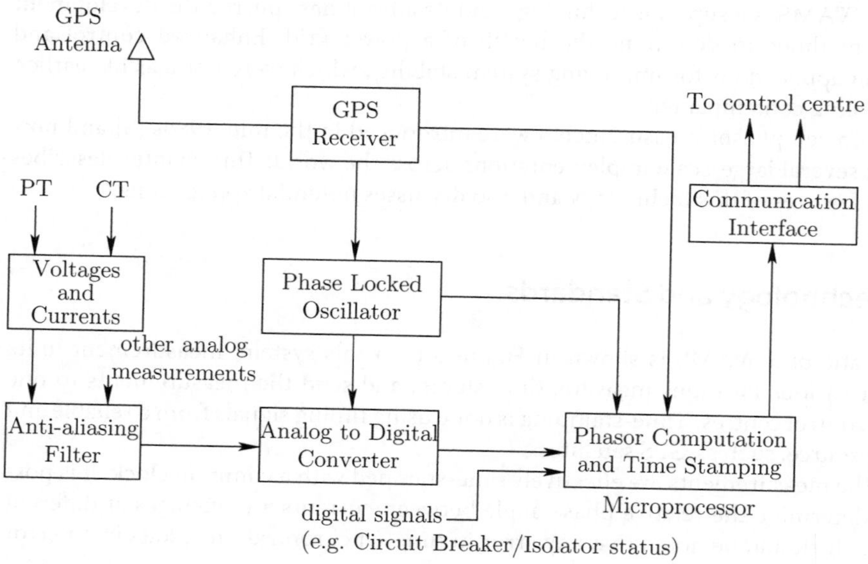

Figure 9.2 A phasor measurement unit.

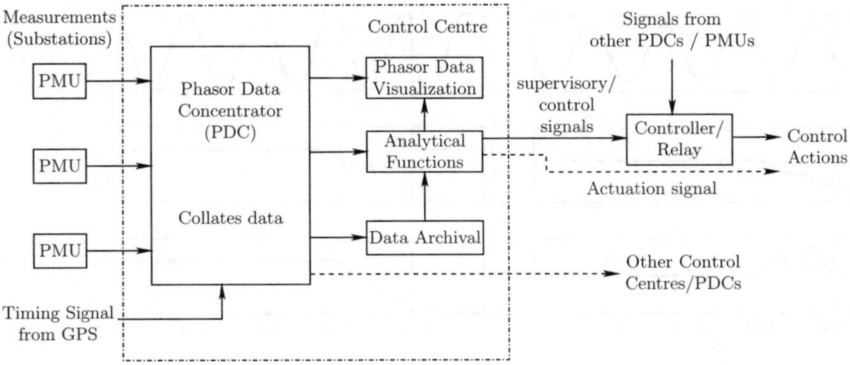

Figure 9.3 Phasor data concentrator and control centre functions.

as shown in Figure 9.3. These applications range from monitoring and event analysis to control and system protection. The data, or a filtered and down-sampled subset of it, may be also be forwarded to other control centres, where other applications may reside. The architecture of WAMS could be designed to be hierarchical, with data flowing from individual substations to local control centres, then to the regional control centres, and finally to an apex control centre. The hierarchy could be bypassed for some control and protection applications for which the data is required with minimal time delay.

In order to facilitate inter-operability among PMUs from different manufacturers, and the systematic development of applications, a WAMS must adhere to certain standards. The IEEE Synchrophasor Standard C37.118.1-2011 [4] and the amendment C37.118.1a [5] address issues like the definition of a synchronized phasor, application of time-tags, message formats for communication, and verification of compliance with the standard.

9.2.1 Synchrophasor Definition

Consider a signal $x(t)$ which is given by $x(t) = X_m \cos(\omega t + \xi)$. The synchrophasor representation of $x(t)$ is the value $\mathbf{X}$ given by the equation

$$\mathbf{X} = \frac{X_m}{\sqrt{2}} e^{j\phi} = X_r + jX_i \tag{9.1}$$

ϕ is the phase angle of $x(t)$ relative to a cosine function at the nominal system frequency (ω_o) and synchronized to coordinated universal time (UTC). Note that all phasors are computed using a common time and frequency reference. If $x(t)$ is a sinusoidal signal at an off-nominal frequency then ϕ, which equals $(\omega - \omega_o)t + \xi$, is not constant.

A synchrophasor is usually computed by applying a discrete Fourier transform (DFT) algorithm on the samples of the waveform $x(t)$. The one-cycle DFT of the sampled signal $x(k)$ is given below:

$$\hat{\mathbf{X}}(l) = \frac{\sqrt{2}}{N} \sum_{k=l-N+1}^{l} x(k) e^{-j\omega_o kh}$$

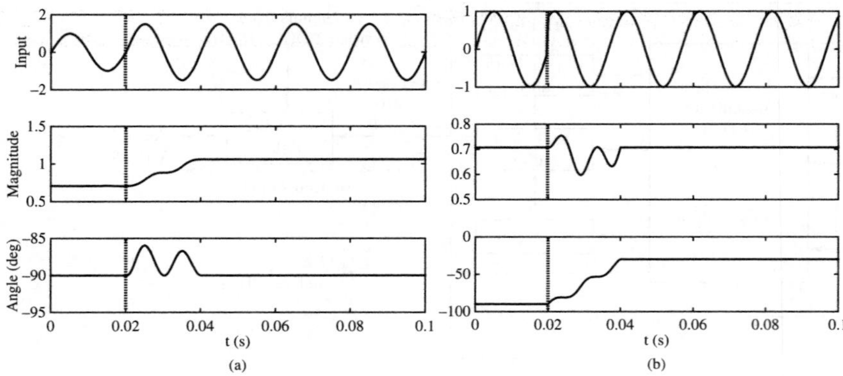

Figure 9.4 Response of a synchrophasor computed using DFT. (a) Step change in amplitude (b) Step change in angle

where N denotes the number of samples in one cycle of nominal frequency ω_o and h is the sampling time step. Note that filtering is necessary (prior to sampling the raw input signal) to prevent aliasing due to harmonics, fast electromagnetic transients, and noise. During transient conditions, the computed synchrophasor $\hat{\mathbf{X}}$ differs from the actual synchrophasor $\mathbf{X}$ due to filtering and the DFT response time.

The synchrophasor standard prescribes the limits on the error $(\mathbf{X} - \hat{\mathbf{X}})$ and the response time under mandated test conditions. The response time is the period of time for which the PMU measurement is outside specified error limits after a step change in the input. For example, the responses of a synchrophasor to a step change in the amplitude and the phase angle of a 50 Hz sinusoidal signal $x(t)$ are shown in Figure 9.4. The sampling time step is $h = 1$ ms and the nominal frequency is 50 Hz. The effect of anti-aliasing filters has not been considered here. A one-cycle delay to reach steady state is clearly seen in the responses, which is due to the one-cycle DFT algorithm. The response time can be improved using a shorter DFT window (e.g., half-cycle DFT), but it is prone to errors in the presence of noise, harmonics, and fast-transients. On the other hand, a multi-cycle DFT is less prone to these errors but has a longer response time.

DFT algorithms generally introduce a steady-state amplitude error if the input is at an off-nominal frequency [2]. Appropriate modifications may be required to make it conform to the synchrophasor standard. However, no specific algorithm for synchrophasor computation is mandated in the standard.

Since there is a trade-off between response time and error, the synchrophasor standard has specified two classes of PMUs. Protection (P-class) PMUs are intended for applications requiring fast response with minimal filtering, while measurement (M-class) PMUs are for applications which require good accuracy.

9.2.2 Reporting Rates

If a synchrophasor is computed using a moving window algorithm like DFT, then an updated value is available at every sampling instant. These synchrophasor samples are generally down-sampled and sent to a control centre. A reporting rate of one sample per cycle or one sample per two cycles of the nominal frequency is usually adequate for monitoring the quasi-steady state conditions and electro-mechanical transients. Harmonic phasor calculations and subcycle reporting rates are not implemented in most practical deployments of WAMS, but are conceivable in the future, depending on application requirements.

9.2.3 Latency and Data Loss

PMU reporting latency is defined as the maximum time interval between the data report time as indicated by the data time stamp and the time when the data becomes available at the PMU output. Besides this latency, which occurs due to PMU filtering and processing, additional latencies are caused in the communication of the digital information and due to processing at the data concentrators. Communication latency includes the propagation delay of the communication channel and the delays occurring at the equipment being used for signaling, like ethernet switches and routers [6].

Latency requirements depend on the particular application, and they have to be determined on a case-by-case basis. Transients which may lead to a loss of synchronism between areas typically evolve over 1–5 s. Inter-area swings may have frequencies lower than 0.5 Hz. Therefore, control and protection schemes for enhancing inter-area angular stability should be able to tolerate round-trip latencies of about 100 ms. Primary equipment protection is not feasible with latencies of 100 ms and a 50 Hz/60 Hz reporting rate, but enhancements in slower-acting back-up protection schemes are feasible. For specific applications which require only a few PMU measurements, the required measurements can be directly communicated to the location where the applications reside, without going through a centralized PDC. This avoids the overall delay associated with the collation of a large number of PMU measurements arriving with different time delays at the PDCs.

Data loss due to errors, dropouts or unavailability should generally be 0.1% or less per minute depending on the application [7], and is achievable using standard communication systems. Applications like visualization are not significantly affected by 1–2 dropouts/minute, while control applications should be able to tolerate an occasional single-point data loss. The loss of data in a contiguous block is more serious for control applications, even if the long-term average data loss is small.

9.3 Modeling of WAMS in Angular Stability Programs

Generally, only the positive sequence network is represented in angular stability programs (commonly known as transient stability programs) and the faster network transients are neglected. The formulation of equations is done in the *D-Q* variables in a common synchronously rotating frame of reference (see Chapter 2).

The *D-Q* components of the voltage (v_D, v_Q) when combined to form a complex number $v_Q + jv_D$ conveniently represent the voltage synchrophasors. $\sqrt{v_D^2 + v_Q^2}$ and $\theta = \tan^{-1}(v_D/v_Q)$ represent the corresponding magnitude and phase, respectively. Current synchrophasor measurements can be represented in a similar fashion.

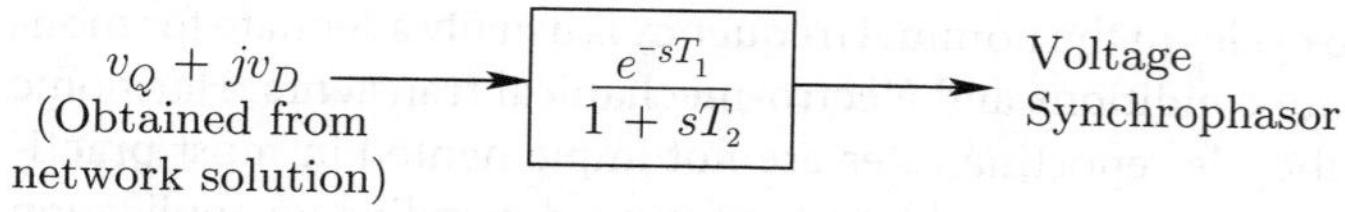

Figure 9.5 Modeling of PMU measurements in angular stability programs.

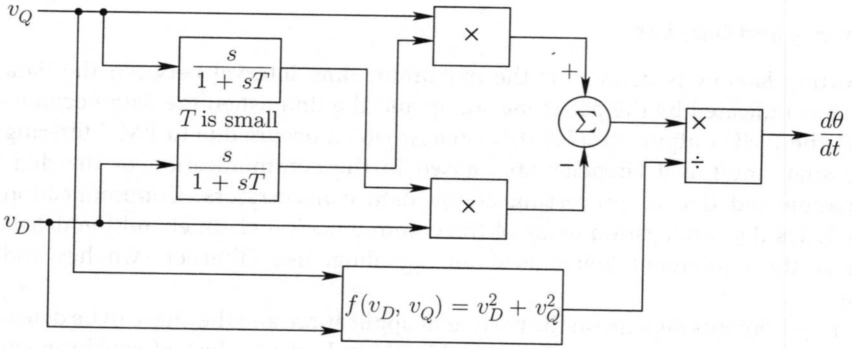

Figure 9.6 Frequency calculation in a transient stability program.

The model of WAMS in angular stability programs should capture the filtering and latencies associated with communication and processing. A simplified representation is shown in Figure 9.5. Measured quantities other than synchrophasors which may be a part of the payload of the PMU may also be modeled in a similar fashion.

In transient stability programs, the bus frequency is obtained by computing the approximate derivative of the phase angle. Generally, the phase angle keeps increasing or decreasing because the frequency may not be at the nominal value. The angles are wrapped at −180° and +180° (or 0° and 360°). A spurious spike is seen in the frequency computation when this wrapping occurs. This can be avoided by computing the derivative indirectly from the derivatives of v_D and v_Q, as given below:

$$\frac{d\theta}{dt} = \frac{d}{dt}\tan^{-1}\frac{v_D}{v_Q} = \frac{v_Q\frac{dv_D}{dt} - v_D\frac{dv_Q}{dt}}{v_D^2 + v_Q^2} \tag{9.2}$$

The derivatives of v_Q and v_D, required in (9.2), are computed using an approximate derivative block, as shown in Figure 9.6.

9.4 Online Monitoring of Power Swing Damping

The phenomenon of inter-machine rotor oscillations, also known as swings, is a characteristic of synchronous grids. These oscillations can be analyzed using a linearized model of the power system, which is valid for small deviations around the equilibrium point. Using such a model, it has been shown in Chapter 6 that a n-machine grid has $n - 1$ swing modes. Typically, these lie in the range 0.2–2 Hz. Since a power system is actually a nonlinear system, the modal characteristics (eigenvalues and eigenvectors) of the linearized system are dependent on the operating conditions. Some of the swing modes may be lightly damped for certain operating conditions. Occasionally, a swing mode may even be negatively damped, resulting in growing oscillations. The growing oscillations may lead to loss of synchronism or a sustained large amplitude oscillation (see Figure 6.1).

Despite the use of power swing damping controllers in FACTS, HVDC, and generator excitation systems, poorly damped swings are occasionally seen in many power systems. This is because the damping controllers may not be able to provide sufficient damping for all swing modes under all operating conditions. Therefore, online monitoring of power oscillations is essential so that preventive/corrective actions may be taken.

It is possible to assess damping of swings by analyzing the signals which are measured by WAMS. Power swings are generally observable quite well in quantities like generator speeds and rotor angles, real power flows, bus voltage frequencies, and bus voltage phase angles. Since the oscillations are in the 0.2–2 Hz range, PMU measurements (typical reporting rates 50 Hz or 25 Hz) can capture these oscillations without aliasing.

A power system is occasionally subjected to disturbances like line or generator trippings due to events like faults and protective relay operations. These excite the natural response of the system ("ringdown"), which can be used to estimate the modal properties, that is, modal frequency, modal damping, and modal observability. Alternatively, one can also inject known probing signals into the set-points of the generator excitation system, HVDC, and FACTS controllers, or carry out staged disturbances to excite a response. In these analyses, the responses are assumed to be that of a linear system, which is true if the transient deviations from the equilibrium condition are small. At the same time, the deviations should be large enough to be visible over the ambient noise caused due to the continually occurring random load variations.

Interestingly, if the random load variations have certain stochastic properties, then it is also possible to determine the dominant modes of the system by analyzing the noisy ambient measurements. This obviates the need for disturbances, staged or otherwise, or the injection of probing signals. Thus, there are three approaches to modal estimation: (i) analysis of ringdown waveforms following disturbances, (ii) analysis of the response to probing signal inputs, and (iii) analysis of ambient data. These are briefly described in the following sections.

9.4.1 Modal Estimation based on Ringdown Analysis

Power systems are occasionally subjected to disturbances like faults, line trippings, and generator trippings, which excite the oscillatory swing modes. If the disturbance is large enough, then the oscillatory response is visible in the measured waveforms, above the ambient noise. The deviation of the jth measured signal from its post-disturbance equilibrium value can be modeled as follows:

$$y_j(t) = \sum_{i=1}^{L} B_{ij} e^{\lambda_i t} \tag{9.3}$$

where λ_i is the ith eigenvalue of the model, B_{ij} is the modal observability of the ith mode in the signal $y_j(t)$, and L is the total number of eigenvalues considered in the model.

The objective of ringdown analysis is to identify the eigenvalues and modal observabilities from the N evenly-spaced samples $y_j(kT) = y_j[k]$, $k = 0, 1, 2, ...N - 1$. Note that

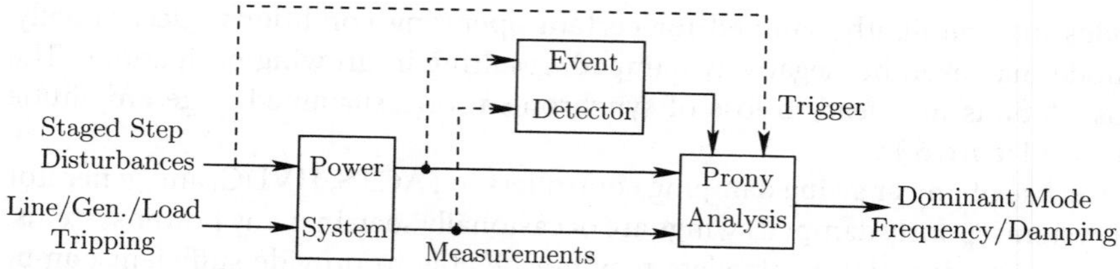

Figure 9.7 Ringdown analysis.

the sampling period T should be less than the Nyquist period to avoid aliasing. The sampled signal can be written as

$$y_j[k] = \sum_{i=1}^{L} B_{ij}\mu_i^k \tag{9.4}$$

where $\mu_i = e^{\lambda_i T}$ is the discrete-time eigenvalue. Note that if $y_j[k]$ satisfies (9.4), then it also satisfies an equation of the form

$$y_j[k] = a_{1j}y_j[k-1] + a_{2j}y_j[k-2] + \dots\dots + a_{Lj}y_j[k-L] \tag{9.5}$$

where the coefficients $a_{1j}, \dots a_{Lj}$, are independent of k.

Since the data are generally corrupted by noise, the objective is to fit the observed data samples to (9.5), that is, to determine the coefficients $a_{1j}, \dots a_{Lj}$. This may be done by using methods like the Prony method [8] and the matrix pencil method [9]. Subsequently the discrete-time eigenvalues μ_i are obtained by computing the roots of the characteristic polynomial $1 - (a_1 z^{-1} + a_2 z^{-2} \dots a_L z^{-L})$. Furthermore, the B_{ij}s are obtained using (9.4).

The schematic of modal estimation by ringdown analysis using the Prony method is shown in Figure 9.7. Note that if the disturbance is unintentional (not staged) then a separate algorithm is required to detect the disturbance (event detector) and trigger the modal estimation algorithm [10].

In a practical situation, the ringdown waveforms are contaminated by ambient noise caused by random load variations. The window of the recorded waveform will generally be cropped in order to exclude the initial part, which may contain nonlinear effects (if the initial deviations are large). Similarly, the tail of the ringdown is excluded as it may decay below the ambient noise level. High-frequency components are filtered out using low-pass filters. The measurements also contain slow trends that are drifts and changes in the steady state caused by load-generation imbalances, switching events or set-point changes. If the oscillatory swing modes are of interest, then the removal of slow trends improves the relative strength of the swing-mode components in the measured signal. Frequency and phase angular *difference* signals may be used in order to exclude trends that are due to the centre-of-inertia movement (caused by load-generation imbalances). For more general situations, trend removal algorithms like the one proposed in [11] may be used.

From ringdown analysis, it is usually possible to accurately determine the dominant modes which are excited by a disturbance. Dominant modes are the ones having low damping and high modal observability in the measured signals. Although the order of practical power systems is very high, the number of dominant modes excited by a disturbance and visible in a measured signal is typically around six or less [12]. The order of the fitted model L needs to be higher than this number for good numerical conditioning. This results in the estimation of a few modes which are weakly excited or are numerical artifacts. An automated mode-selection algorithm may be used to select only those modes which have high observability and poor damping as they are the ones of interest.

Note that when multiple signals $y_1, y_2, ..., y_n$ are measured for the same disturbance, the relative modal observabilities (or mode-shape) for the ith mode, that is, $B_{i1}, B_{i2}, ..., B_{in}$, are available from Prony analysis. The use of multiple signals generally improves the accuracy of the modal estimates [13]. The mode-shape information is useful in correlating the estimated swing modes from the measurements with the ones obtained from a small-signal model of the system.

9.4.2 Modal Estimation based on Probing Signals

The modal characteristics of the system can also be determined from the sinusoidal steady-state response of the system to known sinusoidal input signals. The main idea is to modulate the set-point of a controllable device using sinusoidal signals and measure the response of various output signals. From this, the frequency response of the corresponding transfer function can be obtained. The details of the scheme are shown in Figure 9.8. The frequency response can be obtained one frequency at a time, but in each case one has to wait for the natural response to die out, which is time-consuming. Instead one may inject a *multi-sine* signal. The steady-state response to this input is a superposition of the individual frequency components (assuming linearity). A multi-sine signal can be mathematically represented as follows:

$$u(t) = \sum_{l=N_1}^{N_2} a_l \sin(2\pi l f_d t + \phi_l) \tag{9.6}$$

Note that this signal has $N_2 - N_1 + 1$ frequency components (from $N_1 f_d$ to $N_2 f_d$) each spaced by f_d Hz. Therefore, the overall period of $u(t)$ is $(1/f_d)$ s. The maximum amplitude of $u(t)$ has to be limited to avoid causing large deviations from the equilibrium. a_l and ϕ_l may be chosen appropriately to reduce the maximum amplitude. For example, consider

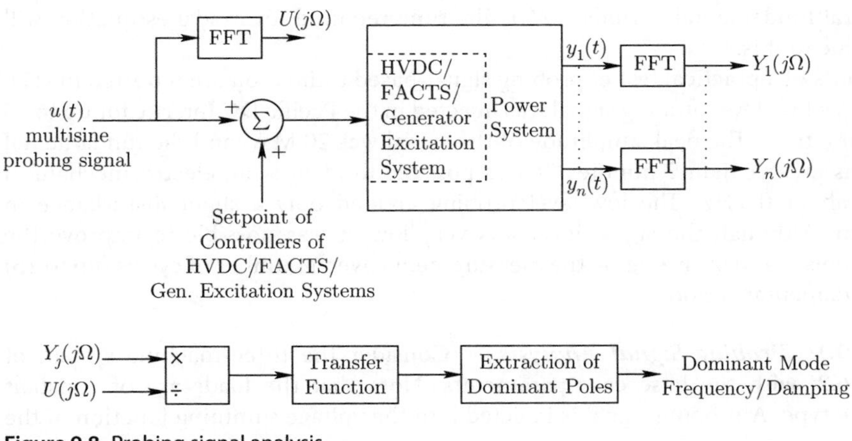

Figure 9.8 Probing signal analysis.

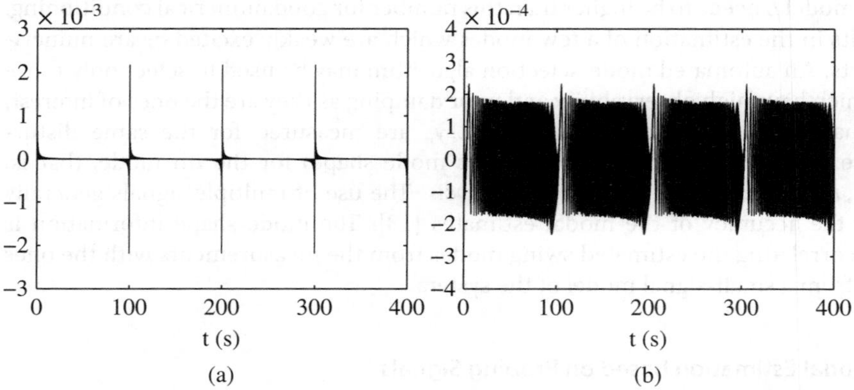

Figure 9.9 Comparison of multi-sine signals. (a) With $\phi_l = 0$ (b) With ϕ_l as given in (9.7)

the multi-sine signal which has the same a_l for all frequencies, but ϕ_l is chosen as per the following equation [14]:

$$\phi_l = -\frac{(l - N_1)(l - N_1 + 1)}{N_2 - N_1 + 1}\pi \tag{9.7}$$

The signal that uses ϕ_l as given in (9.7) and the signal that uses $\phi_l = 0$ are shown in Figure 9.9. The other parameters are identical ($a_l = 10^{-5}$, $N_1 = 1$, $N_2 = 300$ and $f_d = 0.01$ Hz).

The fast Fourier transform (FFT) algorithm [15] may be used to determine the frequency components of the input and the outputs. In the absence of noise, the steady-state response is periodic with a frequency f_d and the FFT calculation is carried out over this period. A rational transfer function is fitted to the obtained frequency response using methods such as vector fitting [16]. The poles and residues of this transfer function reveal the modes, which are controllable using the selected input and observable in the measured signals. In practice, the output signals will contain not only the response to the probing signal, but also the response to the random load variations. The signal-to-noise ratio can be improved by taking the average of the waveforms over several fundamental periods ($1/f_d$). The time required for modal estimation will increase due to this.

The results of a practical test of probing signal-based estimation are reported in [17]. In this test, a low-level probing signal was injected in the Pacific DC Intertie for up to 35 minutes at a time. The peak amplitude of the signal was 20 MW, and the fundamental period was greater than 2 minutes. It was possible to identify an electro-mechanical mode of about 0.3 Hz. The low-level probing created only a slight disturbance in the system. Although the signal level was very low, it was possible to improve the signal-to-noise ratio by averaging the measurements over a number of cycles (up to 15) of the fundamental period.

Example 9.1 ***Probing Signal Analysis*** Consider the three-machine system of Example 6.2, with the base case parameters. Note that the loads are of constant impedance type. A probing signal is injected into the voltage summing junction of the AVR of Generator 2.

The parameters of the probing signal are as follows: $f_d = 0.01$ Hz, $a_l = 10^{-4}$ pu, $N_1 = 1$, and $N_2 = 300$. ϕ_l is chosen as given in (9.7).

Load noise is modeled by varying the load impedance Z_{Li} (at each load bus i) every 20 ms in the following fashion: $Z_{Li} = Z_{Li0}(1 + 0.05\gamma)$, where γ is a random number with normal distribution (mean 0 and standard deviation 1). Obtain the swing modes of the system from the simulated steady-state response of the bus–frequency differences between buses 6 and 4, 8 and 9, and 4 and 5.

Solution

The probing signal is added to the set-point of the AVR of Generator 2 and is shown in Figure 9.10a. The simulated bus frequency differences are shown in Figures 9.10b–9.10d. Since the response to the load noise practically drowns the response to the probing signal, we average the signal over several cycles of the fundamental period to improve the signal-to-noise ratio. In this example, the frequency measurements are averaged over 10 windows, each of 100 s duration (corresponding to $1/f_d$). The frequency response is then obtained for the averaged waveforms. The modal estimates from these measurements are obtained by fitting the data to three rational transfer functions using vector fitting [16]:

$$G_1(s) = \sum_{i=1}^{N} \frac{R_{1i}}{s - \lambda_i} = \frac{\Delta f_6 - \Delta f_4(s)}{\Delta V_{ref2}(s)} \tag{9.8}$$

$$G_2(s) = \sum_{i=1}^{N} \frac{R_{2i}}{s - \lambda_i} = \frac{\Delta f_5 - \Delta f_4(s)}{\Delta V_{ref2}(s)} \tag{9.9}$$

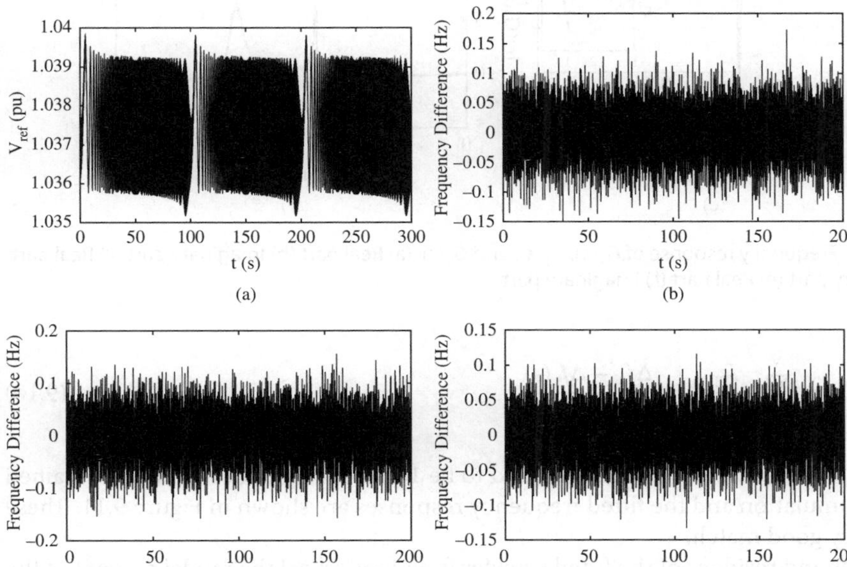

Figure 9.10 AVR inputs and the measured frequency-difference signals. (a) Set-point of the AVR of Generator 2 (b) $\Delta f_6 - \Delta f_4$ (c) $\Delta f_5 - \Delta f_4$ (d) $\Delta f_8 - \Delta f_9$

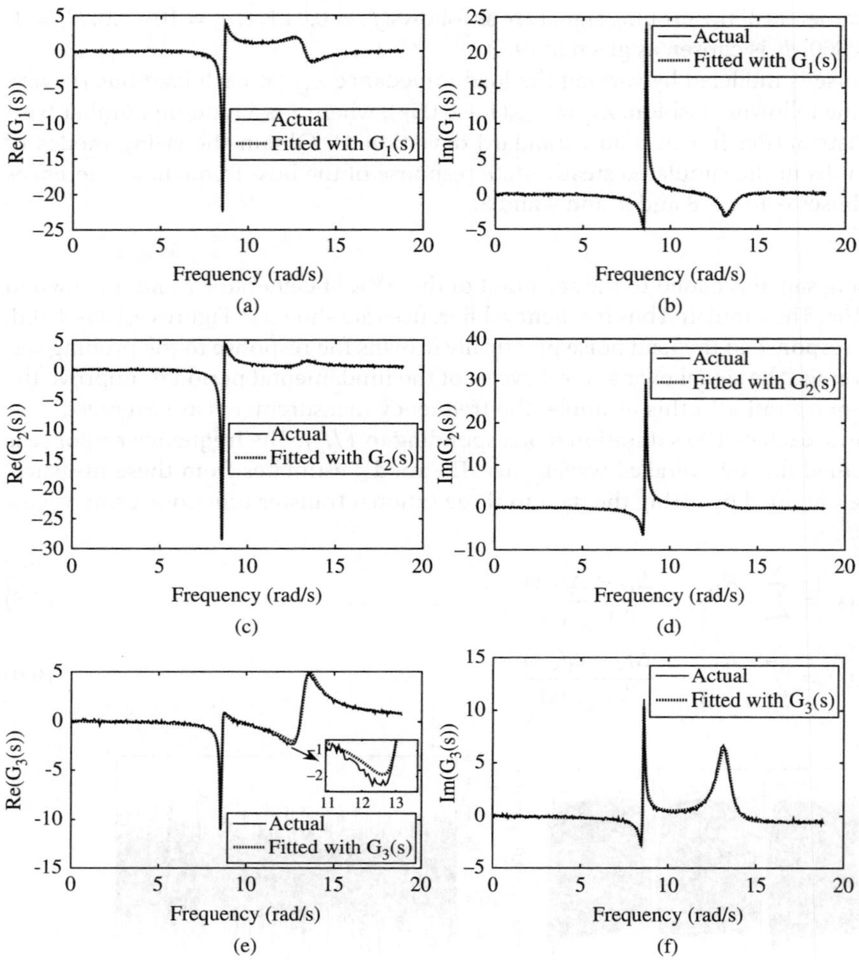

Figure 9.11 Frequency response of $G_1(s)$, $G_2(s)$, and $G_3(s)$. (a) Real part (b) Imaginary part (c) Real part (d) Imaginary part (e) Real part (f) Imaginary part

$$G_3(s) = \sum_{i=1}^{N} \frac{R_{3i}}{s - \lambda_i} = \frac{\Delta f_8 - \Delta f_9(s)}{\Delta V_{ref2}(s)}. \tag{9.10}$$

A suitable value for N in this case is found to be 10. The frequency responses obtained from the simulation and the fitted frequency responses are shown in Figure 9.11. These show a very good match.

The poles and residues of the fitted transfer functions reveal the modal content of the system. Note that for a system with state-space matrices A, B, C, and D, the residue of the transfer function R_i corresponding to the ith eigenvalue of A is the product of the modal observability (Cv_i) and the modal controllability ($u_i^t B$): $R_i = (Cv_i)(u_i^t B)$, where v_i and u_i are the normalized right and left eigenvectors of A (normalized implies that $u_i^t v_i = 1$).

Table 9.1 Poles and the residues corresponding to the frequency-difference signals (probing signal analysis).

Poles	Residues
$\mathbf{-0.049 \pm j8.58}$	$[1.38\angle \pm 139.50°,\ 1.74\angle \pm 140.88°,\ 0.65\angle \pm 146.01°]$
$\mathbf{-0.462 \pm j13.26}$	$[1.55\angle \mp 99.06°,\ 0.47\angle \pm 33.66°,\ 3.17\angle \pm 70.97°]$
$-23.668 \pm j37.56$	$[4.04\angle \mp 40.79°,\ 13.87\angle \mp 101.09°,\ 25.16\angle \mp 121.93°]$
$-0.742 \pm j2.37$	$[0.19\angle \mp 153.54°,\ 0.15\angle \mp 157.28°,\ 0.04\angle \mp 72.38°]$
-0.725	$[0.29\angle 180°,\ 0.23\angle 180°,\ 0.08\angle 0°]$
-0.051	$[0.02\angle 0°,\ 0.02\angle 0°,\ 0.01\angle 180°]$

Table 9.2 Poles and the residues corresponding to the frequency-difference signals (small-signal analysis of the linearized system model).

Poles	Residues
$\mathbf{-0.039 \pm j8.59}$	$[1.36\angle \pm 139.88°,\ 1.71\angle \pm 141.08°,\ 0.64\angle \pm 145.94°]$
$\mathbf{-0.45 \pm j13.16}$	$[1.60\angle \mp 86.97°,\ 0.43\angle \pm 46.07°,\ 3.21\angle \pm 83.42°]$

Since the transfer functions are obtained for the same input (ΔV_{ref2}), the residues also reveal the relative modal observabilities of the swing modes in the frequency difference measurements. The extracted modal information is shown in Table 9.1. The swing modes and the corresponding residues that are obtained from the small-signal analysis of the linearized model are shown in Table 9.2 for comparison. The swing modes can generally be identified by the frequency range in which they lie (approximately 0.2–2 Hz), their under-damped nature, and their mode shape (see Example 6.2). Other modes are also identified by the probing signal analysis depending on the order of the transfer function that is chosen for the fitting. Some of these could be numerical artifacts. In this example, both swing modes of this system are identified by probing signal analysis. There is a good match between the extracted swing modes and the modes obtained from small-signal analysis.

9.4.3 Modal Estimation based on Ambient Data Analysis

The analyses in the previous sections are based on either the natural response triggered due to discrete events or the steady-state response to known probing signals. The ambient noise due to random load variations is an undesirable contaminant in these analyses. On the other hand, random load variations may also be treated as a small-amplitude input to the system, which may be used to determine the modal content of the output signals. The schematic of this modal estimation scheme is shown in Figure 9.12.

Ambient data analysis methods for modal extraction are based on certain hypotheses. It is hypothesized that under ambient conditions the random load variations are

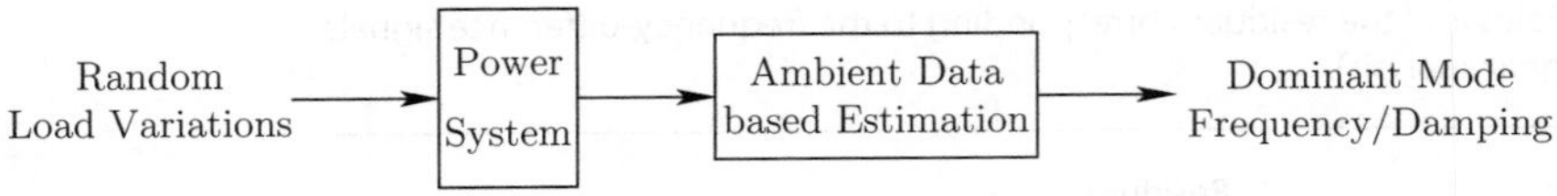

Figure 9.12 Ambient data analysis.

stationary, which is a notion similar to steady state in deterministic signals. For a stationary signal, properties like the mean and autocorrelation are time-invariant [18]. It is further hypothesized that the random load variations can be represented as white noise or integrated white noise, with the variations at individual buses being independent [19]. Note that white noise signals have the following properties:

1) The *autocorrelation* $R_x(\tau)$ of a white noise signal $x(t)$ which has a variance a satisfies the following relationship

$$R_x(\tau) = \lim_{T\to\infty} \frac{1}{T} \int_0^T x(t)x(t+\tau)\, dt = a\delta(\tau) \tag{9.11}$$

where δ is the Dirac delta function.

2) The one-sided power spectral density function $G_x(f)$ of a white noise signal is given by

$$G_x(f) = 4 \int_0^\infty R_x(\tau)\cos(2\pi f\tau)\, d\tau = 2a \tag{9.12}$$

If the frequency response of a linear system is $H(f)$ and a stationary random signal $x(t)$, with the power spectral density function $G_x(f)$, is applied as an input, then the output of the system will be a stationary random signal, $y(t)$, with a power spectral density function given by

$$G_y(f) = |H(f)|^2 G_x(f) \tag{9.13}$$

and

$$R_y(\tau) = \lim_{T\to\infty} \frac{1}{T} \int_0^T y(t)y(t+\tau)\, dt = (h(\tau) * h(-\tau)) * R_x(\tau) \tag{9.14}$$

where $h(\tau)$ denotes the system impulse response and * denotes the convolution operation.

Clearly, if $R_x(\tau)$ is known (e.g., if $x(t)$ is white noise) then it is possible to infer $(h(\tau) * h(-\tau))$ from $R_y(\tau)$. Since $h(\tau)$ denotes the impulse response, $R_y(\tau)$ contains the signatures of the modes of the system. For a first-order system, if $x(t)$ is white noise and $h(t) = e^{\lambda t}u(t)$, $u(t)$ being the unit step function, then $R_y(\tau) = \frac{1}{2}e^{\lambda|\tau|}$ [18]. Therefore, methods similar to Prony analysis can be used to extract λ from the samples of $R_y(\tau)$. The calculation of autocorrelation $R_y(\tau)$ from $y(t)$ using (9.14) requires an integration over an infinite time duration. In practice, discrete samples of $y(t)$ in a finite but long-enough time window are used to approximately compute $R_y(\tau)$ as given below:

$$R_y(\tau) \approx \frac{h}{T} \sum_{n=l}^{N+l-1} y(nh)y(nh+\tau) \tag{9.15}$$

where h is the sampling period and $N = \frac{T}{h}$ is the number of samples in the time-window T.

The foregoing discussion brings out the principle of modal estimation based on ambient measurements. Several algorithms have been developed to extract the modal information for practical applications. In block-processing algorithms like the Yule–Walker [20] method, the modes are estimated from a window of data. For each new window of data, a new estimate is calculated. In recursive methods like the least mean square (LMS) method [21], the modal estimates are updated at each new sample of the data. The new estimate is obtained using a combination of the new data point and the previous modal estimate.

Modal information can also be obtained from frequency domain analysis [19]. Since the nature of load noise and therefore $G_x(f)$ is assumed to be known, $|H(f)|^2$ can be obtained by evaluating $G_y(f)$ as given below:

$$G_y(f) = |Y(f)|^2$$

where

$$Y(f) = \lim_{T\to\infty} \frac{1}{T} \int_{t-T}^{t} y(\tau)e^{-j2\pi f\tau}\, d\tau$$

In practice, the DFT is computed over a finite time-window T using samples of $y(t)$. Such DFT computations are done over several successive time-windows, and these are then averaged. In the Welch method [22], the time-windows are allowed to overlap. The modal frequencies are then estimated from the peaks of the spectral density plots.

In practice, assumptions regarding the stationarity of the data may be violated due to transients under switching conditions, missing data due to a temporary loss of the data link and outliers due to sensor or processing errors. A study of three block-processing algorithms [12] concluded that the accuracy of the algorithms improves as a mode becomes less stable. The presence of transient ring-down waveforms due to disturbances was found to improve the performance of the algorithms. Data corruption due to missing data and outliers did not cause significant degradation in the estimator performance if the corrupt data was in the range of 6% or less.

Example 9.2 ***Ambient Data Analysis*** In the three-machine system of Example 6.2, the loads at buses 5, 6 and 8 are assumed to have random variations as given in Example 9.1.

1) Obtain the autocorrelation of the generator speed differences from the simulated waveforms.
2) From these autocorrelation signals, estimate the modal content of the signals using multiple-signal Prony analysis [13].

Solution:

Note that speed differences between generators are considered here in order to eliminate the centre-of-inertia motion component, since the main objective is to estimate the swing-mode properties. The generator speed-difference waveforms $(\omega_1 - \omega_2)$, $(\omega_2 - \omega_3)$, and $(\omega_3 - \omega_1)$, which are obtained by simulating the system with the load noise, are shown in Figure 9.13.

The autocorrelations are obtained for $\tau = 0$ to 40 s, using data samples ($h = 10$ ms) of the speed-difference signals in a finite time-window of $T = 300$ s (refer to (9.15)). The plots of the autocorrelations obtained from one window of data samples are shown

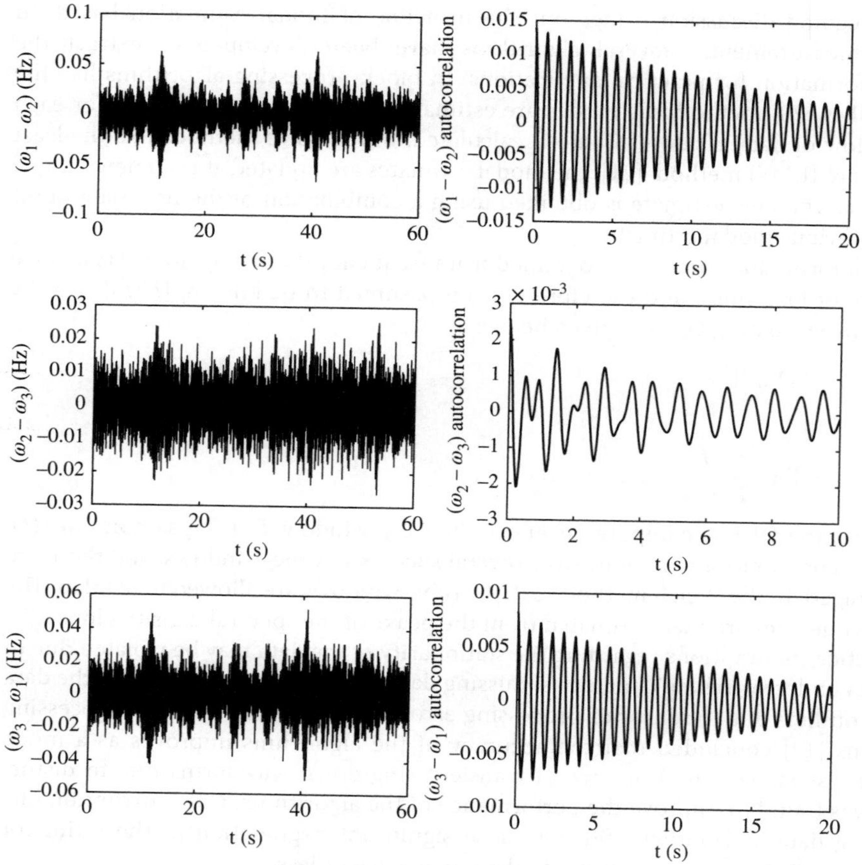

Figure 9.13 Generator speed differences and their autocorrelations.

in the figure. Note that signatures of the swing modes are revealed in the autocorrelation of the speed-difference signals, although the speed-difference signals themselves are noisy.

Multiple-signal Prony analysis is performed on the three autocorrelation signals in order to obtain the modal content of the signals. A 20 s window of the autocorrelation plots are used for Prony analysis. This is repeated for several autocorrelation plots, each obtained using overlapping 300 s windows of the recorded speed-difference signals. The windows are shifted by 50 s. Multiple-signal Prony analysis is performed on each set of three autocorrelation waveforms thus obtained. The scatter of eigenvalues obtained for 20 such time windows is shown in Figure 9.14. The total length of the data record that is used for obtaining the scatter plot is, therefore, $(300 + 40) + 50 \times 19 = 1290$ s.

It can be seen that the modal frequency and damping estimates of both swing modes are scattered around and quite close to the actual values that are obtained from small-signal analysis. The mean values of the estimates are only slightly off the mark for both swing modes. The numerical experiments carried out in [12] show that the scatter

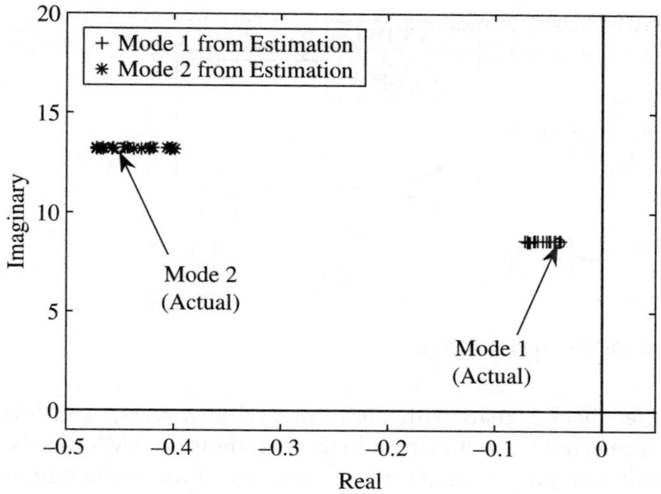

Figure 9.14 Scatter of the modal estimates obtained from several overlapping windows.

around the mean can be reduced by using larger time windows in the computation of the autocorrelations. However, this means that the time required for modal estimation will be longer.

9.5 WAMS Applications in Power Swing Damping Controllers

Non-local measurements from WAMS can be used in excitation systems, FACTS, and HVDC-based power swing damping controllers (PSDCs) to enhance their performance in the following ways:

1) WAMS-based supervisory actions: These include the triggering of discrete actions like the the arming/disarming of pre-determined features in controllers. The objective is to enhance the damping of modes that are found to be poorly damped by online-damping estimators (discussed in the previous section). The actions may also include tripping of reactors, capacitors or loads. The nature and quantum of actions are decided based on (i) off-line studies or (ii) evaluations carried out online using a small-signal stability analysis program.
2) WAMS-based feedback signals: This involves the use of a signal synthesized using WAMS for continuous feedback control. Figure 9.15 depicts a possible WAMS-based PSDC scheme. By appropriately choosing the weighting factors, W_1 and W_2, the local or the non-local channel (which uses a WAMS-based feedback signal) may be given more prominence. To disarm a channel the corresponding weighting factor is set to zero.

The motivation for using wide-area measurements in feedback control schemes is that there is a wider choice of feedback signals from which the ones with desirable properties can be chosen. Refer to the multi-modal decomposition shown in Figure 6.22

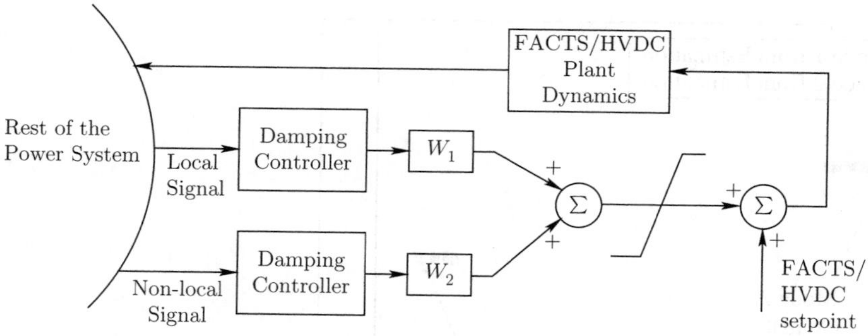

Figure 9.15 PSDC using local and WAMS-based channels.

in Section 6.6. Recall that feedback signals with high observability, $K_{oi}(s)$, and low inner-loop gain, $K_{ILi}(s)$, are preferred for achieving a higher damping. WAMS may be used to synthesize signals with such properties. Another advantage of wide-area signals is that the observability of the local non-swing modes is likely to be small, therefore the destabilization of these modes is unlikely to be a major concern in a WAMS-based controller.

Modal Speed Signal-Based PSDC

The modal speed ω_{mi} corresponding to the ith swing mode is defined as $\omega_{mi} = f_i^T\omega$ where ω is the vector of generator speeds and f_i^T is the ith row of the matrix V^{-1}, where V is the right eigenvector matrix of A_{21} (refer Section 6.6).

If the feedback signal y in Figure 6.22 is chosen to be ω_{mi}, then $K_{ILi} = 0$, therefore the achievable damping influence for this mode is high. Since other swing modes are not observable in this signal, only the ith mode is affected by the damping controller. These are attractive properties if we are seeking the targeted damping of a poorly damped swing mode. Note that a poorly damped mode and its mode-shape can be identified using the WAMS-based online mode estimation techniques described in the previous section. The modal signal itself is synthesized from wide-area measurements of the speed signals [23]. The use of a WAMS-based PSDC using modal speed as a feedback signal is illustrated using the following example.

Example 9.3 ***Modal Speed Signal-Based PSDC*** Consider the three-machine system shown in Figure 6.34. There are two swing modes in this system. A thyristor-controlled series compensator (TCSC) is placed in lines 4–6 and has a quiescent reactance equal to 0.01 pu, as in Example 6.5. The TCSC is modeled as a variable reactance, X_{TCSC}, and is equipped with a damping controller which has three channels, as shown in Figure 9.16. The non-local channels use modal speed signals ω_{m1} and ω_{m2} corresponding to the two swing modes. The controller allows for selectively changing the strength of a channel through the weighting factors W_1, W_2, and W_3.

1) Synthesize the modal speed signals using generator speed measurements.
2) Obtain the root loci with increasing gain for the non-local channels (used one at a time) by using the small-signal model of the system.

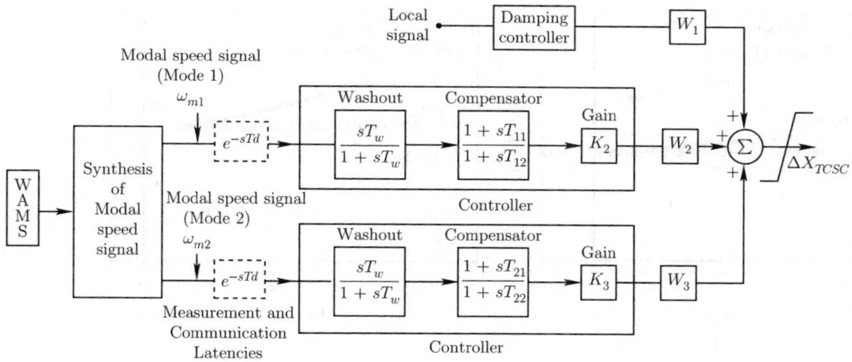

Figure 9.16 Use of controller channels using local feedback signal and WAMS-based modal signals.

Solution:

The two swing modes in this system have frequencies of 8.61 rad/s and 13.18 rad/s. The modal speed signals corresponding to the swing modes are computed from the following equations:

$$\omega_{m1} = 0.768\omega_1 - 0.624\omega_2 - 0.144\omega_3$$
$$\omega_{m2} = -0.241\omega_1 - 0.555\omega_2 + 0.796\omega_3 \tag{9.16}$$

ω_1, ω_2, and ω_3 are the rotor speed deviations of machines 1, 2, and 3. The coefficients used in the above equations are obtained from the left eigenvectors of matrix A_{21}, as discussed previously.

The phase compensation requirements for achieving damping of the relevant swing mode when either of these modal signals is used can be determined from the controller phase index (CPI) defined in Section 6.6.1. These are found to be almost zero for both modes if the modal speed signals defined in (9.16) are used. Therefore, if the communication latency is neglected, just a gain (without phase compensation) will be able to achieve the damping affect.

The root loci when the ω_{m1} and ω_{m2} channels are armed (one at a time) are shown in Figure 9.17. The plots are shown for the case in which no phase compensation is provided ($T_{11} = T_{12} = T_{21} = T_{22} = 1$ s). The washout time constant is taken to be $T_w = 5.0$ s. When the ω_{m1} channel alone is activated ($W_1 = W_3 = 0$, $W_2 = 1$), there is an almost exclusive movement of the first swing mode. Similarly, when the ω_{m2} channel alone is activated ($W_1 = W_2 = 0$, $W_3 = 1$) it mostly moves the second mode. In each case, slight movements in the other swing mode are also seen because the synthesis of the modal signal is approximate.

Latencies associated with WAMS cause the signals to have a time delay T_d, which translates to a phase delay of $e^{-j\Omega T_d}$ at the modal frequency Ω. The movement of eigenvalues with $T_d = 100$ ms is shown in Figure 9.17. Note that the first-order Padé approximation is used to model the delay block for the small-signal analysis, that is, $e^{-sT_d} \approx \frac{1-s\frac{T_d}{2}}{1+s\frac{T_d}{2}}$. The introduction of a delay causes an increase in swing mode frequency with an increase in gain, as seen in the figure. A lesser damping improvement for the same controller gain is also seen when the delay is present.

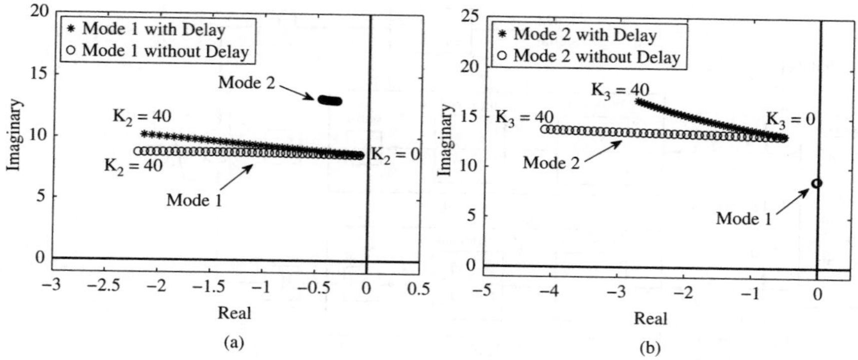

Figure 9.17 Root locus with the modal signals. (a) ω_{m1} channel alone is activated (b) ω_{m2} channel alone is activated

Note that the phase delay caused by the latencies is not compensated by this controller since T_{11}, T_{12}, T_{21}, and T_{22} are equal. An appropriate phase lead at the modal frequencies may be provided based on the average delay of the signal to improve the damping influence. The controller should be disarmed if excessive signal delays occur, especially over contiguous blocks of time.

In a practical situation, all the generator speeds may not be available for the synthesis of the modal speed signals. Therefore, approximate calculations using bus frequencies near the generators may be used. It should be understood that modal speed signals are only one example of the use of wide-area measurements and many alternative signals can be considered in a WAMS-based PSDC.

The application of WAMS-based feedback signals for damping control is being practically tested using the Pacific DC Intertie as the actuator [24]. The frequency measurements from selected PMUs are being used as inputs to this controller. The damping controller uses bus frequency-difference signals using four PMUs in the north and four PMUs in the south for redundancy and diversity, allowing for 16 controller instances in parallel. Supervisory functions are implemented which disarm a controller instance should an issue arise, and ensure a bump-less transition to another controller instance. The overall time delay (from the measurement to the time when the command is sent to actuator) is 71 ms on an average with a worst case value of 103 ms [24].

9.6 WAMS Applications in Emergency Control

Conventional software in energy management centres performs state estimation, power flow/optimal power flow, and security assessment. This helps a system operator to determine the nature and quantum of preventive control actions during an alert state. Preventive action is anticipatory in nature and is necessary to take the system to a secure state before any major disturbance occurs. Despite efforts by a system operator to maintain the system in a secure state, a series of unforeseen large disturbances, often exacerbated by the unexpected behavior of protective relays due to hidden deficiencies in design or settings, may lead the system to an emergency state. In an emergency state, the system tends to move rapidly to an unstable condition, leading to large fluctuations in voltages,

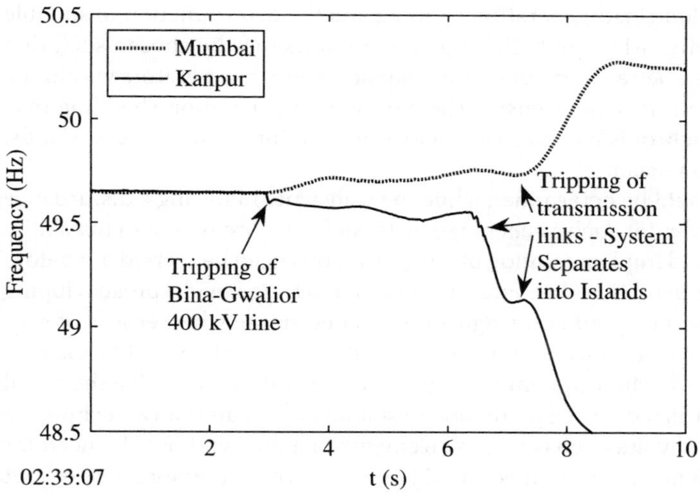

Figure 9.18 Frequencies during the blackout in the Indian grid (July 30, 2012).

currents, torques, and speeds which will trigger protective relay actions. This may lead to a partial or total blackout due to widespread equipment trippings. The available time to take any corrective action is limited during these emergencies, often to just a few seconds. For example, the loss of synchronism event experienced in the Indian grid led to the separation of the system into two islands just 5 s after the initiating event (see Figure 9.18). One of the islands subsequently blacked out just a few seconds after the separation due to very low frequency.

To deal with such situations, system operators rely on pre-designed, automatic emergency control schemes (also known as system protection schemes, remedial action schemes or system integrity protection schemes). The key concerns in any emergency control scheme are the avoidance of false alarms, false dismissals, and inadequate, excessive or incorrect control actions. Industry experiences with existing emergency control schemes are documented in two major surveys [25, 26].

Under-frequency, undervoltage, and rate-of-change of frequency load shedding relays are common system protection measures. These measures are (i) not tailor-made for any specific disturbance, (ii) use local information, (iii) perform minimal computations, and (iv) are generally not adapted based on the prevailing operating conditions. Some emergency schemes, however, can be highly customized and use information from specific remote locations. For example, the Pacific Northwest scheme, implemented in the late 1970s [27], uses sensors at various locations to detect events that could cause instability and system separation. The scheme initiates a transfer-trip of up to 3,000 MW of industrial load upon the detection of such events. In addition, to compensate for the loss of the Pacific AC Intertie, fast power change controls are implemented on the parallel Pacific HVDC Intertie. The level of control actions is determined from off-line transient stability simulations. The controls are armed only during critical system conditions to avoid malfunctions.

WAMS can facilitate the development of sophisticated centralized stability controllers which can be adapted based on the prevailing conditions [28]. For angular instability, the control actions may include load/generation tripping and the determination of suitable separation points (adaptive islanding). The separation points may be chosen such that the islands formed have a better load-generation balance, and the generators within the island are likely to remain in synchronism. The aim of load/generation shedding may be to prevent loss of synchronism or to ensure a load-generation balance in the islands, should system separation occur.

The control actions could be determined while the transients following a disturbance are still evolving. However, for such a scheme to be feasible, the prediction of instability and the determination and implementation of control actions need be carried out before the instability-induced fluctuations become large enough to trigger widespread tripping of equipment. Therefore, fast prediction algorithms to determine whether loss of synchronism is likely to occur are essential for implementing such schemes. For systems which behave like a two-machine system during system separation, tools like the Equal Area Criterion may be utilized for the prediction of stability [29]. Alternative techniques based on the current and voltage phasor measurements in a line, without the need for external system equivalents, are presented in [30] and [31]. These methods also do not require knowledge of the mode of instability, and no assumptions are made regarding the power-angle relationship. The results reported in these papers are encouraging; it was possible to predict instability with sufficient time being available to carry out emergency measures like a controlled system separation.

Like angular instability, the early detection of an impending voltage instability situation is also essential, so that pre-emptive measures like disabling of on-load tap-changers and activation of reactive power sources, and emergency measures like load tripping may be taken. Voltage instability detection can be based on local phasor measurements implemented in a distributed manner without the exchange of information [32]. The local phasor measurements are used to estimate the apparent impedance of the load and the Thevenin equivalent of the power system at the load bus, as shown in Figure 9.19. These are estimated by applying curve-fitting techniques on the measurements taken at several time instants.

The time-window of the measurements should be wide enough for operating conditions to change, but narrow enough to ensure that the Thevenin equivalent of the network does not change significantly. The proximity to voltage instability is inferred from the proximity to the maximum power point. This is a “static” indicator based on the assumption of constant power loads. The method cannot anticipate non-local events like generator reactive power limits being hit, which quickly change the parameters of

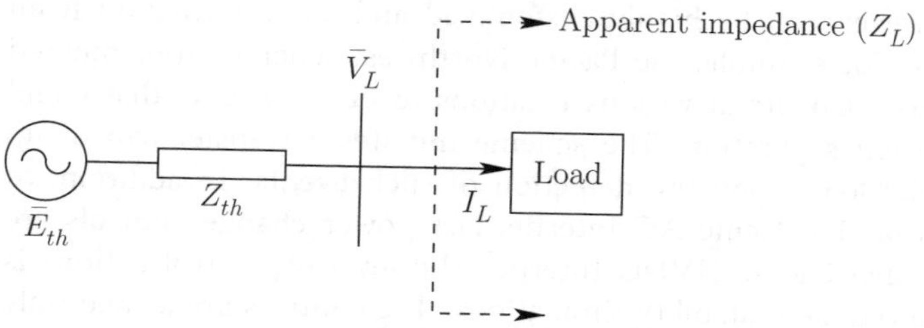

Figure 9.19 Thevenin equivalent of power system at the load bus.

the system equivalent. This problem can be addressed by using the measurements from WAMS to estimate the generator reactive power reserves and the reserve depletion trend [33], as well as the voltage magnitude reduction trend at key buses. Through this, the time and margin to instability can be estimated and appropriate control measures can be initiated.

WAMS-Based Supervision of Relays

The signals derived from WAMS could also facilitate supervision of many protection and control schemes, which otherwise use local signals. Supervisory signals derived from WAMS could help a relay distinguish between faults, heavy line loading, and power swings, thereby avoiding false trips, especially of backup relays [2]. Another possibility, discussed in [2], is the use of supervisory signals to adapt the "voting logic" in cases where multiple relays are used to detect a fault. For example, the logic could be changed adaptively from "trip the breaker if any of the relays see a fault" to "trip the breaker if at least two out of three relays see a fault" or, possibly, "trip the breaker if all relays see a fault". The logic may be chosen based on whether failure to trip (low dependability) is a bigger concern than a false trip (low security), or vice versa. Under healthy system conditions (an adequate stability margin exists), a protection system may be biased towards dependability because a false trip is an acceptable risk given the advantage of speedy fault clearing. Under insecure conditions the logic may be biased towards better security since a false trip may worsen the conditions and lead to cascading outages and instability.

Another possible use of system-wide signals is in under-frequency load shedding. Under-frequency relays trip blocks of loads when the bus frequency falls below preset thresholds. In addition, rate of change of frequency (ROCOF) relays are used to avoid large changes in frequency due to sudden and large load-generation imbalances. The relays trigger load-shedding when the ROCOF exceeds a preset threshold value. ROCOF may be calculated using local bus frequency or generator speed signals. The local bus frequency or generator speed signals not only contain the centre-of-inertia (COI) component, which is a measure of the overall load-generation balance in the system, but are also contaminated by "swings" due to inter-machine oscillations.

Consider the three-machine system given in Example 6.2 (base case). The bus frequencies and ROCOF at the load buses if Generator 3 is suddenly tripped are shown in Figure 9.20. The frequency and ROCOFs are calculated by passing the bus phase angles and the frequencies, respectively, through an approximate derivative block $\frac{s}{1+0.005s}$. A spike is visible in the frequency and ROCOF at the instant of tripping since the bus voltage phase angles undergo a sudden change due to the tripping. Note that in the transient stability simulation model used here the D-Q components of the bus voltages are not state variables, therefore the phase angles undergo a sudden change following the tripping. In a practical relay, the phase angle transition will depend on the filtering and phasor computation algorithms which are employed by the relay. The initial spike in the frequency and ROCOF should be ignored.

Subsequently, however, there is a drop in the frequencies, which is as expected, but the drop is not uniform due to the presence of swings. The swings are amplified in the ROCOF due to the derivative action. This may cause a mis-operation of ROCOF relays

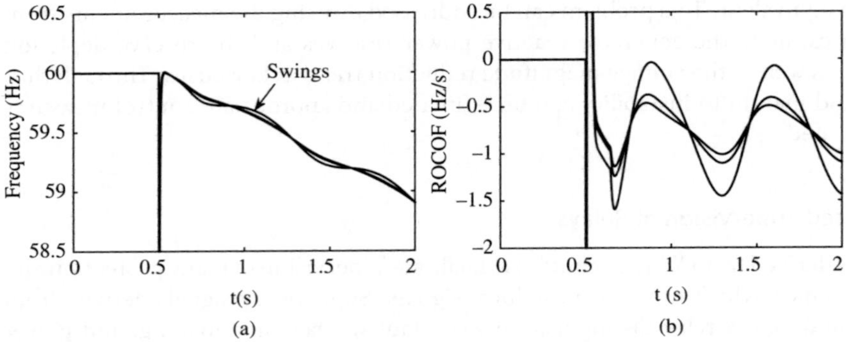

Figure 9.20 Frequency and ROCOF at the load buses following the tripping of Generator 3. (a) Bus frequencies (b) Rate of change of frequencies.

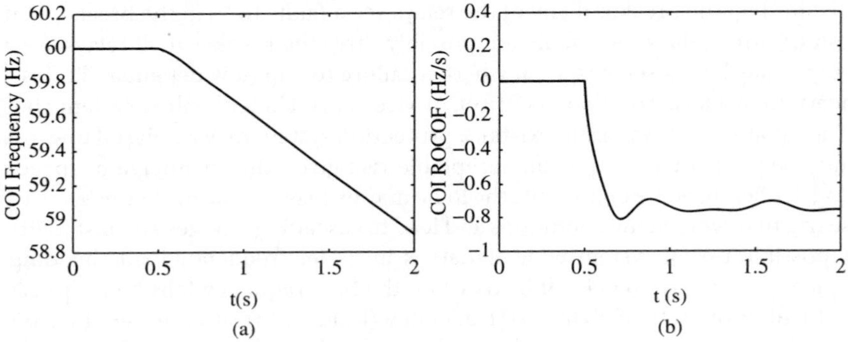

Figure 9.21 COI frequency and its rate of change. (a) COI frequency (b) Rate of change of COI frequency.

due to the large variations. On the other hand, the COI frequency $\frac{H_1 f_{g1}+H_2 f_{g2}}{H_1+H2}$ and its rate of change are relatively free of swings, as shown in Figure 9.21, therefore these COI signals are good candidates for frequency-based load-shedding. Note that f_{g1} and f_{g2} denote the electrical frequencies of Generator 1 and Generator 2, respectively, and H_1 and H_2 are the corresponding inertia constants.

The generation load-imbalance after the tripping is affected by the voltage and frequency dependence of the load. In this example, although the load-generation imbalance is expected to be 1.123 pu (the power output of Generator 3 before it tripped), it is actually less than 0.8 pu. This is because the power drawn by the constant impedance loads is reduced due to the lower voltages following the generator tripping. This is seen in Figure 9.22.

System-wide measurements obtained from WAMS can be used to estimate the actual load-generation imbalance and the total inertia in the grid after a disturbance. This information can be used to block or unblock under-frequency or ROCOF relays in the system, or for switching their settings. Since the COI frequency can be determined using WAMS, it may be used in a centralized load-shedding scheme. However, this scheme will involve time delays due to the communication latencies associated with WAMS signals.

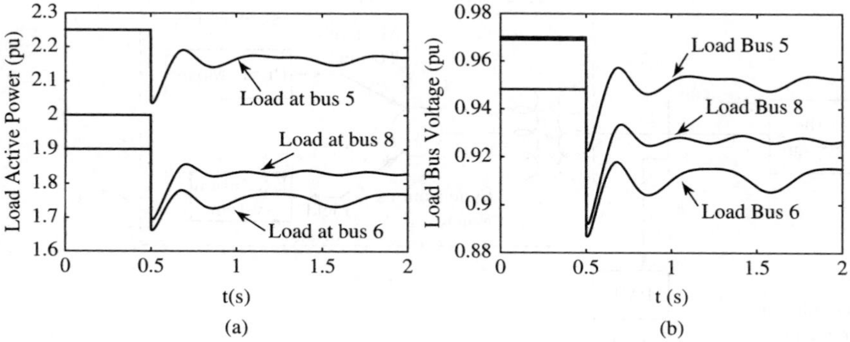

Figure 9.22 Load active power and bus voltages following the tripping of Generator 3. (a) Load active powers (b) Load bus voltages.

9.7 Generator Parameter Estimation

A synchronous generator and its controllers are critical components of a power system from the point of view of stability. Hence, accurate representation of these components in stability studies is essential. While parameter estimation may be done when the generator is off-line, online estimation may be done by recording responses to staged or naturally occurring disturbances in a power system. Since PMUs are generally installed at major power plants, the responses to small and large power system disturbances are readily available. Generator parameter estimation using PMUs may be carried out using the following steps [34]:

1) Model validation: A hybrid simulation is carried out in which the measured signals, like the voltage phasor at the terminals of a generator, are used as inputs to the simulation model of a generator. The model's response (usually currents or power flows) is compared to the measured signals of the same quantities. This analysis is carried out for several disturbances. If the error between the simulated and measured responses is very small, the model can be considered to be accurate.
2) Parameter estimation: If the error is large in the model validation step, the parameters of the models should be tuned using an optimization program which minimizes the error between the measured and simulated responses. This can be challenging because the models of interest are typically nonlinear and involve many parameters.

A parameter estimation scheme is shown in Figure 9.23. While the figure depicts the parameter estimation scheme for a single generator, it can be extended to multiple generators and other devices and subsystems as well.

9.8 Electro-Mechanical Wave Propagation and Other Observations in Large Grids

The large number of time-synchronized measurements already available from WAMS have revealed several important characteristics of power systems. Analysis of disturbance data has confirmed the wave-like propagation of electro-mechanical disturbances

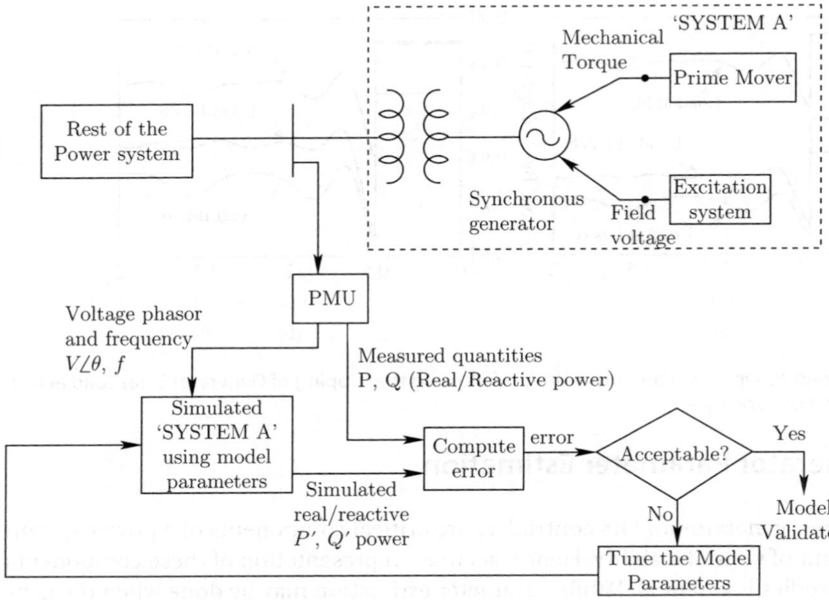

Figure 9.23 Generator parameter estimation scheme.

in large grids. This was theorized much earlier [35] using a simplified continuum model of a power system; some of the simplifying assumptions were relaxed in the subsequent work on this topic [36]. A large AC power system containing a large number of small but well-distributed generators connected by AC transmission lines can be approximated by a continuum of machine inertias and transmission capacities. Phenomena like

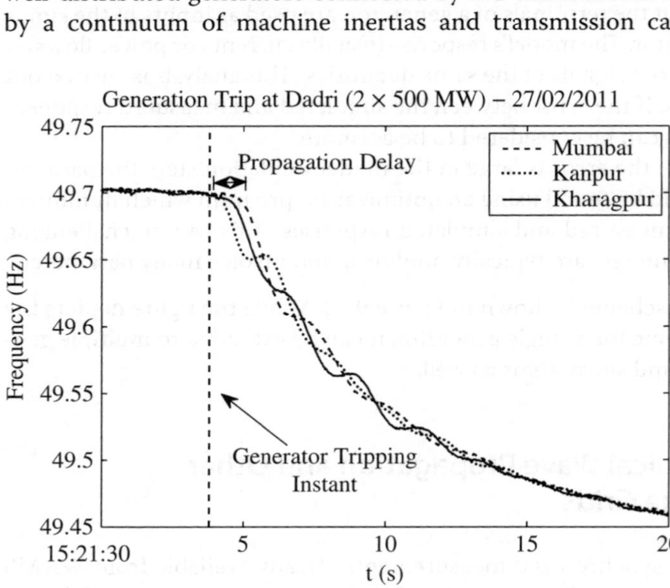

Figure 9.24 Propagation delay after generator trip in the Indian grid.

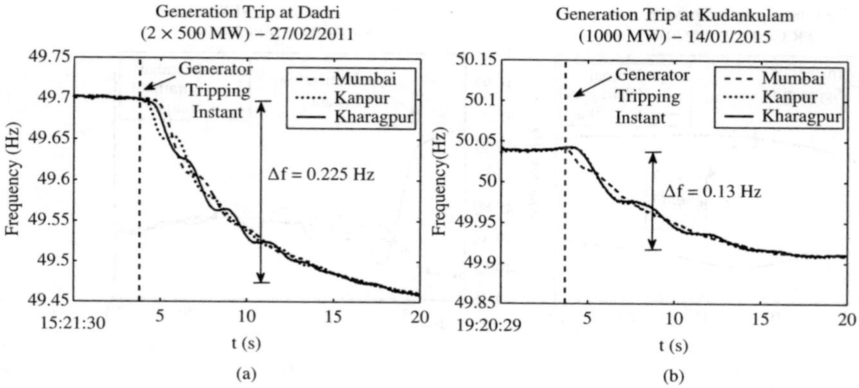

Figure 9.25 Frequency drop following generation loss of approximately 1000 MW. (a) 2011 (b) 2015

propagation delay, a characteristic of distributed parameter systems, emerge from this model. Large continent-size synchronous grids come close to the idealized continuum model, although the distribution of generators and transmission capacities is not uniform.

Propagation speeds ranging from hundreds of miles per second to thousands of miles per second have been observed in North America [37]. Propagation delays have also been observed in the Indian grid. Figure 9.24 shows the frequencies measured at several locations in the Indian grid following a generation trip at Dadri (in northern India) using a wide-area frequency measurement system [38]. There is a noticeable time delay between the instant at which the bus frequency starts changing at Kanpur (approximately 450 km from Dadri) and the instant at which the change is experienced at Mumbai and Kharagpur (both approximately 1400 km from Dadri).

Using the wide-area measurements recorded over several years, it is possible to observe changes in the system characteristics over the years. Consider the Indian grid whose size has grown significantly in the last decade. The southern grid of India (SR) was synchronized to the North-East-West (NEW) grid in early 2014. The system responses to generation tripping events, one in 2011 and the other in 2015, are shown in Figure 9.25. Although the loss of generation in both cases is approximately the same (1000 MW), the steady-state frequency deviation and the initial rate of frequency drop is lower in 2015 compared to 2011. The settling time has also reduced. This indicates an increase in the system inertia as well as the frequency-dependence of load and generation (MW/Hz) because of the growth in system size.

The examination of the swings in the frequency signal following the re-synchronization of the southern grid with the rest of the Indian grid, a few hours after they were separated due to the tie-line tripping (see Figure 9.26), clearly indicates the emergence of a low-frequency inter-area oscillation of approximately 0.25 Hz. This is a north-south mode with Kanpur (northern India) and Surathkal (southern India) displaying a relatively large swing as compared to Mumbai (western India). Prior to synchronization, the observed lowest frequency swing mode in the system was approximately 0.4 Hz, involving the machines of the west swinging against those in the east and the north [38].

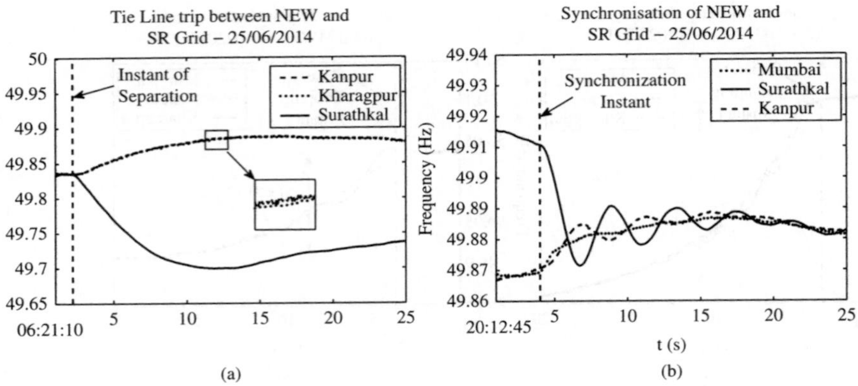

Figure 9.26 Separation and re-synchronization of two large grids [38]. (a) Separation of the grids due to tie-line tripping (b) Re-synchronization of the grids by reconnecting the tie-line.

References

1 Novosel, D., Madani, V., Bhargava, B. *et al.* (2008) Dawn of the grid synchronization. *IEEE Power and Energy Magazine*, **6** (1), 49–60.

2 Phadke, A.G. and Thorp, J.S. (2008) *Synchronized Phasor Measurements and Their Applications*, Springer.

3 Phadke, A.G. (2002) Synchronized phasor measurements – A historical overview, in *IEEE/PES Transmission and Distribution Conference and Exhibition*, Yokohama, pp. 476–479.

4 (2011) IEEE Standard for Synchrophasor Measurements for Power Systems. *IEEE Std C37. 118. 1-2011 (Revision of IEEE Std C37. 118-2005)*, pp. 1–61.

5 (2014) IEEE Standard for Synchrophasor Measurements for Power Systems – Amendment 1: Modification of Selected Performance Requirements. *IEEE Std C37. 118. 1a-2014 (Amendment to IEEE Std C37. 118. 1-2011)*, pp. 1–25.

6 Naduvathuparambil, B., Valenti, M.C., and Feliachi, A. (2002) Communication delays in wide area measurement systems, in *Proceedings of the Thirty-Fourth Southeastern Symposium on System Theory*, Huntsville, pp. 118–122.

7 Working Group C-14 of the System Protection Subcommittee, Power System Relaying Committee (2014) Use of synchrophasor measurements in protective relaying applications, in *67th Annual Conference for Protective Relay Engineers*, College Station, Texas, pp. 23–29.

8 Hauer, J.F. (1991) Application of Prony analysis to the determination of modal content and equivalent models for measured power system response. *IEEE Transactions on Power Systems*, **6** (3), 1062–1068.

9 Crow, M.L. and Singh, A. (2005) The matrix pencil for power system modal extraction. *IEEE Transactions on Power Systems*, **20** (1), 501–502.

10 Gajjar, G. and Soman, S.A. (2014) Auto detection of power system events using wide area frequency measurements, in *18th National Power System Conference*, Guwahati.

11 Zhou, N., Trudnowski, D., Pierre, J.W. *et al.* (2008) An algorithm for removing trends from power-system oscillation data, in *IEEE PES General Meeting*, Pittsburgh, pp. 1–7.
12 Trudnowski, D.J., Pierre, J.W., Zhou, N. *et al.* (2008) Performance of three mode-meter block-processing algorithms for automated dynamic stability assessment. *IEEE Transactions on Power Systems,* **23** (2), 680–690.
13 Trudnowski, D.J., Johnson, J.M., and Hauer, J.F. (1999) Making Prony analysis more accurate using multiple signals. *IEEE Transactions on Power Systems,* **14** (1), 226–231.
14 Pintelon, R. and Schoukens, J. (2012) *System Identification: A Frequency Domain Approach,* John Wiley & Sons, Hoboken, New Jersey, 2nd edn.
15 Oppenheim, A.V., Schafer, R.W., and Buck, J.R. (1999) *Discrete-time Signal Processing*, Prentice Hall, New Jersey, 2nd edn.
16 Gustavsen, B. and Semlyen, A. (1999) Rational approximation of frequency domain responses by vector fitting. *IEEE Transactions on Power Delivery,* **14** (3), 1052–1061.
17 Pierre, J.W., Zhou, N., Tuffner, F.K. *et al.* (2010) Probing signal design for power system identification. *IEEE Transactions on Power Systems,* **25** (2), 835–843.
18 Bendat, J.S. and Piersol, A.G. (2010) *Random Data: Analysis and Measurement Procedures*, John Wiley & Sons, Hoboken, New Jersey, 4th edn.
19 Hauer, J. and Cresap, R. (1981) Measurement and modeling of Pacific AC Intertie response to random load switching. *IEEE Transactions on Power Apparatus and Systems,* **PAS-100** (1), 353–359.
20 Pierre, J., Trudnowski, D., and Donnelly, M. (1997) Initial results in electromechanical mode identification from ambient data. *IEEE Transactions on Power Systems,* **12** (3), 1245–1251.
21 Wies, R.W., Pierre, J.W., and Trudnowski, D.J. (2004) Use of least mean squares (LMS) adaptive filtering technique for estimating low-frequency electromechanical modes in power systems, in *IEEE PES General Meeting*, Denver, pp. 1863–1870 Vol. 2.
22 Proakis, J.G. and Manolakis, D.G. (1996) *Digital Signal Processing: Principles, Algorithms and Applications,* Prentice Hall, New Jersey, 3rd edn.
23 Pradhan, V., Kulkarni, A.M., and Khaparde, S.A. (2018) A model-free approach for emergency damping control using wide area measurements. *IEEE Transactions on Power Systems,* **33** (5), 4902–4912.
24 Pierre, B.J., Wilches-Bernal, F., Schoenwald, D.A. *et al.* (2017) Open-loop testing results for the Pacific DC Intertie wide area damping controller, in *IEEE Manchester PowerTech,* Manchester, pp. 1–6.
25 Anderson, P.M. and LeReverend, B.K. (1996) Industry experience with special protection schemes. *IEEE Transactions on Power Systems,* **11** (3), 1166–1179.
26 Madani, V., Novosel, D., Horowitz, S. *et al.* (2010) IEEE PSRC report on global industry experiences with system integrity protection schemes (SIPS). *IEEE Transactions on Power Delivery,* **25** (4), 2143–2155.
27 Taylor, C.W., Nassief, F.R., and Cresap, R.L. (1981) Northwest power pool transient stability and load shedding controls for generation-load imbalances. *IEEE Transactions on Power Apparatus and Systems,* **PAS-100** (7), 3486–3495.

28 Begovic, M., Novosel, D., Karlsson, D. *et al.* (2005) Wide-area protection and emergency control. *Proceedings of the IEEE*, **93** (5), 876–891.

29 Centeno, V., Phadke, A.G., Edris, A. *et al.* (1997) An adaptive out-of-step relay. *IEEE Transactions on Power Delivery*, **12** (1), 61–71.

30 Padiyar, K.R. and Krishna, S. (2006) Online detection of loss of synchronism using energy function criterion. *IEEE Transactions on Power Delivery*, **21** (1), 46–55.

31 Lavand, S.A. and Soman, S.A. (2016) Predictive analytic to supervise zone 1 of distance relay using synchrophasors. *IEEE Transactions on Power Delivery*, **31** (4), 1844–1854.

32 Vu, K., Begovic, M.M., Novosel, D., and Saha, M.M. (1999) Use of local measurements to estimate voltage-stability margin. *IEEE Transactions on Power Systems*, **14** (3), 1029–1035.

33 Milosevic, B. and Begovic, M. (2003) Voltage-stability protection and control using a wide-area network of phasor measurements. *IEEE Transactions on Power Systems*, **18** (1), 121–127.

34 Huang, Z., Du, P., Kosterev, D., and Yang, S. (2013) Generator dynamic model validation and parameter calibration using phasor measurements at the point of connection. *IEEE Transactions on Power Systems*, **28** (2), 1939–1949.

35 Semlyen, A. (1974) Analysis of disturbance propagation in power systems based on a homogeneous dynamic model. *IEEE Transactions on Power Apparatus and Systems*, **PAS-93** (2), 676–684.

36 Parashar, M., Thorp, J.S., and Seyler, C.E. (2004) Continuum modeling of electromechanical dynamics in large-scale power systems. *IEEE Transactions on Circuits and Systems I: Regular Papers*, **51** (9), 1848–1858.

37 Zhong, Z., Xu, C., Billian, B.J. *et al.* (2005) Power system frequency monitoring network (FNET) implementation. *IEEE Transactions on Power Systems*, **20** (4), 1914–1921.

38 Salunkhe, K.A., Gajjar, G., Soman, S.A., and Kulkarni, A.M. (2014) Implementation and applications of a wide area frequency measurement system synchronized using network time protocol, in *IEEE PES General Meeting*, National Harbor, Maryland, pp. 1–5.

10

Analysis of Subsynchronous Resonance

10.1 Introduction

Chapters 6 and 7 primarily dealt with the stability of relative motion between the rotors of different synchronous generators (angular stability). These transients are much slower than the electro-magnetic transients associated with the electrical network and generator stator fluxes. For the study of these electro-mechanical transients, the fast transients may be neglected, that is, the differential equations associated with the stator fluxes and the states of the network (in the d-q frame) are converted into algebraic equations by neglecting the time-derivative terms of these quantities. In these slow-transient studies, the rotating system of the turbine-generator is assumed to be a single rigid body. Therefore, the inertia constant H considered in these studies is the cumulative inertia of the turbine-generator system.

Another class of dynamic phenomena in a power system involves the relatively faster transient interactions between (i) turbine-generator systems, (ii) the electrical network, and (iii) power-electronic systems and their controllers (HVDC, FACTS, renewable energy/storage interfaces, generator excitation systems). The interactions may involve all of these components together or any pair of them. Adverse interactions between these systems are generally manifested as poorly damped or unstable oscillations in the subsynchronous range of frequencies, but could be higher in some cases that involve electrical network–controller interactions. Note that these interactionsare at a higher frequency than the swing modes (0.2–2 Hz) studied in Chapter 6.

The adverse interactions were brought into focus by the occurrence of unanticipated generator-turbine shaft failures in the Mohave power plant in the western United States in 1970 and 1971 [1]. These failures were due to the interaction of a series-capacitor compensated transmission system with the torsional modes of the generator-turbine system. It was observed that if the subsynchronous natural frequency of the compensated electrical network (in the d-q frame) is near the natural frequency corresponding to a torsional mode, then this may lead to small-signal instability of that torsional mode. Hence this phenomenon is known as subsynchronous resonance (SSR). Adverse interaction between the generator electrical system and a series-compensated network may also lead to a small-signal instability phenomenon known as the induction generator effect [2, 3]. This phenomenon does not involve the generator-turbine-shaft dynamics. The controllers of power electronics systems could also potentially destabilize the torsional modes and/or the electrical network modes of a system. Instances of such

Dynamics and Control of Electric Transmission and Microgrids, First Edition.
K. R. Padiyar and Anil M. Kulkarni.

adverse interactions, known as subsynchronous torsional interactions (SSTIs) and subsynchronous control interactions (SSCIs), have been also been reported in the literature [4]. SSTIs and SSCIs can often be mitigated by changes in the controller design or the controller parameters. Troublesome torsional oscillations may also be "actively" damped out by auxiliary subsynchronous damping controllers (SSDCs), which modulate the firing angles of grid-connected power electronic systems in the vicinity of the generator.

The key modeling differences in the analysis of these phenomena and the relatively slower power-swing phenomenon considered in Chapter 6 are (i) the detailed dynamic representation of the electrical network elements and generator stator fluxes, (ii) the modeling of the generator-turbine mechanical system as a multiple-mass-shaft system, and (iii) modeling of the higher-bandwidth controllers of power-electronic systems. In this chapter, we first consider the nature of transients in an electrical network and the induction generator effect. Thereafter, the modeling of the generator-turbine system and the nature of torsional modes is described. The adverse interaction between the torsional system and the series-compensated electrical network is then brought out. A case study of a TCSC-based SSDC is also presented. Modeling techniques for power-electronic systems, including dynamic phasors and numerical scanning techniques, are also described.

10.2 Analysis of Electrical Network Dynamics

An electrical network consists of elements such as transformers, transmission lines, shunt/series-connected capacitors, and shunt reactors. For transients in the subsynchronous frequency range, a transformer is modeled as a series-connected leakage inductance. Transmission lines are modeled either as series R-L branches, or as one or more cascaded π circuits consisting of series R-L and shunt capacitor branches, depending on the transmission line length. Distributed parameter modeling is generally not necessary for transients in the subsynchronous range of frequencies.

In the study of slow transients such as power swings, an electrical network is assumed to be in the quasi-sinusoidal steady state, with the amplitude and phase determined by the dynamics of the slower-moving states such as the generator rotor angle and fluxes. In those studies, the voltages and currents in the synchronously rotating frame of reference are assumed to be related algebraically. This modeling is inadequate for the faster transients considered in this chapter, and a differential-equation model of the network is necessary.

Consider a short, symmetrical three-phase transmission line connected in series with capacitors, as shown in Figure 10.1. The three-phase transmission line is represented as an R-L branch with mutual coupling between the phases. The shunt capacitance of the line is neglected. The differential equations representing this system are:

$$\begin{bmatrix} \dfrac{di_L}{dt} \\ \dfrac{dv_C}{dt} \end{bmatrix} = [A] \begin{bmatrix} i_L \\ v_C \end{bmatrix} + [L]^{-1} \begin{bmatrix} v_S \\ [0] \end{bmatrix} \tag{10.1}$$

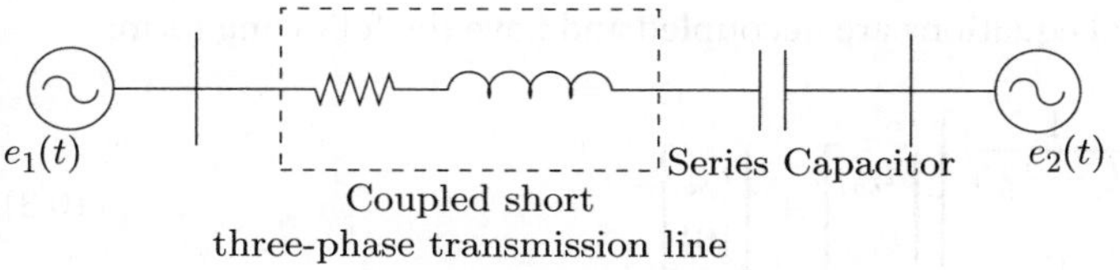

Figure 10.1 A series-capacitor compensated transmission line.

where

$$[A] = \begin{bmatrix} -[L]^{-1}[R] & -[L]^{-1} \\ [C]^{-1} & [0] \end{bmatrix}, \quad [L] = \begin{bmatrix} L_s & L_m & L_m \\ L_m & L_s & L_m \\ L_m & L_m & L_s \end{bmatrix},$$

$$[C] = \begin{bmatrix} C & 0 & 0 \\ 0 & C & 0 \\ 0 & 0 & C \end{bmatrix} \text{ and } [R] = \begin{bmatrix} R_s & R_m & R_m \\ R_m & R_s & R_m \\ R_m & R_m & R_s \end{bmatrix}$$

$i_L^t = \begin{bmatrix} i_{La} & i_{Lb} & i_{Lc} \end{bmatrix}$ represents the three-phase inductor currents.

$v_C^t = \begin{bmatrix} v_{Ca} & v_{Cb} & v_{Cc} \end{bmatrix}$ represents the three-phase capacitor voltages.

$v_S^t = \begin{bmatrix} v_{Sa} & v_{Sb} & v_{Sc} \end{bmatrix}$ represents the difference between the two voltage sources.

The eigenvalues of the matrix $[A]$, assuming that $R_s = R_m = 0$, are

$$\lambda_{1,2} = \pm j \frac{1}{\sqrt{(L_s - L_m)C}}$$

$$\lambda_{3,4} = \pm j \frac{1}{\sqrt{(L_s - L_m)C}}$$

$$\lambda_{5,6} = \pm j \frac{1}{\sqrt{(L_s + 2L_m)C}}$$

The eigenvalue pair $\lambda_{5,6}$ is exclusively associated with the zero sequence component of the phase quantities. This is readily evident if we transform the phase variables into Clarke's variables using Clarke's transformation, which was defined in Section 2.4:

$$\begin{bmatrix} f_a \\ f_b \\ f_c \end{bmatrix} = \begin{bmatrix} \sqrt{\frac{2}{3}} & 0 & \frac{1}{\sqrt{3}} \\ -\frac{1}{\sqrt{6}} & -\frac{1}{\sqrt{2}} & \frac{1}{\sqrt{3}} \\ -\frac{1}{\sqrt{6}} & \frac{1}{\sqrt{2}} & \frac{1}{\sqrt{3}} \end{bmatrix} \begin{bmatrix} f_\alpha \\ f_\beta \\ f_o \end{bmatrix} \tag{10.2}$$

The resulting differential equations are decoupled and have the following form:

$$\begin{bmatrix} \frac{di_{L\alpha}}{dt} \\ \frac{dv_{C\alpha}}{dt} \end{bmatrix} = \begin{bmatrix} 0 & -\frac{1}{(L_s - L_m)} \\ \frac{1}{C} & 0 \end{bmatrix} \begin{bmatrix} i_{L\alpha} \\ v_{C\alpha} \end{bmatrix} + \begin{bmatrix} v_{S\alpha} \\ [0] \end{bmatrix} \tag{10.3}$$

$$\begin{bmatrix} \frac{di_{L\beta}}{dt} \\ \frac{dv_{C\beta}}{dt} \end{bmatrix} = \begin{bmatrix} 0 & -\frac{1}{(L_s - L_m)} \\ \frac{1}{C} & 0 \end{bmatrix} \begin{bmatrix} i_{L\beta} \\ v_{C\beta} \end{bmatrix} + \begin{bmatrix} v_{S\beta} \\ [0] \end{bmatrix} \tag{10.4}$$

$$\begin{bmatrix} \frac{di_{Lo}}{dt} \\ \frac{dv_{Co}}{dt} \end{bmatrix} = \begin{bmatrix} 0 & -\frac{1}{(L_s + 2L_m)} \\ \frac{1}{C} & 0 \end{bmatrix} \begin{bmatrix} i_{Lo} \\ v_{Co} \end{bmatrix} + \begin{bmatrix} v_{So} \\ [0] \end{bmatrix} \tag{10.5}$$

10.2.1 Equations in *DQo* Variables

The equations may also be expressed in the *DQo* variables, obtained by applying Kron's transformation defined in Section 2.4 with $\theta = \omega_o t$, where ω_o is constant:

$$\begin{bmatrix} f_a \\ f_b \\ f_c \end{bmatrix} = \sqrt{\frac{2}{3}} \begin{bmatrix} \cos \omega_o t & \sin \omega_o t & \frac{1}{\sqrt{2}} \\ \cos\left(\omega_o t - \frac{2\pi}{3}\right) & \sin\left(\omega_o t - \frac{2\pi}{3}\right) & \frac{1}{\sqrt{2}} \\ \cos\left(\omega_o t + \frac{2\pi}{3}\right) & \sin\left(\omega_o t + \frac{2\pi}{3}\right) & \frac{1}{\sqrt{2}} \end{bmatrix} \begin{bmatrix} f_D \\ f_Q \\ f_o \end{bmatrix} \tag{10.6}$$

The equations of a symmetrical L-C network in the *DQo* variables are of the following form:

$$\frac{dx_{DQ}}{dt} = [A_{DQ}]x_{DQ} + [B_{DQ}]u_{DQ} \tag{10.7}$$

where

$$x_{DQ}^t = \left[i_{LD} \; i_{LQ} \; i_{Lo} \; v_{CD} \; v_{CQ} \; v_{Co} \right],$$

$$u_{DQ}^t = \left[v_{SD} \; v_{SQ} \; v_{So} \right],$$

$$[A_{DQ}] = \begin{bmatrix} 0 & -\omega_o & 0 & -\frac{1}{(L_s - L_m)} & 0 & 0 \\ \omega_o & 0 & 0 & 0 & -\frac{1}{(L_s - L_m)} & 0 \\ 0 & 0 & 0 & 0 & 0 & -\frac{1}{(L_s + 2L_m)} \\ \frac{1}{C} & 0 & 0 & 0 & -\omega_o & 0 \\ 0 & \frac{1}{C} & 0 & \omega_o & 0 & 0 \\ 0 & 0 & \frac{1}{C} & 0 & 0 & 0 \end{bmatrix},$$

$$[B_{DQ}]^t = \begin{bmatrix} 1 & 0 & 0 & 0 & 0 & 0 \\ 0 & 1 & 0 & 0 & 0 & 0 \\ 0 & 0 & 1 & 0 & 0 & 0 \end{bmatrix}$$

Remarks

1) The eigenvalues obtained from the differential equation formulations using phase variables and Clarke's variables are the same, as the variables are related by a time-invariant transformation. The eigenvalues obtained from the *DQo* formulation are

$$\lambda'_{1,2} = \pm j\omega_o + j\frac{1}{\sqrt{(L_s - L_m)C}}$$

$$\lambda'_{3,4} = \pm j\omega_o - j\frac{1}{\sqrt{(L_s - L_m)C}}$$

$$\lambda'_{5,6} = \pm j\frac{1}{\sqrt{(L_s + 2L_m)C}}$$

Note that $\lambda'_{1,2} = \lambda_1 \pm j\omega_o$ and $\lambda'_{3,4} = \lambda_2 \pm j\omega_o$. This is not surprising as Kron's transformation is a time-periodic transformation (with the frequency ω_o).

2) The differential equations of the R-L-C network are *time-invariant* when expressed in the phase variables, Clarke variables, as well as the *DQo* variables. For a symmetrical R-L-C circuit, Clarke's formulation leads to decoupled equations in the $\alpha\beta o$ variables. An advantage of the *DQo* formulation is that the differential equations of components such as synchronous generators are also time-invariant in the *DQo* variables. Therefore, this formulation is generally used for interfacing the generators with the electrical network.

3) $L_s - L_m$ is also the *positive sequence reactance* of the transmission line. If $\omega_o(L_s - L_m) > \frac{1}{\omega_o C}$ then $\frac{1}{\sqrt{(L_s - L_m)C}} < \omega_o$. Thus, a *subsynchronous* frequency network mode is

present in a transmission line whose (positive sequence) series reactance is partially compensated by the reactance of the series capacitor.

4) The zero-sequence variables (subscript o) are decoupled from the D-Q and α-β variables unless imbalances involving the neutral and ground occur. It should be noted that even within a network some of the zero-sequence variables are decoupled from the others due to star-delta transformer connections. In this chapter, we neglect the zero-sequence variables, as they are not generally associated with the subsynchronous instabilities that are considered here.
5) A network may also be represented by a matrix transfer function in the α-β and D-Q variables. Such a representation is useful for analytical studies as well as studies in the frequency domain. For the network considered in this section,

$$\begin{bmatrix} V_{S\alpha}(s) \\ V_{S\beta}(s) \end{bmatrix} = \begin{bmatrix} Z_\alpha(s) & 0 \\ 0 & Z_\beta(s) \end{bmatrix} \begin{bmatrix} I_{L\alpha}(s) \\ I_{L\beta}(s) \end{bmatrix} \tag{10.8}$$

where $Z(s) = Z_\alpha(s) = Z_\beta(s) = s(L_s - L_m) + \frac{1}{sC} = \frac{(L_s - L_m)C\, s^2 + 1}{sC}$. If the impedance matrix in the D-Q variables is represented as follows,

$$\begin{bmatrix} V_{SD}(s) \\ V_{SQ}(s) \end{bmatrix} = \begin{bmatrix} Z_{DD}(s) & Z_{DQ}(s) \\ Z_{QD}(s) & Z_{QQ}(s) \end{bmatrix} \begin{bmatrix} I_{LD}(s) \\ I_{LQ}(s) \end{bmatrix} \tag{10.9}$$

then, from the relationship between the α-β and the D-Q variables, we obtain the following relationship:

$$Z_{DD}(s) = Z_{QQ}(s) = \frac{Z(s + j\omega_o) + Z(s - j\omega_o)}{2} \tag{10.10}$$

$$Z_{DQ}(s) = -Z_{QD}(s) = \frac{Z(s + j\omega_o) - Z(s - j\omega_o)}{2j} \tag{10.11}$$

The derivation of these expressions is given in Appendix 10.A.

6) In larger networks, there are many network modes. If multiple series-compensated lines are present, several subsynchronous network modes may exist. If shunt capacitors are present, then supersynchronous network modes may also exist, as shown in the following subsection.
7) Power electronics controllers such as HVDC systems, FACTS, and grid-connected interfaces for renewable energy systems, which are also a part of the electrical network, have characteristics which are different from the passive R-L-C circuits associated with transmission lines, transformers, shunt reactors, and series capacitors. The modeling of these devices is taken up later in this chapter.

10.2.2 Interfacing a DQ Network Model with a Generator Model

The state-space model of a synchronous generator (excluding the dynamics of the mechanical system) can be written as follows:

$$\dot{X}_e = [A_e(\omega)]X_e + [B_{e1}]E_{fd} + [B_{e2}]\begin{bmatrix} v_d \\ v_q \end{bmatrix} \tag{10.12}$$

$$\begin{bmatrix} i_d \\ i_q \end{bmatrix} = [C_e]X_e \tag{10.13}$$

For Model (2.2) of the synchronous generator (see Chapter 3), X_e consists of six states: four states corresponding to the rotor field and damper windings, and two states corresponding to the stator fluxes. E_{fd} is the field voltage per unit (pu), which is assumed to be constant. Note that in this study, the differential equations corresponding to the stator fluxes (ψ_d and ψ_q) are *not* approximated by algebraic equations.

In this section, we focus on the electrical system of the generator only, therefore let us assume that the speed of the synchronous machine is constant and equal to ω_o. The generator currents and terminal voltages (i_d, i_q, v_d, v_q) are expressed in Park's reference frame (*d*-*q* variables) (see Section 2.4), which is tied to the generator rotor position $\theta = \omega_o t + \delta$. These variables need to be transformed to Kron's reference frame (*D*-*Q* variables), which is a reference frame rotating at a fixed speed ω_o. This facilitates interfacing of the generator equations with the network equations, which are expressed in the *D*-*Q* variables. The transformation between the *d*-*q* and *D*-*Q* variables is given by:

$$\begin{bmatrix} v_d \\ v_q \end{bmatrix} = P \begin{bmatrix} v_D \\ v_Q \end{bmatrix} \quad \text{and} \quad \begin{bmatrix} i_D \\ i_Q \end{bmatrix} = P^{-1} \begin{bmatrix} i_d \\ i_q \end{bmatrix} \tag{10.14}$$

where $P = \begin{bmatrix} \cos\delta & \sin\delta \\ -\sin\delta & \cos\delta \end{bmatrix}$. Note that δ is the rotor position of the generator relative to the synchronously rotating frame. The generator currents and voltages in the *D*-*Q* variables can now be interfaced to those of the network through Kirchoff's current and voltage laws. This is illustrated using two examples.

System 1: SMIB system with a capacitor at the generator terminals

Consider the system shown in Figure 10.2, which has a star-connected shunt capacitor at the generator terminals. The generator is connected to the infinite bus via a transformer and a transmission line. The equations of the network external to the generator are:

$$\frac{d}{dt}\begin{bmatrix} v_{CD} \\ v_{CQ} \end{bmatrix} = [A_C]\begin{bmatrix} v_{CD} \\ v_{CQ} \end{bmatrix} + \frac{1}{C_{sh}}\begin{bmatrix} i_{CD} \\ i_{CQ} \end{bmatrix} \tag{10.15}$$

$$\frac{d}{dt}\begin{bmatrix} i_{LD} \\ i_{LQ} \end{bmatrix} = [A_L]\begin{bmatrix} i_{LD} \\ i_{LQ} \end{bmatrix} + \frac{1}{L_{eq}}\left(\begin{bmatrix} v_{CD} \\ v_{CQ} \end{bmatrix} - \begin{bmatrix} e_D \\ e_Q \end{bmatrix}\right) \tag{10.16}$$

where

$$[A_L] = \begin{bmatrix} -\dfrac{R_L}{L_{eq}} & -\omega_o \\ \omega_o & -\dfrac{R_L}{L_{eq}} \end{bmatrix}$$

$$[A_C] = \begin{bmatrix} 0 & -\omega_o \\ \omega_o & 0 \end{bmatrix}$$

L_{eq} is the equivalent positive sequence series reactance of the network, that is, $L_{eq} = \frac{X_T+X_L+X_{sys}}{\omega_B}$, and $C_{sh} = \frac{b_c}{\omega_B}$. These equations are combined with the generator equations using the following relationships to yield the overall state-space model:

$$\begin{bmatrix} v_D \\ v_Q \end{bmatrix} = \begin{bmatrix} v_{CD} \\ v_{CQ} \end{bmatrix} \tag{10.17}$$

$$\begin{bmatrix} i_{CD} \\ i_{CQ} \end{bmatrix} = \begin{bmatrix} i_D \\ i_Q \end{bmatrix} - \begin{bmatrix} i_{LD} \\ i_{LQ} \end{bmatrix} \tag{10.18}$$

Example 10.1 ***Eigenvalue analysis of the system shown in Figure 10.2*** In the system shown in Figure 10.2, the generator is supplying 1.0 pu real power and its terminal voltage magnitude is 1.0 pu. The infinite bus voltage is $1.0\angle 0°$, that is, $e_D = 0.0$ and $e_Q = 1.0$ pu. The base speed $\omega_B = 2\pi \times 60$ rad/s. The generator speed is constant and equal to ω_B. The frequency of infinite bus ω_b is also equal to ω_B. The data for the system is adapted from [5] and given below:

Network

$R_L = 0.02$ pu, $X_T = 0.14$ pu, $X_L = 0.50$ pu, $X_{sys} = 0.06$ pu

Generator

The base frequency ω_B is $2\pi \times 60$ rad/s.

d-axis parameters

$x_d = 1.790$ pu, $x'_d = 0.169$ pu, $x''_d = 0.135$ pu,
$T'_{do} = 4.300$ s, $T''_{do} = 0.032$ s,
$T''_{dc} = 0.0259$ s

q-axis parameters:

$x_q = 1.710$ pu, $x'_q = 0.228$ pu, $x''_q = 0.200$ pu,
$T'_{qo} = 0.85$ s, $T''_{qo} = 0.050$ s

The reactances are in pu on the generator base (890 MVA) and the time-constants are in seconds. Generator resistance is assumed to be negligible. The shunt capacitor has a susceptance, $b_c = 0.112$ pu (equivalent to 100 MVAr shunt compensation at the generator terminals). Obtain the eigenvalues of the system if:

1) **Case 1:** The generator is represented as a constant voltage source behind a sub-transient reactance $x'' = 0.135$ pu.
2) **Case 2:** The generator is represented by the two-axis Model (2.2).
3) **Case 3:** The generator is represented by the two-axis Model (2.2), with modified q-axis time-constants: $T'_{q0} = 0.17$s, $T''_{q0} = 0.01$s.

In all these cases, the field voltage E_{fd} is assumed to be constant.

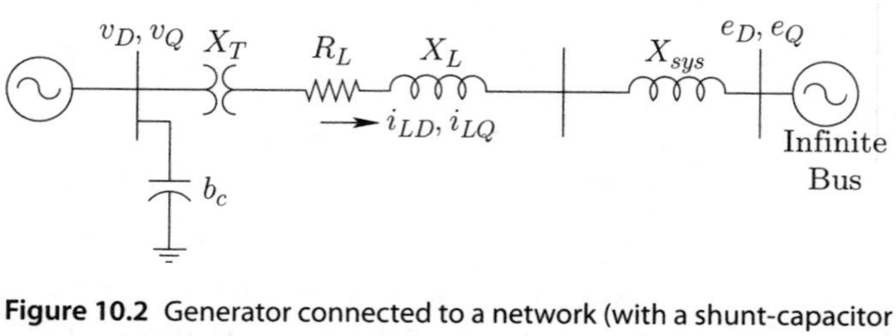

Figure 10.2 Generator connected to a network (with a shunt-capacitor at the generator terminals).

Solution

The system in this example is linear since the rotor speed ω and rotor angle δ are assumed to be constant. The state-space representation of the generator electrical system is obtained from the given parameter values using the operational-inductance approach given in [6]. The overall state matrix is obtained using the procedure outlined earlier in this section. The eigenvalues of the system for the different cases are shown in Table 10.1. The following inferences can be drawn from the results:

1) There are three network modes, two of which are *supersynchronous*. The damping of the supersynchronous modes is significantly affected by the rotor time-constants, but both modes are stable.
2) For the classical model, the supersynchronous modal frequencies are spaced apart by $2 \times \omega_b$, that is, $3725.5 - 2971.5 = 2 \times 377$ rad/s. The damping of the two supersynchronous modes is also the same for the classical model. This is only approximately true for the detailed model. This is because the d- and q-axis parameters of a generator are not identical for the detailed model (dynamic saliency).
3) The damping of the network mode with the frequency 377 rad/s is given by $\sigma \approx -\frac{R_L\omega_B}{x''+X_L+X_T+X_{sys}} = -9.02\ s^{-1}$. This mode is manifested during transients as an exponentially decaying DC-offset in the phase (abc) variables.

System 2: Generator connected to a series-compensated line

In the system shown in Figure 10.3, the differential equations corresponding to the series capacitor $C = \frac{1}{\omega_B X_C}$ are given by

$$\frac{d}{dt}\begin{bmatrix} v_{CD} \\ v_{CQ} \end{bmatrix} = [A_C]\begin{bmatrix} v_{CD} \\ v_{CQ} \end{bmatrix} + \frac{1}{C}\begin{bmatrix} i_{LD} \\ i_{LQ} \end{bmatrix} \tag{10.19}$$

The differential equations of the line current can be written down as follows:

$$\frac{d}{dt}\begin{bmatrix} i_{LD} \\ i_{LQ} \end{bmatrix} = [A_L]\begin{bmatrix} i_{LD} \\ i_{LQ} \end{bmatrix} + \frac{1}{L_{eq}}\left(\begin{bmatrix} v_D \\ v_Q \end{bmatrix} - \begin{bmatrix} e_{bD} \\ e_{bQ} \end{bmatrix} - \begin{bmatrix} v_{CD} \\ v_{CQ} \end{bmatrix}\right) \tag{10.20}$$

Table 10.1 Eigenvalues for the system of Figure 10.2.

Case 1	Case 2	Case 3	Remarks
$-0.871 \pm j3725.5$	$-4.552 \pm j3553.4$	$-7.270 \pm j3553.3$	Network Mode 1
$-0.871 \pm j2971.5$	$-5.431 \pm j2650.9$	$-18.121 \pm j2650.7$	Network Mode 2
$-9.030 \pm j376.99$	$-8.700 \pm j376.95$	$-8.695 \pm j376.85$	Network Mode 3
	-32.286	-101.06	
—	-20.209	-14.987	Rotor
	-2.996	-32.286	flux modes
	-0.642	-0.642	

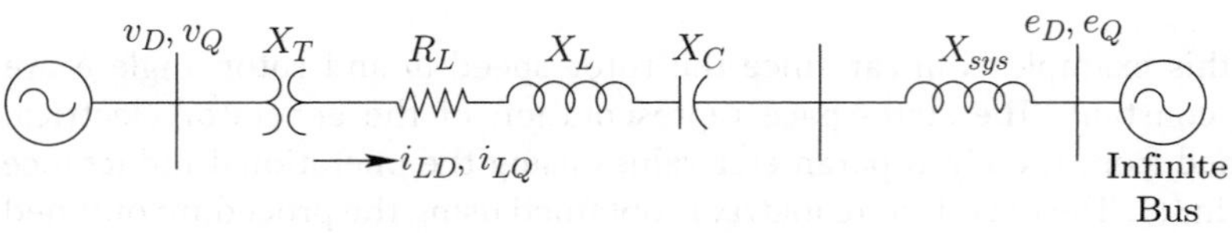

Figure 10.3 Generator connected to a series-compensated line.

However, these equations are redundant since the line current is equal to the generator current, which is already defined in terms of the fluxes (see (10.13)). Therefore, it is also true that

$$\frac{d}{dt}\begin{bmatrix} i_d \\ i_q \end{bmatrix} = [C_e]\dot{X}_e$$
$$= [C_e][A_e(\omega)]X_e + [C_e][B_{e1}]E_{fd} + [C_e][B_{e2}]\begin{bmatrix} v_d \\ v_q \end{bmatrix} \tag{10.21}$$

which yields

$$\frac{d}{dt}\begin{bmatrix} i_D \\ i_Q \end{bmatrix} = [P]^{-1}[C_e]\left([A_e(\omega)]X_e + [B_{e1}]E_{fd} + [B_{e2}]P\begin{bmatrix} v_D \\ v_Q \end{bmatrix}\right)$$
$$-[P]^{-1}[\dot{P}][P]^{-1}[C_e]X_e \tag{10.22}$$

Note that

$$-[P]^{-1}[\dot{P}] = \begin{bmatrix} 0 & -(\omega-\omega_o) \\ (\omega-\omega_o) & 0 \end{bmatrix}$$

Since in this section the speed of the generator ω is assumed to be equal to ω_o, this term is zero (but this is not true in general).

The states of this system can be chosen to be $\left[X_e^t \; v_{CD} \; v_{CQ}\right]$. v_d and v_q, which are required as inputs to (10.12), may be obtained by equating the right-hand sides of (10.20) and (10.22). The line currents i_{LD} and i_{LQ}, which are required for (10.19), are obtained from the generator currents, which in turn are related to the generator fluxes and the rotor angle δ.

In this example, we had to choose the state variables and remove the redundant differential equations, while this was not required in the earlier example. In general, the network topology has to be considered while choosing state variables. If the network has a loop consisting of only capacitors and voltage sources then not all of those capacitor voltages can be state variables. A similar situation arises with a cut-set having only inductors and current sources. In addition, if such a network also contains switches, then the state variables may have to be changed when there is a switching event.

When it comes to *simulating* electrical networks, the need to choose state variables can be avoided by using Dommel's methodology [7], which is incorporated in the widely used Electromagnetic Transient Program (EMTP). The use of the trapezoidal rule for numerical integration in EMTP may, however, cause spurious numerical oscillations during switching operations [8].

Example 10.2 ***Induction generator effect*** Consider the system shown in Figure 10.3. The data for the generator and transmission line is the same as in Example

10.1. Obtain the eigenvalues of the system for $X_c = 0.0$ pu and $X_c = 0.35$ pu, and for the three cases given in Example 10.1. The field voltage E_{fd} is assumed to be constant as before.

Solution

The eigenvalues of the system for all the cases are shown in Tables 10.2 and 10.3.

1) In the case where $X_C = 0$, we have a network mode at 377 rad/s. With $X_C = 0.35$ pu this gets transformed into two modes, one at a supersynchronous frequency and the other at a subsynchronous frequency.
2) The supersynchronous and subsynchronous frequencies add up exactly to $2\omega_o$ when the classical model is considered, and approximately to the same value when the detailed model is considered.
3) The damping of the subsynchronous mode is significantly affected by the q-axis time-constants. The mode becomes unstable when $T'_{q0} = 0.17$ s and $T''_{q0} = 0.01$ s (Case 3). Note that reduced q-axis time-constants are indicative of increased resistance of the rotor q-axis windings. The destabilization of the subsynchronous mode with series compensation is also called the induction generator effect.

An intuitive explanation for this instability can be given as follows. If a subsynchronous oscillation (which is present due to series capacitor compensated electrical network) is excited, then the synchronous machine has a negative slip with respect to this oscillation (since it is rotating at a higher frequency than this oscillation). For this oscillation, the rotor behaves like an induction machine running above the synchronous speed.

Table 10.2 Eigenvalues for the system of Figure 10.3 (with $X_c = 0.0$ pu).

Case 1	Case 2	Case 3	Remarks
$-9.03 \pm j376.99$	$-8.70 \pm j376.95$	$-8.69 \pm j376.85$	Network mode
	-32.36	-101.13	
—	-20.22	-15.58	Rotor
	-3.12	-32.36	flux modes
	-0.67	-0.67	

Table 10.3 Eigenvalues for the system of Figure 10.3 (with $X_C = 0.35$ pu).

Case 1	Case 2	Case 3	Remarks
$-4.51 \pm j621.02$	$-4.71 \pm j616.57$	$-5.51 \pm j616.51$	Network Mode 1
$-4.51 \pm j132.96$	$-2.77 \pm j137.41$	$0.56 \pm j137.97$	Network Mode 2
	-33.14	-101.62	
—	-20.39	-33.13	Rotor
	-4.32	-21.52	flux modes
	-0.97	-0.97	

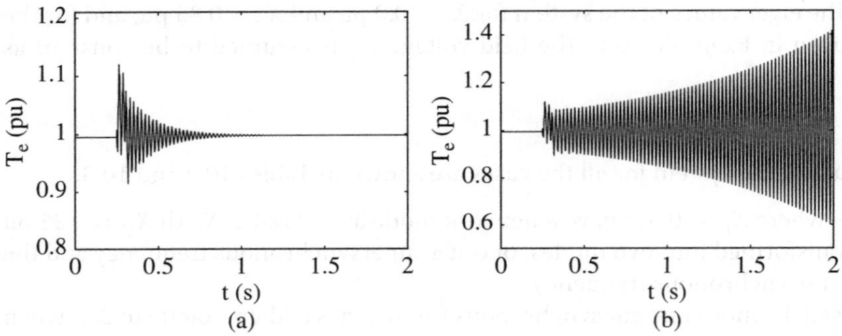

Figure 10.4 Response of the electrical torque to a pulse change in E_b ($X_C = 0.35$ pu). (a) Case 2 (b) Case 3.

Therefore the apparent resistance of the machine at its terminals (at this oscillation frequency) can become negative. If this resistance is greater than the resistance offered by the network, then instability is seen. Note that this explanation neglects the fact that the synchronous machine has dynamic saliency, unlike an induction machine.

It is important to note that the induction generator effect does not involve the mechanical dynamics (generator speed was assumed to be constant in the analysis), that is, it is purely an electrical effect. The instability predicted by eigenvalue analysis is corroborated by the results of a simulation study. The system, which is initially in steady state, is subjected to a 0.1 s positive pulse of 5% in the magnitude of E_b. Case 2 and Case 3 are contrasted in Figure 10.4. As expected, Case 3 exhibits small-signal instability.

Countermeasures for the Induction Generator Effect

Pole-face amortisseur windings can be added to reduce the net negative resistance of a generator at subsynchronous frequencies, thereby reducing the possibility of this instability. It is relatively inexpensive to install pole-face amortisseur windings in new machines, but retrofitting on old machines is not feasible. A bypass damping filter connected in parallel across the series capacitor in each phase (see Figure 10.5) can be useful for countering the induction generator effect. A resistor is connected in series with a parallel combination of a reactor and capacitor which is tuned at the system frequency. Thus the filter has a very high impedance at the system frequency, and the power losses in the resistor are limited under normal conditions. The damping resistor becomes effective at subsynchronous frequencies, and it can introduce significant positive resistance in the circuit for subsynchronous oscillation frequencies.

The induction generator effect is an example of network mode instability. In general, the damping of network modes may also be affected by the controllers of FACTS, HVDC, and other power-electronic systems that are present in the vicinity. The

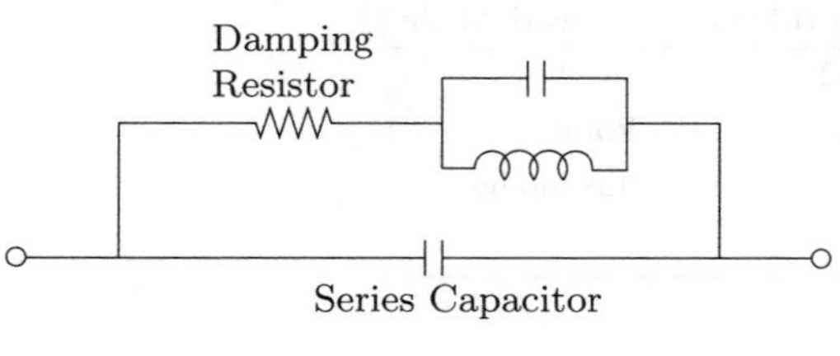

Figure 10.5 Bypass damping filter.

controllers of these systems can be designed to mitigate small signal instabilities such as the induction generator effect. Sometimes the controllers may themselves (inadvertently) contribute to the instability of a network mode at sub- or supersynchronous frequencies. This is usually due to a complex interaction of the power electronic converter, the controller, and passive elements of the electrical network (transmission lines, transformer, and filters). Well-known examples are the interaction near the second harmonic frequency observed in a HVDC link [9–11] and the SSCIs between a type-3 wind turbine-generator system and a series-compensated network [4]. Such instabilities can be alleviated by modifications in the controller and/or its parameters.

10.3 Torsional Dynamics of a Generator-Turbine System

Consider a turbine-generator system modeled as two lumped masses corresponding to the turbine and generator, and connected by an elastic shaft, as shown in Figure 10.6.

The equations which describe this system are

$$\frac{d\delta_T}{dt} = (\omega_T - \omega_o) \tag{10.23}$$

$$\frac{d\delta}{dt} = (\omega - \omega_o) \tag{10.24}$$

$$M_T \frac{d(\omega_T - \omega_o)}{dt} = T_m - K(\delta_T - \delta) \tag{10.25}$$

$$M_G \frac{d(\omega - \omega_o)}{dt} = -T_e - K(\delta - \delta_T) \tag{10.26}$$

where

M_T : inertia of turbine (kg-m^2)
M_G : inertia of generator (kg-m^2)
K : stiffness constant (N-m/rad)
T_m : mechanical torque (N-m)
T_e : electrical torque (N-m)
δ : angular position of the generator with respect to a synchronous rotating frame of frequency ω_o.
δ_T : angular position of the turbine with respect to a synchronous rotating frame of frequency ω_o
ω : generator speed (rad/s)
ω_T : turbine speed (rad/s).

Figure 10.6 Two rotating masses connected by a shaft.

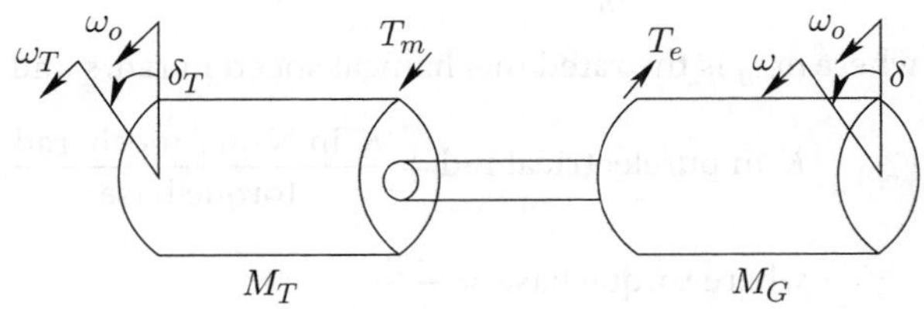

If the torques are expressed in pu, then these equations can be rewritten as follows:

$$\frac{d\delta_T}{dt} = (\omega_T - \omega_o) \tag{10.27}$$

$$\frac{d\delta}{dt} = (\omega - \omega_o) \tag{10.28}$$

$$\frac{2H_T}{\omega_B}\frac{d(\omega_T - \omega_o)}{dt} = T_m - K(\delta_T - \delta) \tag{10.29}$$

$$\frac{2H_G}{\omega_B}\frac{d(\omega - \omega_o)}{dt} = -T_e - K(\delta - \delta_T) \tag{10.30}$$

These equations can be written in the following form:

$$[E_1]\dot{x} = [A_1]x + u \tag{10.31}$$

where

$$[E_1] = \begin{bmatrix} 1 & 0 & 0 & 0 \\ 0 & 1 & 0 & 0 \\ 0 & 0 & \frac{2H_T}{\omega_B} & 0 \\ 0 & 0 & 0 & \frac{2H_G}{\omega_B} \end{bmatrix}, \quad [A_1] = \begin{bmatrix} 0 & 0 & 1 & 0 \\ 0 & 0 & 0 & 1 \\ -K & K & 0 & 0 \\ K & -K & 0 & 0 \end{bmatrix}$$

$$x = \left[\delta_T \ \delta \ (\omega_T - \omega_o) \ (\omega - \omega_o)\right]^t$$

$$u = \left[0 \ 0 \ T_m \ -T_e\right]^t$$

H_T : inertia of turbine (MJ/MVA)
H_G : inertia of generator (MJ/MVA)
K : stiffness constant (pu/electrical rad)
ω_B : base electrical frequency (rad/s)

The angular speeds ω and ω_T are expressed in electrical rad/s and δ and δ_T are expressed in electrical radians.
Note that

$$H_T = \frac{\frac{1}{2}M_T\,\omega_{mB}^2}{S_B}, \qquad H_G = \frac{\frac{1}{2}M_G\,\omega_{mB}^2}{S_B}$$

where ω_{mB} is the rated mechanical speed in rad/s and S_B is the volt-ampere base.

$$K \text{ in pu/electrical rad} = \frac{K \text{ in N-m / mech. rad}}{\text{torque base}} \times \frac{\omega_{mB}}{\omega_B}$$

$$\text{where torque base} = \frac{S_B}{\omega_{mB}}$$

If T_m and T_e are treated as external inputs, then the natural frequencies of the mechanical system are obtained from the eigenvalues of the following matrix:

$$[E_1]^{-1}[A_1] = \begin{bmatrix} 1 & 0 & 0 & 0 \\ 0 & 1 & 0 & 0 \\ 0 & 0 & \frac{2H_T}{\omega_B} & 0 \\ 0 & 0 & 0 & \frac{2H_G}{\omega_B} \end{bmatrix}^{-1} \begin{bmatrix} 0 & 0 & 1 & 0 \\ 0 & 0 & 0 & 1 \\ -K & K & 0 & 0 \\ K & -K & 0 & 0 \end{bmatrix}$$

The eigenvalues of the state matrix are

$$0,\ 0,\ \pm j\sqrt{\frac{K\omega_B}{2H_{eq}}}$$

where $H_{eq} = \frac{H_T H_G}{H_T + H_G}$.

The zero eigenvalues and the complex pair of eigenvalues are associated with the common mode and the torsional mode, respectively.

The Torsional Mode

The complex conjugate pair of eigenvalues is associated with torsional oscillations (i.e., the relative motion between the generator and the turbine masses). The relative motion is highlighted by examining the dynamics of the angle and frequency difference, as given below:

$$\frac{d(\delta_{diff})}{dt} = \omega_{diff} \tag{10.32}$$

$$\frac{2H_{eq}}{\omega_B}\frac{d(\omega_{diff})}{dt} = T_{eq} - K(\delta_{diff}) \tag{10.33}$$

where $\delta_{diff} = (\delta_T - \delta)$, $\omega_{diff} = (\omega_T - \omega)$, and $T_{eq} = \frac{H_G}{H_T + H_G}T_m + \frac{H_T}{H_T + H_G}T_e$.

The relative motion between the generator and turbine, which is oscillatory, causes cyclic torsional stress on the shaft. If the magnitude of the oscillations exceeds a certain value, which depends on the shaft material, then the shaft experiences fatigue. The shaft may fail (crack) if the cumulative number of cycles of the oscillation exceeds a certain limit, which is determined by the amplitude of oscillations. This is depicted by the S-N curve shown in Figure 10.7.

Large disturbances such as faults, auto-reclosure of circuit breakers, bypass and re-insertion of series capacitors, and loss of synchronism may cause large *transient shaft torques*, leading to some loss of life of the shaft. In addition, small-signal instabilities of the torsional mode may cause oscillation amplitudes to grow to large values even if the initiating disturbances are small. Such instabilities may be caused by adverse interactions with the electrical network. This will be discussed in the next section.

The Common Mode

The zero eigenvalues are associated with the common mode, which is highlighted by adding the differential equations corresponding to the turbine and generator, as

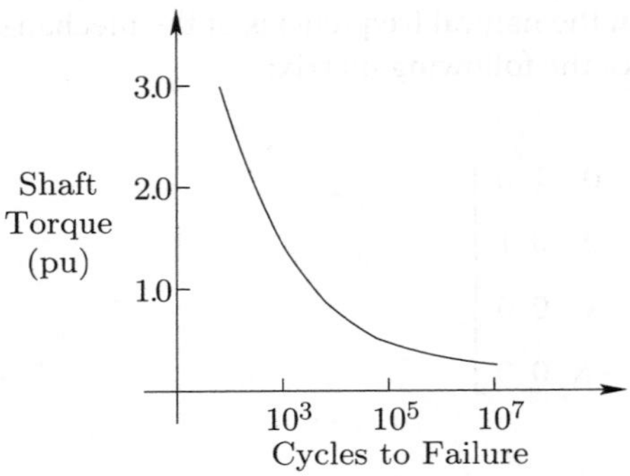

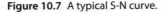

Figure 10.7 A typical S-N curve.

given below:

$$\frac{d(\delta_{com})}{dt} = \omega_{com} - \omega_o \tag{10.34}$$

$$\frac{2H_{com}}{\omega_B}\frac{d(\omega_{com} - \omega_o)}{dt} = T_m - T_e \tag{10.35}$$

where

$$\delta_{com} = \frac{(H_T\,\delta_T + H_G\,\delta_G)}{H_T + H_G}$$

$$\omega_{com} = \frac{(H_T\,\omega_T + H_G\,\omega_G)}{H_T + H_G}$$

The inertia H_{com} associated with this mode is given by $H_{com} = H_T + H_G$. The common mode is associated with the zero eigenvalues in the analysis of this section only because the electrical torque is considered as an "input" in the equation. In reality, the torque is itself affected by the relative rotor positions of different generators. The common modes of different turbine-generators get coupled to each other through their electrical torques, which are dependent on the relative position of the generator angles. This coupling gives rise to low-frequency oscillations (see Chapter 6), the loss of synchronism phenomenon (see Chapter 7), and the centre-of-inertia movement of the system frequency (see Chapter 6).

Since the dynamics of the common mode of the generator-turbine is generally much slower compared to the torsional oscillations, it is possible to study the common mode and torsional modes in a decoupled fashion without a significant loss of accuracy. In the decoupled study of the common mode, the torsional oscillations may be neglected, which means that $\omega = \omega_T$ and $\delta = \delta_T$. The common mode equations then reduce to the "swing equations" studied in Chapters 6 and 7. Therefore, one does not have to unlearn the earlier analyses of power swings, loss of synchronism, and system centre-of-inertia movement. These are simply the decoupled analyses of the common mode of the turbine-generator system and are expected to give reasonably accurate results due to the different timescales associated with the common and torsional modes.

Multi-Turbine Systems

In many cases, various turbine sections, such as high pressure (HP), intermediate pressure (IP), and low pressure (LP), may have to be modeled as separate lumped masses, which are coupled with shafts. A rotating excitation system, if present, may also be modeled as a separate lumped mass. For an n-mass turbine-generator system, $[H]$ and $[K]$ are $n \times n$ inertia and spring constant matrices, respectively.

The state equations are given by

$$\begin{bmatrix} [I] & [0] \\ [0] & \frac{2}{\omega_B}[H] \end{bmatrix} \frac{d}{dt} \begin{bmatrix} \delta_{TG} \\ (\omega_{TG} - \omega_o) \end{bmatrix} = \begin{bmatrix} [0] & [I] \\ -[K] & [0] \end{bmatrix} \begin{bmatrix} \delta_{TG} \\ (\omega_{TG} - \omega_o) \end{bmatrix} + \begin{bmatrix} [0] \\ e_m \end{bmatrix} T_m - \begin{bmatrix} [0] \\ e \end{bmatrix} T_e \tag{10.36}$$

where $[I]$ and $[0]$ are the identity and zero matrices, respectively, of appropriate dimensions. δ_{TG} and ω_{TG} are the $n \times 1$ vectors of rotor angles and speeds of the rotating masses including the generator. T_m is the total mechanical torque applied at the turbines. e_m is a $n \times 1$ column vector which contains as its elements the fraction of the total mechanical torque applied at each turbine. The rows of e_m corresponding to the masses other than the turbine masses have entries equal to zero. The sum of entries of e_m is 1. e is a $n \times 1$ column vector of zeros, except for an entry corresponding to the generator mass, which is 1. The equation (10.36) may be rewritten as

$$\frac{2}{\omega_B}[H]\frac{d^2\delta_{TG}}{dt^2} = -[K]\delta_{TG} + [e_m]T_m - [e]T_e \tag{10.37}$$

If μ_i denotes an eigenvalue of $\frac{\omega_B}{2}[H]^{-1}[K]$, then this corresponds to a conjugate pair of eigenvalues (λ_i, λ_i^*) of the state matrix $\begin{bmatrix} [I] & [0] \\ [0] & \frac{2}{\omega_B}[H] \end{bmatrix}^{-1} \begin{bmatrix} [0] & [I] \\ -[K] & [0] \end{bmatrix}$ as obtained from (10.36). These eigenvalues are given by

$$(\lambda_i, \lambda_i^*) = \pm j\sqrt{\mu_i} \tag{10.38}$$

One pair of eigenvalues (0, 0) is because $[K]$ is singular. This pair corresponds to the common mode. The other eigenvalues are complex conjugate pairs and correspond to the $n-1$ torsional modes.

The mode-shape v_i of the ith mode is the right eigenvector of the matrix $\frac{\omega_B}{2}[H]^{-1}[K]$, corresponding to that mode. The mode shape indicates the relative observability of the mode in the speeds of the rotating masses.

Note that the angle (δ) and speed (ω) of the generator mass are given by $\delta = e^t\,\delta_{TG}$ and $\omega = e^t\,\omega_{TG}$, respectively. Therefore, the transfer function $\frac{\delta(s)}{T_e(s)}$ is given by:

$$\frac{\delta(s)}{T_e(s)} = -e^t\left(\frac{2[H]}{\omega_B}s^2 + K\right)^{-1} e \tag{10.39}$$

Example 10.3 ***Torsional Modes of a Turbine-Generator System*** Consider the generator-turbine system shown in Figure 10.8. There are two low-pressure turbines, namely LPA and LPB. The rotating excitation system is denoted EXC. The data for this example have been taken from [5] and are given in the following tables.

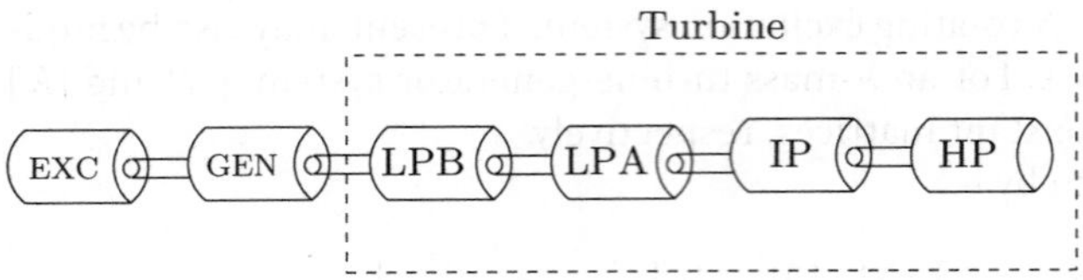

Figure 10.8 Multi-mass system.

Inertia	HP	IP	LPA	LPB	GEN	EXC
(MJ/MVA)	0.092897	0.155589	0.858670	0.884215	0.868495	0.0342165

Obtain the eigenvalues of the multi-mass system ($\omega_B = 2\pi \times 60$ rad/s).

Stiffness constant	HP-IP	IP-LPA	LPA-LPB	LPB-GEN	GEN-EXC
(pu torque/elect. rad)	19.303	34.929	52.038	70.858	2.822

Solution

The states are:

$$X_m = [\delta_{HP} \;\; \delta_{IP} \;\; \delta_{LPA} \;\; \delta_{LPB} \;\; \delta \;\; \delta_{EXC} \;\; \omega_{HP} \;\; \omega_{IP} \;\; \omega_{LPA} \;\; \omega_{LPB} \;\; \omega \;\; \omega_{EXC}]^t$$

The inertia and stiffness matrices are given by:

$$[H] = \begin{bmatrix} 0.093 & 0 & 0 & 0 & 0 & 0 \\ 0 & 0.156 & 0 & 0 & 0 & 0 \\ 0 & 0 & 0.859 & 0 & 0 & 0 \\ 0 & 0 & 0 & 0.884 & 0 & 0 \\ 0 & 0 & 0 & 0 & 0.869 & 0 \\ 0 & 0 & 0 & 0 & 0 & 0.0342 \end{bmatrix}$$

$$[K] = \begin{bmatrix} 19.30 & -19.30 & 0 & 0 & 0 & 0 \\ -19.30 & 54.23 & -34.93 & 0 & 0 & 0 \\ 0 & -34.93 & 86.97 & -52.04 & 0 & 0 \\ 0 & 0 & -52.04 & 122.90 & -70.86 & 0 \\ 0 & 0 & 0 & -70.86 & 73.68 & -2.82 \\ 0 & 0 & 0 & 0 & -2.82 & 2.82 \end{bmatrix}$$

Since there are six masses, there are five torsional modes and one common mode. The eigenvalues corresponding to the torsional mode are given in Table 10.4, while the mode shapes (the components of v_i) are given in Table 10.5. The torsional mode shapes indicate that Torsional Mode 5 has very low observability in the generator speeds relative to the turbine masses. It is associated with the HP and IP turbines swinging against each other with low participation of the other rotating masses. Torsional Mode 1 is the lowest frequency torsional mode which involves the HP, IP, and LPA turbine swinging against

Table 10.4 Eigenvalues corresponding to the torsional modes and the common mode.

Mode	Eigenvalue (λ_i)
Torsional Mode 1	$\pm j98.72$
Torsional Mode 2	$\pm j126.99$
Torsional Mode 3	$\pm j160.52$
Torsional Mode 4	$\pm j202.85$
Torsional Mode 5	$\pm j298.18$
Common Mode	(0,0)

Table 10.5 Mode shapes of the torsional modes and the common mode.

Rotor mass	Torsional Mode 1	Torsional Mode 2	Torsional Mode 3	Torsional Mode 4	Torsional Mode 5	Common Mode
HP	−2.083	− 2.942	6.025	−1.392	176.519	1.0
IP	−1.564	− 1.731	2.062	0.070	−224.176	1.0
LPA	−0.918	− 0.402	−1.384	0.810	25.394	1.0
LPB	0.299	1.057	−0.575	−1.612	− 4.733	1.0
GEN	1.000	1.000	1.000	1.000	1.000	1.0
EXC	2.680	−26.771	−1.521	−0.607	− 0.212	1.0

the rest of the rotating masses. The common mode is observable to the same extent in all the generator speeds.

10.3.1 Damping of Torsional Oscillations

Damping has not been considered in the analysis of the turbine-generator system so far. As a result, all the eigenvalues (λ) of the turbine-generator system have zero real parts. In reality, damping is present due to several factors, which are explained below.

1) Damping due to factors other than the external electrical network. This consists of two components:
   (a) Mechanical damping, which is due to components such as steam forces on turbine blades, windage, fretting, and material hysteresis.
   (b) Damping torque of electro-magnetic origin due to rotating excitation systems, and eddy current losses in the generator stator laminations due to flux rotation.

   While these factors contribute to damping when the generator is synchronized with the external electrical network, they are also present when the generator is not connected to the external network. The damping from these sources is generally minimum during no-load conditions. The damping characteristics due to these factors should be obtained from field tests, as it is difficult to obtain them theoretically. In [12] it is mentioned that "in the absence of test data from similar units, an estimated value for no-load damping for modes not involving exciter dynamics will range between 0.02–0.05 rad/sec. For modes involving exciter dynamics, values of 0.05–0.1 rad/sec are more indicative." These values are applicable to steam-turbine-generator units.

2) Damping due to connection with the external electrical network. The electrical torque T_e of the generator, when it is connected to the external electrical network, also affects the damping of torsional modes. This component of damping is present only when the generator is synchronized to the external system. The damping is affected by the nature of elements connected in the network, such as series capacitors as well as controllers of power-electronic actuators. Importantly, the damping due to the external network can be negative in some situations. This aspect will be discussed in detail shortly.

10.3.2 Sensitivity of the Torsional Modes to the External Electrical System

When the generator is connected to an electrical network, the generator speed ω and angle δ affect the electrical torque, which in turn affects the mechanical system through (10.36). Therefore, the feedback loop is closed as shown in Figure 10.9. The electrical system is described by the equations given in Section 10.2.2. Note that δ and ω are not constant in this study and are obtained from equations of the mechanical system. The electrical torque is a nonlinear function of the fluxes and currents. δ and ω appear in the state equations of the electrical system in a manner that causes additional nonlinearities. Therefore, the overall electro-mechanical system is nonlinear, and the combined system that is shown in Figure 10.9 is obtained by linearizing the system around an equilibrium point. ΔT_m is assumed to be zero.

The feedback system shown in Figure 10.9b can be described by the relation $\Delta T_e = \kappa\Delta\delta$. Furthermore, using $\Delta\delta = e^t\Delta\delta_{TG}$, we obtain $\Delta T_e = \kappa e^t\Delta\delta_{TG}$. Substituting for ΔT_e in (10.37), we obtain $\frac{2}{\omega_B}[H]\frac{d^2\Delta\delta_{TG}}{dt^2} = -([K] + \kappa ee^t)\Delta\delta_{TG}$. Therefore, κee^t represents a perturbation to the original matrix $[K]$. The sensitivity μ_i, which is the eigenvalue of $\frac{\omega_B}{2}[H]^{-1}[K]$, to κ is therefore given by [13]

$$\frac{d\mu_i}{d\kappa} = u_i^t\,\frac{d}{d\kappa}\left(\frac{\omega_B}{2}[H]^{-1}([K] + \kappa ee^t)\right)\,v_i = u_i^t\,\frac{\omega_B}{2}[H]^{-1}ee^t\,v_i$$

where u_i^t and v_i are the normalized left and right eigenvectors of $\frac{\omega_B}{2}[H]^{-1}[K]$ corresponding to the eigenvalue μ_i (i.e., $u_i^t v_i = 1$). The relation between μ_i and the ith eigenvalue of the mechanical system, λ_i, is given by $\lambda_i^2 = -\mu_i$. Therefore, the sensitivity of λ_i

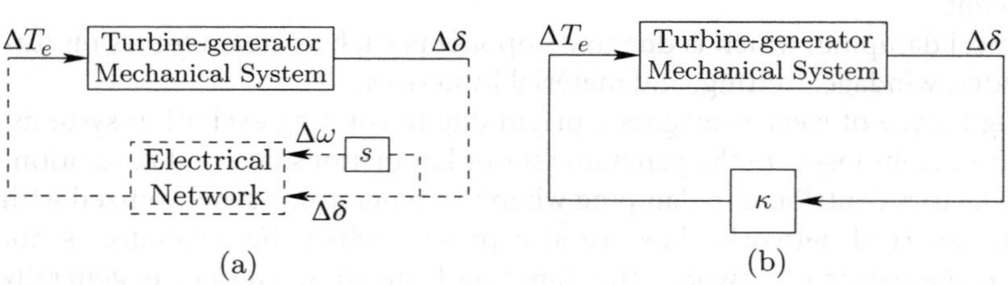

Figure 10.9 Combined system (linearized around an equilibrium point). (a) Interface between the electrical and mechanical systems (b) System for evaluation of sensitivity of torsional modes to the external network.

to κ (the external electrical system) is given by:

$$\frac{d\lambda_i}{d\kappa} = -\frac{1}{2\lambda_i}\left(u_i^t \frac{\omega_B}{2}[H]^{-1} e\, e^t\right) v_i = -\frac{1}{2\lambda_i} \times \frac{\omega_B}{2\,H_{mi}} \tag{10.40}$$

where $H_{mi} = \frac{1}{u_i^t[H]^{-1}ee^t v_i}$ is called the *modal inertia* of the mechanical system corresponding to the ith torsional mode. A higher modal inertia results in lower sensitivity of the corresponding torsional mode to the external electrical network. Note that $\frac{d\lambda_i}{d\kappa}$ is also the residue of the transfer function of the mechanical system $\frac{\delta(s)}{T_e(s)}$ as given in (10.39), corresponding to the eigenvalue λ_i.

The sensitivity expression captures the potential effect of the electrical network on the mechanical modes of the system when the generator is connected to the electrical network. For the two-mass system discussed earlier in this section

$$V = [v_1\ v_2] = \begin{bmatrix} 1 & \dfrac{H_G}{H_T + H_G} \\ -1 & -\dfrac{H_T}{H_T + H_G} \end{bmatrix}$$

$$V^{-1} = \begin{bmatrix} u_1^t \\ u_2^t \end{bmatrix} = \begin{bmatrix} \dfrac{H_T}{H_T + H_G} & \dfrac{H_G}{H_T + H_G} \\ 1 & -1 \end{bmatrix}$$

where v_1 and u_1 correspond to the common mode and v_2 and u_2 correspond to the torsional mode. The sensitivity of $\lambda_2 = +j\sqrt{\frac{K\omega_B}{2H_{eq}}}$, which is an eigenvalue corresponding to the torsional mode, is given by:

$$\frac{d\lambda_2}{d\kappa} = -\frac{1}{2\lambda_2} u_2^t \frac{\omega_B}{2}[H]^{-1} e\, e^t\, v_2$$

$$= -\frac{1}{2\lambda_2}[1\ -1]\begin{bmatrix} \dfrac{\omega_B}{2H_T} & 0 \\ 0 & \dfrac{\omega_B}{2H_G} \end{bmatrix}\begin{bmatrix} 0 \\ 1 \end{bmatrix}[0\ 1]\begin{bmatrix} \dfrac{H_G}{H_T + H_G} \\ -\dfrac{H_T}{H_T + H_G} \end{bmatrix}$$

$$= -\frac{1}{2\lambda_2}\frac{\omega_B}{2}\frac{\frac{H_T}{H_G}}{H_T + H_G} \tag{10.41}$$

In hydro turbine-generator systems, the inertia of the generator is generally much larger than the inertia of the turbine; $\frac{H_G}{H_T}$ varies from 7 to 30 for hydro units with Kaplan turbines, while for Francis and Pelton wheel turbines it varies from 10 to 40 [14]. This is unlike steam turbine-generator systems, where the generator inertia is generally smaller than the overall inertia of the turbine. Therefore, the sensitivity of a torsional mode to the electrical network is much lower in hydro-turbines. The mechanical damping of hydro-turbines is also larger than that in steam-turbines. Therefore, adverse network-torsional interactions have not been observed for hydro-turbines.

10.4 Generator-Turbine and Network Interactions: Subsynchronous Resonance

Generator-turbine and network interactions can be studied by performing eigenvalue analysis of the combined electro-mechanical system. The equations of the mechanical system and the electrical network are interfaced as shown in Figure 10.9a. The state-space equations of the electrical subsystem (generator and network) are formulated as explained in Section 10.2.2. The electrical torque T_e and (δ, ω) are the interface variables between the two subsystems.

The study of interactions of the torsional system and the electrical network is of interest because of the possibility of the small-signal instability of torsional modes when a generator is connected to a series-capacitor compensated line. This is illustrated using the following example.

Example 10.4 ***Subsynchronous Resonance*** Consider the system shown in Figure 10.10. The data for the electrical system is the same as in Example 10.2 (Case 2), while the data for the mechanical system is the same as in Example 10.3. The generator terminal voltage magnitude and the infinite bus voltage magnitude E_b are both 1.0 pu. ω_o is the frequency of the infinite bus. The field voltage E_{fd} is assumed to be constant (AVR is not active). Damping sources other than the external electrical network are neglected.

1) Compute the eigenvalues of the linearized system for $X_C = 0.2, 0.3, 0.35$, and 0.45 pu, and for two values of quiescent real power output of the generator, 0.0 pu and 1.0 pu.
2) Simulate the response of the system when the series capacitor, which is initially bypassed, is inserted at $t = 0.1$ s. The mechanical torque applied to the turbine masses (HP, IP, LPA, and LPB) are 30%, 26%, 22%, and 22% of the total mechanical torque (1.0 pu), respectively, that is, $e_m^t = [0.3\ \ 0.26\ \ 0.22\ \ 0.22\ \ 0\ \ 0]$.
3) The concept of synchronizing and damping torque analysis was discussed in Section 6.4. Evaluate the transfer function $\frac{\Delta T_e(j\Omega)}{\Delta\delta(j\Omega)}$ of the electrical subsystem in the subsynchronous range of frequencies. Plot the damping torque (T_{De}) versus the frequency Ω, which is given by:

$$T_{De}(\Omega) = \Re\left(\frac{\Delta T_e(j\Omega)}{\Delta\omega(j\Omega)}\right) = \frac{1}{\Omega}\Im\left(\frac{\Delta T_e(j\Omega)}{\Delta\delta(j\Omega)}\right) \tag{10.42}$$

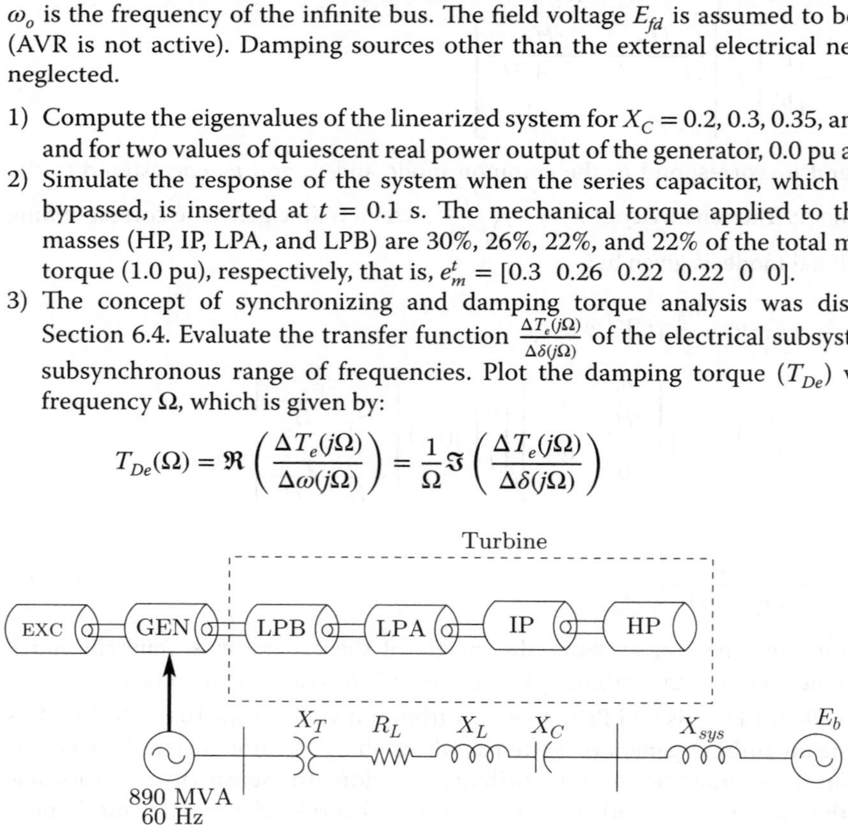

Figure 10.10 Turbine-generator system connected to a series compensated transmission line.

Solution

Eigenvalue analysis

The eigenvalues of this system are obtained by linearizing the differential equations of the entire system around the equilibrium point. The equilibrium conditions of the generator (including δ) are computed from the prevailing load-flow conditions as given in Section 3.15. The equilibrium values of the rotor angles of the turbine/exciter masses are obtained by setting the LHS in (10.36) to zero and plugging in the equilibrium value of δ. Note that the speeds of all masses are equal to the infinite bus frequency ω_o under equilibrium conditions.

Without series compensation

The eigenvalues of the system without compensation are shown in Table 10.6 for different loading levels. The real part of the torsional modes is extremely small, indicating negligible interactions between the torsional modes and the network. Since the damping sources other than the electrical network are not modeled here, the results for the torsional mode damping are pessimistic. Considering the typical values of the damping contribution from the other sources (see Section 10.3.1), it is likely that the overall damping would be positive and the modes would be stable in an actual scenario.

When the generator is connected to the infinite bus through the ac transmission system, the common mode of the mechanical system (associated with the zero eigen-values in Example 10.2) transforms into a low-frequency, oscillatory "swing mode". While the damping and frequency of the swing mode is a strong function of the operating condition, this is not true for the torsional and network mode frequencies and their damping. The torsional mode frequencies are not affected significantly due to the connection of the generator to the external network (compare the torsional mode frequencies in Tables 10.4 and 10.6). Since the field voltage (E_{fd}) is constant, one of the real eigenvalues moves towards the origin as loading is increased. This indicates that the system is susceptible to monotonic instability with increased loading (which may lead to a loss of synchronism). Note that this is a manifestation of reduced steady-state synchronizing torque (see the discussion in Section 6.4). This can be alleviated by decreasing the external reactance using series compensation.

Table 10.6 Eigenvalues without series compensation.

P_m = 0.0 pu	P_m = 0.5 pu	P_m = 1.0 pu	Remarks
$-0.955 \pm j8.23$	$-0.476 \pm j8.15$	$-0.372 \pm j6.95$	Swing Mode
$-0.0061 \pm j99.11$	$-0.0041 \pm j99.10$	$-0.0029 \pm j99.00$	Torsional Mode 1
$-0.0003 \pm j126.98$	$-0.0001 \pm j126.98$	$0.0000 \pm j126.97$	Torsional Mode 2
$-0.0002 \pm j160.61$	$0.0002 \pm j160.61$	$0.0005 \pm j160.58$	Torsional Mode 3
$0.0008 \pm j202.98$	$0.0013 \pm j202.98$	$0.0019 \pm j202.94$	Torsional Mode 4
$0.0000 \pm j298.18$	$0.0000 \pm j298.18$	$0.0000 \pm j298.18$	Torsional Mode 5
$-8.711 \pm j376.95$	$-8.712 \pm j376.95$	$-8.714 \pm j376.95$	Network Mode 1
-32.361	-32.298	-32.269	
-20.186	-20.213	-20.224	Rotor
-1.216	-2.642	-3.115	flux modes
-0.670	-0.241	0.006	

With Series Compensation

The eigenvalues of the system with various levels of series compensation are shown in Tables 10.7 and 10.8. There are now two network modes, one subsynchronous and the other supersynchronous, as discussed in Section 10.2. The swing mode frequency increases due to series compensation, which is indicative of increased synchronizing torque at low frequencies. With series compensation, the real eigenvalue of lowest magnitude is negative for all conditions. However, destabilization of the torsional modes due to the series compensation is evident, the real parts of Torsional Modes 1–4 are positive.

Table 10.7 Eigenvalues with series compensation (with $P_m = 0.0$ pu).

$X_c = 0.2$ pu	$X_c = 0.3$ pu	$X_c = 0.35$ pu	$X_c = 0.45$ pu
$-1.229 \pm j9.29$	$-1.433 \pm j10.00$	$-1.563 \pm j10.43$	$-1.910 \pm j11.49$
$-0.0047 \pm j99.28$	$0.0035 \pm j99.49$	$0.0205 \pm j99.72$	$\mathbf{3.0428 \pm j101.58}$
$0.0005 \pm j127$	$0.0056 \pm j127.04$	$0.0466 \pm j127.15$	$0.0078 \pm j126.9$
$0.0125 \pm j160.72$	$\mathbf{0.4597 \pm j159.93}$	$0.0237 \pm j160.38$	$0.0017 \pm j160.51$
$\mathbf{0.3515 \pm j202.32}$	$0.0046 \pm j202.83$	$0.0010 \pm j202.86$	$-0.0014 \pm j202.89$
$0.0000 \pm j298.18$	$0.0000 \pm j298.18$	$0.0000 \pm j298.18$	$0.0000 \pm j298.18$
$\mathbf{-3.887 \pm j196.24}$	$\mathbf{-3.554 \pm j155.54}$	$-2.899 \pm j136.93$	$\mathbf{-5.143 \pm j103.01}$
$-4.653 \pm j558.1$	$-4.695 \pm j598.83$	$-4.712 \pm j616.61$	$-4.740 \pm j648.7$
-32.707	-32.972	-33.141	-33.590
-20.233	-20.267	-20.288	-20.342
-1.2159	-1.2160	-1.2160	-1.2161
-0.8012	-0.9017	-0.9666	-1.1430

Table 10.8 Eigenvalues with series compensation (with $P_m = 1.0$ pu).

$X_c = 0.2$ pu	$X_c = 0.3$ pu	$X_c = 0.35$ pu	$X_c = 0.45$ pu
$-0.428 \pm j8.90$	$-0.512 \pm j9.98$	$-0.571 \pm j10.58$	$-0.744 \pm j12$
$0.0028 \pm j99.25$	$0.0184 \pm j99.53$	$0.0462 \pm j99.82$	$\mathbf{3.7857 \pm j101.63}$
$0.0013 \pm j127$	$0.0087 \pm j127.05$	$0.0654 \pm j127.19$	$0.0085 \pm j126.87$
$0.0185 \pm j160.73$	$\mathbf{0.5988 \pm j159.75}$	$0.0272 \pm j160.31$	$0.0011 \pm j160.47$
$\mathbf{0.4613 \pm j202.11}$	$0.0045 \pm j202.77$	$0.0004 \pm j202.81$	$-0.0022 \pm j202.86$
$0.0000 \pm j298.18$	$0.0000 \pm j298.18$	$0.0000 \pm j298.18$	$0.0000 \pm j298.18$
$\mathbf{-4.009 \pm j196.34}$	$\mathbf{-3.710 \pm j155.63}$	$-2.946 \pm j136.80$	$\mathbf{-5.889 \pm j102.95}$
$-4.653 \pm j558.11$	$-4.695 \pm j598.84$	$-4.712 \pm j616.62$	$-4.740 \pm j648.71$
-32.562	-32.782	-32.92	-33.28
-20.295	-20.349	-20.384	-20.478
-3.593	-3.920	-4.124	-4.658
-0.112	-0.151	-0.168	-0.201

As discussed previously, in an actual situation some positive damping contribution from mechanical/other sources is expected. Since this contribution is not modeled here, we have somewhat pessimistic results.

The torsional modes are affected to different extents depending on the level of series-capacitor compensation. A profound destabilizing effect is seen in cases where the frequency of the subsynchronous network mode nearly coincides with the frequency of a torsional mode (see the network mode and torsional modes shown in bold in the tables). This phenomenon is called *subsynchronous resonance.*

When the network subsynchronous resonant frequency is not tuned to the frequency of any torsional mode (e.g., when $X_C = 0.35$ pu), the effect on the torsional modes is less. Torsional Mode 5 (298 rad/s) is hardly affected by the electrical system. This can be understood from the sensitivity analysis, which will be discussed later.

Effects of Modeling

It is instructive at this stage to consider the effect of the various models as given below:

Model 1: Multi-mass model of mechanical system with network and stator transients included.

Model 2: Single-mass model of mechanical system with network and stator transients included.

Model 3: Single-mass model of mechanical system with network and stator transients neglected.

The eigenvalues are shown in Table 10.9. It is clear that that Model 3 is acceptable for the study of the low-frequency swing mode, although the damping estimate is slightly inaccurate. This justifies the approximations made in Chapter 6 for the study of the power swing phenomena.

Simulation Study

The response to the insertion of the series capacitor at $t = 0.1$ s is shown in Figure 10.11 for $X_C = 0.2$ pu and $X_C = 0.3$ pu, for a quiescent power output of 1.0 pu. Instability of a

Table 10.9 Eigenvalues obtained from different models ($X_C = 0.35$ pu, $P_m = 1.0$ pu).

Eigenvalues (Model 1)	Eigenvalues (Model 2)	Eigenvalues (Model 3)	Remarks
$-0.571 \pm j10.58$	$-0.583 \pm j10.67$	$-0.445 \pm j10.64$	Swing Mode
$0.0462 \pm j99.82$			Torsional Mode 1
$0.0654 \pm j127.19$			Torsional Mode 2
$0.0272 \pm j160.31$	—	—	Torsional Mode 3
$0.0004 \pm j202.81$			Torsional Mode 4
$0.0000 \pm j298.18$			Torsional Mode 5
$-2.946 \pm j136.80$	$-2.783 \pm j137.14$	—	Network Mode 1
$-4.712 \pm j616.62$	$-4.711 \pm j616.58$		Network Mode 2
−32.920	−32.946	−33.043	
−20.384	−20.384	−20.383	Rotor
−4.124	−4.124	−4.096	flux modes
−0.168	−0.168	−0.169	

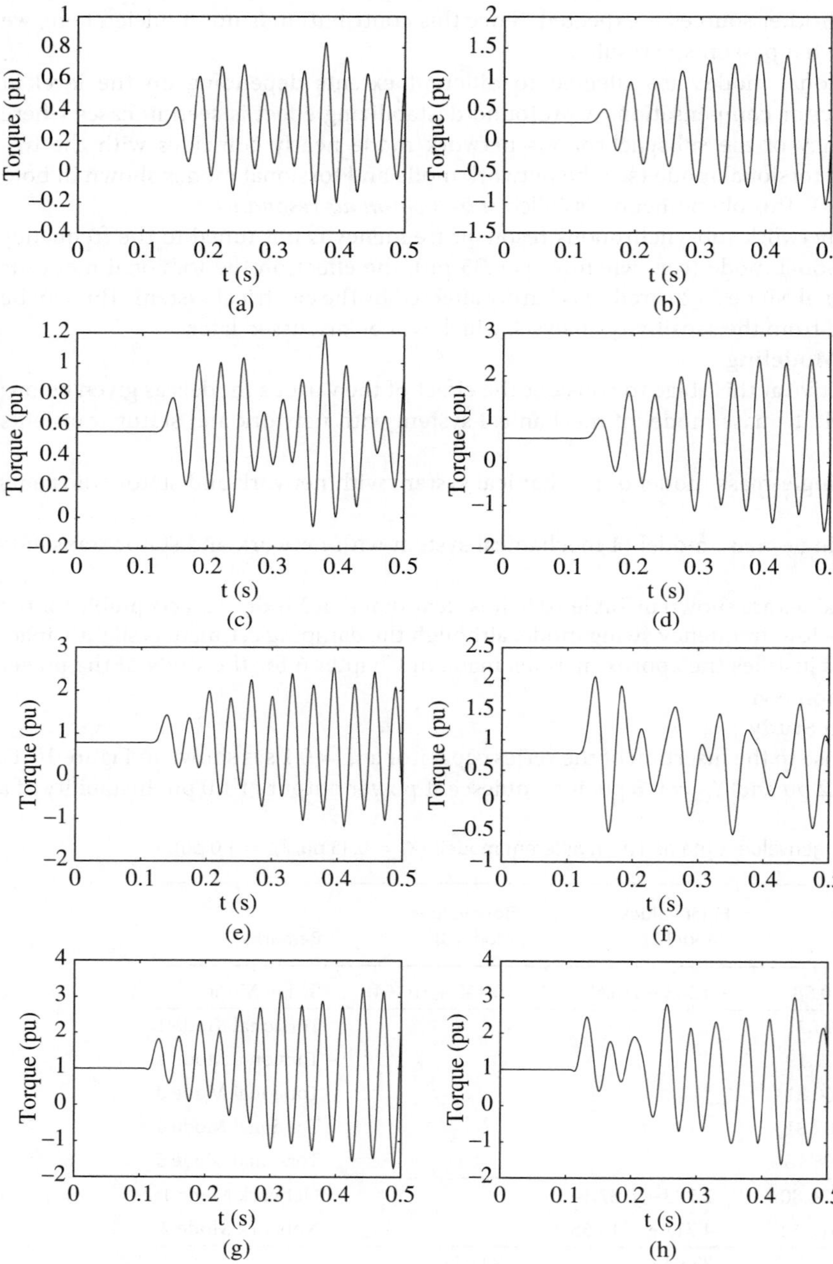

Figure 10.11 Response when the series capacitors are inserted in the transmission line. (a) HP–IP shaft torque: $X_C = 0.2$ pu (b) HP–IP shaft torque: $X_C = 0.3$ pu (c) IP–LPA shaft torque: $X_C = 0.2$ pu (d) IP–LPA shaft torque: $X_C = 0.3$ pu (e) LPA–LPB shaft torque: $X_C = 0.2$ pu (f) LPA–LPB shaft torque: $X_C = 0.3$ pu (g) LPB–GEN shaft torque: $X_C = 0.2$ pu (h) LPB–GEN shaft torque: $X_C = 0.3$ pu.

torsional mode at about 202 rad/s when $X_C = 0.2$ pu and about 159 rad/s when $X_C = 0.3$ pu is clearly seen in the shaft torque waveforms. Note that the HP–IP shaft torque is given by $T_{sh} = K_{HP-IP} \times (\delta_{HP} - \delta_{IP})$, where δ_{HP} and δ_{IP} are the rotor angular positions of the HP and IP turbines, respectively, and K_{HP-IP} is the stiffness constant of the HP–IP shaft. The other shaft torques are similarly obtained.

Damping Torque Analysis

The damping torques for $X_C = 0.35$ pu and $X_C = 0.45$ pu for quiescent power output of 1.0 pu are shown in Figure 10.12. The vertical (dashed) lines in Figure 10.12 correspond to the swing mode and torsional mode frequencies. The damping torque has a negative peak at the subsynchronous network frequency. The peak nearly coincides with the lowest torsional frequency when $X_C = 0.45$ pu, while it is "detuned" from the torsional modes when $X_C = 0.35$ pu.

Note that the value $\frac{\Delta T_e(j\Omega_i)}{\Delta\delta(j\Omega_i)}$ can be approximately taken as the feedback gain κ in Figure 10.9b. Assuming the connection of the electrical network causes only a small change in the torsional mode characteristics, we can use the eigenvalue sensitivity formula given in the previous section to approximate the shift in the eigenvalue. Therefore

$$\Delta\lambda_i = \Delta\sigma_i + j\Delta\Omega_i \approx \frac{d\lambda_i}{d\kappa} \times \frac{\Delta T_e(j\Omega_i)}{\Delta\delta(j\Omega_i)}$$

As the mechanical damping is neglected, eigenvalues of the mechanical system can be taken as purely imaginary, that is, $\lambda_i = j\Omega_i$, where Ω_i is the radian frequency of the ith torsional mode. Substituting for $\frac{d\lambda_i}{d\kappa}$ using (10.40), we obtain

$$\Delta\sigma_i + j\Delta\Omega_i \approx -\frac{1}{j2\Omega_i} \times \frac{\omega_B}{2H_{mi}} \times \frac{\Delta T_e(j\Omega_i)}{\Delta\delta(j\Omega_i)}$$

The shift in the real part of the eigenvalue can be computed by

$$\Delta\sigma_i \approx -\frac{1}{2\Omega_i} \times \frac{\omega_B}{2H_{mi}} \times \Im\left(\frac{\Delta T_e(j\Omega_i)}{\Delta\delta(j\Omega_i)}\right)$$

This can be simplified further using (10.42). Therefore, we obtain

$$\Delta\sigma_i \approx -\frac{\omega_B}{4H_{mi}} \times T_{De}(\Omega_i) \tag{10.43}$$

The eigenvalue sensitivities corresponding to the torsional modes are calculated using the expression (10.41) and are shown in Table 10.10. The $\Delta\sigma_i$ values calculated

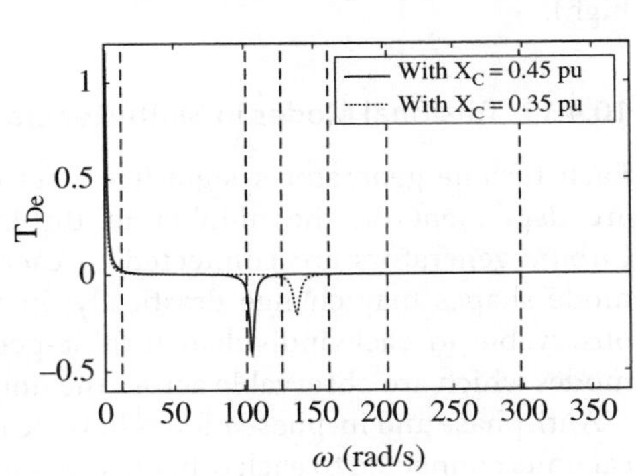

Figure 10.12 Damping torque (T_{De}) in pu/(rad/s) for $P_m = 1.0$ pu.

Table 10.10 Sensitivity of eigenvalues corresponding to the torsional modes, and the damping torque evaluated at the torsional mode frequencies for $P_m = 1.0$ pu, mechanical damping neglected, AVR not considered.

Mode	Eigenvalue ($\lambda_i = j\Omega_i$)	Modal inertia H_{mi}	$T_{De}(\Omega_i)$ pu/(rad/s) $X_C = 0.35$ pu	$T_{De}(\Omega_i)$ pu/(rad/s) $X_C = 0.45$ pu
Torsional Mode 1	$j98.723$	2.7004	-1.2809×10^{-3}	-9.6494×10^{-2}
Torsional Mode 2	$j126.99$	27.788	-1.5926×10^{-2}	-2.6249×10^{-3}
Torsional Mode 3	$j160.52$	6.9186	-2.0179×10^{-3}	-8.0715×10^{-5}
Torsional Mode 4	$j202.85$	3.9222	-1.6655×10^{-5}	9.1600×10^{-5}
Torsional Mode 5	$j298.18$	11288	0.0000	0.0000

Table 10.11 Real parts of the torsional modes ($P_m = 1.0$ pu).

Mode	Real part ($\Delta\sigma_i$) (With $X_C = 0.35$ pu) Actual	Calculated	(With $X_C = 0.45$ pu) Actual	Calculated
Torsional Mode 1	0.0462	0.0447	3.7857	3.3675
Torsional Mode 2	0.0654	0.0540	0.0085	0.0089
Torsional Mode 3	0.0272	0.0275	0.0011	0.0011
Torsional Mode 4	0.0004	0.0004	−0.0022	−0.0022
Torsional Mode 5	0.0000	0.0000	0.0000	0.0000

approximately as above, along with the actual values obtained from the eigenvalues of the combined system, are shown in Table 10.11. These show a close match, which indicates that the damping torque of the electrical system, which is obtained independent of the mechanical system, and the eigenvalue sensitivity of the mechanical system can be used to approximate the real part of the eigenvalues of the combined system. In this example, the series-compensated electrical network causes the real part of the eigenvalues (corresponding to the torsional modes) to become positive, that is, it destabilizes the torsional modes. Torsional Mode 5, however, is practically unaffected, as the corresponding eigenvalue sensitivity is very low (modal inertia is very high).

10.4.1 Torsional Modes in Multi-Generator Systems

Each turbine-generator system has a set of torsional modes. The number of modes are dependent on the number of turbine and generator masses. When identical turbine-generators are connected to each other through an electrical network, the mode shapes may change drastically: instead of sets of torsional modes exclusively observable in each individual turbine-generator, there are anti-phase and in-phase modes which are observable across the units.

Anti-phase and in-phase modes have nearly the same frequency. In the case of identical units connected to each other in a power plant, anti-phase modes are not observable

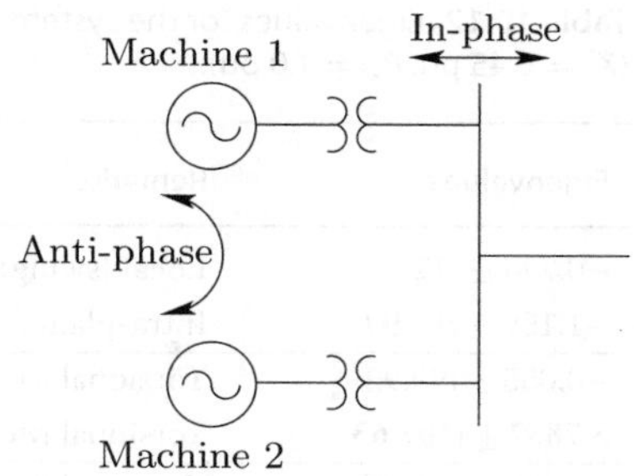

Figure 10.13 Parallel connection of two identical generators: anti-phase and in-phase modes.

outside the power plant. On the other hand, in-phase modes are observable externally and affected by series compensation in the external transmission system. This is illustrated using the following example.

Example 10.5 ***In-Phase and Anti-Phase Modes*** Consider the system given in Example 10.4, but with the synchronous machine and the generator transformer replaced by two identical generators and their transformers having half the MVA rating of the original generator and transformer. The generator and their unit transformers are connected in parallel on the transmission line side, as shown in Figure 10.13.

The reactances of the individual generators and transformers are doubled when evaluated on the original MVA base, while the inertias (H) and stiffness constants (K) are halved. The power output of each generator is 0.5 pu.

Compute the eigenvalues of this system for $X_C = 0.45$ pu, assuming the same conditions of the external system as in Example 10.4. Also, comment on whether a particular torsional mode is in-phase or anti-phase by looking at the mode shapes.

Solution

The eigenvalues of the system are given in Table 10.12. Note that for a power plant with two identical generating units system the torsional modes are in pairs (in-phase and anti-phase) which are of nearly the same frequency. The in-phase modes are the modes in which the turbine-generator systems of both units move together. The in-phase modes are affected by the series-compensated line (Torsional Mode 2 experiences SSR). The units behave like one unit of double the individual rating when this mode is excited. Not surprisingly, the eigenvalue corresponding to the in-phase mode is identical to the one in Table 10.10. The anti-phase modes involve the inter-unit oscillations within the power plant and are not affected by the (external) series-compensated line.

The mode-shapes for Torsional Modes 1 and 2 are given in Table 10.13. Note that the mode shape denotes the observability of the torsional mode in the speeds of the various rotor masses, and are obtained from the eigenvector of the overall state matrix. For Torsional Mode 2, the turbine, generator, and exciter masses of both machines move together (the mode-shape components have the same sign), while they swing against each other for the anti-phase mode (Torsional Mode 1).

SSR in Meshed Networks

The examples considered so far have a turbine-generator system connected radially via a series-compensated line to an infinite bus. The subsynchronous network mode in these

Table 10.12 Eigenvalues for the system in Example 10.5 ($X_C = 0.45$ pu, $P_m = 1.0$ pu).

Eigenvalues	Remarks
$-0.744 \pm j12$	Local (swing) mode
$-1.159 \pm j13.99$	Intra-plant (swing) mode
$-0.055 \pm j99.92$	Torsional Mode 1 (anti-phase)
$\mathbf{3.7857 \pm j101.63}$	Torsional Mode 2 (in-phase)
$0.0085 \pm j126.87$	Torsional Mode 3 (anti-phase)
$-0.0036 \pm j127.04$	Torsional Mode 4 (in-phase)
$0.0011 \pm j160.47$	Torsional Mode 5 (anti-phase)
$-0.0088 \pm j160.81$	Torsional Mode 6 (in-phase)
$-0.0022 \pm j202.86$	Torsional Mode 7 (anti-phase)
$-0.1000 \pm j203.26$	Torsional Mode 8 (in-phase)
$0.0000 \pm j298.18$	Torsional Mode 9 (anti-phase)
$0.0000 \pm j298.18$	Torsional Mode 10 (in-phase)
$-5.8886 \pm j102.95$	Network Mode 1 (subsynchronous)
$-4.7401 \pm j648.71$	Network Mode 2 (supersynchronous)
$0.0000 \pm j376.99$	Network Mode 3 (intra-plant)
$-34.178, -33.280$	
$-20.661, -20.478$	Rotor
$-5.499, -4.658$	flux modes
$-0.269, -0.201$	

Table 10.13 Mode-shapes of Torsional Modes 1 and 2 ($X_C = 0.45$ pu, $P_m = 1.0$ pu).

Mass	HP		IP		LPA		LPB		GEN		EXC	
Machine	1	2	1	2	1	2	1	2	1	2	1	2
Torsional Mode 1	−2.06	2.06	−1.54	1.54	−0.88	0.88	0.33	−0.33	1.00	−1.00	2.80	−2.80
Torsional Mode 2	2.02	2.02	1.49	1.49	0.83	0.83	−0.38	−0.38	−1.00	−1.00	−2.94	−2.94

cases is observable in the generator current and the electrical torque. In a meshed transmission system with one or more series-compensated lines and many generators, the observability of network modes at different generators depends on the network topology and parameters. There may be more than one subsynchronous network mode if multiple series-compensated lines are present. Mere proximity of a subsynchronous network mode frequency (in the *D*-*Q* frame) with a torsional mode frequency may not result in SSR. SSR may occur only if all of the following conditions are satisfied:

1) The subsynchronous network mode frequency is near the frequency of a torsional mode.

2) The subsynchronous network mode is observable at the generator(s) associated with that torsional mode. Damping torque analysis is suitable for evaluating the extent of the influence of individual network modes on a generator. A significant negative peak in damping torque at a generator at the network-mode frequency is indicative of the observability of the network mode at that generator.
3) The torsional mode is controllable by the electrical torque. Recall that in Example 10.3, Torsional Mode 5 is practically uncontrollable by the electrical torque (eigenvalue sensitivity to the parameters of the electrical network is very small).

10.4.2 Adverse Interactions with Turbine-Generator Controllers

Adverse torsional interaction was reported with a speed-governor system in [15]. In general, the torsional modes observable in feedback signals (turbine or generator speed) can cause adverse interaction if the modal components in these signals are not sufficiently attenuated by the governor controller and servo system.

A power system stabilizer (PSS) is a part of a generator's excitation system controller. It modulates the set-point of an automatic voltage regulator (see Section 6.6.3 in Chapter 6). The aim of a PSS is to damp power swings, which are of relatively low frequency (0.1–2 Hz).

A PSS which uses an input signal in which a torsional mode is significantly observable may affect the damping of that mode. It has been found that torsional modes are susceptible to instability with speed-input stabilizers [16]. The use of torsional filters has been suggested to alleviate the problem [17]. As an alternative, signals which are similar to the speed signal at the swing-mode frequencies, but in which the observability of the torsional modes is lower, may be used. The delta-P-omega signal [18] is an example of such a signal. This signal mimics the speed signal at low frequencies, but attentuates the high-frequency torsional components. The synthesis of this signal is based on the following equations in the Laplace domain:

$$\Delta\omega(s) = \frac{\omega_B}{2Hs}[\Delta T_m(s) - \Delta T_e(s)] \approx \frac{\omega_B}{2Hs}[\Delta P_m(s) - \Delta P_e(s)] \tag{10.44}$$

Therefore

$$\frac{\omega_B}{2Hs}\Delta P_m(s) \approx \frac{\omega_B}{2Hs}\Delta P_e(s) + \Delta\omega(s) \tag{10.45}$$

$$\approx \left(\frac{\omega_B}{2Hs}\Delta P_e(s) + \Delta\omega(s)\right) G_f(s) \tag{10.46}$$

$G_f(s)$ is a low-pass filter which filters out the torsional frequencies.

Since the changes in the mechanical power P_m are generally slow and the the gain of $G_f(s)$ is approximately 1 at low frequencies, the synthesis of $\frac{\omega_B}{2Hs}\Delta P_m(s)$ using (10.46) is expected to be accurate. The delta-P-omega signal is then obtained as follows:

$$\Delta\omega'(s) = \left(\frac{\omega_B}{2Hs}\Delta P_e(s) + \Delta\omega(s)\right) G_f(s) - \frac{\omega_B}{2Hs}\Delta P_e(s) \tag{10.47}$$

The synthesis of this signal is shown in Figure 10.14. Note that at low frequencies $\Delta\omega' \approx \Delta\omega$.

Example 10.6 ***Effect of Speed Input and delta-P-omega input PSS*** Consider the system given in Example 10.4, but with no series compensation ($X_C = 0.0$). The parameters of the generator, network, and multi-mass turbine model are as given in

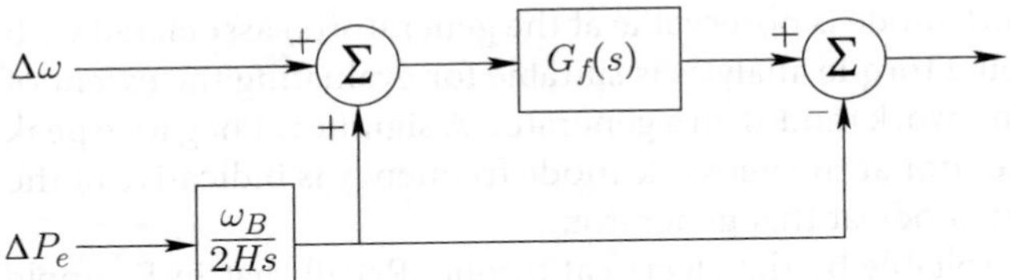

Figure 10.14 Synthesis of the delta-P-omega signal.

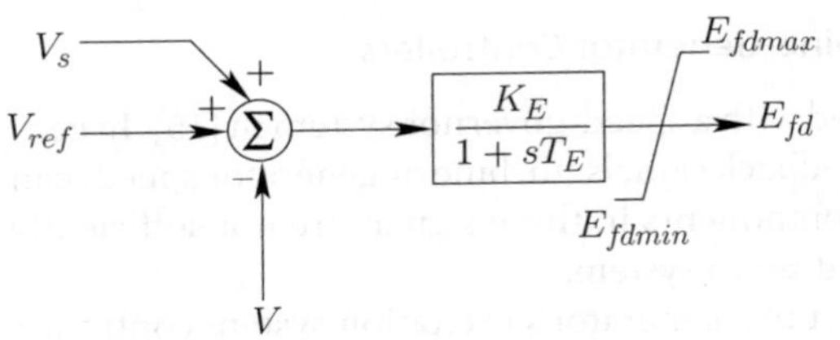

Figure 10.15 Model of excitation system (PSS output added at the AVR summing junction).

Example 10.4. In addition, the generator has an AVR to control the field voltage. A PSS is also included to damp out low-frequency power swings. The block diagrams of AVR and PSS considered in this example are shown in Figures 10.15 and 10.16, respectively.

The magnitudes of the generator terminal voltage and the infinite bus voltage are 1.0 pu. The quiescent power output is 1.0 pu. The parameters of the AVR and PSS are as follows:

AVR

$K_E = 200$ pu/pu, $T_E = 0.025$ s

PSS

Washout block: $T_w = 10$ s

Compensators: $T_1 = T_3 = 0.1$ s, $T_2 = T_4 = 0.05$ s

PSS gain: $K_s = \frac{5}{\omega_B}$ pu/(rad/s)

Obtain the eigenvalues of the system and plot the damping torque (T_{De}) versus the frequency Ω for the following cases:

1) No PSS is present.
2) Speed-input PSS is present without a torsional filter.
3) Speed-input PSS is present with a torsional filter, as shown in Figure 10.16. The torsional filter parameters are $\zeta = 0.5$ and $\omega_n = 20$.
4) Delta-P-omega PSS is present with the same parameters as the speed-input PSS. However, the torsional filter shown in Figure 10.16 is removed. The filter used in the delta-P-omega PSS has the transfer function $G_f(s) = (\frac{1}{1+0.2s})^4$. The integration operation $\frac{\omega_B}{2Hs}\Delta P_e(s)$ is approximated by $\frac{\omega_B}{2H}\frac{10}{1+10s}\Delta P_e(s)$.

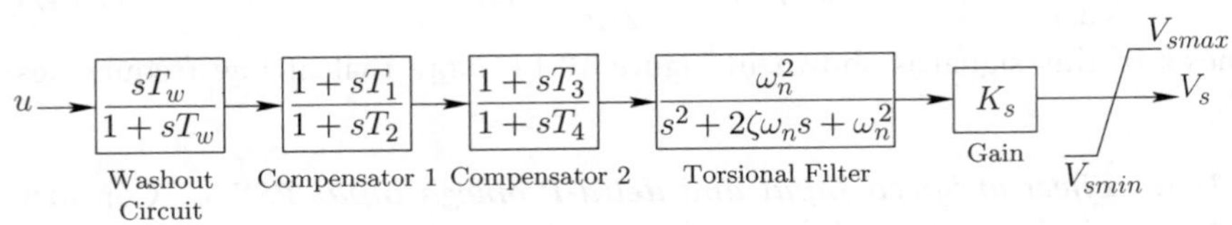

Figure 10.16 Block diagram of a PSS with a torsional filter.

Solution

Observations

1) The eigenvalues and the damping torque plots (T_{De}) for all cases are shown in Table 10.14, and Figure 10.17, respectively. The vertical dashed lines correspond to the swing-mode and torsional-mode frequencies.
2) It can be seen that the damping torque is negative at the frequency corresponding to the swing mode when there is no PSS present. The damping torque is positive at the frequencies corresponding to the torsional modes.
   Introduction of the PSS (without a torsional filter) improves the damping of the swing mode but the torsional modes are destabilized.
3) Introduction of a torsional filter when used in cascade with the stabilizer alleviates the problem. This is corroborated by the plot of damping torque. It is positive in the frequency range of the swing mode as well as torsional modes. An additional oscillatory mode is introduced due to the second-order torsional filter.
4) The use of the delta-P-omega signal instead of the speed input signal is able to give the same performance as a speed input PSS for the swing mode, but does not destabilize the torsional modes.

10.4.3 Detection of SSR/Torsional Monitoring

SSR can be detected by monitoring variables in which the torsional modes are observable, such as the generator and turbine speeds. Typically, the speeds are measured at the

Table 10.14 Eigenvalues with and without PSS (no series compensation present).

With AVR without PSS	Speed-input PSS with no torsional filter	Speed-input PSS with torsional filter	Delta-P-omega PSS
$\mathbf{0.434 \pm j7.86}$	$-0.635 \pm j7.49$	$-0.944 \pm j7.94$	$-0.631 \pm j7.47$
$-0.0151 \pm j98.99$	$\mathbf{0.2728 \pm j99.18}$	$-0.0253 \pm jj98.98$	$-0.0137 \pm j98.99$
$-0.0009 \pm j126.97$	$0.0181 \pm j126.98$	$-0.0013 \pm j126.97$	$-0.0008 \pm j126.97$
$-0.0017 \pm j160.58$	$0.0473 \pm j160.60$	$-0.0024 \pm j160.58$	$-0.0016 \pm j160.58$
$-0.0006 \pm j202.94$	$0.0553 \pm j202.96$	$-0.0011 \pm j202.94$	$-0.0005 \pm j202.94$
$0.0000 \pm j298.18$	$0.0000 \pm j298.18$	$0.0000 \pm j298.18$	$0.0000 \pm j298.18$
$-8.716 \pm j376.95$	$-8.718 \pm j376.95$	$-8.716 \pm j376.95$	$-8.716 \pm j376.95$
$-18.664 \pm j24.21$	$-14.475 \pm j24.82$	$-19.676 \pm j26.72$	$-14.477 \pm j25.34$
		$-5.680 \pm j16.33$	$-14.243 \pm j1.19$
−38.887	−39.139	−38.949	−39.077
	−31.031	−28.236	−30.337
−20.231	−20.224	−20.224	−20.224
	−15.786	−15.546	−10.487
−3.108	−3.108	−3.108	−3.108
	−0.100	−0.100	$-0.100 \pm j0.0009$
			$-10.600 \pm j0.7590$

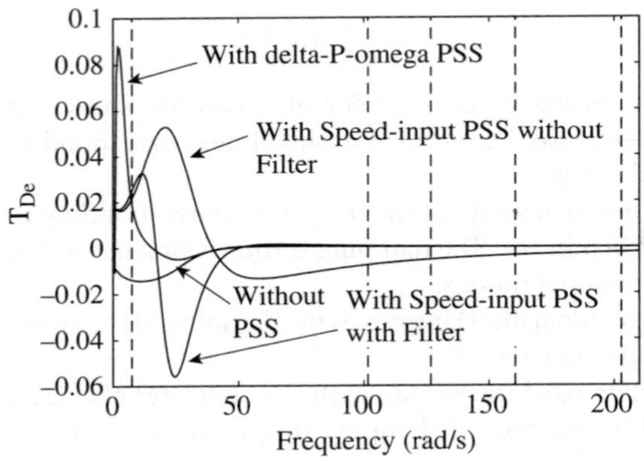

Figure 10.17 Damping torque in pu/(rad/s) with and without a PSS (Example 10.6).

two ends of the turbine-generator shaft. Electrical variables such as armature current may also be monitored for the presence of poorly damped or unstable sub-synchronous oscillations. The torsional frequencies and mode shapes of a turbine-generator system are generally known from prior measurements and analysis, and do not change significantly with changes in the external electrical network. This information (torsional mode frequencies and mode shapes) is useful, since in an SSR situation the offending torsional mode has to be detected early, before its amplitude in the measured signal becomes large. The amplitude of the excited torsional modes may be used to estimate the torsional stresses and thereby estimate the fatigue (expenditure of shaft life). Alarms and/or tripping may be initiated if this exceeds the pre-set thresholds.

10.4.4 Countermeasures for Subsynchronous Resonance and Subsynchronous Torsional Interactions

Series versus Shunt Compensation

At the planning stage, shunt compensation could be considered as an alternative to series compensation. The network modes due to shunt capacitors generally are not at subsynchronous frequencies but are at supersynchronous frequencies (see Example 10.1). At supersynchronous frequencies damping torque is positive for such systems [19]. While both series and shunt compensation facilitate greater power transfer in an AC network, series compensation may be more economical and flexible.

Turbine-Generator Modifications

The torsional mode frequencies of the turbine-generator can be varied only within small limits. The constraints on shaft and bearing size make it impractical to design machines for which the lowest torsional mode frequency is greater than the synchronous frequency. Changing the torsional mode frequency by a small amount to avoid resonance with the subsynchronous network mode frequency is not of much use as changes in the network due to future growth/line outages would affect the subsynchronous network mode frequency. Increasing mechanical damping, which is normally present due to friction, hysteresis, and fretting, is also considered impractical.

System Modifications

The steady-state SSR problem is mainly due to the coincidence of electrical resonant frequency with the complement of a torsional mode frequency. It is thus possible to adjust the series compensation to avoid this coincidence. As the problem is more severe at lower frequency torsional modes, reducing the level of series compensation helps. This can also be done during system operation by bypassing some capacitor segments. It is also possible to have coordinated series capacitor use with loading. As the mechanical damping increases with the load (power output of the generator), it practical to insert capacitors only when the loading exceeds the minimum loading. As series capacitors are mainly required at higher loadings, this procedure of coordinating the capacitor use with load is consistent with the dual objectives of improved stability limit and avoidance of the steady-state SSR problem.

Filtering Schemes

Filtering involves the insertion of passive elements in the electrical network in order to significantly reduce the observability of subsynchronous network modes at the generator. A static blocking filter is a three-phase filter made up of separate filters connected in series [20]. Each section of the filter is a high Q, parallel resonant circuit tuned to block electric currents at frequencies complementary to each of the torsional modes. It may be inserted in series with the generator step-up transformer winding on the neutral end of the transformer high-voltage winding. It can also be placed on the high-voltage side of the transformer winding. While the performance of the filter is not significantly affected by the external network, it is affected by drift in the parameter values of the capacitors.

Phase Imbalance Schemes

Edris [21] suggested two schemes of passive phase imbalance for the mitigation of SSR, as shown in Figure 10.18. Imbalance is created by inserting additional capacitors and inductors in the circuit in such a manner that the system is balanced at the nominal frequency (ω_o), although it is unbalanced at other frequencies. This alters the damping torque characteristics of the electrical network. It is possible to detune the network from the torsional frequencies and reduce the negative peak of the damping torque [22, 23] using this scheme.

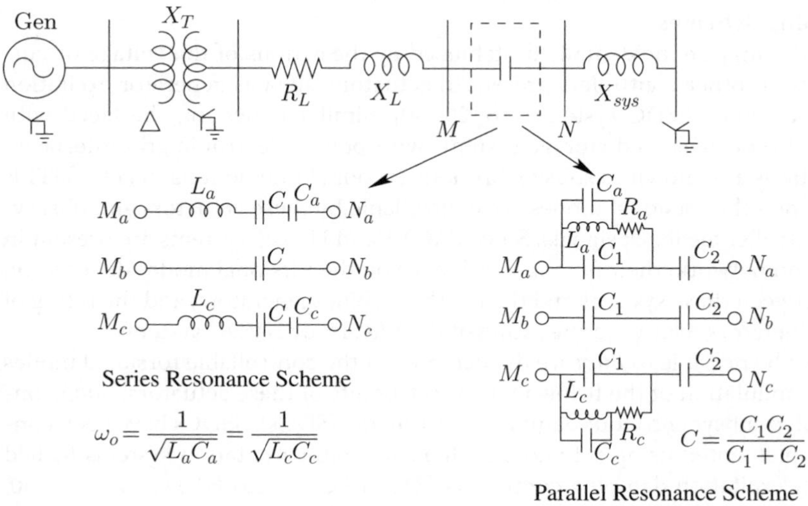

Figure 10.18 Series and parallel phase imbalance schemes.

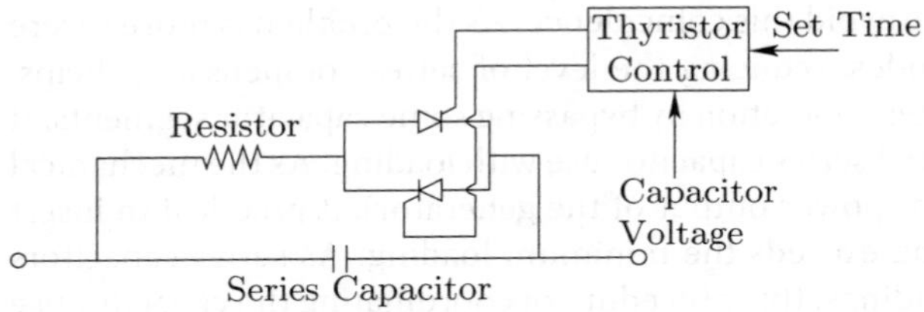

Figure 10.19 NGH damping scheme.

NGH Damping Scheme

The NGH-SSR damping scheme was invented by N.G. Hingorani in 1980 [24, 25]. It was successfully applied at Mohave generating station in the United States and South California Edison's Lugo substation [26, 27]. The basic NGH damping scheme for one phase is shown in Figure 10.19. It involves a linear resistor in series with antiparallel thyristors connected across the capacitor.

When a zero crossing of the capacitor voltage is detected, the succeeding half-cycle period is timed. As soon as the half cycle exceeds the set time, the corresponding thyristor is turned on to discharge the capacitor through the resistor, thereby discharging it sooner than it would otherwise. If there is a problem of small-signal instability of torsional modes, the set period is set to slightly less than a half-cycle period. The thyristors conduct during steady state at the tail end of each half cycle of the capacitor voltage. This is expected to provide a detuning effect against the gradual build-up of oscillations. During practical tests, the NGH damper was found to damp the on-tune torsional mode. However, the damping of a few other off-tune subsynchronous torsional modes were reduced by the device [25, 28]. This problem was alleviated by utilizing the NGH damper in conjunction with an excitation-system-based SSR damping controller. The NGH damping scheme as discussed above is a "passive" scheme as it does not require feedback controller design. The excitation-based damping controller, on the other hand, is an example of an "active" SSR damping scheme.

Active Damping Schemes

Torsional modes may be incidentally destabilized by the actions of the voltage or current regulators, or other controllers present in actuators, such as generator excitation systems, FACTS, and HVDC systems [16, 29, 30]. Similar issues may be faced with grid-connected renewable and storage systems with power-electronic grid interfaces. These interactions are known as subsynchronous torsional interactions (SSTI). SSTI is a possibility when the torsional modes are controllable by these actuators and observable in the controller feedback signals. Since FACTS and HVDC systems are present in the transmission network, their potential influence on the torsional modes depends on impedance between these systems and the nearby turbine-generators, and the rating of the turbine-generators relative to the rating of the FACTS or HVDC systems.

It should also be possible to *improve* the damping of the controllable torsional modes by deliberate modulation of the firing angle or set-points of these actuators. Such controllers are called subsynchronous damping controllers (SSDCs). SSDCs have been conceived and implemented using actuators such as generator excitation systems (called supplementary excitation damping controllers [31]), FACTS, and HVDC systems [30, 32–37]. To illustrate the damping action of a SSDC, a case study of an SSDC based on a thyristor-controlled series compensator (TCSC) is presented in the following section.

10.4.5 Case Study: TCSC-Based SSDC

A TCSC consists of a fixed capacitor in parallel with a thyristor-controlled reactor (TCR). A TCR is a series combination of back-to-back connected thyristors and a fixed reactor. The detailed description of a TCSC is given in Section 5.4. The TCSC circuit and its steady-state waveforms are shown in Figure 10.20. Note that the effective reactance of a TCSC can be controlled by controlling the firing angle α.

Now consider the series compensated system shown in Figure 10.21, wherein a TCSC is used in conjunction with a fixed capacitor (X_C). A phase-locked loop (PLL) which is locked on to the line current is used to generate the reference signal for the firing pulse generator, as shown in Figure 10.22. The firing angle order α_0 is augmented by the output of the SSDC shown in Figure 10.23. The input for the SSDC is the generator speed (ω). The steady-state output of the SSDC is zero because of the washout block.

The parameters of the generator, network, and multi-mass turbine model are as given in Example 10.4. An AVR with the same parameters as in Example 10.6 is present, but the PSS is not present. The magnitudes of the generator terminal voltage and infinite bus voltage are both 1.0 pu, the quiescent power output is 0.5 pu, and $\alpha_0 = 70°$.

The other parameter values are $X_C = 0.15$ pu, $X_{TC} = 0.05$ pu, and $X_{TL} = 0.006$ pu. The parameters of the speed-input SSDC are $T_f = 1$ ms, $T_w = 5$ s, and $G_c(s) = \frac{1+sT_1}{1+sT_2}$, with $T_1 = 20$ ms and $T_2 = 1$.

To illustrate the damping effect of the TCSC, we simulate this system for a +5 % pulse in the infinite bus voltage magnitude for 0.01 s, with and without a SSDC. The response

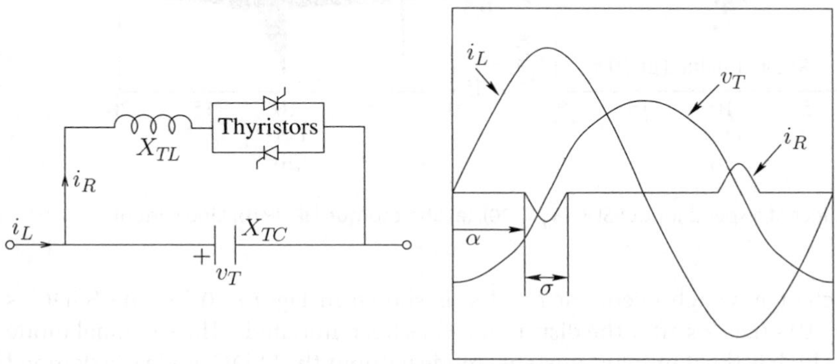

Figure 10.20 A TCSC circuit and its steady-state waveforms.

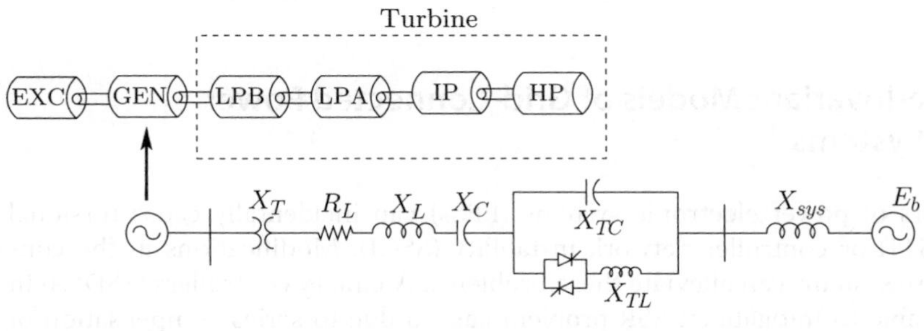

Figure 10.21 A single machine connected to an infinite bus by a TCSC compensated line.

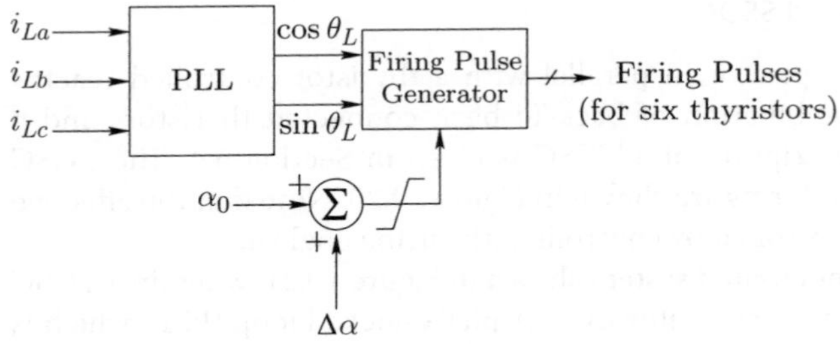

Figure 10.22 TCSC firing system.

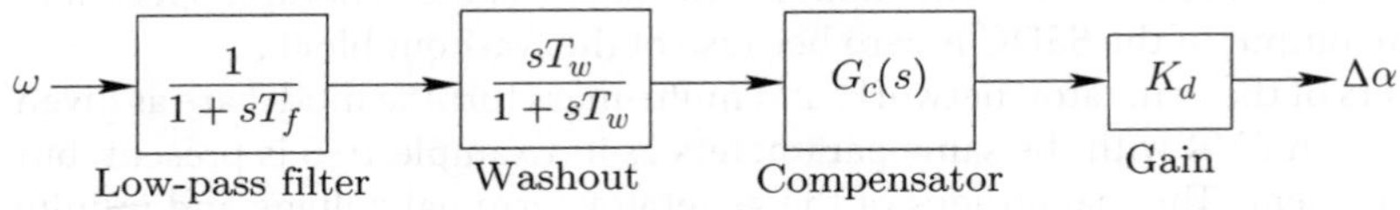

Figure 10.23 Block diagram of a SSDC.

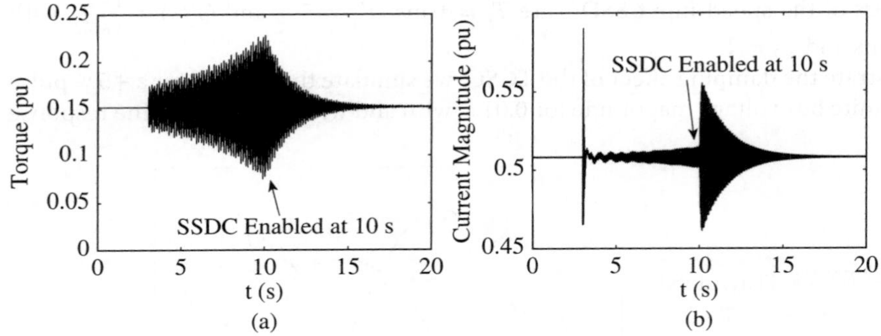

Figure 10.24 Effect of a speed-input SSDC ($K_d = 30$). (a) Shaft torque HP–IP (b) Line current magnitude.

to the pulse change, which occurs at $t = 3$ s, is shown in Figure 10.24. The SSDC is enabled at $t = 10$ s (i.e., 7 s after the disturbance has been initiated). The torsional mode at 202 rad/s, which is the dominant unstable mode without the SSDC, is clearly damped out when the SSDC is enabled. The other torsional modes are also stabilized due to the SSDC.

10.5 Time-Invariant Models of Grid-Connected Power Electronic Systems

The controllers of power electronic systems (PESs) can incidentally cause torsional instability (SSTI) or controller-network instability (SSCI). Modifications in the controllers of these systems can alleviate these problems. Auxiliary controllers (SSDCs) in PESs may be able to mitigate an SSR problem caused due to series compensation of lines. In the case of grid-connected converters (FACTS, HVDC, renewables, storage systems) it is necessary to model the power-electronic converter and controller action in more detail than in a generator excitation system. This is because these systems are of generally of higher power rating and act through the fast-acting electrical transmission network.

Typically, the model of a PES in terms of sequence/phase variables is time-varying due to the switching action of the power electronic devices. The switching instants themselves are determined by a controller, which includes the synchronization and firing system. While it is possible to simulate a detailed model of the entire system in the time domain to evaluate the response of the system (as done in the SSDC case study), characterization of the system using frequency responses and eigenvalues is useful for parametric studies and controller design. Frequency domain analysis and eigenvalue analysis can be performed if the system is linearized and is time-invariant. Since PESs are generally time-periodic systems in the phase variables, it should be possible to derive a time-invariant model of the system using a suitable time-periodic transformation.

The derivation of time-invariant models suitable for the analysis of subsynchronous interactions may be based on the following approaches:

1) *D-Q* variable-based models: These models are based on the assumption that DC-AC conversion is ideal, that is, the switching functions which are used to model converter action can be approximated by fundamental frequency sinusoidal functions. The approach is suitable for voltage source converter (VSC) based systems that use vector-control strategies and have insignificant lower-order switching harmonics. See, for example, the STATCOM model given in Chapter 5.
2) Discrete-time models: These are based on the Poincaré mapping technique. The relationships between successive samples of the variables of the system for small deviations around the periodic steady state are derived using this technique.
3) Dynamic phasor models: These are based on the DC, fundamental, and harmonic frequency components of the dynamic response evaluated using a sliding, finite time-window.
4) Numerical models: LTI models can also be *numerically* extracted from the simulated time-domain responses of the system to low-amplitude probing inputs. Numerical frequency scanning is an example of this approach, wherein a multi-sine probing input is injected and the frequency response of the system is evaluated.

The dynamic phasor and discrete-time approaches are suitable for thyristor-based PESs such as line-commutated converter HVDC systems, thyristor-controlled series compensators (TCSCs), and Static VAr Compensators (SVCs). These systems are switched at a relatively low frequency compared to typical VSC-based systems. Moreover, the turn-off instants of thyristors are dependent on the external circuit and cannot be independently determined by a controller. Therefore, the first approach is generally not suitable for such PESs.

The numerical approach is suitable for both thyristor-based as well as VSC-based systems. The PES can be modeled in detail in a time-domain simulation program. The extracted model will capture all the details represented in this simulation model. In fact, a user need not have the complete details of the underlying simulation model since the model is probed/observed only at available inputs/outputs. Therefore, this approach is also suitable when manufacturers of PESs provide only the black-box simulation models to their clients.

The following subsections describe the basic principle of discrete-time dynamic modeling, dynamic phasor modeling, and numerical modeling. An analytical fundamental frequency dynamic phasor model is derived for a TCSC. The use of the numerical approach (frequency scanning) is also described.

10.5.1 Discrete-Time Dynamic Models using the Poincaré Mapping Technique

Consider a power system with embedded PESs. The steady-state trajectories (of the phase or D-Q variables) of such a system are generally periodic, although they may not be sinusoidal. If the continuous-time system model is linearized around the steady-state trajectories, then the resulting equations are of the form

$$\Delta\dot{x} = A(t)\Delta x,$$

where x are the states of the system. The response of the time-variant system can be represented as follows [38]:

$$\Delta x(t) = \Phi(t, t_o)\Delta x(t_o)$$

where Φ is called the state transition matrix of the system. Φ satisfies the following properties:

1) $\Phi(t, t) = I$, where I is a unity matrix
2) $\Phi(t_1, t_2) = \Phi(t_1, t_3)\Phi(t_3, t_2)$.

Since the system is time-periodic, $A(t) = A(t + T)$, where T is the time period. It follows from Floquet theory [39] that $\Phi(t + T, t_o) = \Phi(t, t_o)A_d$, where A_d is a non-singular matrix. Consider the periodically taken samples of x: $(\Delta x(t_o), \Delta x(t_o + T) \cdots \Delta x(t_o + kT) \cdots)$, which are denoted by $(\Delta x_o, \Delta x_1 \cdots \Delta x_k \cdots)$. For a time-periodic system, the mapping between the successive samples can therefore be expressed as follows:

$$\Delta x_{k+1} = A_d \Delta x_k$$

Note that A_d is sample-invariant (not dependent on k).

The eigenvalues of this discrete-time, sample-invariant model can be used to infer the stability of the system. Note that the system is stable if the magnitude of the eigenvalues of A_d are less than unity. This technique has been applied for computing the damping of subsynchronous oscillations by a TCSC [40]. Considerable simplicity and modularity can be achieved if a time-invariant TCSC model can be derived independently, since the computation of the state matrix of the entire system can be avoided. This is possible if it is assumed that the line current through the TCSC (in the D-Q frame) is constant between samples [41]. This TCSC model can then be interfaced with the analytical model of the rest of the system to obtain the overall system model.

10.5.2 Dynamic Phasor-Based Modeling

The kth Fourier coefficient or dynamic phasor, $\langle x\rangle_k$, corresponding to a signal $x(\tau)$ is defined as follows:

$$\langle x\rangle_k(t) = \frac{1}{T}\int_{t-T}^{t} x(\tau)e^{-jk\omega_o\tau}\, d\tau \tag{10.48}$$

where $\omega_o = \frac{2\pi}{T}$ is the fundamental frequency in rad/s. $x(\tau)$ can then be represented in terms of complex Fourier coefficients over a window length T as given below:

$$x(\tau) = \sum_{k=-\infty}^{\infty} \langle x \rangle_k(t) e^{jk\omega_o \tau}, \qquad \tau \in [t-T, t) \tag{10.49}$$

The important properties of dynamic phasors are given below.

(a) The derivative of a dynamic phasor satisfies the following equation:

$$\frac{d\langle x \rangle_k}{dt} = \left\langle \frac{dx}{dt} \right\rangle_k - jk\omega_o \langle x \rangle_k \tag{10.50}$$

This may be re-written as follows by separating out the real and imaginary terms:

$$\begin{aligned} \frac{d\langle x \rangle_k^{re}}{dt} &= \left\langle \frac{dx}{dt} \right\rangle_k^{re} + k\omega_o \langle x \rangle_k^{im} \\ \frac{d\langle x \rangle_k^{im}}{dt} &= \left\langle \frac{dx}{dt} \right\rangle_k^{im} - k\omega_o \langle x \rangle_k^{re} \end{aligned} \tag{10.51}$$

(b) The dynamic phasor of the product of two signals $u(\tau)$ and $v(\tau)$ can be obtained by the convolution of the corresponding dynamic phasors as follows:

$$\langle uv \rangle_k = \sum_{l=-\infty}^{\infty} \langle u \rangle_{k-l} \langle v \rangle_l \tag{10.52}$$

(c) If $x(\tau)$ is real, then $\langle x \rangle_{-l} = \langle x \rangle_l^*$, where * denotes the complex conjugate.

For time-periodic systems, dynamic phasors are constant in the steady state. The dynamic model of a system in terms of the dynamic phasor variables $\langle x \rangle_k$ can be obtained from the original model, which is in terms of x, using (10.50) [42]. The relationship given in (10.52) is useful if multiplicative switching functions that are time-periodic in steady state are present in the original model. The presence of these multiplicative terms causes a coupling between the dynamic phasors corresponding to the different harmonic components (k). Since k varies from $-\infty$ to ∞, a dynamic phasor model is a coupled system of infinite order. However, a lower order model (which uses only the lower-order harmonic coefficients) is usually sufficient to capture sub-synchronous interactions.

A dynamic phasor model of a power system with PESs is time-invariant and lends itself to conventional eigenvalue analysis. Dynamic phasor models of synchronous machines, networks, and PESs such as TCSCs have been derived in the literature [42–44]. SSR analysis of a TCSC-compensated system using the dynamic phasor approach has been presented in [45]. Following their approach, we present a fundamental frequency dynamic phasor model of a TCSC.

Fundamental Frequency Dynamic Phasor Model of a TCSC

The dynamical equations for one phase of a TCSC in pu form are

$$\frac{1}{\omega_B X_{TC}} \frac{dv_T}{dt} = (i_L - i_R) \tag{10.53}$$

$$\frac{X_{TL}}{\omega_B} \frac{di_R}{dt} = q v_T \tag{10.54}$$

where q is a switching function whose value is 1 when either of the thyristors is conducting and zero otherwise. Note that the switching instants are dependent on the firing angle α and the turn-off instants, as shown in Figure 10.20. ω_B is the base frequency in rad/s. X_{TL} and X_{TC} represent inductive and capacitive reactances, respectively, while i_L, i_R, and v_T denote the line current, reactor current, and TCSC voltage, respectively, as shown in Figure 10.20.

The dynamic phasor model of TCSC for $k = 1$ is then given by the following equations:

$$\frac{1}{\omega_B X_{TC}} \frac{d\langle v_T \rangle_1}{dt} = (\langle i_L \rangle_1 - \langle i_R \rangle_1) - \frac{j\omega_o}{\omega_B X_{TC}} \langle v_T \rangle_1 \tag{10.55}$$

$$\frac{X_{TL}}{\omega_B} \frac{d\langle i_R \rangle_1}{dt} = \langle q v_T \rangle_1 - \frac{j\omega_o X_{TL}}{\omega_B} \langle i_R \rangle_1 \tag{10.56}$$

The expansion of the term $\langle q v_T \rangle_1$ in accordance with (10.52) causes a coupling of these phasor equations ($k = 1$) with those corresponding to other k. A simplified fundamental-frequency-only model may, however, be derived by making the following approximations/assumptions [45].

1) It is expected that for the transients of interest (SSR), $\frac{X_{TL}}{\omega_B} \frac{d\langle i_R \rangle_1}{dt}$ is relatively much smaller than the other terms in (10.56). Moreover, the frequency ω_o is normally very close to ω_B or equal to it. Therefore, $\langle q v_T \rangle_1 \approx \frac{j\omega_o X_{TL}}{\omega_B} \langle i_R \rangle_1$ and we are left with a single differential equation as given below:

$$\frac{1}{\omega_B X_{TC}} \frac{d\langle v_T \rangle_1}{dt} = \langle i_L \rangle_1 - \left(\frac{j\langle v_T \rangle_1}{X_{TC}} + \frac{\langle q v_T \rangle_1}{jX_{TL}} \right) \tag{10.57}$$

2) It is assumed that $\frac{\langle q v_T \rangle_1}{jX_{TL}}$ can be written as follows: $\frac{\langle q v_T \rangle_1}{jX_{TL}} = \frac{\langle v_T \rangle_1}{jX_{eq}(\sigma)}$. σ is the prevailing conduction angle (see Figure 10.20), which is given by

$$\sigma = \sigma_{ref} + 2\angle(-j\langle i_L \rangle_1 \, \langle v_T \rangle_1^*) \tag{10.58}$$

where $\langle v_T \rangle_1^*$ denotes the complex conjugate of $\langle v_T \rangle_1$. This equation assumes that the firing is synchronized with the line current zeros.

The conduction angle reference σ_{ref} is related to the firing angle reference α_{ref} by the relation $\sigma_{ref} = \pi - 2\alpha_{ref}$.

The fundamental frequency state-space model can now be written as follows:

$$\begin{aligned} \frac{1}{\omega_B X_{TC}} \frac{d\langle v_T \rangle_1}{dt} &= \langle i_L \rangle_1 - \left(\frac{j\langle v_T \rangle_1}{X_{TC}} + \frac{\langle v_T \rangle_1}{jX_{eq}(\sigma)} \right) \\ &= \langle i_L \rangle_1 - \frac{j\langle v_T \rangle_1}{X_{TCSC}(\sigma)} \end{aligned} \tag{10.59}$$

In steady state, that is, when $\frac{d\langle v_T \rangle_1}{dt} = 0$, X_{TCSC} represents the fundamental-frequency capacitive reactance of the TCSC. $X_{TCSC}(\sigma)$ should therefore be consistent with the TCSC fundamental-frequency capacitive reactance expression which was derived earlier in Chapter 5 (see equation (5.28)):

$$\frac{X_{TCSC}}{X_{TC}} = 1 + \frac{2}{\pi} \frac{\lambda^2}{(\lambda^2 - 1)} \left[\frac{2\cos^2 \frac{\sigma}{2}}{(\lambda^2 - 1)} \left(\lambda \tan \frac{\lambda\sigma}{2} - \tan \frac{\sigma}{2} \right) - \frac{\sigma}{2} - \frac{\sin \sigma}{2} \right] \tag{10.60}$$

where $\lambda = \sqrt{\frac{X_{TC}}{X_{TL}}}$. $X_{TCSC}(\sigma)$ given in (10.60) is now used in the differential equation (10.59), thereby completing the derivation of the model.

The dynamic phasor model (10.59) is obtained here on a per-phase basis. The single complex equation corresponding to one phase can be split into two real differential equations by separating out the real and imaginary parts. These equations can then be interfaced with the dynamic phasor model of the rest of the system, which may include synchronous generators [23]. An alternative interfacing approach is given below.

It is shown in [23] that the fundamental-frequency dynamic phasors ($k = 1$) of the phase variables can be related to the dynamic phasors of the D-Q variables corresponding to $k = 0$ by the following equations:

$$\langle v_{TQ}\rangle_0 + j\langle v_{TD}\rangle_0 = j\sqrt{\frac{2}{3}}[\langle v_{Ta}\rangle_1 + \alpha\langle v_{Tb}\rangle_1 + \alpha^2\langle v_{Tc}\rangle_1] \tag{10.61}$$

$$\langle v_{TD}\rangle_0 + j\langle v_{TQ}\rangle_0 = \sqrt{\frac{2}{3}}[\langle v_{Ta}\rangle_1^* + \alpha^2\langle v_{Tb}\rangle_1^* + \alpha\langle v_{Tc}^*\rangle_1] \tag{10.62}$$

where $\alpha = e^{j120^\circ}$. Similar equations can be written for the line current. The fundamental frequency zero-sequence components may be assumed to be zero. Using this additional constraint it is possible to convert the D-Q dynamic phasor variables ($k = 0$) to the dynamic phasors in the phase variables ($k = 1$), and vice versa.

Note that the dynamic phasors corresponding to $k = 0$ are real and are the moving average values of the corresponding variables over a cycle. If we assume that the variations within a cycle are small, then $\langle v_{TD}\rangle_0 \approx v_{TD}$ and $\langle v_{TQ}\rangle_0 \approx v_{TQ}$, which holds true similarly for the line current. With this assumption, the equations of the TCSC can be directly interfaced with the D-Q model of the rest of the system (note that i_{LD} and i_{LQ} are the inputs and v_{TD} and v_{TQ} are the outputs of the state-space model of the TCSC).

10.5.3 Numerical Derivation of PES Models: A Frequency Scanning Approach

Frequency scanning involves the use of a time-domain simulation program to numerically obtain the small-signal frequency response of a system [10]. In this technique, a small-amplitude wide-band periodic signal is injected as an input in the time-domain simulation of the system. This injection is superimposed upon the existing sources which are required to set up the equilibrium conditions around which the frequency response is desired. The small-signal frequency response of the system is obtained in the periodic steady state by computing the frequency components of the injected signal and the output signals. Examples of input and output signals are current and voltage (to obtain the response of the impedance), and generator speed and electrical torque (to obtain synchronizing/damping torques). The basic principle of this method is given in Section 9.4.2, where it was used for the estimation of the low-frequency swing modes of the system from synchronized phasor measurements. For the study of network–controller–torsional interactions we need to obtain the frequency response in a larger frequency range (typically from 0 to 150 Hz). The scanning can be done conveniently in the D-Q variables, as described next.

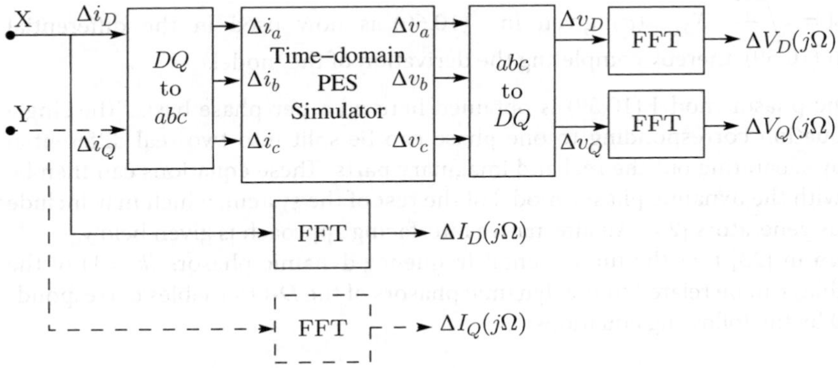

Figure 10.25 *D-Q*-based injection for scanning the impedance. The multi-sine signal is injected one at a time at X and Y.

***D-Q*-Based Frequency Scanning [46]**

A frequency scanning scheme to obtain the frequency response of the impedance of a PES in the *D-Q* variables is shown in Figure 10.25. The implicit assumption is that the underlying model is LTI in the *D-Q* variables, and the PES which is being probed is stable. The zero-sequence variables are not shown here and are assumed to be negligible.

For obtaining the impedance scan, a wide-band, small-magnitude periodic signal is injected one at a time at the *D-Q* current ports as shown in the figure. The system is simulated using a time-domain simulator and Δv_D and Δv_Q are measured. Once we have the individual measurements of all outputs and the inputs, we can find the frequency response using the fast Fourier transform (FFT) algorithm. A vector-fitting tool may then be used to fit a rational transfer function matrix to this frequency response [47]. The transfer function matrix in the Laplace domain may then be obtained as follows:

$$\begin{bmatrix}\Delta V_D(s)\\ \Delta V_Q(s)\end{bmatrix} = \begin{bmatrix}Z_{DD}(s) & Z_{DQ}(s)\\ Z_{QD}(s) & Z_{QQ}(s)\end{bmatrix}\begin{bmatrix}\Delta I_D(s)\\ \Delta I_Q(s)\end{bmatrix} \tag{10.63}$$

where

$$Z_{DD}(s) = \left.\frac{\Delta V_D(s)}{\Delta I_D(s)}\right|_{\Delta I_Q(s)=0}, \quad Z_{DQ}(s) = \left.\frac{\Delta V_D(s)}{\Delta I_Q(s)}\right|_{\Delta I_D(s)=0},$$

$$Z_{QD}(s) = \left.\frac{\Delta V_Q(s)}{\Delta I_D(s)}\right|_{\Delta I_Q(s)=0}, \quad Z_{QQ}(s) = \left.\frac{\Delta V_Q(s)}{\Delta I_Q(s)}\right|_{\Delta I_D(s)=0}$$

In addition to the impedance matrix, the transfer function between other inputs and outputs can be obtained using the same approach. For example, the set-point of a controller could be an input and the outputs could be the feedback signals for the controller. For each additional input, a separate frequency scan has to be performed.

A convenient way of utilizing this multi-input multi-output transfer function model is to convert it to state-space form. Since state-space models in the *D-Q* variables of most other components (e.g., a synchronous generator and the electrical network) are available, it is easy to combine the state-space model obtained from frequency scanning of the

PES with the rest of the system. Eigenvalue analysis of the overall system can then be performed to check the stability of the system [48].

10.A Transfer Function Representation of the Network

The transfer function model of a network (impedance or admittance) is useful in the understanding of the subsynchronous resonance phenomenon. The impedance/admittance of a three-phase network is in the form of a matrix transfer function as there are multiple inputs and outputs.

For example, the impedance matrix of the L-C branch considered in the previous section may be written in α-β variables as follows:

$$\begin{bmatrix} V_{S\alpha}(s) \\ V_{S\beta}(s) \end{bmatrix} = \begin{bmatrix} Z_\alpha(s) & 0 \\ 0 & Z_\beta(s) \end{bmatrix} \begin{bmatrix} I_{L\alpha}(s) \\ I_{L\beta}(s) \end{bmatrix} \tag{10.64}$$

where $Z(s) = Z_\alpha(s) = Z_\beta(s) = s(L_s - L_m) - \frac{1}{sC} = \frac{(L_s - L_m)C\,s^2 - 1}{sC}$. In the time domain, the D-Q and α-β variables are related as follows (see Section 2.4):

$$\begin{bmatrix} f_D(t) \\ f_Q(t) \end{bmatrix} = \begin{bmatrix} \cos\omega_o t & -\sin\omega_o t \\ \sin\omega_o t & \cos\omega_o t \end{bmatrix} \begin{bmatrix} f_\alpha(t) \\ f_\beta(t) \end{bmatrix} \tag{10.65}$$

Therefore, in the frequency domain they are related as follows:

$$\begin{aligned} \begin{bmatrix} F_D(s) \\ F_Q(s) \end{bmatrix} &= \frac{1}{\sqrt{2}} \begin{bmatrix} 1 & 1 \\ j & -j \end{bmatrix} \begin{bmatrix} \dfrac{F_\alpha(s+j\omega_o) - jF_\beta(s+j\omega_o)}{\sqrt{2}} \\ \dfrac{F_\alpha(s-j\omega_o) + jF_\beta(s-j\omega_o)}{\sqrt{2}} \end{bmatrix} \\ &= T_f \begin{bmatrix} \dfrac{F_\beta(s+j\omega_o) + jF_\alpha(s+j\omega_o)}{\sqrt{2}} \\ \dfrac{F_\beta(s-j\omega_o) - jF_\alpha(s-j\omega_o)}{\sqrt{2}} \end{bmatrix} = T_f \begin{bmatrix} F_p(s+j\omega_o) \\ F_n(s-j\omega_o) \end{bmatrix} \end{aligned} \tag{10.66}$$

where $F_p(s) = \frac{F_\alpha(s) - jF_\beta(s)}{\sqrt{2}}$ and $F_n(s) = \frac{F_\alpha(s) + jF_\beta(s)}{\sqrt{2}}$. These expressions are derived using the following identities:

$$f(t) \leftrightarrow F(s) \Rightarrow f(t)e^{j\omega_o t} \leftrightarrow F(s - j\omega_o),$$

$$\cos(\omega_o t) = \frac{e^{j\omega_o t} + e^{-j\omega_o t}}{2} \quad \text{and} \quad \sin(\omega_o t) = \frac{e^{j\omega_o t} - e^{-j\omega_o t}}{j2}$$

Since $Z(s) = Z_\alpha(s) = Z_\beta(s)$, we can infer the following:

$$\begin{bmatrix} V_p(s+j\omega_o) \\ V_n(s-j\omega_o) \end{bmatrix} = \begin{bmatrix} Z(s+j\omega) & 0 \\ 0 & Z(s+j\omega) \end{bmatrix} \begin{bmatrix} I_p(s+j\omega_o) \\ I_n(s-j\omega_o) \end{bmatrix} \tag{10.67}$$

Therefore, if the impedance matrix in the D-Q variables is represented as follows,

$$\begin{bmatrix} V_{SD}(s) \\ V_{SQ}(s) \end{bmatrix} = \begin{bmatrix} Z_{DD}(s) & Z_{DQ}(s) \\ Z_{QD}(s) & Z_{QQ}(s) \end{bmatrix} \begin{bmatrix} I_{LD}(s) \\ I_{LQ}(s) \end{bmatrix} \tag{10.68}$$

then for the L-C network

$$\begin{bmatrix} Z_{DD}(s) & Z_{DQ}(s) \\ Z_{QD}(s) & Z_{QQ}(s) \end{bmatrix} = T_f \begin{bmatrix} Z(s+j\omega) & 0 \\ 0 & Z(s-j\omega) \end{bmatrix} T_f^{-1} \tag{10.69}$$

From this we obtain

$$Z_{DD}(s) = Z_{QQ}(s) = \frac{Z_\alpha(s+j\omega_o) + Z_\alpha(s-j\omega_o)}{2} \tag{10.70}$$

$$Z_{DQ}(s) = -Z_{QD}(s) = \frac{Z_\alpha(s+j\omega_o) - Z_\alpha(s-j\omega_o)}{2j} \tag{10.71}$$

Remarks

1) The transfer function representation is illustrated using a simple L-C network, but can be extended to larger symmetrical three-phase networks.
2) The relationships $Z(s) = Z_\alpha(s) = Z_\beta(s)$, $Z_{DD}(s) = Z_{QQ}(s)$, and $Z_{QD}(s) = -Z_{DQ}(s)$ hold true for a balanced transmission network, but this may not be true for all power system components.
3) The variables $F_p(s)$ and $F_n(s)$ are, in fact, positive and negative sequence variables, respectively, since it is evident from Clarke's transformation that:

$$\begin{bmatrix} F_p(s) \\ F_n(s) \\ F_o(s) \end{bmatrix} = \begin{bmatrix} \frac{1}{\sqrt{2}} & \frac{-j}{\sqrt{2}} & 0 \\ \frac{1}{\sqrt{2}} & \frac{j}{\sqrt{2}} & 0 \\ 0 & 0 & 1 \end{bmatrix} \begin{bmatrix} F_\alpha(s) \\ F_\beta(s) \\ F_o(s) \end{bmatrix} = \frac{1}{\sqrt{3}} \begin{bmatrix} 1 & e^{j120^\circ} & e^{-j120^\circ} \\ 1 & e^{-j120^\circ} & e^{j120^\circ} \\ 1 & 1 & 1 \end{bmatrix} \begin{bmatrix} F_a(s) \\ F_b(s) \\ F_c(s) \end{bmatrix}$$

References

1 Hall, M.C. and Hodges, D.A. (1976) Experience with 500 kV subsynchronous resonance and resulting turbine generator shaft damage at Mohave generating station, in *IEEE PES Winter Meeting*, New York.

2 Concordia, C. and Carter, G.K. (1941) Negative damping of electrical machinery. *Transactions of the American Institute of Electrical Engineers*, **60** (3), 116–119.

3 Kilgore, L.A., Elliott, L.C., and Taylor, E.R. (1971) The prediction and control of self-excited oscillations due to series capacitors in power systems. *IEEE Transactions on Power Apparatus and Systems*, **PAS-90** (3), 1305–1311.

4 Irwin, G.D., Jindal, A.K., and Isaacs, A.L. (2011) Sub-synchronous control interactions between Type 3 wind turbines and series compensated AC transmission systems, in *IEEE PES General Meeting*, San Diego, pp. 1–6.

5 IEEE Subsynchronous Resonance Task Force (1977) First benchmark model for computer simulation of subsynchronous resonance. *IEEE Transactions on Power Apparatus and Systems*, **96** (5), 1565–1572.

6 Kotian, S.M. and Shubhanga, K.N. (2014) Performance of synchronous machine models in a series-capacitor compensated system. *IEEE Transactions on Power Systems*, **29** (3), 1023–1032.

7 Dommel, H.W. (1969) Digital computer solution of electromagnetic transients in single- and multiphase networks. *IEEE Transactions on Power Apparatus and Systems*, **PAS-88** (4), 388–399.

8 Gole, A.M., Fernando, I.T., Irwin, G.D., and Nayak, O.B. (1997) Modeling of power electronic apparatus: Additional interpolation issues, in *International Conference on Power System Transients*, Seattle, pp. 23–28.

9 Hammad, A.E. (1992) Analysis of second harmonic instability for the Chateauguay HVDC/SVC scheme. *IEEE Transactions on Power Delivery*, **7** (1), 410–415.

10 Jiang, X. and Gole, A.M. (1995) A frequency scanning method for the identification of harmonic instabilities in HVDC systems. *IEEE Transactions on Power Delivery*, **10** (4), 1875–1881.

11 Ainsworth, J.D. (1967) Harmonic instability between controlled static convertors and AC networks. *Proceedings of the Institution of Electrical Engineers*, **114** (7), 949–957.

12 Katz, E., Tang, J., Bowler, C.E.J. *et al.* (1989) Comparison of SSR calculations and test results. *IEEE Transactions on Power Systems*, **4** (1), 336–344.

13 Pagola, F.L., Perez-Arriaga, I.J., and Verghese, G.C. (1989) On sensitivities, residues and participations: applications to oscillatory stability analysis and control. *IEEE Transactions on Power Systems*, **4** (1), 278–285.

14 Andersson, G., Atmuri, R., Rosenqvist, R., and Torseng, S. (1984) Influence of hydro units' generator-to-turbine inertia ratio on damping of subsynchronous oscillations. *IEEE Transactions on Power Apparatus and Systems*, **PAS-103** (8), 2352–2361.

15 Lee, D.C., Beaulieu, R.E., and Rogers, G.J. (1985) Effects of governor characteristics on turbo-generator shaft torsionals. *IEEE Transactions on Power Apparatus and Systems*, **PAS-16** (6), 1254–1261.

16 Watson, W. and Coultes, M.E. (1973) Static exciter stabilizing signals on large generators – mechanical problems. *IEEE Transactions on Power Apparatus and Systems*, **PAS-92** (1), 204–211.

17 Larsen, E.V. and Swann, D.A. (1981) Applying power system stabilizers Part I: General concepts, Part II: Performance objectives and tuning concepts, and Part III: Practical considerations. *IEEE Transactions on Power Apparatus and Systems*, **PAS-100** (6), 3017–3046.

18 Lee, D.C., Beaulieu, R.E., and Service, J.R.R. (1981) A power system stabilizer using speed and electrical power inputs – design and field experience. *IEEE Transactions on Power Apparatus and Systems*, **PAS-100** (9), 4151–4157.

19 Padiyar, K.R. (1999) *Analysis of Subsynchronous Resonance in Power Systems*, Kluwer, Boston.

20 Farmer, R.G., Schwalb, A.L., and Katz, E. (1977) Navajo project report on subsynchronous resonance analysis and solutions. *IEEE Transactions on Power Apparatus and Systems*, **96** (4), 1226–1232.

21 Edris, A.A. (1993) Subsynchronous resonance countermeasure using phase imbalance. *IEEE Transactions on Power Systems*, **8** (4), 1438–1447.

22 Subhash, S., Sarkar, B.N., and Padiyar, K.R. (2001) A novel control strategy for TCSC to damp subsynchronous oscillations, in *Seventh International Conference on AC-DC Power Transmission*, London, pp. 181–186.

23 Chudasama, M.C. and Kulkarni, A.M. (2011) Dynamic phasor analysis of SSR mitigation schemes based on passive phase imbalance. *IEEE Transactions on Power Systems*, **26** (3), 1668–1676.

24 Hingorani, N.G. (1981) A new scheme for subsynchronous resonance damping of torsional oscillations and transient torque – Part I. *IEEE Transactions on Power Apparatus and Systems*, **PAS-100** (4), 1852–1855.

25 Hedin, R.A., Stump, K.B., and Hingorani, N.G. (1981) A new scheme for subsynchronous resonance damping of torsional oscillations and transient torque – Part II, Performance. *IEEE Transactions on Power Apparatus and Systems*, **PAS-100** (4), 1856–1863.

26 Hingorani, N.G., Bhargava, B., Garrigue, G.F., and Rodriguez, G.D. (1987) Prototype NGH subsynchronous resonance damping scheme Part I – Field installation and operating experience. *IEEE Transactions on Power Systems*, **2** (4), 1034–1039.

27 Benko, I.S., Bhargava, B., and Rothenbuhler, W.N. (1987) Prototype NGH subsynchronous resonance damping scheme Part II – Switching and short circuit tests. *IEEE Transactions on Power Systems*, **2** (4), 1040–1047.

28 Cheriyan, E.P. and Kulkarni, A.M. (2007) Discrete-time dynamic model of NGH damper. *IEEE Transactions on Power Systems*, **22** (4), 1888–1897.

29 Bahrman, M., Larsen, E.V., Piwko, R.J., and Patel, H.S. (1980) Experience with HVDC – turbine-generator torsional interaction at Square Butte. *IEEE Transactions on Power Apparatus and Systems*, **PAS-99** (3), 966–975.

30 Rostamkolai, N., Piwko, R.J., Larsen, E.V. *et al.* (1990) Subsynchronous interactions with static var compensators-concepts and practical implications. *IEEE Transactions on Power Systems*, **5** (4), 1324–1332.

31 Bowler, C.E.J. and Baker, D.H. (1981) Concepts of countermeasures for subsynchronous supplementary torsional damping by excitation modulation. *IEEE Special Publication, Symposium on Countermeasures for Subsynchronous Resonance, IEEE Publication 81TH0086-9-PWR*, pp. 64–69.

32 Putman, T.H. and Ramey, D.G. (1982) Theory of the modulated reactance solution for subsynchronous resonance. *IEEE Transactions on Power Apparatus and Systems*, **PAS-101** (6), 1527–1535.

33 Wasynczuk, O. (1981) Damping subsynchronous resonance using reactive power control. *IEEE Transactions on Power Apparatus and Systems*, **PAS-100** (3), 1096–1104.

34 Ramey, D.G., Kimmel, D.S., Dorney, J.W., and Kroening, F.H. (1981) Dynamic stabilizer verification tests at the San Juan station. *IEEE Transactions on Power Apparatus and Systems*, **PAS-100** (12), 5011–5019.

35 Hammad, A.E. and El-Sadek, M. (1984) Application of a thyristor controlled var compensator for damping subsynchronous oscillations in power systems. *IEEE Transactions on Power Apparatus and Systems*, **PAS-103** (1), 198–212.

36 Svensson, S. and Mortensen, K. (1981) Damping of subsynchronous oscillations by an HVDC link: An HVDC simulator study. *IEEE Power Engineering Review*, **PER-1** (3), 45–46.

37 Piwko, R.J. and Larsen, E.V. (1982) HVDC system control for damping of subsynchronous oscillations. *IEEE Transactions on Power Apparatus and Systems*, **PAS-101** (7), 2203–2211.

38 Kailath, T. (1980) *Linear Systems*, Prentice Hall, Englewood Cliffs.

39 Chen, C.T. (1995) *Linear System Theory and Design*, Oxford University Press, New York, 2nd edn.

40 Rajaraman, R., Dobson, I., Lasseter, R.H., and Shern, Y. (1996) Computing the damping of subsynchronous oscillations due to a thyristor controlled series capacitor. *IEEE Transactions on Power Delivery*, **11** (2), 1120–1127.

41 Othman, H.A. and Angquist, L. (1996) Analytical modeling of thyristor-controlled series capacitors for SSR studies. *IEEE Transactions on Power Systems*, **11** (1), 119–127.

42 Stankovic, A.M., Sanders, S.R., and Aydin, T. (2002) Dynamic phasors in modeling and analysis of unbalanced polyphase AC machines. *IEEE Transactions on Energy Conversion*, **17** (1), 107–113.

43 Demiray, T.H. (2008) *Simulation of power system dynamics using dynamic phasor models*, Ph.D. thesis, Department of Information Technology and Electrical Engineering, Swiss Federal Institute of Technology, Zurich.

44 Chudasama, M.C. (2012) *Application of dynamic phasor models for the analysis of power system with phase imbalance*, Ph.D. thesis, Department of Electrical Engineering, Indian Institute of Technology Bombay, Mumbai.

45 Mattavelli, P., Stankovic, A.M., and Verghese, G.C. (1999) SSR analysis with dynamic phasor model of thyristor-controlled series capacitor. *IEEE Transactions on Power Systems*, **14** (1), 200–208.

46 Kulkarni, A.M., Das, M.K., and Gole, A.M. (2016) Frequency scanning analysis of STATCOM – network interactions, in *6th International Conference on Power Systems (ICPS)*, New Delhi, pp. 1–6.

47 Gustavsen, B. and Semlyen, A. (1999) Rational approximation of frequency domain responses by vector fitting. *IEEE Transactions on Power Delivery*, **14** (3), 1052–1061.

48 Das, M.K. and Kulkarni, A.M. (2017) Inclusion of frequency scanning based DQ models of FACTS and HVDC systems in large scale eigen-value analysis programs for analysis of torsional interactions, in *13th IET International Conference on AC and DC Power Transmission*, Manchester, pp. 1–6.

11
Solar Power Generation and Energy Storage

11.1 Introduction

Solar energy can be converted to electricity in the following ways:

1) Solar thermal power conversion: Solar radiation (photons) interacts with a material to increase the kinetic energy of atoms (heat). This heat is used to produce steam, which is then used to drive a turbine. The turbine is coupled with an electrical generator, which produces electricity.
2) Solar photovoltaic (PV) conversion: Photons falling on a semiconductor P-N junction cause a voltage to be generated across the junction (photovoltaic effect). This can drive current into an external circuit and deliver power to it.

Electricity generation systems based on both solar thermal and solar PV technologies have been deployed in the world. The global installed capacity of solar PV in 2015 was about 220 GW [1], while for solar thermal conversion systems it was less than 5 GW. Solar PV systems have seen the highest annual growth rates among the renewable energy systems due to numerous advantages such as quiet conversion, easy design and installation, and falling costs.

The size of solar PV plants may vary from a few kilowatts (e.g., roof-top solar plants) to large farms of hundreds of megawatts. Depending on their capacity, the plants may be connected to the low or medium voltage distribution grid or the high-voltage grid. They may also be operated in the stand-alone mode (no grid connection). The largest solar PV plant in 2015 had a total power rating of 579 MW [1], but plants much larger than this are now being installed. A significant proportion of solar PV generation is in the form of a large number of smaller capacity units connected close to the points of consumption (distributed generation). This is in contrast to the large centralized power stations in a conventional grid. Solar PV generation is static, that is, it does not involve any rotating parts. The most significant dynamics are due to the control of the power-electronic converters, which are needed to interface the solar panels to the grid.

Solar power is available only during daytime and when it is not cloudy. Therefore, the power generated by solar PV systems is intermittent and variable. This means that either the conventional generation has to adjust its power output to maintain the grid frequency or energy storage devices (if present) store/supply the balance power. While pumped hydro-storage plants provide this functionality to a limited extent in some power grids, other storage technologies like batteries are also being considered for smoothening short-term variations. Besides the large-scale grid applications, energy

Dynamics and Control of Electric Transmission and Microgrids, First Edition.
K. R. Padiyar and Anil M. Kulkarni.

storage systems are needed at lower power levels for improving the power quality, for emergency power supplies, and for matching power demand in isolated grids. In this chapter, we present the important features of solar PV generation and an overview of electrical storage technologies.

11.2 Solar Thermal Power Generation

In solar thermal electric power generation systems, solar radiation is concentrated onto a small area in order to achieve high temperatures. This is achieved by using cylindrical-parabolic collectors (also referred to as parabolic troughs or linear parabolic collectors), which concentrate the radiation along a line, or by point-focus parabolic concentrators. Point-focus parabolic concentration may be achieved by using thousands of distributed tracking mirrors (heliostats) which focus light on to a point on a central tower. The parabolic shape ensures that incoming light rays which are parallel to the concentrator's axis will be reflected toward the focus. Other technologies include the linear Fresnel reflector, dish-stirling, and solar chimneys. Over 60% of applications use cylindrical-parabolic collectors [1]. The concentrating systems require sun tracking to maintain the focus at the collector as the power output from such systems falls drastically under diffused light conditions.

The high temperatures are used to generate steam, which runs a conventional steam-turbine synchronous generator. The heat from the concentrator may be directly used to produce steam. Alternatively, it may be used to heat a material such as molten salt, which acts as a thermal energy storage medium. The stored energy can be transferred to produce steam as per the power demand. This allows a solar thermal plant to produce electricity at night and on cloudy days. A schematic of a solar thermal generation system is shown in Figure 11.1. The electrical system is similar to a conventional power plant (synchronous generator connected to a grid). Parabolic trough-based solar thermal plants may achieve at least 25% efficiency [1]. As compared to solar PV systems, solar thermal systems are suitable for higher power generation, but they have disadvantages, such as significantly higher installation and running costs, and a higher sensitivity to dust and humidity.

11.3 Solar PV Power Generation

The basic unit of a solar PV generation system is a solar cell, which is a P-N junction diode. Solar radiation causes a large number of electron-hole pairs to be created in

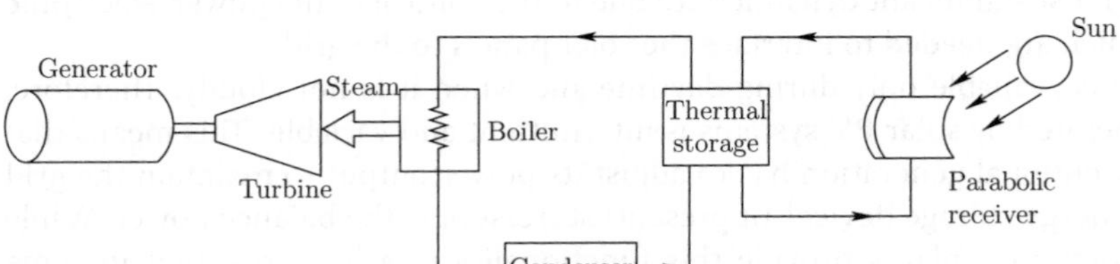

Figure 11.1 A solar-thermal electrical power generation with thermal storage.

the semiconductor material. The asymmetry in a P-N junction provides a built-in electric field at the junction, therefore the generated electrons and holes flow from the P-side to the N-side and from the N-side to the P-side, respectively. A voltage appears across the diode, which can drive current into an external circuit and deliver power to it.

While crystalline silicon cells (mono-crystalline and multi-crystalline) are the most prevalent solar cells (about 93% in 2015), thin-film-based solar cells (amorphous silicon (a-Si), cadmium telluride (Cd-Te), and copper indium gallium selenide (CIGS)) also have a significant proportion of the market (about 7% in 2015) [1]. Other PV technologies that are being explored include organic, dye-sensitized, and gallium arsenide (GaAs) solar cells [2]. Over the past ten years, the efficiency of average commercially available wafer-based modules has increased from 12% to 17%, while it has increased from 9% to 16% for Cd-Te thin-film modules [3].

Solar PV modules are made up of many solar cells which are connected in series and parallel in order to reach convenient values of DC voltage and power. While the typical power output of a solar cell is around 3–4 W, a module may be available from 3 W to 300 W (peak power). Several modules may then be connected in series and parallel to form solar PV arrays. Several such arrays may be used in a solar power plant.

11.3.1 Solar Module I-V Characteristics

From an external circuit perspective, a solar PV module is characterized by its current-voltage (I-V) curves, which depend on the amount of solar radiation falling over a unit area (irradiance) and the temperature. Typical I-V curves for different values of solar irradiation at a given temperature are shown in Figure 11.2a. These may be obtained from the information given in the datasheets of the module, using the method given in [4]. A manufacturer generally specifies the peak output power under standard test conditions (1000 W/m^2 solar irradiance with the specified spectral characteristics and 25°C module temperature).

The power delivered to an external circuit ($v \times i$) as a function of the voltage across the circuit is shown in Figure 11.2b. Maximum power extraction is possible if the load (the external circuit) draws a current corresponding to the maximum power point shown in the figure. Note that the output voltage, current, and power are determined by the intersection of I-V characteristic of the external circuit with that of the PV module. For example, consider the case when a fixed resistance R is connected to the PV module, as shown in Figure 11.3. The power extracted is a maximum if $R = 3.83\ \Omega$ for the case that is considered here.

In PV arrays, where several modules are connected in series and parallel to achieve higher power ratings, the overall I-V characteristic and the load characteristic determine the output power. A mismatch in the parameters of the modules and/or non-identical operating conditions (e.g., partial shading of modules) may result in suboptimal power extraction and may cause hot spots in the modules [2].

11.3.2 Solar PV Connections and Power Extraction Strategies

The output of the solar array(s) can be fed to a load, energy storage devices like batteries, an external grid, or a combination of these. Solar systems may be classified

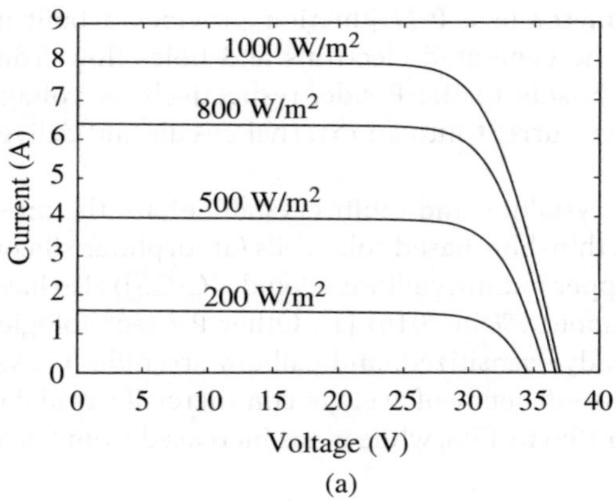

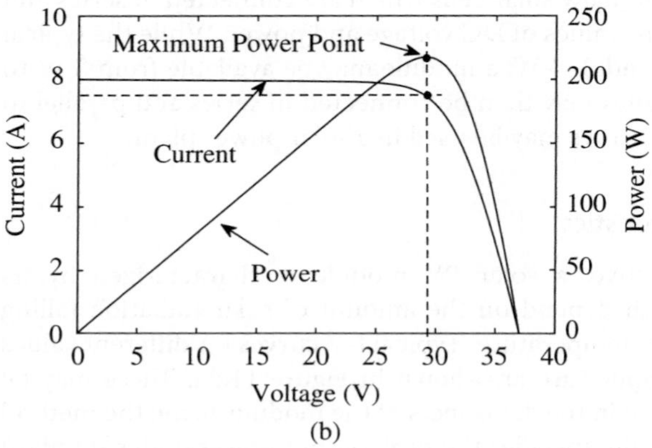

Figure 11.2 I-V curves for a solar module. (a) I-V curves for different solar radiation levels (b) I-V curve and the corresponding power output.

as stand-alone or grid-connected systems. In an unregulated stand-alone system, the output of a solar array is directly fed to a load, as shown in Figure 11.3.

In other systems (see Figure 11.4), power electronic converters are necessary to regulate the load voltage magnitude (AC and DC loads) and frequency (for AC loads), and/or facilitate controlled extraction of power. Since solar energy is a free resource, it makes

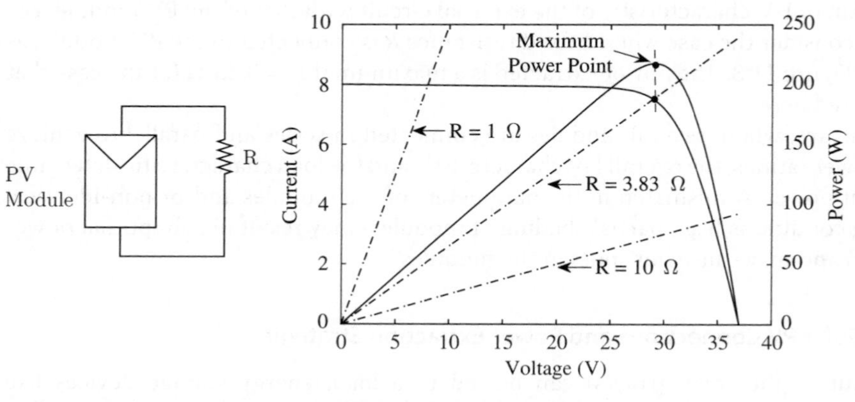

Figure 11.3 Operating point when a PV module is connected to a resistance *R*.

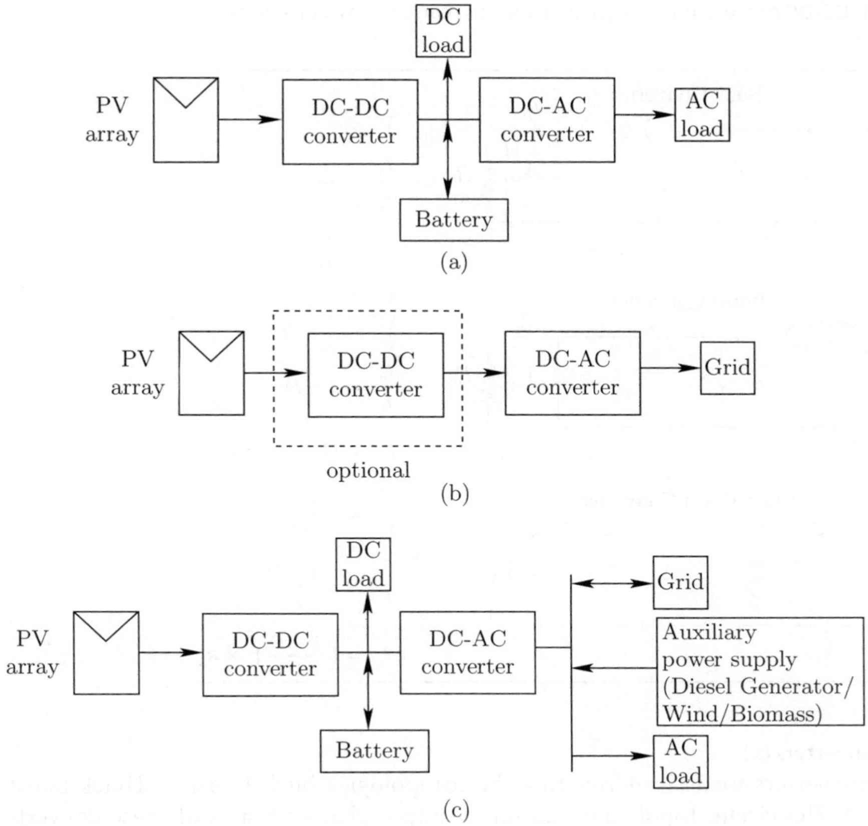

Figure 11.4 Solar-PV load connections. (a) Regulated stand-alone system with energy storage (b) Grid-connected system (c) Hybrid system.

sense to extract maximum power from the solar arrays. This may be pumped into a large grid or stored in devices such as batteries (when availability exceeds the demand). The connection of solar panels to a relatively large capacity grid facilitates the absorption of the variable or intermittent power from the solar arrays without significant deviations in voltage and frequency.

If the solar plant is connected to a weak or small grid, or is a stand-alone system without much storage capability, then it is necessary to control the extracted power based on the prevailing demand. The control strategy has to be coordinated with the other generators, loads, and storage devices present in the grid.

11.3.3 Power Electronic Converters for Solar PV Applications

The power electronic converters used in solar systems are usually DC-DC converters and DC-AC converters. Either or both these converters may be necessary depending on whether the solar panel is connected to a DC load, an AC load or an AC grid (or all of these). For a grid-connected PV system, single-stage or two-stage conversion may be employed, as shown in Figure 11.4b.

Table 11.1 DC-DC converters characteristics, the load current is assumed to be continuous.

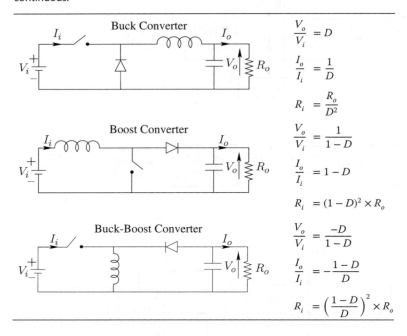

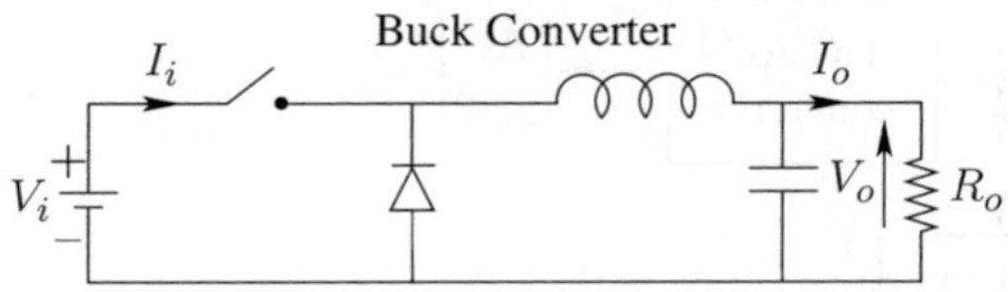	$\frac{V_o}{V_i} = D$ $\frac{I_o}{I_i} = \frac{1}{D}$ $R_i = \frac{R_o}{D^2}$
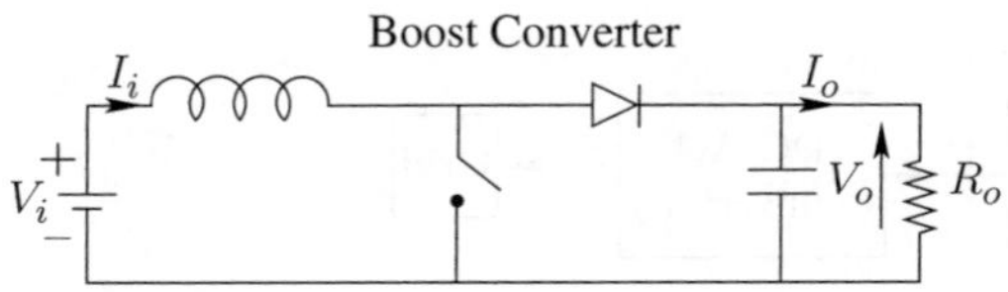	$\frac{V_o}{V_i} = \frac{1}{1-D}$ $\frac{I_o}{I_i} = 1 - D$ $R_i = (1-D)^2 \times R_o$
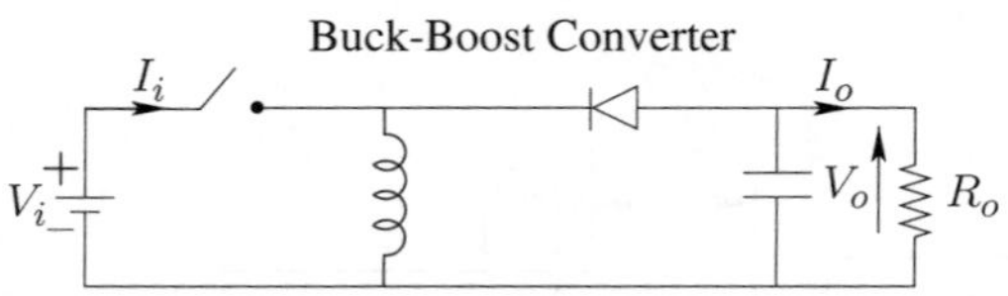	$\frac{V_o}{V_i} = \frac{-D}{1-D}$ $\frac{I_o}{I_i} = -\frac{1-D}{D}$ $R_i = \left(\frac{1-D}{D}\right)^2 \times R_o$

DC-DC Converters [5]

DC-DC converters are derived from three basic topologies: buck, boost, and buck-boost converters. The circuit topologies and input-output characteristics of these converters are presented in Table 11.1. Switching is done periodically and at a high frequency using switches like MOSFETs and IGBTs. The duty cycle D is defined by $D = \frac{T_{on}}{T}$, where T is the period corresponding to one switching cycle. T_{on} is the period during which the switch is in the "on" position within this cycle, therefore D can vary between 0 and 1. The inductors and capacitors are chosen so that the output voltage ripple is small. The input-output characteristics (as a function of D) are derived assuming that the components are ideal and the load currents are continuous. As the names suggest, a buck converter is able to provide lower voltage than that of the input, a boost converter can provide higher voltage, while a buck-boost converter can provide either a higher or a lower voltage than the input.

Since the equivalent load resistance seen at the source (R_i) is a function of the duty cycle, it follows that duty cycle control can be used to extract maximum power from the solar modules. The control of the duty cycle in this fashion is known as maximum power point tracking (MPPT). MPPT should not be confused with mechanical sun-tracking, which is used to adjust the solar panels so that they face the sun (maximizing the irradiance).

For small power grid-connected solar PV applications, galvanic isolation between the grounds on the solar panel side and the grid side may be achieved at the DC-DC conversion stage itself, using a small-sized high frequency transformer. The push-pull

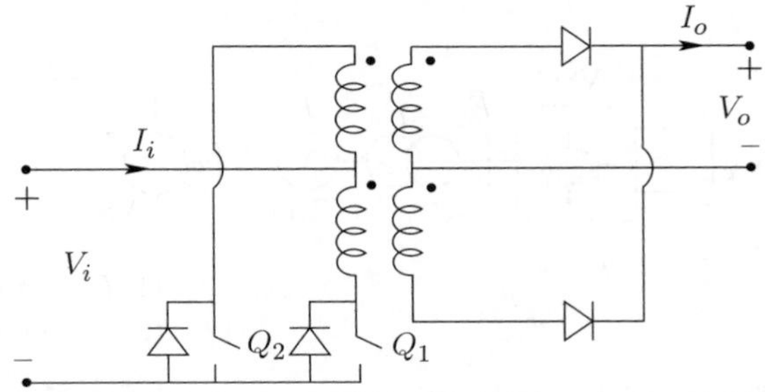

Figure 11.5 Isolated DC-DC converter: push-pull topology.

DC-DC converter shown in Figure 11.5 is an example of such an isolated converter topology. Note that the switches Q_1 and Q_2 are switched in alternate switching cycles, with each switch having a duty D in its switching cycle. The output voltage is given by the expression $\frac{V_o}{V_i} = 2aD$, where a is the transformer turns ratio.

Example 11.1 ***Maximum Power Extraction Using a Buck-Boost Converter*** Consider a solar module with the I-V characteristics shown in Figure 11.3. This PV module is connected to a load resistance, $R_o = 20\ \Omega$, through a buck-boost converter operating in continuous mode. What is the duty cycle needed to extract maximum power from the module?

Solution
From the I-V characteristic it is clear that for maximum power $V_i = 28.86$ V, $I_i = 7.53$ A and the equivalent resistance $R_i = \frac{V_i}{I_i} = \frac{28.86}{7.53} = 3.83\ \Omega$. Assuming continuous current operation, the input resistance seen by the converter is:

$$R_i = \frac{V_i}{I_i} = R_o \times \left(\frac{1-D}{D}\right)^2$$

Hence, if R_i has to be 3.83 Ω, then the duty cycle should be $D = 0.6956$.

DC-AC Converters
DC-AC converters can be implemented using a voltage source converter (VSC) topology as described in Chapters 4 and 5. Recall that the magnitude and phase of the fundamental component of the voltages at the AC terminals can be changed by changing the modulation-index (m) and the phase α, as depicted in Figure 11.6. For grid-connected systems, the controllability of the AC voltage allows us to control the real and reactive power output of the converter. Since output real power can be controlled, MPPT can also be achieved by a VSC.

11.3.4 Maximum Power Point Tracking Algorithms

The maximum power available for extraction depends on the module characteristics, solar irradiance, and temperature. Since solar irradiance and temperature vary with time, it is necessary to track the maximum power point and adjust the duty ratio accordingly. In a large power plant, a large number of modules and arrays may be connected

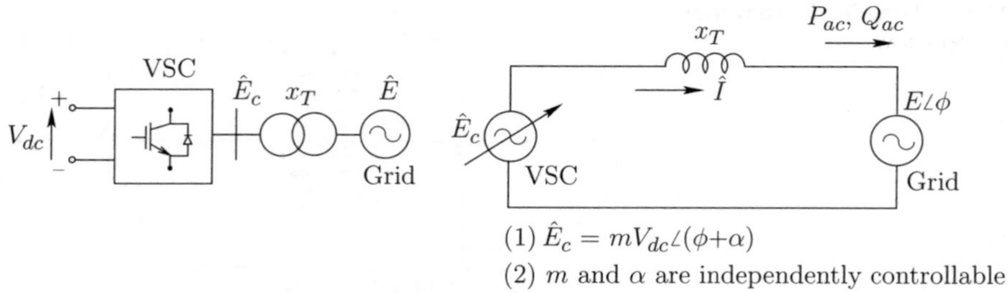

Figure 11.6 Real and reactive power control using a VSC.

in series and parallel. They are spread over a large area and may experience different solar irradiance. They may also have slightly different I-V characteristics. Therefore, a feedback-type approach may be used to track the overall maximum power point. This requires the current and voltage measurements at the output of the array.

The Hill climbing method is an example of a perturb-and-observe MPPT method, in which the duty cycle of the DC-DC converter is perturbed at regular intervals. If the output power increases with an increase in the output voltage, then adjustments in duty ratio are tried in that direction. On the other hand, if power decreases with an increase in the output voltage the duty ratio is adjusted so that the voltage is decreased. This perturbation, observation, and adjustment is continued until the power no longer increases/decreases. The adjustments may be reduced when the change in power with the voltage perturbation becomes smaller in order to prevent oscillations around the maximum power point.

Several other MPPT algorithms have also been proposed, such as the incremental conductance method [6] and the constant voltage method [7]. The MPPT algorithm may be chosen based on parameters such as implementation complexity, convergence speed, need for customization/periodic tuning, need for temperature/irradiance measurements, and whether the true maximum point is achieved or not [8] .

11.3.5 Control of Grid-Connected Solar PV Plants

The topologies and control of solar PV plants that are connected to a grid depend on the size of the solar power plant. While residential and industrial solar plants vary from a few hundred watts to a few megawatts, high power solar plants may have power ratings of hundreds of megawatts. The factors which determine the choice of topology are cost, efficiency, energy yield, and DC voltage levels.

Grid-Connected Residential and Industrial Solar Plants

For grid-connected residential and industrial PV applications, the typical structure is the series connection of PV arrays ("string" configuration). For low power levels (typically 25 kW or less), each string has its own dedicated two-stage power-electronic interface (DC-DC converter and DC-AC converter).

The control strategy for this type of grid-connected PV system is shown in Figure 11.7. The duty ratio of the DC-DC converter is usually set to extract maximum power (MPPT), but it is also possible to reduce the power from this level (using the power demand controller) if there is low demand or over-frequency in the grid. The

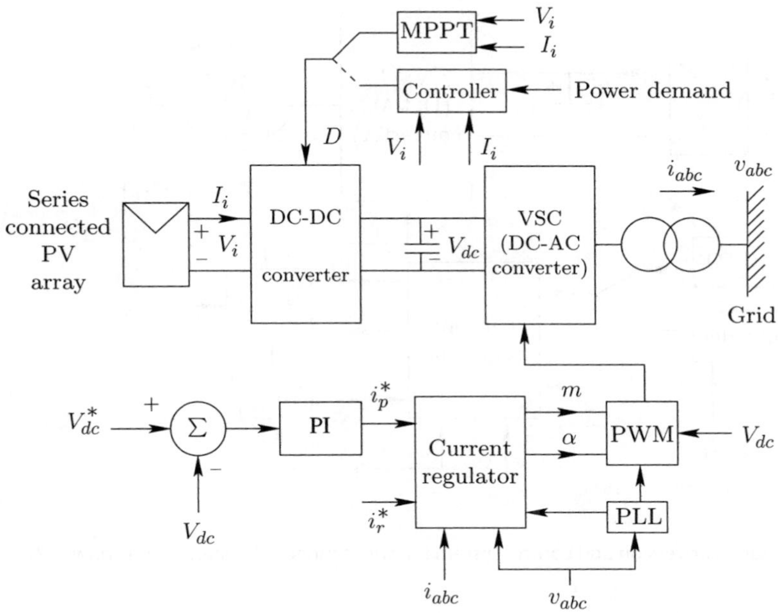

Figure 11.7 Two-stage conversion and control system for grid-connected systems (residential and industrial plants).

VSC transfers the power arriving at the DC link from the DC-DC converter to the AC grid. It does so indirectly, by regulating the DC link capacitor voltage, which would otherwise increase/decrease. The output of the capacitor voltage regulator sets the reference (i_p^*) for the component of AC-side current that is in phase with the AC voltages of the VSC (also called the real current). The quadrature (reactive) current reference (i_r^*) may be set based on the grid reactive power/voltage requirements. The currents are regulated using a vector control scheme similar to the one used for a STATCOM (see Section 5.5.5).

A multi-string structure may be used for higher power levels where each string has its own DC-DC converter, but with a common DC-AC inverter. MPPT may then be applied locally at each DC-DC converter. Smaller residential rooftop systems that are in the range of hundreds of watts to a few kilowatts are generally connected to a single phase of the grid. Typically, these systems use a step-up isolated DC-DC converter like the one shown in Figure 11.5, and a single-phase DC-AC converter. Since the isolation is provided at the DC-DC converter itself, a bulky low-frequency transformer at the point of connection to the grid can be avoided.

Grid-Connected High-Power Solar Plants

In high-power PV plants, solar panels are commonly connected in series and parallel to reach the convenient levels of voltage and power. Instead of using two stages of conversion (DC-DC and DC-AC conversion), a single centralized DC-AC converter may be used, as shown in Figure 11.8. In this configuration, the real-current reference (i_p^*) of the VSC is set by a MPPT controller or a power demand controller. Note that the power

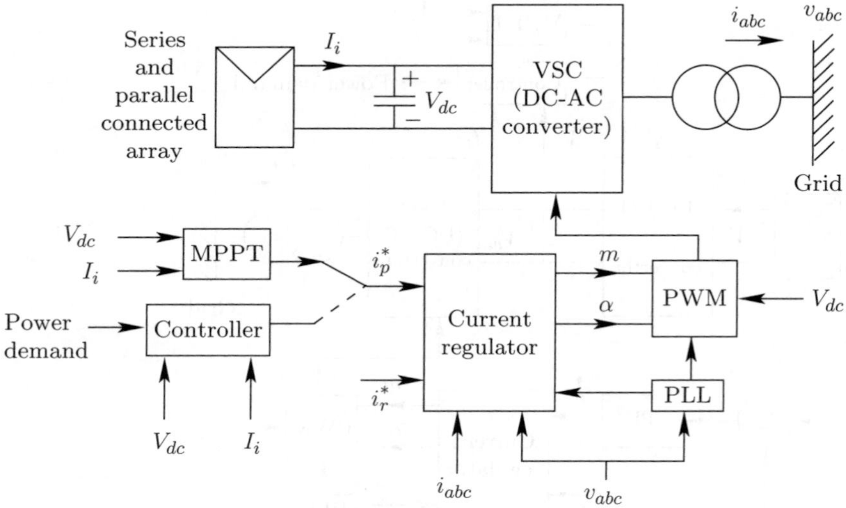

Figure 11.8 Single-stage conversion and control system for grid-connected systems (high-power PV plants).

demand controller is activated if the power required by the grid is less than the MPPT level. The DC link capacitor voltage is determined by the solar array I-V characteristic and the real current. The plant may be divided into several areas, each managed by a three-phase VSC. These may then be paralleled on the AC side and connected to the grid using a low-frequency transformer. A limitation of this scheme is that since there is only one centralized VSC in one area, MPPT is applied to the overall power output in that area. The power output consists of the contribution of many arrays connected in parallel. Since the conditions at each array may be different, and the output of each array is not controlled individually by a converter, the overall power output is less than what can be achieved if MPPT were to be applied individually to each array.

When connected to a weak or small grid, the real and reactive power injected by a large solar plant may significantly affect the frequency and voltage of the grid. In such a situation, the control strategy has to be decided in coordination with other generators, loads, and storage devices in the grid. Even in larger grids, as the penetration of solar PV and other renewable energy systems increases, it would become necessary for a grid-connected solar PV system to contribute to voltage and frequency control. In addition, the fast-acting and vernier ability of the controllers could also be harnessed for power swing damping [9]. However, unlike a wind energy system, which has an inherent store of kinetic energy due to its rotating parts, a solar PV array cannot (even temporarily) supply real power more than the maximum power level determined by its I-V characteristic (at the prevailing solar irradiance and temperature).

11.3.6 Low-Voltage Ride Through and Voltage Support Capability

Solar PV systems are connected to the grid through power electronic interfaces (DC-AC converters). Usually these are VSCs, which can rapidly control the real and

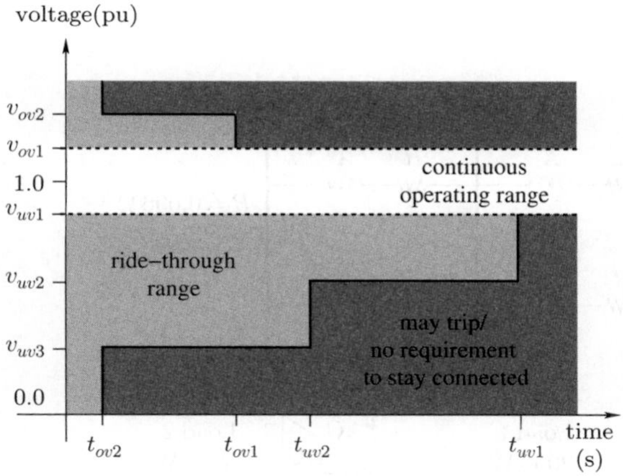

Parameter	Typical Value
v_{ov2}	1.20 pu
v_{ov1}	1.10 pu
v_{uv1}	0.88 pu
v_{uv2}	0.60 pu
v_{uv3}	0.45 pu
t_{ov2}	0.16 s
t_{ov1}	1 s
t_{uv2}	1 s
t_{uv1}	2 s

Figure 11.9 An example of a low-voltage ride-through characteristic. The voltage is at the point of connection with the grid.

reactive power injection into the grid. This ability can be used by a grid to recover quickly following disturbances such as faults. However, faults result in low voltages and over-currents, which may lead to tripping of the DC-AC converter. In such a scenario, the grid will lose the power generated by the solar PV system, and the system also cannot assist in grid recovery. Therefore, grid regulators require that the system remain connected for short duration faults and low voltage, and resume normal feed-in immediately afterwards. This is known as the low-voltage ride-through (LVRT) capability. The LVRT requirement is often specified by the grid regulators in the form of a characteristic (see Figure 11.9). Note that the thresholds and the nature of the characteristic may be specified differently in different grids. A solar PV system may be required/permitted to disconnect if the voltage remains outside the specified magnitude-time limits, but should remain connected otherwise. Note that similar requirements are applicable for other grid-connected renewable energy systems such as wind generators.

To understand this, consider the system shown in Figure 11.10, which consists of a 100 kW grid-connected solar PV system connected in parallel with unity power factor loads. The PV system uses the control system given in Figure 11.7. The reactive current reference i_r^* is set by a voltage regulator (PI type) which regulates the terminal voltage of the PV system at the reference value. In this case study we assume that the solar PV system operates at the maximum power point (100 kW). The initial reactive power exchange with the grid is zero. The DC-AC converter rating is taken to be 112 kVA, which gives it some spare capacity for reactive power exchange with the grid.

A single-line ground fault occurs in one of the cables close to the solar PV plant, which is cleared in 0.06 s by disconnecting the faulted cable. The response of this system to the disturbance is shown in Figure 11.11. The power and current are per unit on the inverter kVA base (112 kVA), while the voltage is per unit on a 415 V base. It is seen that although the currents are not strictly regulated during the unbalanced fault, the current magnitude is mostly below 1.4 pu for the duration of the fault. The d-q-based current regulators cannot strictly regulate the current to the reference values during the fault

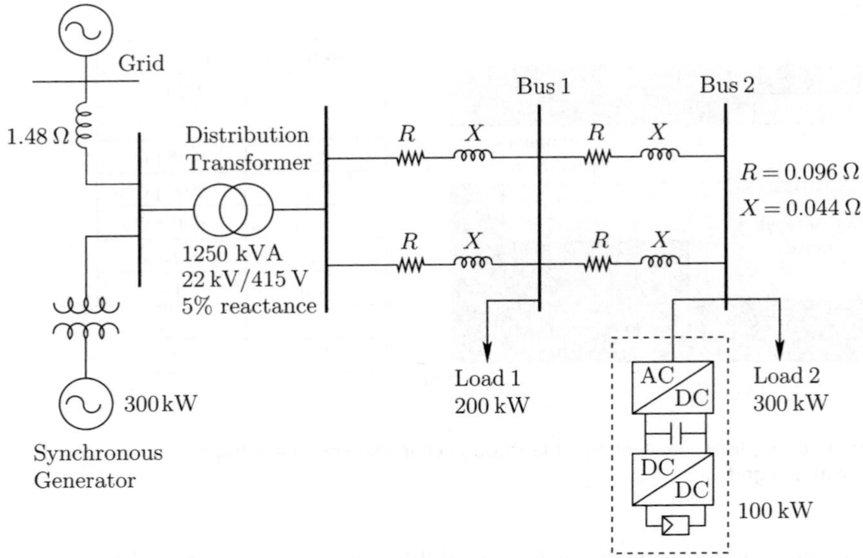

Figure 11.10 A distribution system with solar PV input.

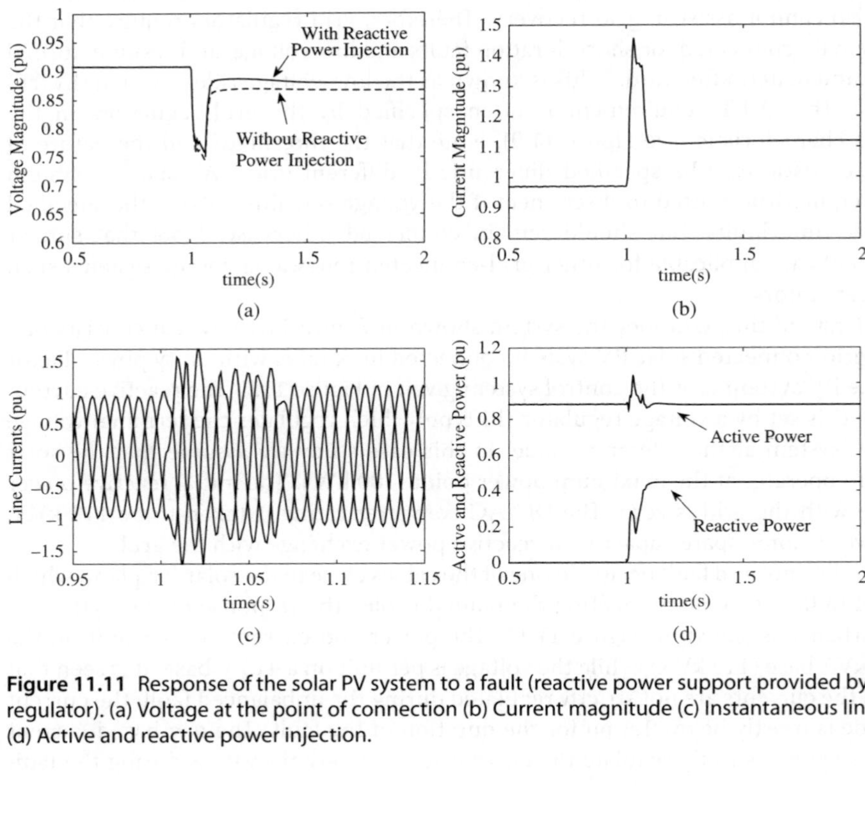

Figure 11.11 Response of the solar PV system to a fault (reactive power support provided by a voltage regulator). (a) Voltage at the point of connection (b) Current magnitude (c) Instantaneous line currents (d) Active and reactive power injection.

because (i) the finite bandwidth of the controllers and (ii) unbalanced currents cannot be regulated by the conventional *d-q* controllers. Negative sequence controllers have been proposed to overcome this problem [10], and these can reduce the over-currents for unbalanced faults. If the currents are within the transient rating of the inverter, then the solar PV system can ride through the disturbance.

After the fault is cleared by tripping the faulted cable, the voltage reduces due to the larger voltage drop on the remaining cable. The solar PV system, by the action of its voltage regulator, can inject reactive power in order to support the voltage of the system, as shown in the figure. Note that the voltage regulator has a dead-band between 0.88 and 1.1 pu; therefore the regulator prevents the steady-state voltage from dropping below 0.88 pu.

11.4 Energy Storage

Most large conventional electrical grids can operate without significant storage of energy after it has been converted to electric energy. This is because the load-generation balance is maintained in near real time through the control of the generated power, with frequency as the feedback signal. The load is forecast in the short term and the long term. This facilitates the prior scheduling of generators. For satisfactory operation, there should be sufficient availability of generation capacity to cater to the peak demand, a sufficient generation margin to ride through unexpected generation/load imbalances, and the ability to ramp up and ramp down generation at the required rate.

The power output from renewable energy sources such as wind and solar is variable and intermittent. Since generation capacity has to be able to cater to the peak demand, a significant overcapacity has to be built to avoid an inadequate generation situation. Alternatively, the peak demand may be reduced by time-shifting some of the load demand to off-peak hours (leveling the load curve). This can be achieved through demand management and/or energy storage. When the load demand is low, the energy may be stored and this may be used to support the peak demand when it occurs. The availability of energy storage also allows maximum power extraction from renewable sources even when the demand is low.

Besides load-leveling applications, energy storage may be used to:

1) smoothen short-term variations in the power output of renewable energy systems
2) serve as a supply reserve
3) improve grid-stability by:
   (a) reducing the sudden drop of frequency following a sudden loss of generation by providing a temporary boost in power injection ("synthetic inertia")
   (b) modulation of the power output to improve power-swing damping
   (c) avoiding transmission line congestion (especially if storage devices are placed near the loads)
   (d) supporting the voltage by control of reactive power
4) tide over short disruptions in power in the distribution system due to upstream disturbances or medium-term disruptions due to disconnection of supply
5) improve the power quality by mitigation of voltage sags and swells, harmonics, and compensation of reactive power.

Note that power electronic converters are used to interface the storage devices to the grid in many applications [11]. The fast control ability of these converters facilitates many of the aforementioned functions.

11.4.1 Attributes of Energy Storage Devices

Some important considerations for the evaluation of energy storage technologies (besides the cost and maturity of the technology) are as follows:

1) Discharge time: This is the time required to discharge the rated stored energy at the rated power.
2) Self-discharge losses: These are the losses during storage. Storage devices with large self-discharge losses are unsuitable for long-duration storage.
3) Cycle efficiency: This is the ratio between the released energy and stored energy for one charge–discharge cycle, that is, the efficiency associated with energy conversion.
4) Cycle lifetime: The number of charge–discharge cycles which the device can have in its lifetime. The overall life (in years) depends on the number of cycles required by the application and the aging processes in the storage device.
5) Power and energy densities per unit volume: This determines the space requirement of the storage device. The densities in terms of per-unit weight are important for some applications which require mobile storage units.
6) Response time: This indicates how fast the storage device can reach its full power rating from zero power.

11.4.2 Energy Storage Technologies

Energy storage technologies can be classified as [12]:

1) mechanical, for example pumped hydro, compressed air energy storage (CAES), and flywheels
2) electro-chemical:
   (a) secondary batteries, for example lead acid, nickel cadmium (NiCd), nickel metal hydride (NiMH), lithium ion (LI) and sodium sulphur (NaS)
   (b) flow batteries, for example vanadium redox batteries (VRBs), Zn-Br batteries, and polysulphide bromide batteries (PSBs)
3) chemical, for example fuel cells/gas turbines based on hydrogen or synthetic natural gas (SNG)
4) thermal, for example low-temperature thermal energy systems (TES) such as cryogenic energy storage (CES) and high-temperature TES such as molten salt storage and phase-change materials
5) electrical, for example super-conducting magnetic energy storage (SMES) and super-capacitors such as the double-layer capacitor (DLC).

A brief outline of a few of these technologies is given below [12–24].

Pumped Storage Hydro-Plants (PHS): This is a large-scale energy storage system. It is a mature technology that involves the pumping of water from a lower reservoir to a higher reservoir during off-peak hours and discharging it through a turbine when power has to be extracted. The discharge time ranges from one hour to more than one day. The cycle efficiency is about 70–85%, while the response time is of the order of minutes. The overall lifetime is very long, with practically unlimited cycling. The main drawback is the large land use. PHS constituted almost 99% of the installed storage capacity in the world in 2011 [12].

Secondary Batteries: A battery consists of one or more electro-chemical cell. Each cell consists of a liquid, paste or solid electrolyte along with positive (anode) and negative (cathode) electrodes. Electro-chemical reactions occur at the two electrodes during discharge. This causes a flow of electrons through an external circuit. These reactions are reversible, which allows recharging by application of an external voltage. Lead-acid batteries are a common choice for power quality and uninterruptible power supplies (UPSs), and have cycle efficiencies ranging from 70% to 90%. They suffer from disadvantages like short cycle life and low energy density. There are several types of lead-acid batteries, including the flooded battery (which requires topping up with distilled water), the sealed maintenance-free battery with a gelled/absorbed electrolyte, and the valve regulated lead-acid battery (VRLA). Lithium-ion (LI) batteries are relatively recent and have achieved higher energy densities, higher cycle life, low self-discharge, and higher efficiency, but at a higher cost than lead-acid batteries. Sodium-sulphur (NaS) is mature technology for high-power and high-energy applications. An NaS battery consists of molten sulphur at the positive electrode and molten sodium at the negative electrode, separated by a solid ceramic electrolyte. The battery operating temperature is in the range of 300–360°C. NaNiCl batteries (also called ZEBRA batteries [12]) are also high-temperature batteries like NaS batteries and are commercially available for medium power and energy applications.

Flow Batteries: Flow batteries consume electrolytes stored in different tanks and are separated by a microporous membrane. These batteries have low self-discharge. Unlike classical batteries, where there is a strong coupling between the power and the energy rating, in flow batteries the power rating is proportional to the area of the membrane and the flow of the reactants whereas the energy rating is proportional to the electrolyte tank capacity. These batteries therefore have a good potential for both power and energy scale-up. At present, these batteries are in the R&D/demonstration stage.

Compressed Air Energy Storage (CAES): Like PHS, CAES is a technology for large-scale energy storage. The energy is stored in the form of elastic potential energy of the compressed air. The excess or off-peak power is used to compress air. The compressed air is pumped into a storage vessel for later use. When power is needed, the stored air is used to run a gas-fired turbine generator, and the power is delivered back to the grid. The storage vessels for compressed air can be in the form of underground rock caverns, salt caverns or porous media reservoirs. Thus CAES can be built only in certain geographical areas. CAES is also less attractive as it involves combustion with fossil fuels for conversion to electricity. Improved CAES systems are under investigation which are hybridized with thermal energy storage systems.

Fuel Cells: A fuel cell is an electro-chemical energy conversion device, but differs from a battery in that it requires a continuous replenishment of reactants (fuel and oxidant). An electrolyser is used which splits water into hydrogen and oxygen with the help of electricity. This is an endothermal process requiring external heating. A hydrogen fuel cell uses hydrogen and oxygen to produce electricity and water. Hydrogen fuel cells have a high energy density but suffer from low cycle efficiency. Apart from fuel cells, gas motors, gas turbines, and combined cycles of gas and steam turbines are also under consideration for power generation from hydrogen. Synthetic natural gas (methane) may be produced by combining CO_2 and the hydrogen obtained from electrolysis. This gas may be used thereafter for running a gas turbine. This technology is still in the development stage for high power and energy applications [12].

Flywheels: Flywheels store energy in the inertia of the rotating mass. A flywheel is rotated by using a motor, which acts like a generator when discharging. For a given inertia, the storage of energy can be large if it is rotated at very high speed. This calls for a high vacuum environment to minimize windage losses and a low loss-bearing assembly. Nonetheless, self-discharge due to running losses is relatively high. Flywheels have a long cycling life, which is determined by the life of the mechanical components. Flywheels are suitable for high-power, short-duration applications, for example mitigation of power quality problems and short-duration interruptions.

Super-Conducting Magnetic Energy Storage (SMES): In SMES storage, energy is stored as direct current in an inductor. The inductor coil is made of super-conducting material so the loss in the inductor is practically zero. SMES systems have a fast response and a high cycle life, and are therefore suitable for applications requiring continual and frequent cycling (charge-discharge).

Super-capacitors: Conventional capacitors can be charged or discharged very quickly, but have very low energy density. A larger capacitance is achievable with super-capacitors, which use electrodes with a very high surface area/per-unit weight while maintaining a small electrode distance. Like flywheels, super-capacitors have a high self-discharge loss. They can be used for quick injection of energy when required and are therefore useful in application areas such as disturbance ride-through and power quality.

Thermal Energy Systems (TES): These systems store materials at high/low temperature in insulated containers. Electrical energy is generated using heat engine cycles. TESs can be classified into low- and high-temperature TESs. Low-temperature TESs includes aquiferous low-temperature TES (ALTES) in which water is iced during off-peak hours and used to meet cooling needs during peak time. Another type of low-temperature TES is cryogenic energy storage (CES), where air or nitrogen is liquefied during off-peak hours. This is used to generate electricity when required using a cryogenic heat engine. High-temperature TESs include molten-salt storage, concrete storage, and latent heat storage (using phase-change materials).

The key technical attributes of energy storage technologies are summarized in Tables 11.2–11.4. The data has been compiled from different papers (2010–2017) that are cited in the table against the data.

Note: Most systems have a modular design (except for some such as PHS and underground CAES) and could be implemented with even larger power output and energy capacities than indicated in the tables. Therefore, if a larger power range or higher energy capacity is not realized, it will be mainly due to economic reasons (cost per kW and cost per kWh, respectively) and/or application requirements.

11.4.3 Mapping to Applications

PHS and lead-acid batteries are mature technologies. CAES, NiCd, NaS, NaNiCl, Li-ion, flow batteries, SMES, flywheel, super-capacitors, AL-TES, and HT-TES are developed

Table 11.2 Technical attributes of energy storage systems (Part A).

	Available/projected module sizes†		Discharge	τ_{rs} ‡		Self-§	Life in years	Power and energy densities	
Technology	Power (MW)	Energy (MWh)	time	[12]	η	discharge	(in cycles)	W/L (W/kg)	Wh/L (Wh/kg)
PHS	100–5000 [16]	5400–14000	1–24 h [12, 16]				40–60		0.5–1.5 [13, 16],
	900–1400 [19, 23]	[12, 19],	6–10 h [19]	Minutes	~81 [19, 21],	Very	[13, 16, 20, 23]	0.1–0.2 [12]	0.2–2 [12]
	250–530 [21]	1680–5300 [21],	Few days [23]		65–87 [16, 20]	small [16]	(>13000		(0.5–1.5 [13, 16],
	100–1000 [22]	<24000 [23]					[19, 21])		0.2–2 [12, 20])
CAES	110–290 [13, 14, 23]	400–7000 [23]	4–6 h [13]-[14]						
	5–300 [16, 20]		1–24 h [16, 20]						
	20–2000 [21, 22, 24]								1–6 [12, 13, 16]
CAES-AG	50 [19]	250 [19]	5 h [19]	Minutes	85 [14],	Small [16]	20–40 [16]	0.5–2 [16]	(10–30 [20],
CAES-UG					60–70 [21]		(>13000 [19])	0.2–0.6 [12]	30–60 [16])
Adiabatic	10 [12]	20–250 [12]	5–24 h [12]						
Diabatic	500 [12]	2000–6000 [12]	5–24 h [12]						
FES	10–800 kW [21, 22]		Few seconds				20–30 [17, 23]	1000–2000 [16]	20–80 [12, 13, 16]
	20–100 kW [12]	0.1–80 kWh [12]	to few minutes [14, 16, 23]	<1 s	80–87 [14, 19],	100% [16, 17]	(>20000	4500–5500 [12]	(50–100 [17, 20],
	100–250 kW [14, 16]				90–95 [20, 23]		[13, 17, 20])	(1000, 5000 [20])	5–30 [12, 16, 17, 20])
SMES	0.1–2 [12, 21, 22]	0.1–20 kWh [12]	1–3 s [14]				>20 [16],	300 [13]	6 [12]
	0.1–10 [16, 19]	<250 kWh [23]	Milliseconds to 8 s [13, 16, 19]	<1 s	98 [14],	10–15% [16]	40 [23]	1000–4000 [12, 16]	(0.2–5 [13, 16]
	0.2–30 [20, 23]		6 s to 10 min [12]		85–90 [12, 16]		(>10^5 [16])	(500–2000 [16])	5.5–6.1 [12])
Super-	<0.1 [14], 0–0.3 [16]	<3 [23]	Few seconds				>20 [16]	36–120 kW/L	10–20 [12, 13]
capacitor (DLC)	0.01–1 [19]–[22]	0.1–2 kWh [12]	to minutes [19–22]	<1 s	90–98	20–40% [16]	30–40 [23]	[12, 16], 30 [13]	(1–15 [12, 16],
	<0.25 [23]				[12, 20, 23]		(>10^5 [16, 17])	(500–5000 [16, 20])	<50 [20])

PHS, pumped hydro storage; CAES, compressed air energy storage (AG, above ground; UG, under ground); FES, flywheel energy storage; SMES, super-conducting magnetic energy storage; DLC, double layer capacitor; η, cycle efficiency. ‡Response time. §Per day; †See the note on page 406.

Table 11.3 Technical attributes of energy storage systems (Part B).

Technology	Available/projected module sizes† Power (MW)	Energy (MWh)	Discharge time	τ_{rs}‡ [12]	η	Self-§ discharge	Life in years (in cycles)	Power and energy densities W/L (W/kg)	Wh/L (Wh/kg)
TES	3–200 [20],		3–20 h [20]		80 [23]				
	0.1–200 [23]		Few h [23]						
AL-TES	0–5 [16]		1–8 h [16]		60 [16]	0.5% [16]	10–20 [16]		80–120 [16]
CES	0.1–0.3 [16]		1–8 h [16]		50 [16]	0.5–1% [16]	20–40 [16]	(10–30 [16])	120–200 [16]
HT-TES	0–60 [16]		1–24 h [16]		60 [16]	0.05–1% [16]	5–10 [16]		120–500 [16]
H_2	0.5–800 [12]							>500 [16]	500–3000 [16]
FC/GT/GM	0–50 [16] (in FC),	10–1000 [12]	>24 h [16]	Seconds to	60–85 [14]	Negligible	5–15 [16]	0.2–2, 2–20 [12]	600 [12]
	<20 [23] (in (FC)		>1 h [12]	minutes	70–90 [23]	[16]	2–10 [23]	(>500 [16])	(800–10000 [16],
	<10 [12] (GT, GM)								33330 [12])
SNG	1 [12]	100–1000 [12]		Minutes	<35 [12]			0.2–2 [12]	1800 [12]
GT/GM									10000 [12]
Lead-acid							5–15 [13, 16]		
flooded	0–20 [12, 16, 18, 20]	40 [18]	Few seconds	Milliseconds	70–90	Very small	(500–2200	10–400 [13, 16]	50–80 [12, 13, 16]
type	20–100 [19, 21]	250–400 [19, 21]	to hours [16, 18, 19]		[16, 18, 19, 21]	[17, 21]	[13], [16–18], [21],	(75–300 [16])	(25–50 [12, 16, 18])
							4500 [19])		
VRLA	0.3 [18], 10 [22]	0.58 [18]	<1 h [22]	Milliseconds	72–78 [18]	Very small	5–15 [13, 16]	90–700 [12]	
	100 [22]					[17, 21]	(200–300 [18])	(180 [18])	(20–50 [17, 18, 20])
Lithium ion	0.1–100 kW	0.25–25 [19, 21],	Minutes to hours				5–15 [16]	1500–10000 [12]	200–500 [12, 16]
	[16, 19, 21], 1–30 [12],	1–30 kWh [12],	[12, 16, 19, 25]	<1 s	80–93 [19],	0.1–0.3% [16]	(>10^5 [19, 21],	(150–315 [16],	(75–200 [16, 18],
	1–100 [19, 21],	1–100 [12],			70–85 [20]		500–2000 [20],	360 [20])	100–200 [20],
	30 [12], 100 [25]	129 [25]					1000–10000 [16])		60–200 [12])
Ni-Cd	0.01–40 [16, 24],						10-20 [16, 17]	75–700 [12]	15–80 [12],
	1–300 kW		Few seconds	<1 s	60–70 [16],	Very	(2000–2500	(150–300	60–150 [12, 16]
	[19, 21, 22],		to minutes		60–90 [20]	small [16, 17]	[16, 17],	[16, 20],	(40–80 [18, 20]
	0.01–10 [20]		[16, 19, 20, 24]				500–2000 [20])	50–1000 [17])	15–50 [12, 17])

TES, thermal energy storage; AL-TES, aquiferous low-temperature TES; CES, cryogenic energy storage; HT-TES, high-temperature TES; FC, fuel cell; GT, GM, gas turbines, gas motor; SNG, synthetic natural gas; VRLA, valve regulated lead-acid; η, cycle efficiency. †Response time. §Per day. ‡See the note on page 406

Table 11.4 Technical attributes of energy storage systems (Part C).

Technology	Available/projected module sizes† Power (MW)	Energy (MWh)	Discharge time	τ_{rs}‡ [12]	η	Self-§ discharge	Life in years (in cycles)	Power and energy densities W/L (W/kg)	Wh/L (Wh/kg)
Ni-MH	1–2000 kW [19, 21, 22], 3–100 kW [12]		Few seconds to few minutes [19]	<1 s	50–80 [20]		(<3000 [20])	500–3000 [12] (220 [20])	80–200 [12] (40–80 [12, 20])
NaS	0.05–8 [16] 0.01–30 [12, 20] 0.5–50 [19], 1 [21]	20–500 [12] 300 [19], 7.2 [21]	1 h to 1 day [12, 20]	<1 s	75 [19, 21]	~20% [16]	10–15 [13, 16] (2000–2500 [16, 20], 4500 [19, 21])	40 [13], 120–160 [12] (150–230 [16], 120 [20])	130–300 [12, 13, 16] (100–250 [12, 16, 18, 20])
NaNiCl#	0.01–300 kW [16] 0.07–3 [19] 30 kW, 30 [12]	0.1–2 [12] 20–500 [12]	Few mins [19] 1 h to 1 day [12]	<1 s		~15% [16]		220–300 [12, 16] (150–200 [16])	150–200 [12, 16] (100–120 [16], 100–200 [12])
Zn-air	0–10 kW [16] 50 [19]	250 [19]	Few seconds to 1 day [16] 5 h [19]	<1 s	75 [19]	Very small [16]	(100–300 [16], >10000 [19])	50–100 [12]	130–200 [12] (150–3000 [16], 450–650 [18])
RFB	5–1200 kW [12] > 20 [23]	>200 [19, 22]	Few hours [23]		65–85, [16, 23]		2 [13], 10 [23]	1–2 [12]	20–70 [12]
Fe-Ti/Fe-Cr	0.02–20 [19, 22]	250 [19]	Few hours [22]		75 [19, 21]		(>10000 [19])		
VRB	0.03-3 [16, 18] 50 [19], 1 [21] 0.1–3 [20], 4 [24]	1.5 [18], 250 [19], 4 [21]	3 h to few days [20] 1.5 h [24]	ms to s	75–85 [16, 18], 65–75 [19, 21]	Small [16]	5–10 [16] (>10000 [19, 21] >16000 [20])	0.5–2 [12, 13] (80–150 [20])	20–70 [12, 16] (10–50 [12, 16, 18, 20])
PSB	1–15 [16]		Few seconds to 10 h [16]	s	60–75 [16]	Small [16]	10–15 [16]		
HFB (Zn-Br)	0.05–2 [16] 15 [18], 50 [19]	120 [18] 250 [19]	Few seconds to hours [16, 19]	ms to s	60–75 [16, 18]	Small [16]	5–10 [13, 16] (>10000 [19])	1–25 [12, 13]	55–65 [12] (75–85 [12])

#Also known as ZEBRA battery. ‡Response time. §Per day; †See the note on page 406.
HFB, hybrid flow battery; RFB, redox flow battery; VRB, vanadium redox battery; PSB, polysulphide bromide battery; η, cycle efficiency.

technologies that are commercially available but deployments are not yet widespread for utility-scale applications. On the other hand, technologies such as fuel cells, metal-air batteries, and CES, are under development. The details and ratings of a few real-life installations of some energy storage technologies are given in [13, 16, 18, 24, 26].

Energy storage technologies can be broadly mapped to the various applications as given below [12, 16, 19, 27]:

1) *Bulk energy management*: PHS and CAES are suitable for large-scale and long-duration requirements. Large-scale batteries, flow batteries, fuel cells and thermal energy storage systems are suitable for medium-scale energy management systems (10–100 MW). SNG-based storage also has the potential for application in energy management, but is still at the research and development stage.
2) *Peaking service, bridging power, intermittent support for renewable sources*: Secondary batteries, flow batteries, and fuel cells are suitable due to their fast response, longer discharge time (hours) and power rating requirements (10 kW to 10 MW). An early application of a battery-based energy storage system for providing 10 MW of peaking power is described in [28].
3) *Grid stability*: Improvement of small-signal stability of power swings can be achieved with medium-power modulation levels and does not require large energy storage. For example, the experimental 30 MJ (8.4 kWh) SMES installation at Bonneville Power Administration's Tacoma Substation [29] has a peak power amplitude capability of 11 MW at 0.1 to 1.0 Hz modulation frequency. Reducing the rate of frequency drop in a grid following a sudden power imbalance requires temporary power injection from a storage device for a few seconds (inertial response). Overall, the requirements for grid stability applications are (i) a quick response time and (ii) an ability to inject megawatts of power for a short duration (typically 1–20 s).
4) *Power quality*: Power quality requirements include a fast response time and the ability to inject real and reactive power for a short duration. This is required to mitigate voltage sags/swells, flicker, and short power disruptions which occur in the distribution system. Therefore, the power requirements are typically lower than 1 MW. Flywheels, SMES, and super-capacitors which have a fast response time (in milliseconds) are suitable for such applications. Batteries may be used for supplying power during longer power interruptions (hours).

Remarks

1) Some application requirements can be met by a combination of energy storage devices. For example, a super-capacitor and battery energy storage can be combined to achieve both a fast discharge capability and a higher sustained power capability.
2) The probable large-scale use of electrical vehicles for transportation [30–32] in the future may have a significant impact on the grid. At any given time a significant fraction of these would be plugged-in to the grid for charging of batteries. Coordinated charging and discharging schemes are being contemplated to optimize the usage of the grid infrastructure for this purpose and to utilize the plugged-in batteries for power and energy management in the grid [32].

11.4.4 Battery Modeling

Electrical energy is produced by the reaction of chemicals in the battery. The rate of the chemical reaction depends on (i) the state of the charge (SOC), (ii) the battery storage

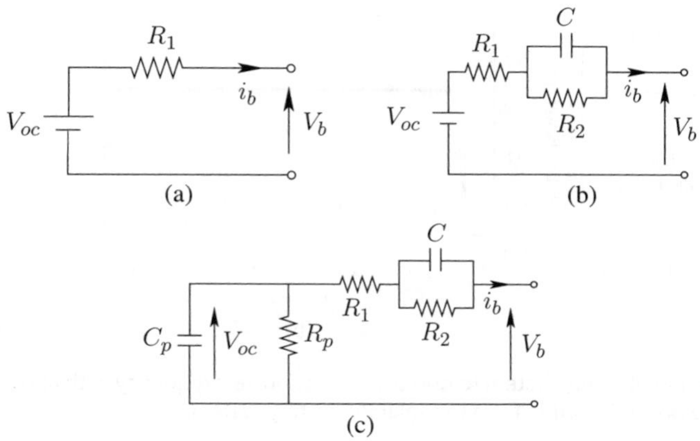

Figure 11.12 Basic battery models. (a) Thevenin model (b) Modeling of transient behavior (c) Modeling of self-discharge.

capacity, (iii) the rate of charge and discharge, (iv) the environment/temperature, and (v) the age/shelf life. SOC is defined as the available capacity (in Ah) expressed as a percentage of its rated capacity. Therefore, $\mathrm{SOC}(t) = 100 \times (1 - \frac{q}{Q})$, where Q is the rated capacity in Ah, $q = \int_0^t i_b(t)\,dt$ in Ah, and i_b is the battery current. In other words $\mathrm{SOC}(t) = \mathrm{SOC}(t_o) - \int_{t_o}^t i_b(t)\,dt$.

It is difficult to arrive at a truly generic model which takes all factors of a battery into account. Several modeling approaches for lead-acid batteries have been reported in the literature [33–36]. The simplest model of a battery is a voltage source V_{oc} behind an internal resistance R_1, as shown in Figure 11.12a. The battery model shown in Figure 11.12b models the capacitance of the battery (the so-called double layer capacitance C) and the resistance contributed by the contact resistance of the plates and the electrolyte R_2 [33]. This model captures the voltage drop at switch-on in a lead-acid battery. This model may be further expanded to take into account the self-discharge, which is modeled by a resistance R_p and a capacitance C_p, as shown in Figure 11.12c. This model has been used in the dynamical modeling of a battery energy storage system in [37]. Note that the differences in the charge and discharge process may be modeled by using different values of R_1 and R_2 during charging and discharging. The model shown in Figure 11.12a may be enhanced by making R_1 and/or V_{oc} functions of the SOC. Such models have been proposed in [38] for lead-acid, Li-ion, NiMH, and NiCd batteries.

Application of a Battery System for Emergency Support

Consider the system shown in Figure 11.10. The system is modified to include a battery, which is connected across the DC link of the solar PV system through a DC-DC converter. The solar PV DC-AC converter rating is increased to 158 kVA to accommodate the additional active power exchange due to the battery. The battery rating is 280 V, 420 Ah and it has a peak discharge current of 210 A. The battery can supply 50 kW for two hours. For this study, the battery is modeled as a constant voltage source behind an internal resistance 0.2 Ω.

If the grid gets suddenly disconnected then an island is formed which consists of the 300 kW synchronous generator, the solar PV system (with a battery), and the loads.

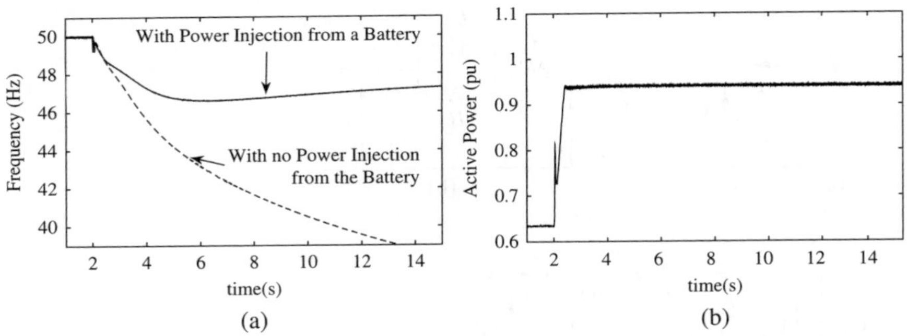

Figure 11.13 Response of the solar PV–battery system to grid disconnection. (a) Frequency with and without battery support (b) Active power injection from the solar PV-battery system.

Many utilities disallow islanded operation with distributed generation systems like this one due to safety concerns for personnel (unexpected live circuits) and large deviations in voltage and frequency. However, it is possible to operate as an island if appropriate protection and control strategies have been deployed, and adequate real and reactive power margins are available to cover any deficit.

In this case study, the synchronous generator is initially operating near its rated capacity. Therefore, it cannot provide for the generation-load deficit of the islanded system, although it has a speed governing system. The battery can complement the synchronous generator by injecting power into the grid when the frequency falls (using power-frequency droop control). This situation is simulated and the responses of system frequency after sudden disconnection of the grid are shown in Figure 11.13. The active power injected by the solar PV–battery system is shown in per unit on the DC-AC converter VA base. With the active power support provided by the battery, the precipitous drop in frequency is prevented and the island attains an equilibrium frequency which is above 47 Hz. Some part of the load may be shed to improve the frequency further.

References

1 Malinowski, M., Leon, J.I., and Abu-Rub, H. (2017) Solar photovoltaic and thermal energy systems: Current technology and future trends. *Proceedings of the IEEE*, **105** (11), 2132–2146.

2 Solanki, C.S. (2011) *Solar Photovoltaics: Fundamentals, Technologies and Applications*, PHI Learning, New Delhi, 2nd edn.

3 Fraunhofer and Institute for Solar Energy Systems (2018) Photovoltaics report, https://www.ise.fraunhofer.de/content/dam/ise/de/documents/publications/studies/Photovoltaics-Report.pdf. (accessed 2 April 2018).

4 Villalva, M.G., Gazoli, J.R., and Filho, E.R. (2009) Modeling and circuit-based simulation of photovoltaic arrays, in *2009 Brazilian Power Electronics Conference*, Bonito-Mato Grosso do Sul, pp. 1244–1254.

5 Rashid, M.H. (2004) *Power Electronics: Circuits, Devices and Applications*, Pearson Education, New Delhi, 3rd edn.

6 Wasynezuk, O. (1983) Dynamic behavior of a class of photovoltaic power systems. *IEEE Transactions on Power Apparatus and Systems*, **PAS-102** (9), 3031–3037.
7 Kobayashi, K., Matsuo, H., and Sekine, Y. (2004) A novel optimum operating point tracker of the solar cell power supply system, in *35th IEEE Annual Power Electronics Specialists Conference*, Aachen, pp. 2147–2151.
8 Esram, T. and Chapman, P.L. (2007) Comparison of photovoltaic array maximum power point tracking techniques. *IEEE Transactions on Energy Conversion*, **22** (2), 439–449.
9 Singh, M., Allen, A., Muljadi, E., and Gevorgian, V. (2014) Oscillation damping: A comparison of wind and photovoltaic power plant capabilities, in *IEEE Symposium on Power Electronics and Machines for Wind and Water Applications*, Milwaukee, pp. 1–7.
10 Song, H.S. and Nam, K. (1999) Dual current control scheme for PWM converter under unbalanced input voltage conditions. *IEEE Transactions on Industrial Electronics*, **46** (5), 953–959.
11 Grainger, B.M., Reed, G.F., Sparacino, A.R., and Lewis, P.T. (2014) Power electronics for grid-scale energy storage. *Proceedings of the IEEE*, **102** (6), 1000–1013.
12 IEC MSB (Market Strategy Board) (2011), White paper: Electric energy storage. International Electrotechnical Commission, Geneva.
13 Kondoh, J., Ishii, I., Yamaguchi, H. *et al.* (2000) Electrical energy storage systems for energy networks. *Energy Conversion and Management*, **41** (17), 1863–1874.
14 Schainker, R.B. (2004) Executive overview: energy storage options for a sustainable energy future, in *IEEE PES General Meeting*, Denver, pp. 2309–2314.
15 Ibrahim, H., Ilinca, A., and Perron, J. (2008) Energy storage systems – characteristics and comparisons. *Renewable and Sustainable Energy Reviews*, **12** (5), 1221–1250.
16 Chen, H., Cong, T.N., Yang, W. *et al.* (2009) Progress in electrical energy storage system: A critical review. *Progress in Natural Science*, **19** (3), 291–312.
17 Hadjipaschalis, I., Poullikkas, A., and Efthimiou, V. (2009) Overview of current and future energy storage technologies for electric power applications. *Renewable and Sustainable Energy Reviews*, **13** (6-7), 1513–1522.
18 Divya, K. and Østergaard, J. (2009) Battery energy storage technology for power systems – an overview. *Electric Power Systems Research*, **79** (4), 511–520.
19 Electric Power Research Institute (2010), Electric energy storage technology options: A white paper primer on applications, costs, and benefits. EPRI, Palo Alto, CA. 1020676.
20 Vazquez, S., Lukic, S.M., Galvan, E. *et al.* (2010) Energy storage systems for transport and grid applications. *IEEE Transactions on Industrial Electronics*, **57** (12), 3881–3895.
21 Sparacino, A.R., Reed, G.F., Kerestes, R.J. *et al.* (2012) Survey of battery energy storage systems and modeling techniques, in *Power and Energy Society General Meeting, 2012 IEEE*, San Diego, pp. 1–8.
22 Boicea, V.A. (2014) Energy storage technologies: The past and the present. *Proceedings of the IEEE*, **102** (11), 1777–1794.
23 Elliman, R., Gould, C., and Al-Tai, M. (2015) Review of current and future electrical energy storage devices, in *50th International Universities Power Engineering Conference* (*UPEC*), Stoke on Trent, pp. 1–5.

24 Aneke, M. and Wang, M. (2016) Energy storage technologies and real life applications – A state of the art review. *Applied Energy*, **179**, 350–377.

25 Stock, A., Bourne, G., Brailsford, L., and Stock, P. (2018), Fully charged: Renewables and storage powering Australia, https://www.climatecouncil.org.au/uploads/d4a4f17c09c83d03f13234051e3e77d8.pdf. Climate Council of Australia Limited. (accessed 1 April 2018).

26 Luo, X., Wang, J., Dooner, M., and Clarke, J. (2015) Overview of current development in electrical energy storage technologies and the application potential in power system operation. *Applied Energy*, **137**, 511–536.

27 Ribeiro, P.F., Johnson, B.K., Crow, M.L., Arsoy, A., and Liu, Y. (2001) Energy storage systems for advanced power applications. *Proceedings of the IEEE*, **89** (12), 1744–1756.

28 Walker, L.H. (1990) 10-MW GTO converter for battery peaking service. *IEEE Transactions on Industry Applications*, **26** (1), 63–72.

29 Rogers, J.D., Schermer, R.I., Miller, B.L., and Hauer, J.F. (1983) 30 MJ superconducting magnetic energy storage system for electric utility transmission stabilization. *Proceedings of the IEEE*, **71** (9), 1099–1107.

30 Saber, A.Y. and Venayagamoorthy, G.K. (2009) One million plug-in electric vehicles on the road by 2015, in *12th International IEEE Conference on Intelligent Transportation Systems*, St. Louis, pp. 1–7.

31 Yilmaz, M. and Krein, P.T. (2013) Review of battery charger topologies, charging power levels, and infrastructure for plug-in electric and hybrid vehicles. *IEEE Transactions on Power Electronics*, **28** (5), 2151–2169.

32 Lopes, J.A.P., Soares, F.J., and Almeida, P.M.R. (2011) Integration of electric vehicles in the electric power system. *Proceedings of the IEEE*, **99** (1), 168–183.

33 Dürr, M., Cruden, A., Gair, S., and McDonald, J. (2006) Dynamic model of a lead acid battery for use in a domestic fuel cell system. *Journal of Power Sources*, **161** (2), 1400–1411.

34 Ceraolo, M. (2000) New dynamical models of lead-acid batteries. *IEEE Transactions on Power Systems*, **15** (4), 1184–1190.

35 Salameh, Z.M., Casacca, M.A., and Lynch, W.A. (1992) A mathematical model for lead-acid batteries. *IEEE Transactions on Energy Conversion*, **7** (1), 93–98.

36 Casacca, M.A. and Salameh, Z.M. (1992) Determination of lead-acid battery capacity via mathematical modeling techniques. *IEEE Transactions on Energy Conversion*, **7** (3), 442–446.

37 Lu, C.F., Liu, C.C., and Wu, C.J. (1995) Dynamic modelling of battery energy storage system and application to power system stability. *IEE Proceedings – Generation, Transmission and Distribution*, **142** (4), 429–435.

38 Tremblay, O. and Dessaint, L.A. (2009) Experimental validation of a battery dynamic model for EV applications. *World Electric Vehicle Journal*, **3** (2), 289–298.

12
Microgrids: Operation and Control

12.1 Introduction

The penetration of distributed generation (DG) at medium and low voltages in electrical distribution systems has been increasing over the years, particularly in developed countries. The distributed generation encompasses a wide range of prime mover technologies, such as internal combustion (IC) engines, gas turbines, microturbines, photovoltaic (PV), fuel cells, and wind turbines (WT). Most emerging technologies require an inverter to interface with the distribution systems [1]. These technologies, collectively labelled "distributed energy resources" (DERs) can substantially reduce carbon emissions, thereby reducing greenhouse gas emissions (as mandated by the Kyoto Protocol) [2]. Another objective is to increase the contribution of the renewable energy component (wind, solar, and hydro), which will overcome the problems of dwindling energy sources based on coal and gas. The presence of distributed generation close to the loads can enhance power quality and reliability (PQR). The occurrence of major power blackouts in 2003 in North America and Europe has also contributed to the desire of customers to install local distributed generation which can minimize the impact of power blackouts. The application of combined heat and power (CHP) equipment, where required, results in increased energy efficiency. These are the major benefits of DER that have resulted in the new paradigm of developing active distribution networks in contrast with the earlier approach of treating distribution networks as passive.

Another important benefit of deploying DERs is the deferral of transmission and distribution upgrades. It is also feasible to reduce the energy losses in the distribution networks and consequently the cost to the customer. This assumes that economies of scale do not apply in the generation based on natural gas.

The control of a large number of DERs poses a major challenge in the operation of the distribution system that can be partially addressed by microgrids, which are entities that coordinate DERs in a consistently decentralized way, thereby reducing the control burden on the grid. A microgrid comprises a LV (≤ 1 kV) or MV (1–69 kV) locally controlled cluster of DERs that behaves as a single producer or load, both electrically and in energy markets. A microgrid operates safely and efficiently within its local distribution network, but it is also capable of islanding (during power blackouts).

Dynamics and Control of Electric Transmission and Microgrids, First Edition.
K. R. Padiyar and Anil M. Kulkarni.

12.2 Microgrid Concept [1–7]

To realize the potential of distributed generation (DG), it is convenient to apply a systems approach. Here, the DG and associated loads are clubbed together to form a subsystem or "microgrid". This approach allows local control of DG thereby reducing or eliminating the need for central dispatch. During disturbances, the generation and corresponding loads can separate from the distribution network to isolate the microgrid's load from the disturbance (maintaining a high level of service reliability) without harming the integrity of the transmission grid. Intentional islanding of DG and loads has the potential to provide a higher local reliability than that provided by the power system as a whole.

Lasseter [1] proposed the approach of providing DG-based controls that enable a plug-and-play model without communication or custom engineering for each site. This concept was further developed in the Consortium for Electric Reliability Technology Solutions (CERTS) microgrid, which is the world's first, full-scale, inverter-based, distributed generation microgrid. CERTS was formed in 1999 to research, develop, and commercialize new methods, tools, and technologies to protect and enhance the reliability of US electric power systems. CERTS is conducting research for the US Department of Energy (DOE) Office of Electricity Delivery and Energy Reliability. CERTS is a consortium involving universities, national laboratories, and industry.

12.2.1 Definition of a Microgrid

A microgrid is a group of interconnected loads and distributed energy resources within clearly defined electrical boundaries that acts as a single controllable entity with respect to the grid and that connects and disconnects from such a grid to enable it to operate in both grid-connected and island mode.

This definition is adopted by the US DOE and the Electric Power Research Institute (EPRI).

It should be noted that DERs include both DG and distributed storage (DS). The DS is required when the generation and loads cannot exactly be matched. Storage capacity is defined in terms of the time that the nominal energy capacity can cover the load at rated power. Storage capacity can then be defined in terms of energy density requirements (for medium- and long-term needs) or in terms of power density requirement (for short-term and very-short-term needs). The DS can provide the following benefits:

1) It stabilizes and permits DG units to run at a constant and stable output, unaffected by load fluctuations.
2) It provides the ride-through capability when there are dynamic variations of primary energy (such as those of sun, wind, and hydro resources).
3) It permits DG to seamlessly operate as a dispatchable unit.
4) It can damp peak surges in load demand, providing outage ride-through while back-up generators respond.

There are several forms of energy storage, including batteries, supercapacitors, and flywheels. Battery systems store energy in the form of chemical energy and require power electronic converters for connection to an AC power system.

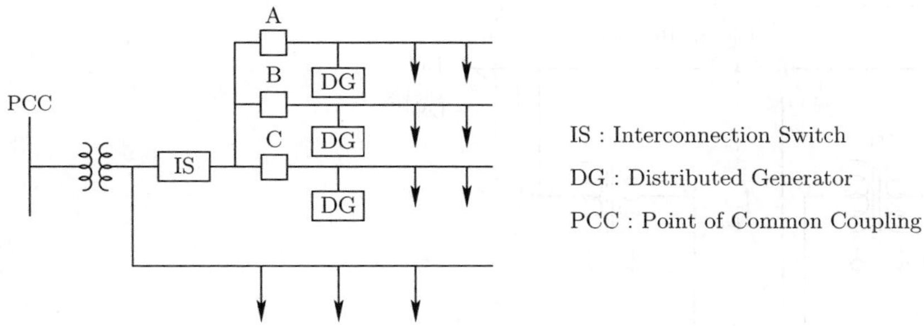

Figure 12.1 Schematic of a microgrid.

A microgrid and its components are shown in Figure 12.1. It consists of a group of radial feeders which could be part of a distribution system or a building's electrical system. There is a single point of connection to the utility called the point of common coupling (PCC). The feeders A, B, and C have sensitive loads that require local generation. Feeders A to C can island from the grid using the interconnection switch (IS), which can be a static switch (that can operate within a cycle). There are three DGs which are controlled using only measurement of local voltages and currents.

When there is a problem with the utility supply, the IS will open, isolating the sensitive loads from the power grid. Non-sensitive loads ride through the event. It is assumed that there is sufficient generation to meet the demand in the microgrid. When the microgrid is connected to the grid, power from the local generation can supply non-sensitive loads.

In the CERTS microgrid concept, each component of the microgrid is assumed to operate in a plug-and-play model with autonomous control in a peer-to-peer concept. This assumes that there is no central controller or storage units in the microgrid. It also implies that the microgrid can continue operating even when one DG fails. Plug-and-play implies that a unit can be placed at any location without re-engineering the controls. The plug-and-play model facilitates placing generators near the thermal loads. An important component is the IS, which connects or disconnects the microgrid from the rest of the distribution network. It consolidates various functions involving power switching, protective relaying, metering, and communications with a digital signal processor (DSP). A schematic diagram of a circuit-breaker-based IS is shown in Figure 12.2. Grid conditions are measured both on utility and microgrid sides of the switch through current transformers (CT) and potential transformers (PT). The ISs are designed to meet grid interconnection standards (IEEE 1547 and UL 1741 for North America) to minimize custom engineering and site-specific approval processes and reduce costs. To maximize applicability and functionality, the controls are also designed to be technology neutral and can be used with a circuit breaker or a faster semiconductor-based static switches like thyristors and insulated-gate bipolar transistors (IGBTs).

12.2.2 Control System [3–5]

A microgrid control system must be designed for safe operation of the system in both grid-connected and islanded modes. The system may be a central controller or

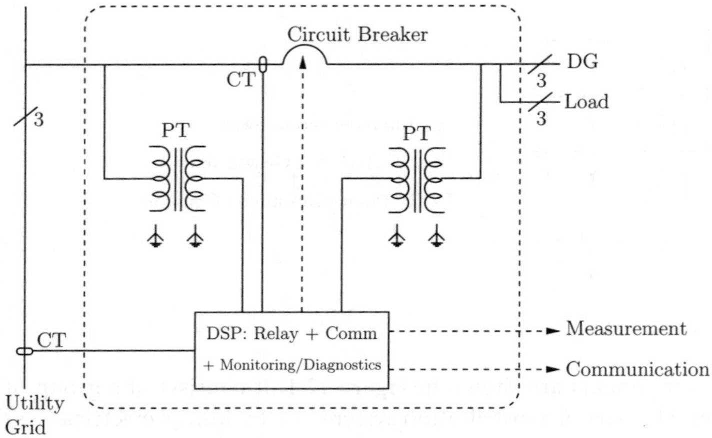

Figure 12.2 Schematic diagram of a circuit-breaker-based IS. Source: [3] Fig.8, p. 44, reproduced with permission ©IEEE 2008.

distributed with each DG. In the islanded mode, the control system must be able to regulate the local voltage and frequency and protect the microgrid. In the absence of rotating masses connected with the generators in a microgrid (containing primarily converter-based DG), frequency control can be complicated. It must coordinate the capabilities of the DERs to change their active power through frequency droops and load shedding as required.

Voltage regulation is also an important facet of the microgrid controller. Without local voltage controllers, systems with high penetration of DERs are prone to oscillations of voltage and/or reactive power.

The voltage control should ensure that there are no large circulating reactive currents between DERs. The problems of voltage regulation are present in both isolated and interconnected modes. In the grid-connected mode, DG units can provide ancillary services of local voltage support. The reactive power support can be provided locally by the adoption of a voltage versus reactive current droop controller, similar to the droop controller for frequency.

The introduction of new technologies in DG and DS in microgrids calls for detailed and comprehensive testing. There are test beds in several countries, including the United States, Canada, Europe, and Japan. The AEP/CERTS test site (with no communications or storage facilities) demonstrated several features of microgrids involving (i) autonomous load following, (ii) seamless separation and automatic resynchronization with the distribution network, (iii) autonomous load transfer from overloaded sources to other sources, (iv) uninterruptible power supply (UPS) level power quality, (v) plug-and-play architecture, and (vi) mechanical and static interface switches [6, 7].

In the development of advanced microgrids, there are three distinct benefits that are anticipated [8]:

1) Economic benefits as customers attempt to lower costs by increasing the efficiency of the microgrid for thermal and electrical requirements.
2) A microgrid provides environmental benefits (reducing greenhouse gases) by adopting renewable generation involving solar PV and wind turbines.

3) Meeting resiliency requirements to retain service during events such as storms or cyclones that can disrupt the electricity supply.

It should be noted that grid resiliency is a new concept or requirement (in addition to PQR) that takes into account the following three aspects when extreme weather and other natural disasters threaten lives, disable communities, and devastate the generation, transmission, and distribution of electrical power:

a) **Prevention:** As far as feasible, preventing damage in the distribution system by changes in the design standards, construction guidelines, maintenance routines, inspection procedures, and recovery practices.
b) **Recovery:** Proper planning to provide rapid damage assessment, prompt crew deployment, and readily available replacement components.
c) **Survivability:** The ability to maintain some basic level of electrical functionality to individual consumers or communities in the event of complete loss of electrical service from the distribution system. It should be noted that microgrids can help here.

It should be noted that balancing the desire for enhanced environmental benefits versus the resiliency needs of a group of critical facilities can be challenging, since the amount and type of generation needed for resiliency may not be able to provide adequate levels of environmental benefits. For example, properly maintained emergency back-up generators with underground distribution network can meet the resiliency requirement without the benefits provided by rooftop solar PV generation.

A robust design of the distribution system operated intelligently to manage disruption as it unfolds and rapid recovery can meet the requirements for resiliency.

12.3 Microgrid Architecture [9]

There are four classes of microgrids:

1) Single facility microgrids
   These include industrial and commercial buildings, residential buildings, and hospitals. The loads are typically under 2 MW. These systems have low inertia and require back-up generation for off-grid operation. Microgrids for these applications need to be designed for better power availability and quality. Loads such as hospitals require seamless transition between grid-connected and islanded operation.
2) Multiple facility microgrids
   This category includes academic campuses and residential, commercial, and industrial complexes which have several buildings spread over an area. The range of load can vary between 2 and 5 MW. The requirements are high PQR.
3) Feeder microgrids
   These microgrids are designed to manage the generation and/or load of all entities within a distribution feeder. The load or generation can vary within a range of 5–10 MW. A feeder microgrid may incorporate smaller microgrids, single or multiple facility. A feeder microgrid has the ability to separate from the bulk grid during disturbances and service its internal loads. Municipal utilities and cooperatives are the potential owners or operators of the feeder microgrids to provide better availability and quality of power supply to the customers.

4) Substation microgrids
   These microgrids manage the generation and/or load of all entities connected to a distribution substation (rated from 5 to 10 MW or more). These microgrids are likely to include some generation connected directly at the substation and distributed generation. Feeder and facility level microgrids could be included in the substation microgrid. The major objective is to improve power quality and availability to customers.

Remote Microgrids [10, 11]

Historically, the development of DG took place in the context of electrification of isolated communities and back-up generation for critical loads. Electrification of remote communities in developing countries and geographical islands is a major application of DERs and the development of microgrids. Energy requirements are met by renewable sources (small hydro, solar PV, and wind) in addition to diesel generators.

A major characteristic of a remote microgrid is that it is sized to serve the entire load along with an adequate level of reserve capacity to manage contingencies. The power/energy balancing between variable load and intermittent generation (solar PV and wind) requires strategies to maintain the desired level of reliability. Isolated microgrids could be based on DC distribution to improve efficiency and reduce costs.

12.4 Distribution Automation and Control [12, 13]

The energy control centres that were developed to improve system security and reliability after the major blackout in 1965 in the northeast United States, essentially introduced the supervisory control and data acquision (SCADA) system for the monitoring and control of generation and transmission systems. Advances in digital technology have resulted in the development of powerful microprocessors which have made it feasible to introduce distribution automation and control (DAC) to monitor the distribution system and the connected load. The objectives of DAC are [12]:

1) to improve overall system efficiency in the use of both capital and energy
2) to increase market penetration of coal, nuclear, and renewable energy sources
3) to reduce the reserve requirements in both transmission and distribution
4) to increase reliability of service for essential loads.

Distribution automation encompasses:

(a) a communication system at the distribution level that can control customer load and reduce peak-load generation through load management
(b) an unattended distribution substation that can (i) continuously monitor the system, (ii) issue commands, (iii) report any change in the status to the distribution dispatch centre (DDC), and (iv) store it on-site for later use based on the application of an on-site microprocessor.

The control functions of DAC include the following:

1) Load management functions:
   (a) Discretionary load switching to reduce the peak load on a feeder or substation that is becoming overloaded

   (b) Peak-load pricing
   (c) Load shedding or pickup
2) Operational management:
   (a) Load reconfiguration involving remote control of switches and breakers to permit routine daily, weekly or seasonal reconfiguration of feeders or feeder segments to take advantage of the load diversity among feeders. This function minimizes feeder reinforcements and enables routine maintenance on feeders without any customer load interruptions.
   (b) Voltage regulation by remote control of selected voltage regulators, together with network capacitor switching.
   (c) Transformer load management (TLM) involving monitoring and continuous reporting of transformer loading data and core temperature to prevent overloads, burnouts or abnormal operation by timely reinforcement, replacement or reconfiguration.
   (d) Feeder load management (FLM), which is similar to TLM, but the loads are monitored on feeders and feeder segments to equalize loads over several feeders.
   (e) Capacitor control to switch distribution capacitors selectively by remote control.
   (f) Fault detection, location, and isolation using sensors located throughout the distribution network.
   (g) Condition monitoring in real time.
   (h) Automatic on-line gathering and recording of load data for specific off-line analysis.
3) Automatic meter reading (AMR) to allow remote reading of customer meters for reading of total energy consumption, peak demand or time-of-day consumption.

12.5 Operation and Control of Microgrids [4, 5]

A microgrid includes a portion of a power distribution network that is located downstream of the distribution substation. It includes a variety of DER units, including DG and DS, with different capacities and characteristics. The microgrid is connected to the distribution system at the low-voltage bus of the distribution transformer. The microgrid normally operates in a grid-connected mode through the substation transformer. However, it is also expected to provide adequate generation capacity and controls to supply a portion of the load after being disconnected from the distribution system (islanded mode of operation). Operation in the islanded mode may require disconnection (curtailment) of some existing loads which are not essential. The existing power utility practice does not normally permit accidental islanding and automatic resynchronization of a microgrid primarily due to human and equipment safety concerns. However, the presence of a large number of DERs (constituting a significant component of the total generation requirement) necessitates the provision for both islanded and grid-connected modes of operation and smooth transition between the two modes.

12.5.1 DER Units

DERs are divided into two groups based on the type of interface with the microgrid. The first group includes conventional (rotary) units based on synchronous or induction

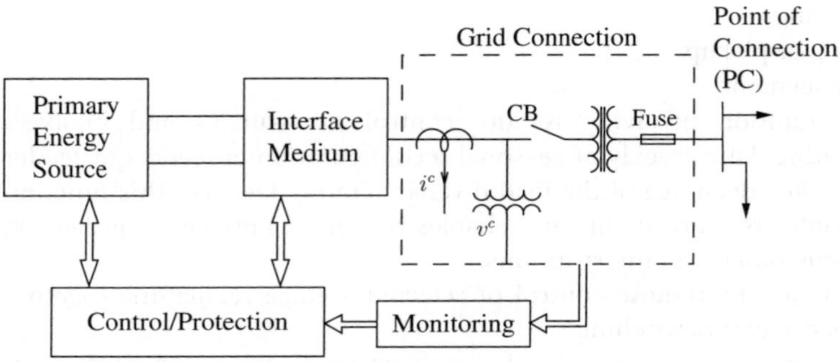

Figure 12.3 Representation of a DG unit. Source: [5] Fig. 3, p. 57, reproduced with permission ©IEEE 2008.

generators that are interfaced directly to the microgrid. The second group are interfaced to the microgrid through power electronic converters which introduce novel control concepts. Figure 12.3 shows the block diagram of a DG unit.

The primary energy source for conventional DG could be (i) reciprocating engines, (ii) small hydro or (iii) fixed-speed wind turbine. The interface for the first two sources is based on a synchronous generator, whereas for the fixed-speed wind turbine it is based on a squirrel cage induction generator (SCIG). When the primary energy source is a variable speed wind turbine or microturbine/solar PV/fuel cell, the interface is based on power-electronic voltage source converters (VSCs). The input power to the interface converter from the source side can be AC at fixed or variable frequency or DC. The microgrid-side converter operates at the frequency of 50 or 60 Hz. Figure 12.4 is a representation of a DS unit for which the "primary source" should be replaced by "storage medium". For long-term storage, the storage medium is battery storage. For short-term storage, the storage medium is either a super-capacitor or a flywheel.

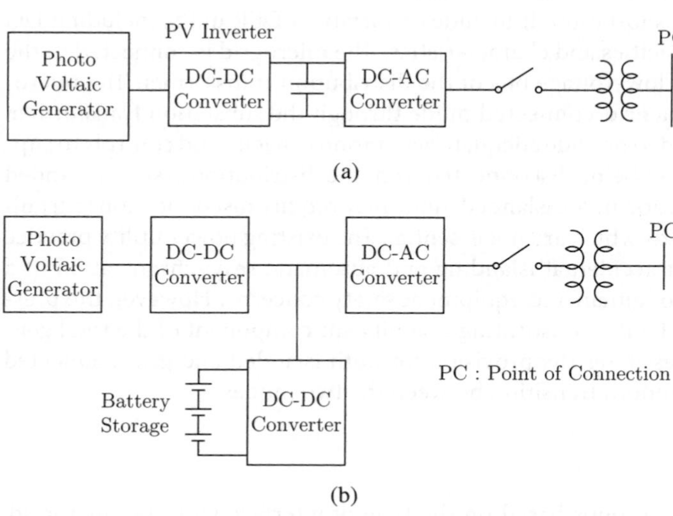

Figure 12.4 Block diagram of a PV-based DER. (a) Non-dispatchable (b) Dispatchable (DG + DS).

In terms of power flow control, a DG unit is either a dispatchable or a non-dispatchable unit. The output power of a dispatchable unit can be controlled externally through set-points provided by a supervisory control system. A dispatchable DG unit can be either a fast-acting or a slow-response type. A DG based on a reciprocating engine is normally equipped with a governor for speed control that adjusts the fuel in-flow. The governor and AVR control the real and reactive power outputs of a dispatchable DG unit.

The output power of a non-dispatchable DG unit is based on the optimal operating condition of its primary energy source. For example, a wind or solar PV-based DG is operated to maximize its output power. The control strategy is based on maximum point of power tracking (MPPT) to deliver maximum power under all operating conditions. Figure 12.4 shows two common structures for VSC-interfaced DER units. The first represents the non-dispatchable PV-based DG. The second represents a dispatchable DER unit using battery storage. It is also possible to replace battery storage by capacitors that provide short-time power requirements during start-up or acceleration/deceleration of the slow genset when it is used as the primary energy source.

12.5.2 Microgrid Loads

A microgrid can serve electrical and thermal loads. In the grid-connected mode, the distribution system can supply or absorb the power deficit or excess generation from the connected microgrid. If the net import or export of power from a microgrid hits hard limits (based on operational strategies or contractual obligations), the load or generation can be shed.

In the islanded mode of operation, load or generation shedding is often required to maintain power balance and stabilize the microgrid. The critical loads in a microgrid should receive priority. The operation of microgrids also considers power quality and reliability enhancement of specified loads. The control of loads is often exercised to optimize the ratings of DS units and dispatchable DG units by reducing the peak load and wide fluctuation of the load with time.

As a matter of fact, part of the non-sensitive load can be considered as a controllable load and entered into a demand response (DR) strategy to reduce the peak and/or load leveling. It is also possible to reschedule the load serving for specified time intervals when power from intermittent DG units (solar PV or wind) is available. The demand response is often initiated by sending the price signal which is supervised by the energy management controller of the microgrid.

12.5.3 DER Controls

The control functions of DER units depend on the operating mode of the microgrid (grid-connected or islanded). The main control functions for a DER unit are (i) voltage and frequency controls, and (ii) active and reactive power control. One could also classify the controls into two categories: grid interactive or non-interactive. For example, in a grid-connected mode of operation, a DER unit may function in a non-interactive mode by exporting power (with or without MPPT). The interactive control approach involves either dispatch of power (where it is feasible) or providing active and reactive power support.

There are two kinds of control strategies used to operate an inverter-based DER:

1) P-Q control: Here the inverter is used to supply a specified active and reactive power. This is an interactive method of control.
2) V-f control: The inverter is controlled to feed the load with pre-defined values of voltage and frequency. Depending on the load, the VSC output (real and reactive power) is defined. This is a non-interactive method.

In P-Q inverter control, the inverter operates by injecting into the microgrid the power available at its input. The reactive power injected corresponds to a pre-specified value, defined locally (using a local control loop) or centrally from the microgrid central controller (MGCC). The P-Q inverter control is implemented as a current-controlled voltage source.

In the second method, the VSC emulates a synchronous generator where voltage (magnitude) and frequency are controlled as described by the following equations:

$$f = f_o - k_P P \tag{12.1}$$

$$V = V_o - k_Q Q \tag{12.2}$$

where P and Q are the inverter active and reactive power outputs, and k_P and k_Q are the droops (positive quantities). f_o and V_o are operating values at no load conditions.

When a DER is connected with a stiff microgrid operating at frequency f_{grid} and with terminal voltage V_{grid}, the f_o and V_o in (12.1) and (12.2) are replaced by

$$f_{o1} = f_{grid} + k_P P_1 \tag{12.3}$$

$$V_{o1} = V_{grid} + k_Q Q_1 \tag{12.4}$$

Figure 12.5 shows the frequency versus active power droop for two different values of k_P. If a cluster of n DERs operates in a stand-alone AC system, frequency variation leads automatically to power sharing, such that

$$\Delta P = \sum_{i=1}^{n} \Delta P_i, \quad \Delta P_i = \frac{\Delta f}{k_{P_i}} \tag{12.5}$$

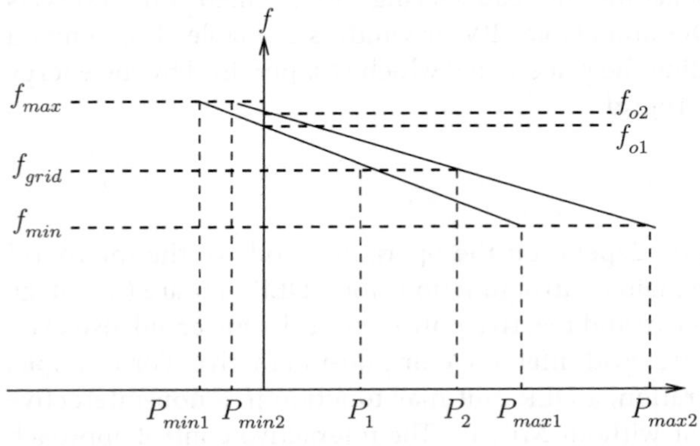

Figure 12.5 Frequency versus active power droop. Source: [4] Fig.4, p. 918, reproduced with permission ©IEEE 2006.

Similar observations can be made for V-Q control droops. However, the impedances of conductors/cables do not permit exact sharing of reactive power according to 12.2. Even active power sharing will be affected by the presence of resistances.

Remarks

1) The operation of DERs in an islanded microgrid does not rely on fast communication with the control method (based on droops).
2) The DERs respond to system disturbances based on information available at its terminals. However, a communication infrastructure will be made available in a microgrid for the purpose of optimal power management.

12.5.4 Control Strategies Under Grid-Connected Operation [4]

When the microgrid is connected to the distribution system, there are two possible modes of control:

(1) *Power export control*: The DER output power is controlled within the voltage and frequency limits determined by the microgrid. For a VSC interfaced DER, current-controlled strategy can be used to determine the reference voltage waveforms for the pulse width modulation (PWM) of the VSC. The reference signals are synchronized to the grid frequency by tracking the voltage waveform at the point of common coupling (PCC). The control strategy can be implemented using a synchronous reference frame (d-q) or a stationary reference frame (α-β). The proportional-integral (PI) controller is widely used in conjunction with d-q control whereas the proportional-resonant (PR) controller is utilized in stationary reference frame control. The d-q current control of a VSC-interfaced DER is shown in Figure 12.6. The blocks d-reference controller, q-reference controller, and d-q current controller are shown separately in Figure 12.7.
In the power export control strategy, the power extracted from the renewable energy source is fed into the DC link, which raises the DC voltage. However, the voltage regulator overcomes the voltage rise by determining the required d-axis current to balance the power flow in the DC link of the VSC. Similarly, the reactive power controller determines the amount of reactive current required to meet the specification of Q. Instead of specifying the reactive power (Q) it is also feasible to specify the required voltage magnitude (V) within limits.

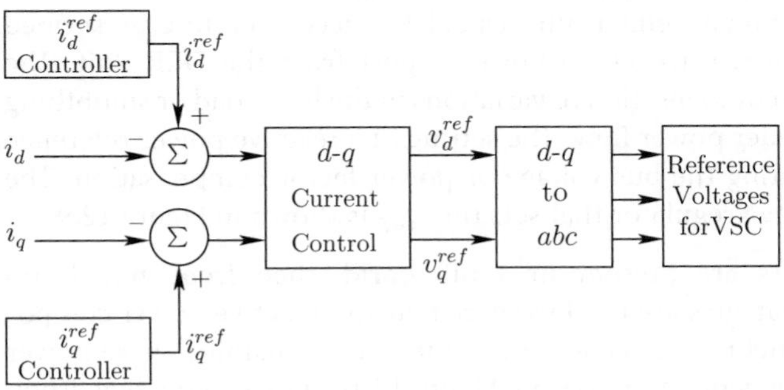

Figure 12.6 Block diagram of a VSC interfaced DER controller.

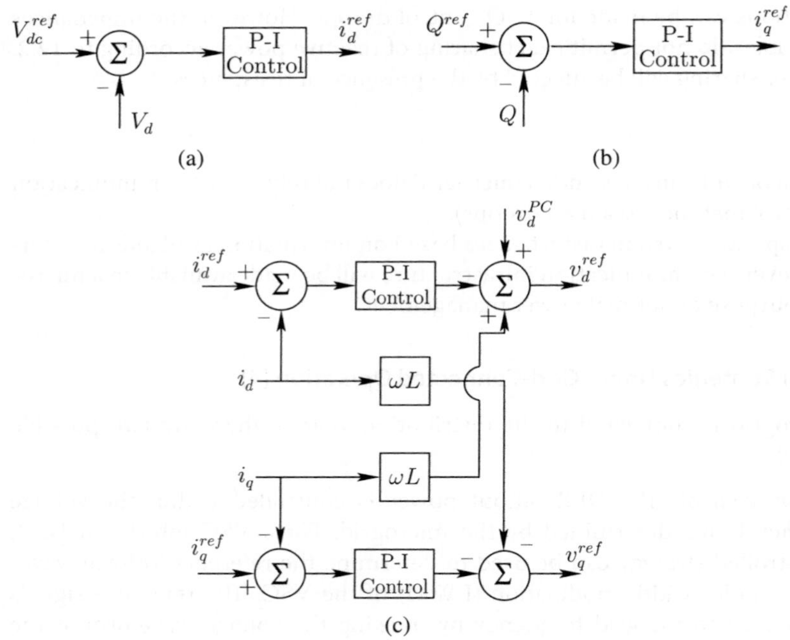

Figure 12.7 Current controller of a DER unit. (a) i_d Reference controller (b) i_q Reference controller (c) *d-q* Current control.

Remark

Figure 12.7 is related to Figure 5.26. The currents i_d and i_q are the currents of a VSC operating as an inverter. They are related to i_p and i_r by the relations

$$i_d = -i_p, \qquad i_q = i_r \tag{12.6}$$

In addition,

$$v_d^{ref} = e_p, \qquad v_q^{ref} = e_r \tag{12.7}$$

(2) *Power dispatch control*: For dispatchable DER units, output power and reactive power control strategies are used based on specified reference values. Figure 12.8 shows the block diagram for this type of control. P_{ref} and Q_{ref} values are set by the supervisory power management unit calculated according to a prescribed power profile to optimize real (reactive) power export from the DER unit. The optimization may involve compensation of variations in the local load or smoothing out fluctuations in the feeder power flow. The setting of a reactive power reference could be based on regulating the bus voltage or power factor compensation. The block diagram of the voltage regulator that sets the Q_{ref} is shown in Figure 12.9.

If two or more DER units are present in a microgrid, then frequency-droop and voltage-droop control strategies are used to share real and reactive power components just as two or more synchronous generators connected in parallel share power and reactive power output of a generating station. Figure 12.10 shows frequency droop (f-P) and voltage droop (V-Q) characteristics where each is specified by its slope

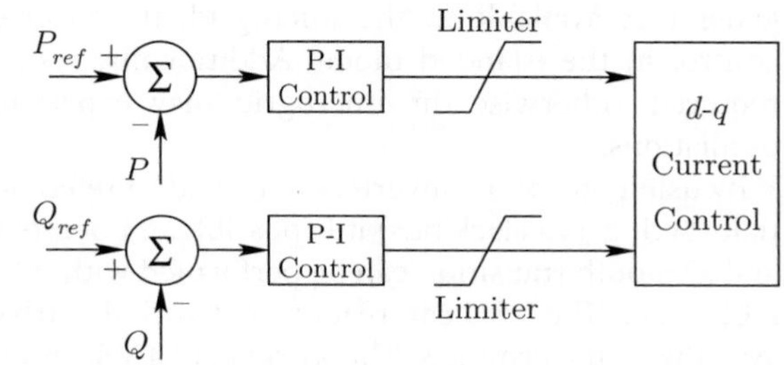

Figure 12.8 Active and reactive power control.

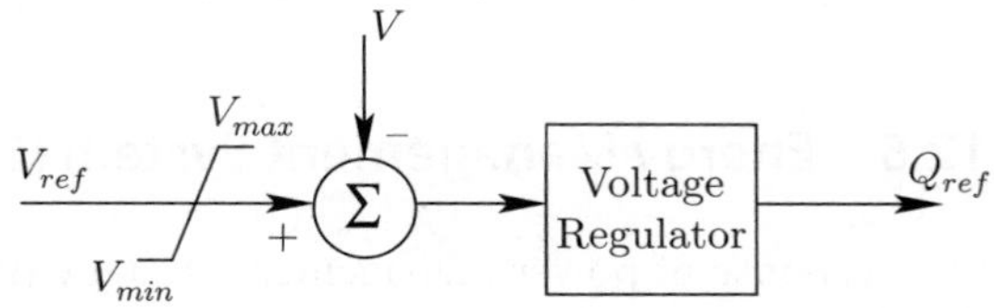

Figure 12.9 Reactive power setting by voltage regulator.

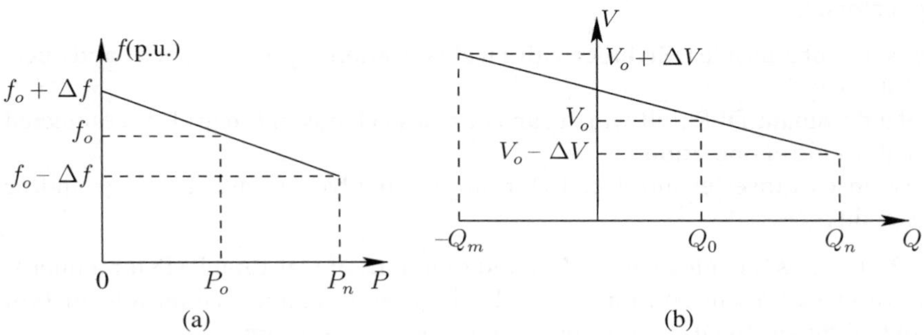

Figure 12.10 Droop characteristics for load sharing among multiple DER units. (a) *f-P* droop (b) *V-Q* droop.

(k_f or k_v) and a base point representing either rated frequency (f_o, P_o) or the nominal voltage (V_o, Q_o), respectively.

The droop parameters and the base points can be controlled to dynamically adjust the operating points of the units. This is achieved by changing the power (and reactive power) sharing levels to set the frequency and voltage at new values. The restoration of the original values is performed slowly, particularly during the resynchronization of the islanded grid. The inputs to the controllers are the locally measured deviations in the frequency and terminal voltage of the DER unit. If DER units have different capacities, the slope of each droop characteristic is selected to be proportional to the rated capacity of the corresponding unit to prevent overloading.

12.5.5 Control Strategy for an Islanded Microgrid [4]

Islanding of a microgrid can occur by unplanned events like faults in the grid or by planned actions such an maintenance work. Under an islanding event, the local generation can be modified to reduce the imbalance between local loads and generation, and minimize the disconnection transients. In the absence of the synchronous generators available in the microgrid, the inverters are tasked with the frequency control in the islanded mode. Additionally, a strategy for voltage regulation is also required, otherwise the microgrid may experience voltage and/or reactive power oscillations.

By using one of the inverters to provide a reference for voltage and frequency (essentially acting as a slack bus) it is possible to operate the microgrid in the islanded mode and a smooth transition can be performed without changing the control mode of other DER units. The VSC can react to network disturbance based only on the information available at its terminals. The working of a VSC provides essentially the primary voltage and frequency regulation in the islanded microgrid.

12.6 Energy Management System [5]

Fast response of power management strategy (PMS) and energy management system (EMS) for microgrids with more than two DER units is a critical requirement due to the following factors:

1) The presence of multiple DER units with a wide variation in power capacity and operational strategy.
2) Lack of a dominant DER unit which can act as a slack bus in the grid-disconnected (islanded) mode of operation.
3) Fast response of inverter interfaced DER units can adversely affect voltage and/or angle stability.

Figure 12.11 shows the information flow and functions of real-time EMS for a microgrid. EMS receives present and forecast of load, generation, and market information to send control signals to DER units, loads, and protection systems.

The EMS signals to DER units determine active and reactive power references. The objectives are:

a) Sharing of active and reactive power among the DER units, in steady state.
b) Enabling of adequate response under microgrid disturbances that give rise to transients.

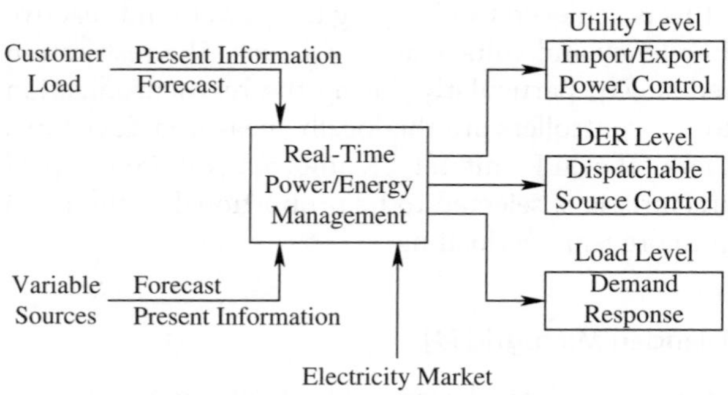

Figure 12.11 Information flow and functions of a real-time EMS. Source: [5] Fig. 11, p. 61, reproduced with permission ©IEEE 2008.

c) Restoration of frequency and resynchronization of the microgrid with the distribution grid.

In the grid-connected mode, DER units operate in a P-Q or P-V control mode. The power output of DER units is controlled to minimize the import of power from the external grid. However, in an islanded mode, the DER units must supply the total load of the microgrid after tripping the non-essential loads. The control strategies for the microgrid are designed to minimize the impact of the microgrid dynamics when subjected to islanding transients. Essentially, the oscillations in the power output and frequency must be well damped. The EMS must cater to both short-term power balancing and long-term energy management.

The short-term power balancing would involve:

a) voltage regulation and frequency control, which may require load shedding
b) voltage and frequency restoration during and subsequent to transients
c) providing adequate power quality for sensitive loads
d) resynchronization subsequent to clearing of the fault that led to the islanding.

The long-term energy management involves:

1) maintaining a satisfactory level of reserve capacity whenever rescheduling the set points of DER units based on optimization techniques, which may be based on (i) minimization of power loss and (ii) maximization of power outputs of renewable energy sources (RES) and/or minimization of the cost of energy production of fuel-based units
2) considerations of the requirements or limitations of each DER unit, which depend on the type of unit, cost of generation, time dependency of the prime source, maintenance intervals, and environment impacts
3) provision for demand response (DR) management and restoration of non-sensitive loads that are disconnected during microgrid transients.

12.6.1 Microgrid Supervisory Control

This must ensure all or a subset of control objectives such as (i) supply of electrical and/or thermal energy, (ii) prespecified service level for critical loads, (iii) black start subsequent to a failure, (iv) provision of ancillary service, and (v) participation in the energy market. These objectives are achieved through a centralized or decentralized supervisory control. A microgrid supervisory control architecture is shown in Figure 12.12. This shows three hierarchical levels:

1) distribution network operator (DNO) and market operator (MO)
2) microgrid central controller (MGCC)
3) local controllers (LCs) associated with each DER unit and/or load.

It should be noted that a DNO provides supervision for several microgrids in an area. One or more MO is responsible for the market functions of each area. The DNO and MO do not belong to the microgrids, but represent the main grid to which the microgrids are connected. The interface between the DNO and the microgrid is the MGCC. The MGCC is responsible for the coordination of the LCs.

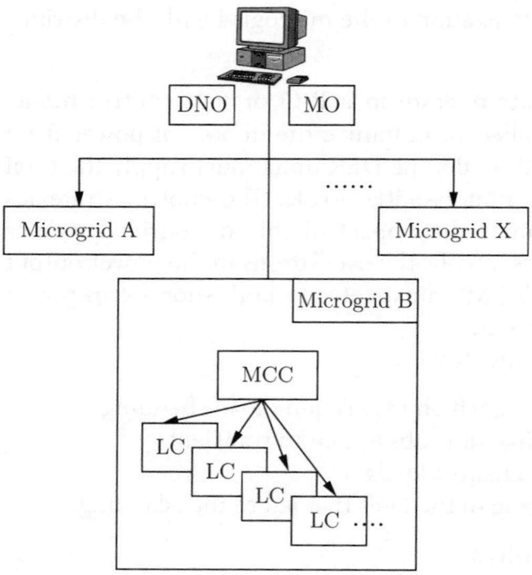

Figure 12.12 A microgrid supervisory control. Source: [14] Fig.1, p. 1447, reproduced with permission ©IEEE 2005.

In a centralized operation, each LC obtains the set-points from the MGCC. In a decentralized operation, the set-points are determined locally. For example, voltage set-points may be obtained from local measurements. In a centralized operation, LCs implement the commands of the MGCC during the grid-connected mode. In the islanded mode, LCs can switch to fast load tracking.

Based on a DER bidding strategy, the MGCC provides set-points to DER units and also determines whether or not to serve low-priority loads. In a decentralized approach, DER LCs take control decisions to meet the local demand and power exports to the main grid based on market prices. LCs also ensure safe operation of controllable loads.

12.6.2 Decentralized Microgrid Control based on a Multi-Agent System [14]

LCs control the DERs (DG and DS) and local loads. Depending on the mode of operation, they have a certain level of intelligence to take decisions locally. There are several levels of decentralization that can be possibly applied, ranging from a centralized approach to a fully decentralized approach where the main responsibility is given to DER controllers, which compete to maximize their production in order to satisfy the demand and probably provide the maximum possible export to the grid based on the market prices. However, LCs should take all suitable decisions to ensure safe and smooth operation of the DER they are controlling.

The organization of a controlled, intelligent entity (whole microgrid) formed by several less intelligent entities (LCs) can be based on a multi-agent system (MAS). The multi-agent technology is the evolution of the classical distributed technology with some specific characteristics that provide capabilities in controlling complex systems. The use of distributed control in a microgrid provides effective solutions for some specific operational problems. For example, local loads and DG/DS units may have different owners and several decisions should be taken locally and independently.

Microgrids operating in a market require that DERs have a certain degree of intelligence. Moreover, the local control of a DER not only controls the power sold to the grid but carries out tasks such as:

a) producing heat for local installations
b) feeding critical loads and preserving sufficient energy or fuel to supply them for a specified time
c) load-shedding as required
d) black start of the microgrid
e) ensuring seamless transition from the grid-connected mode to the islanded mode and vice versa
f) voltage control.

Role of the Agents

An agent can be a physical entity that acts in the environment or a virtual one. For example, an agent that directly controls a microturbine is a physical entity whereas software used for making bids in the energy market is a virtual agent. An agent connected with a DER is capable of influencing the control of other DER units in the microgrid. For example, a diesel generator, by altering its output, changes the set-points of other DG units.

The agents communicate with each other and this can improve the operation of the microgrid. For example, in a system containing a wind generator and a battery the battery system uses power from the wind turbine to charge and to discharge when there is no wind. The agents connected with the wind generator and the battery will be communicating with each other for optimal system operation. The agents also have a certain level of autonomy, which implies they can take decisions without a central control (MGCC). For example, the decision to charge the battery would depend on the price of the energy and the level of charge in the battery.

A MAS is preferable to a system with completely centralized control for the following reasons:

1) It is very difficult and complicated for the MGCC to have access to all available information.
2) It is difficult to implement, at a reasonable cost, a centralized system that can bid in the market every hour and simultaneously has the ability to shut down a specific load or change the set-point of a DER unit within a time frame of a fraction of a second in case of unstable operation.
3) A decentralized operation requires less complex data communication infrastructure.
4) Decentralized approach allows every manufacturer of DER units or loads to embed a programmable agent in the local controller according to some rules. This permits the plug-and-play capability of future DER units and loads. On the contrary, in a centralized system, the installation of any new component would require a change of program of MGCC every time a new component is added.

12.6.3 Industrial Microgrid Controllers

Microgrid controllers are available from manufacturing industries that are designed to manage and automate power generation systems that utilize different energy resources such as diesel, gas, geothermal, hydro, wind, solar, and tidal. It enables and maintains the grid integration of renewable and conventional generators in an energy efficient manner. The controller can typically employ distributed communication protocol. All high-level protocols are based on either standard Ethernet TCP/IP or UDP.

There are two methods of communication:

1) *Broadcasting*: Time-critical messages are broadcast to the system periodically. For example, a generator controller publishes its current status every 100 ms to all other peers in the microgrid. Based on this information, other devices make autonomous decisions to start, stop or adjust their set-points.
2) *Point to point*: Less time-critical information such as data recording and external operator commands to individual controllers are managed by point-to-point communication protocol.

The physical layer used is standard IEEE 802.3 Ethernet. Fibre optics as well as twisted pair cabling can be utilized. Special industrial Ethernet switches are used to create redundancy. In this case, a logical ring Ethernet backbone connects all controllers. During normal operation, the redundant Ethernet switch of the logical ring will block the forwarding traffic. In the case of a fault, this switch detects the problem in less than 20 ms and forwards the traffic to the cut-off part of the network.

One level down, each controller uses a variety of field buses (Modbus, CAN-bus, etc.) or hard-wired I/O (digital, analog) to interface with individual electrical devices. Each controller acts as an interface to the individual electrical device. It translates the data signals into the broadcasting protocol.

In summary, the microgrid controllers available from manufacturers are not meant to be programmable logic controllers (PLCs) and are only amenable to the selection of parameters. The technology used is based on intelligent distributed controller (IDC) hardware. This hardware platform has the flexibility of enabling plant-independent control systems to be configured for interfacing equipment supplied by different manufacturers.

12.7 Adaptive Network Protection in Microgrids [15]

In a low-voltage distribution grid, the feeders are radial with loads tapped-off along feeder sections and thus are designed assuming a unidirectional power flow. The protection is based on overcurrent (OC) relays with time-current discriminating capability. OC protection detects the fault from a high value of the fault current flowing away from the distribution transformer. In modern digital relays, the tripping current can be set for a wide range varying from 0.6 to 15.0 times the circuit breaker (CB) rated current. If a measured current is above the tripping setting, the relay operates to trip the CB on the feeder with a short delay defined by the relay coordination study and compatible with the locking strategy used (no locking, fixed hierarchical locking or directional hierarchical locking).

The power electronic interfaces used in DER units lead to several challenges in microgrid protection, especially in the islanded mode.

12.7.1 Protection Issues

In a microgrid, the fault currents depend also on the connection point of the DER and the feed-in power. (It should be noted that traditionally the distribution grid is a passive network whereas the presence of DERs makes the distribution network active. Also, the operating conditions in a microgrid are constantly changing because of the intermittent nature of the renewable DG (wind and solar)). The network topology is changing regularly in a microgrid due to the power management objectives aimed at economic benefits (such as loss minimization). The loads also vary depending on demand response (DR) or other factors. To summarize, the protection issues can be stated as follows:

1) Protection blinding (under-reach)
   Consider the system shown in Figure 12.13. When a fault occurs at bus 2, the DG 1 supplies part of the fault current, which results in the reduction of the fault current sensed by the relay R1. This phenomenon occurs when a large-scale conventional DG unit is connected to the distribution feeder between the main grid and the fault location. Due to this phenomenon, the relay R1 may not operate to prevent the fault current contribution from the MV grid.
2) Sympathetic tripping
   Consider Figure 12.14, which shows a grid-connected operation mode and a DG unit connected to a feeder, whereas the fault occurs on an adjacent feeder. The fault current contribution from the DG unit might exceed the pick up current setting of the

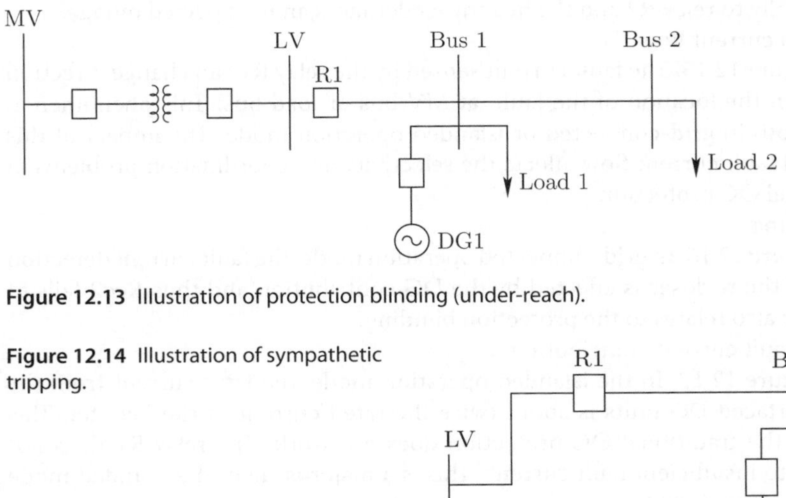

Figure 12.13 Illustration of protection blinding (under-reach).

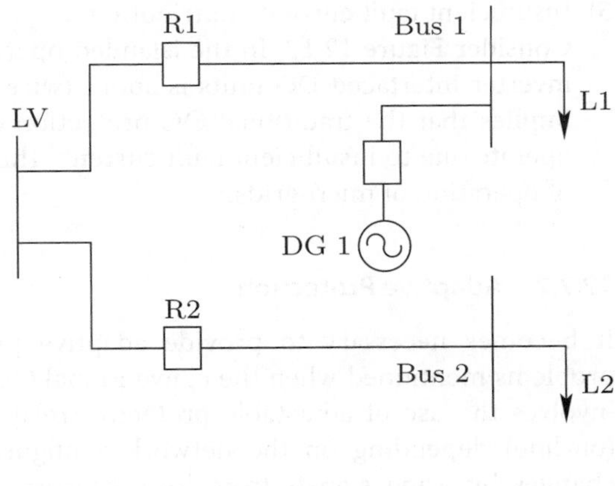

Figure 12.14 Illustration of sympathetic tripping.

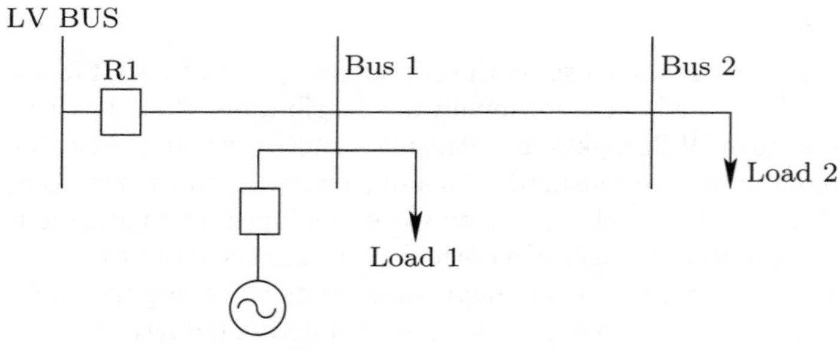

Figure 12.15 Bi-directional flow of fault current.

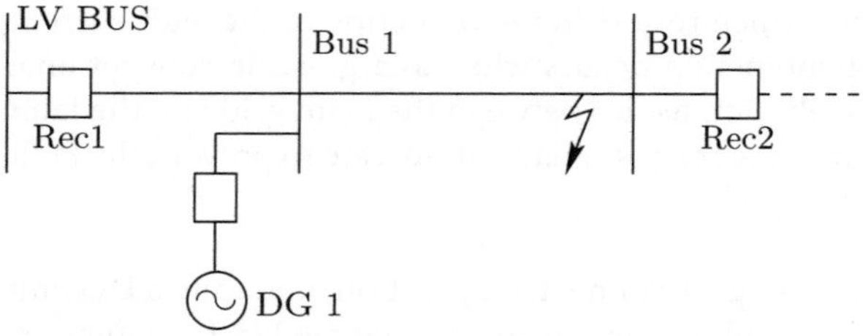

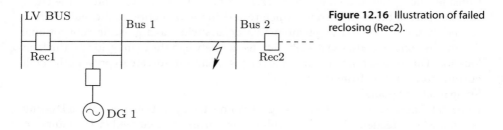

Figure 12.16 Illustration of failed reclosing (Rec2).

relay R1, especially when the DG capacity is sufficiently large. Thus the relay R1 trips sympathetically to relay R2 and the healthy feeder faces an unexpected outage.

3) Bidirectional current flow
Consider Figure 12.15. The fault current sensed by the relay R1 can change direction depending on the location of the fault, at MV bus or load bus. This phenomenon can occur both in grid-connected or islanded operation mode. The impact of this bidirectional fault current flow affects the selectivity and coordination problems in the traditional OC protection.
4) Failed reclosing
Consider Figure 12.16. In grid-connected operation mode, the fault current detection capability of the recloser is affected by the DG contribution and thus Rec2 fails to operate. This also relates to the protection blinding.
5) Insufficient fault current contribution
Consider Figure 12.17. In the islanded operation mode, the fault current from the inverter interfaced DG units is about twice the rated current of the inverter. This implies that the traditional OC protection does not work. The relay R1 does not operate due to insufficient fault current. This is a major issue in the islanded mode of operation of microgrids.

12.7.2 Adaptive Protection

It becomes necessary to provide adaptive protection strategies to overcome the problems mentioned when the conventional OC protection is applied. Essentially, this involves the use of adjustable protective relay settings that can change in real time (on-line) depending on the network configuration (topology and DG connection) changes by using signals from local sensors or a central control system. Adaptive protection systems are based on pre-calculated information where protection settings

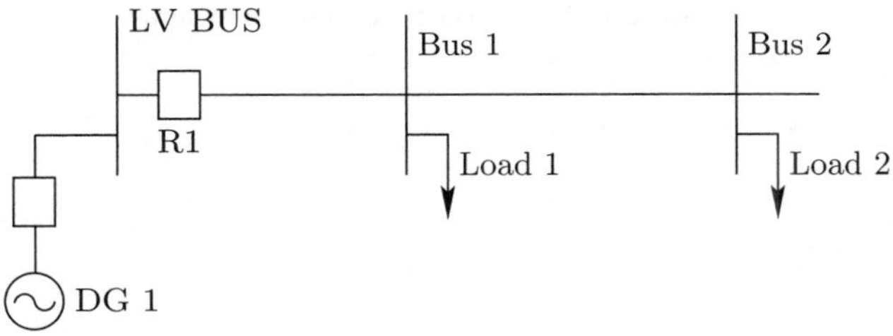

Figure 12.17 Illustration of insufficient fault current.

are updated periodically by the central controller with reference to the operating state of the network. This requires the following technical requirements:

1) Use of digital directional overcurrent relays.
2) Several setting groups must be encapsulated in digital overcurrent relays.
3) Establishment of communication infrastructure and use of industrial communication protocols, such as Modbus, IEC 61850, DNP3. It is necessary for communication between adjacent relays and individual relays with the central control system.

Settings for non-directional or directional overcurrent relays are pre-calculated during off-line fault analysis of the given distribution network, with DG, based on a suitable software.

12.8 Dynamic Modeling of Distributed Energy Resources

The study of any grid system dynamics is essential for the design of suitable controllers for the stable operation of the system under various operating conditions. This involves the modeling of individual components of the system and its simulation. The approach used for a large power system is also applicable here except that there are several types of DG and DS in a microgrid. The low or medium voltage grid is similar to a transmission grid except for the fact that the distribution grids can experience unbalance even in steady state and the distribution lines have a higher R/X ratio. In this section, we present the models of some of the distributed energy resources.

12.8.1 Photovoltaic Array with MPP Tracker [16, 17]

A PV array is made up of solar cells that are grouped in PV modules that are connected in series-parallel configuration. A photovoltaic cell is basically a semiconductor diode whose *p-n* junction is exposed to light. Silicon PV cells are composed of a thin layer of bulk Si or a thin Si film connected to electric terminals. The Si layer is doped to form the *p-n* junction. A thin metallic grid is placed on the sun-facing surface of the semiconductor. The incidence of light on the cell generates charge carriers that result in an electric current if the cell is short circuited. Charges are generated when the energy of the incident photon is sufficient to detach the covalent electrons of the semiconductor and creation of the positive and negative charges (called electron-hole pairs) in the solar cell. The energy (E) of a photon is related to the wavelength (λ) by the equation $E = \frac{hc}{\lambda} = \frac{1.24}{\lambda}$, where h is Planck's constant (6.62×10^{-34} J-s) and c is the velocity of light (3×10^8 m/s). Thus, E is in electron volts (eV) and λ is in μm.

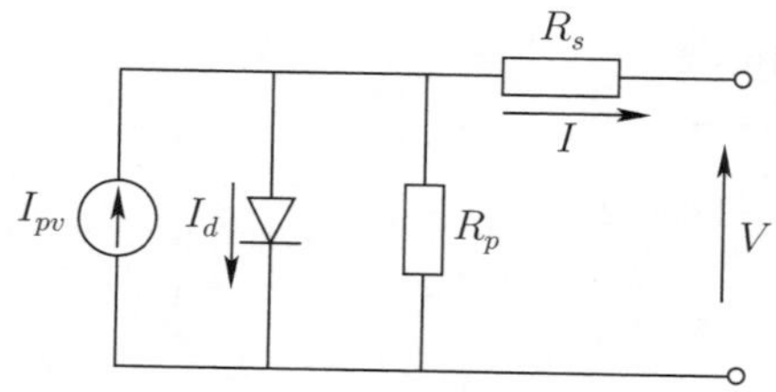

Figure 12.18 Equivalent circuit of a PV array.

The I-V characteristics of the ideal PV cell are given by

$$I = I_{pv} - I_d \tag{12.8}$$

$$I_d = I_o \left[\exp\left(\frac{qV}{akT}\right) - 1\right] \tag{12.9}$$

where I_{pV} is the current generated by the incident light (it is directly proportional to the sun radiation), I_d is the Shockley diode current, and I_o is the leakage current of the diode.

Note q is the electron charge (1.602176×10^{-19} C), k is the Boltzmann constant (1.381065×10^{-23} J/K), T (in Kelvin) is the temperature of the p-n junction, and a is the diode ideality constant.

Modeling of a PV Array

Practical arrays are composed of several series and parallel connected PV cells. The output current of a PV array is given by [16]

$$I = I_{pv}^a - I_o^a \left[\exp\left(\frac{V + R_s I}{V_t a}\right) - 1\right] - \frac{V + R_s I}{R_p} \tag{12.10}$$

where the equivalent circuit of the PV array is shown in Figure 12.18. In (12.10), $I_{pv}^a = N_p I_{pv}$, $I_o^a = I_o N_p$, and $V_t = N_s kT/q$. N_s and N_p are the number of cells connected in series and parallel, respectively. R_s is the equivalent circuit resistance of the array and R_p is the equivalent parallel resistance.

Figure 12.19 shows the characteristic I-V curve of a PV array. The maximum power point (MPP) is shown in the figure where the array gives the maximum power output. Note that $\frac{dP}{dV} = 0$ at the maximum power point and changes sign from positive to negative. It can be shown that $\frac{dI}{dV} = \frac{-I}{V}$ at the MPP. By measuring the incremental and instantaneous array conductance, the PV array voltage can be adjusted to the MPP voltage.

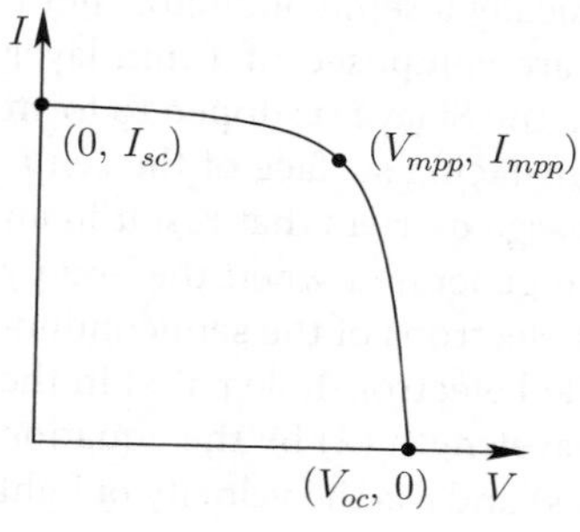

Figure 12.19 I-V curve of a PV array.

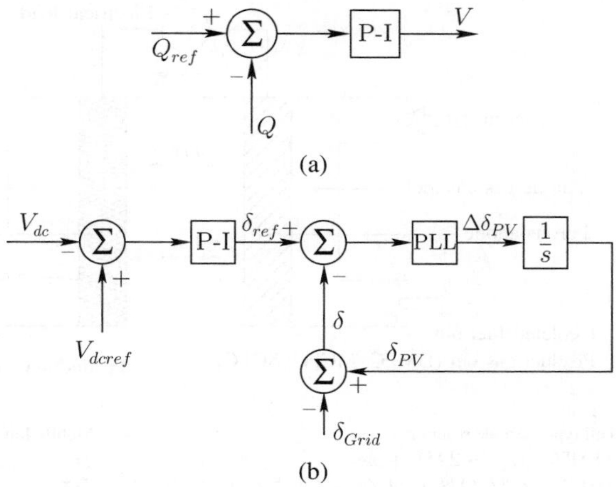

Figure 12.20 Controls for a PV array inverter. (a) Reactive power control (b) DC voltage control.

PV Array Inverter Control

The controller of the inverter of PV array consists of two major control loops: the reactive power and DC voltage control (which controls active power). These controllers are shown in Figures 12.20a and 12.20b, respectively. The reactive power control is a feedback control that regulates the output reactive power to Q_{ref}. The DC voltage controller consists of a PLL whose input is the phase reference δ_{ref} which is generated by the DC voltage error. The DC voltage controller adjusts the phase angle of the inverter AC voltage that affects the power output of the PV array.

12.8.2 Fuel Cells [18, 19]

A fuel cell is an electrochemical device that converts chemical energy directly into electrical energy. A fuel cell is thus similar to a battery in the sense that it also consists of a pair of electrodes and an electrolyte. However, unlike a battery, the species consumed during the electrochemical reactions are continuously replenished so that there is no need to recharge the cell. The basic components of a fuel cell are shown in Figure 12.21. A fuel, usually hydrogen, is supplied to the fuel cell anode. The fuel is oxidized at the anode, yielding electrons, which travel through the external circuit. The anode and cathode reactions, and the composition and direction of the flow of mobile ions vary with the type of fuel cell.

Fuel cells are generally characterized by the type of electrolyte. Although the paper by Ellis [18] considers five types, we consider here only two types that are likely to be used for power generation:

1) molten carbonate fuel cells (MCFC) use a molten mixture of lithium, sodium, and potassium carbonates as the electrolyte.
2) solid oxide fuel cells (SOFC) use a ceramic material as the electrolyte.

For both cells, the net cell reaction is

$$H_2 + \frac{1}{2}O_2 \rightarrow H_2O$$

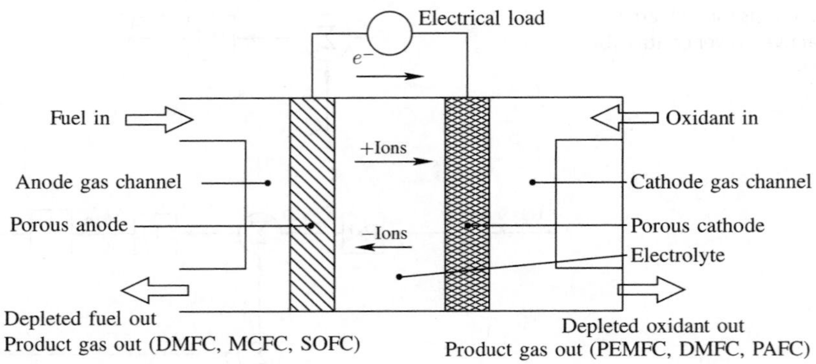

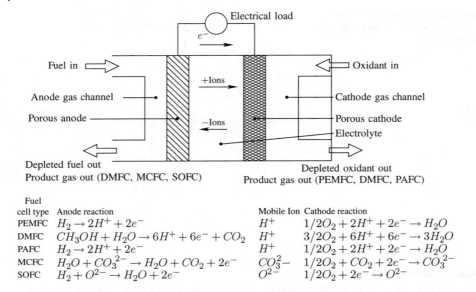

Fuel cell type	Anode reaction	Mobile Ion	Cathode reaction
PEMFC	$H_2 \rightarrow 2H^+ + 2e^-$	H^+	$1/2O_2 + 2H^+ + 2e^- \rightarrow H_2O$
DMFC	$CH_3OH + H_2O \rightarrow 6H^+ + 6e^- + CO_2$	H^+	$3/2O_2 + 6H^+ + 6e^- \rightarrow 3H_2O$
PAFC	$H_2 \rightarrow 2H^+ + 2e^-$	H^+	$1/2O_2 + 2H^+ + 2e^- \rightarrow H_2O$
MCFC	$H_2O + CO_3^{2-} \rightarrow H_2O + CO_2 + 2e^-$	CO_3^{2-}	$1/2O_2 + CO_2 + 2e^- \rightarrow CO_3^{2-}$
SOFC	$H_2 + O^{2-} \rightarrow H_2O + 2e^-$	O^{2-}	$1/2O_2 + 2e^- \rightarrow O^{2-}$

Figure 12.21 Basic components of a fuel cell. Source: [18] Fig.1, p. 1809, reproduced with permission ©IEEE 2001.

A power generation fuel cell system has three main components, namely

1) the fuel processor: converts fuels such as natural gas to hydrogen and byproduct gases
2) the power section: generates electricity
3) the power conditioner: converts DC power to AC power output and includes current, voltage, and frequency control.

Zhu and Tomsovic [19] considered the model of SOFC and mention the following:

1) The dynamic response function of the fuel processor can be modeled as a first-order transfer function with 5 s time constant.
2) The electrical response time in the fuel cells is generally fast and mainly associated with the speed at which the chemical reaction is capable of replacing the charge that has been drained by the load. The dynamic response is also modeled as a first-order transfer function with 0.8 s time constant.
3) The power conditioner enables the control of both active and reactive power. The response time of the power conditioner is less than 10 ms and can be ignored.

12.8.3 Natural Gas Generator Set [20]

Here we consider a natural gas engine driving a synchronous generator. The isochronous governor and the natural gas engine model are shown in Figure 12.22. Here the engine speed reference ω_{ref} is compared with the measured speed and the error is fed to the PID controller of an isochronous governor. The governor controls the flow of fuel to the engine and thus changes the engine torque. However, the change of torque is

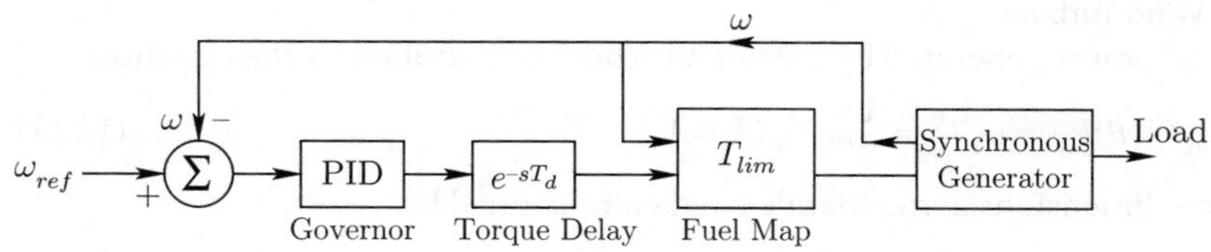

Figure 12.22 Block diagram of a natural gas engine driving a generator set.

Figure 12.23 Engine fuel map.

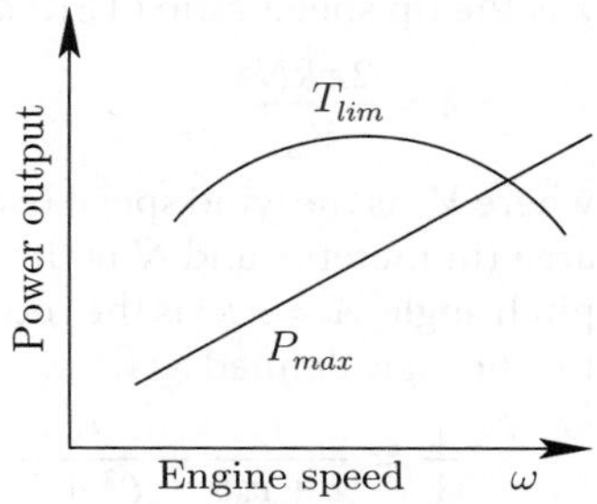

not instantaneous as it is constrained by the fuel transport delay, turbocharger time constant, and availability of air for combustion. The delay time T_d between the fuel injection and torque production for a four-stroke engine can be calculated. The engine fuel map consists of a two-dimensional look-up table which is applied to compute the limit of the mechanical torque of the engine within its maximum capability. The fuel map of a typical natural gas engine is shown in Figure 12.23.

Reciprocating engine-driven generator sets are an important class of DGs having ratings in the range of few kilowatts to hundreds of kilowatts. The utility interface is a synchronous generator or a power electronic inverter. Synchronous generators are prone to stall or introduce voltage and frequency transients under stand-alone operating conditions when subjected to large abrupt load changes. This can bring down an entire industrial grid if they are loaded beyond 80% of the engine capacity. It has been observed that gensets are prone to stalling when they are not provided with an under-frequency (V/Hz) load relief scheme.

12.8.4 Fixed-Speed Wind Turbine Driving SCIG [17–21]

Squirrel cage induction generators (SCIGs) tend to be fixed-speed type when connected to a grid operating at constant voltage in steady state. The block diagram of the system is shown in Figure 12.24.

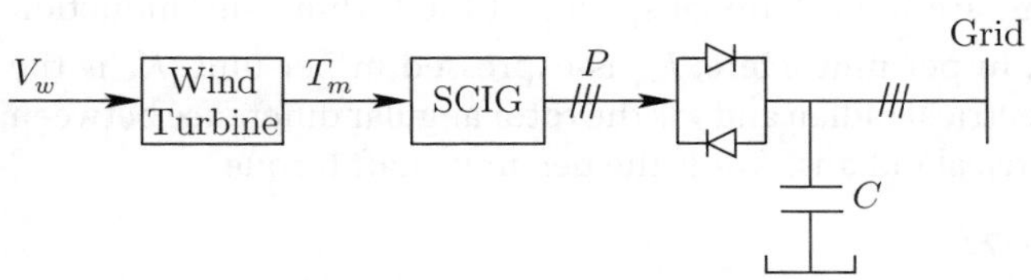

Figure 12.24 Wind power system.

Modeling of a Wind Turbine

The mechanical power generated by the wind turbine is described by the equation

$$P_m = C_p(\lambda,\theta)P_a = \omega_t T_t = \frac{1}{2}\rho A C_p(\lambda,\theta)V_\omega^3 \tag{12.11}$$

where C_p is the dimensions aerodynamic coefficient defined by

$$C_p(\lambda,\theta) = c_1\left(\frac{c_2}{\lambda_i} - c_3\theta - c_4\theta^{c_5} - c_6\right)e^{-\frac{c_7}{\lambda_i}} \tag{12.12}$$

λ is the tip speed ratio (TSR) defined by

$$\lambda = \frac{2\pi RN}{V_\omega} \tag{12.13}$$

where V_w is the wind speed (expressed in meters per second), R is the radius of the swept area (in meters), and N is the rotational speed in revolutions per second (rps). θ is the pitch angle. $A = \pi R^2$ is the rotor swept area in m^2 and ρ is the air density. T_t is the turbine torque. λ_i is defined as

$$\frac{1}{\lambda_i} = \frac{1}{\lambda + c_8\theta} - \frac{c_9}{\theta^3+1}$$

c_1 to c_9 are constants which depend on the turbine blade design. The aerodynamic power (P_a) is actually obtained from the relation given by

$$P_a = \frac{1}{2}(\text{air mass per unit time}) \times (\text{wind velocity})^2$$
$$= \frac{1}{2}(\rho A V_\omega)(V_\omega)^2 = \frac{1}{2}PAV_\omega^3$$

The air density ρ depends on temperature and pressure. At 15°C and normal pressure, $\rho = 1.225\ \text{kg/m}^3$.

It should be noted that C_p is maximum at a specific value of λ. The theoretical maximum value of C_p is approximately 0.59. Thus, TSR is to be maintained at this value for extracting maximum mechanical power.

Turbine Equations

Ignoring the mass of the gear box, the turbine rotor and the induction generator rotor are connected by a shaft which is not rigid and can be modeled as a spring. The rotor equations are

$$\frac{d\omega_t}{dt} = \frac{1}{2H_t}(T_t - K_S\gamma) \tag{12.14}$$

$$\frac{d\gamma}{dt} = 2\pi f(\omega_t - \omega_g) \tag{12.15}$$

where $T_t = \frac{P_m}{\omega_t}$, and ω_t and ω_g are the rotational speeds of the turbine and induction generator rotor, respectively, in per unit. Here, P_m is expressed in per unit. K_S is the shaft stiffness in pu torque/electrical radian and γ is the rotor angular difference between turbine and generator in electrical radians. $K_S\gamma$ is the per-unit shaft torque.

Modeling of Induction Machine [22]

Neglecting stator transients, the stator of a generator can be modeled by the equivalent circuit shown in Figure 12.25. The induced EMF $\hat{E}'$ in the stator results from the fluxes in the rotor winding. There is no loss of generality if the rotor is assumed to have two windings (α and β) rotating at the rotor speed, which can vary. It is convenient to

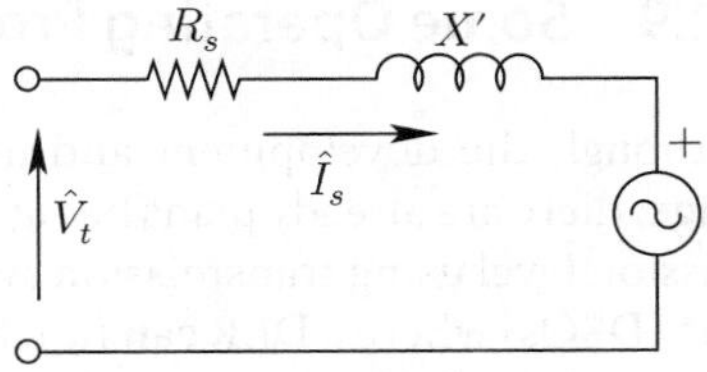

Figure 12.25 Stator equivalent circuit of an induction machine (neglecting stator transients).

transform the voltages in two fictitious axes (D and Q) rotating at the speed related to the system frequency (based on the synchronous reference frame). Thus we obtain

$$\frac{dE'_D}{dt} = -2\pi fSE'_Q + \frac{1}{T_o}[(X - X')i_Q - E'_D] \tag{12.16}$$

$$\frac{dE'_Q}{dt} = 2\pi fSE'_D - \frac{1}{T_o}[(X - X')i_D - E'_Q] \tag{12.17}$$

where S is the slip $(1 - \omega_g)$ and f is the operating system frequency.

The mechanical equations modeling the IG rotor are

$$\frac{d\omega_g}{dt} = \frac{1}{2H}(T_e - T_m) \tag{12.18}$$

where

$$T_e = E'_Q i_Q + E'_D i_D, \quad \hat{E}' = E'_Q + jE'_D$$

$$T_o = \frac{X_m + X_r}{2\pi f R_r}, \quad X = X_m + X_r$$

$$X' = X_s + \frac{X_m X_r}{X_m + X_r}, \quad i_Q + ji_D = \hat{I}_s = \frac{\hat{V}_t - \hat{E}'}{R_s + jX'}$$

The parameters R_s, R_r, X_s, X_m, and X_r are obtained from the steady-state equivalent circuit of the induction machine shown in Figure 12.26. Here, T_m is the mechanical torque at the generator, opposite to the direction of rotation.

It should be noted that both the turbine and IG rotor speeds are expressed in per unit.

Remarks

1) It can be shown that in steady state, $T_e = \frac{I_2^2 R_r}{S}$.
2) Inclusion of stator transients results in a fifth-order model of the induction machine. In Appendix 12.A, a three-phase model of the squirrel cage induction machine is presented in phase variables (similar to the model given in Chapter 3 for synchronous machines). From a three-phase model it is possible to obtain models in D-Q variables and simpler models ignoring stator and rotor transients.

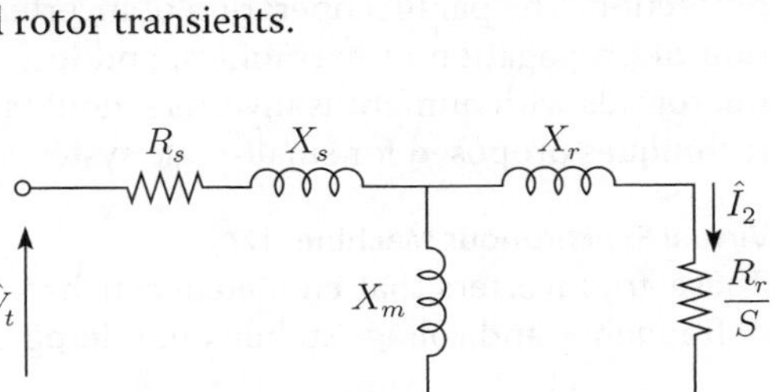

Figure 12.26 Steady-state equivalent circuit of an induction machine.

12.9 Some Operating Problems in Microgirds

Although the development and installation of microgrids is still in the evolutionary stage, there are already plans being made to extend the economic operations at the transmission level using transmission system operators (TSOs) to distribution system operators (DSOs) where a DER can function like a generating station in terms of its operation. In principle, DERs can contribute to the provision of all electrical services provided by TSOs. Perez-Arriaga [23] refers to a report published in 2013 which identified four major regulatory challenges due to DERs: computation of allowed remuneration of distribution networks in the presence of large numbers of DERs, computation of network charges for a diversified ensemble of network users with DERs, an adequate definition of the new role of DSOs with the provision of electricity services, with multiple new actors on suppliers and consumers, and finally managing the new level of complexity in the coordination of DSOs and the TSO.

In this section, the technical issues that were observed during the testing of microgrids will be discussed briefly.

Frequency and Voltage Stability

Frequency stability is a problem faced by islanded microgrids where any individual generator output forms a substantial portion of the total demand. The frequency stability becomes an issue with DERs that are inverter based or have low inertias and damping torques. Under the islanded operation, the use of multiple small DGs can pose problems of synchronization, which is essential for stable operation of the microgrid. There could also be problems of voltage stability.

Conceptually, these problems can be solved by developing intelligent control schemes, but there are trade-offs. They require a relatively fast communications network which is available to link the various microgrid components. In general, the operation of a power system based on many small components rather than one large generator is a more complex task.

Intermittent Generation

The intermittent generation by renewables (wind and solar, caused by wind gusts or clouds passing overhead) implies that not only the power flow is bidirectional, but the power output from these sources varies randomly. The solution is to use DS such as batteries. It would also be desirable to tightly couple the supply to the power demand. Load and generation forecasting can help in the operational management of the microgrid.

Application of Inverters

In microgrids, inverters are used to convert the DC output of the DG to AC, control the frequency and voltage, monitor power flows in the network, and provide basic fault protection. The parallel operation of inverters without loss of synchronization, prevention of propagation of harmonics, and loss of stability is a challenging task. For large microgrids with numerous inverters, doubts remain regarding the scalability of control techniques proposed for small-scale systems.

Virtual Synchronous Machines [24]

These are inverters that emulate a synchronous machine that overcome the problems of frequency and voltage stability in microgrids. A virtual synchronous machine (VSM) can enhance frequency stability by adding inertia and damping to the system. The control of a VSM may require an estimate of the stabilization frequency of the grid.

Stalling of Generator Sets [20]

Reciprocating engine-driven generator sets are a leading class of DERs, from low kilowatts up to hundreds of kilowatts. When these operate under islanded conditions, they are susceptible to stalling and can bring down the entire industrial power system, especially if they are loaded to 80% of the rated engine capacity. The stalling can be prevented by applying an under-frequency load relief (V/Hz) scheme. The stalling problem can be avoided by intelligently controlling the adjustable or sheddable loads. Such load are termed "smart loads".

This approach to stabilize the system is specifically suitable in industrial power systems which are expected to supply fast varying loads like crushers, rolling mills, electrical excavators or even pulsed loads such as spot welders. The profiles of such industrial loads are characterized by sharp step load changes at regular intervals. The use of (V/Hz) load relief mechanisms in the digital voltage regulator can cause a voltage drop at the generator terminals when a large change in the load occurs. Although generator stalling can be prevented by this method, the voltage fluctuations can affect power quality. It may also trip the undervoltage relays. The drop in the voltage can also lead to unwanted reactive power changes with other DERs in the microgrid. Hence, the application of smart loads appears to be an effective means to overcome the problem of stalling in gensets.

Collapse of a Mixed-Source Microgrid [25]

When a mixed-source microgrid operates in the islanded mode of operation near its capacity limits, the survivability of the microgrid is at risk when it experiences a sudden loss of generation from even a single DER. In the example given in [25], there are three DERs: DER1 is a reciprocating engine-driven synchronous generator, DER2 is similar to DER1 except that the engine drives a PMSG whose output is connected to a power conditioner consisting of a rectifier and an inverter, and DER3 is a solar PV system which operates as a grid-following current source whereas DER1 and DER2 act as voltage sources with frequency droop controls. Each DER is rated to deliver 100 kW continuous load. The engine driving DER1 is provided with an isochronous governor whereas DER2 is controlled with P-f droop of 1%.

DER3 is treated as a negative load as it is controlled as a grid-following current source. Thus, the power flow equations are

$$P_1 + P_2 = P_L - P_3, \quad Q_1 + Q_2 = 0$$

Test Case 1

When the net load ($P_L - P_3$) was increased from 75 kW to 150 kW, it was observed that DER2 experienced a voltage collapse initially as it absorbed a large amount of reactive power from DER1 as its capacity is only 100 kW. Therefore, DER1 also collapsed. The initial net load of 75 kW was shared equally between DER1 and DER2.

Test Case 2

Here the initial loading of DER2, $P_2 = 0$. The microgrid survived in this case without collapsing. The steady-state frequency was below the rated frequency at which the microgrid was operating before the disturbance (doubling of the net load from 75 kW).

The problem of system collapse in the first test case is due to the low inertia of the inverter-based DER2, which results in faster loading of DER2 compared to DER1. This shows that the microgrid is susceptible to collapse if the fast-acting DERs do not have adequate reserves when subjected to a sudden increase in the load.

12.10 Integration of DG and DS in a Microgrid [26]

The concept of "plug-and-play" implies decentralized control of DERs in a microgrid. Each DER is controlled by providing frequency/active power (f-P) and voltage/reactive power (V-Q) droop controls to regulate the voltage magnitude and frequency (see Figure 12.10).

Small-sized DERs are characterized by low inertia and cannot absorb any sudden increase in the load demand in a stable manner. For the reliable operation of an islanded microgrid, DGs are supplemented with DS that is suitably sized. A microturbine may take up to 35 s to respond to a 50% step change in power demand. Fuel cells may require 10 s to respond to a 15% change in the power output. The shortfall in the energy provided by a DG can be met by connecting a DS unit at the DC link of the DG connected through an inverter interface (see Figure 12.27). It is shown in [26] that by organizing distributed energy and storage resources (FDERS), reliable supply of power to fast varying loads from a network of multiple small rated resources is feasible. The benefits of this scheme are increased resource life time, optimal energy storage deployment, enhanced controllability, and improved system robustness. The flexibility is achieved by synthesizing virtual reactances that are connected in series with the synthesized voltage sources, virtual inertias, and adaptive frequency-active power (f-P) controls. The hierarchy of distributed energy and storage resources is based on the availability of DERs, their response characteristics, and lifetime costs.

12.11 DC Microgrids [10, 11]

DC microgrids are attractive in residential and commercial buildings where they could service many loads that use DC, such as LED lighting, charging of batteries for electric vehicles, heating, ventilating and air-conditioning (HVAC) equipment, and household appliances. It should be noted that most energy-efficient HVAC equipment and appliances incorporate variable-speed motor drives which operate on DC. Since PV panels mounted on the roofs of residential and commercial buildings generate DC power, it is efficient to avoid conversion of DC power to AC (using inverters) and then to DC where it is utilized. The inverter used to transform power from PV panels to AC typically results in a 10% loss of power generated from solar energy [11]. In off-grid homes (not connected to distribution grids), the DC storage (batteries) is used to proved lighting when there is no solar power. It is estimated that for low-power systems (typically 250 W) deployed in homes, each conversion of power (DC to AC and AC to DC) results in 15% power loss, amounting to a total of (up to) 45% when the local generation is synchronized with the local grid [11].

DC microgrids may be attractive for certain energy-intensive manufacturing plants. These include paper and pulp production and the smelting of aluminium, which now wastes more than 6% of the total energy consumed in the conversion of AC to DC.

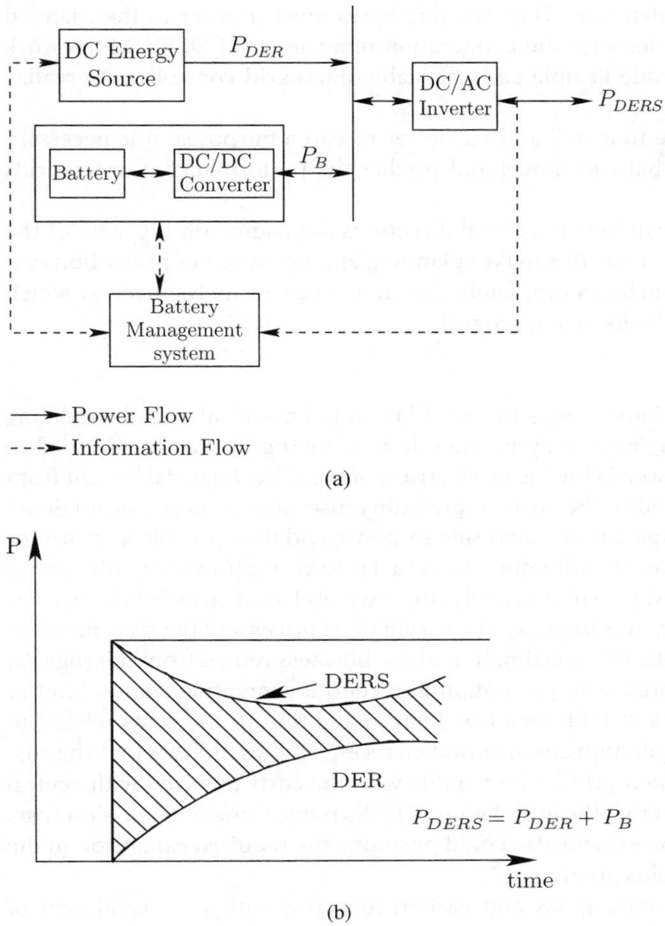

Figure 12.27 Flexible distribution of energy and storage resources (FDERS). Source: [26] Fig. 8, 9, p. 37, 38, reproduced with permission ©IEEE 2015. (a) Block diagram (b) Active power vs time.

12.12 Future Trends and Conclusions [10]

Microgrid is a rapidly developing technology and it is difficult to predict future developments. The design of microgrids can be quite complex as it involves a large number of disparate devices. Issues of installation cost, environmental impact, line loss, grid connectivity, reliability, resource longevity, reuse of waste heat, capacity for intentional islanding, and physical constraints all affect the decision-making process. Sophisticated design tools may be required which take into account the uncertainties.

A central benefit of microgrids, particularly in disaster relief, is the capacity to ride through failures which occur in the utility grid with limited loss of localized service. By rapidly disconnecting from the faulted system (intentional islanding), adjusting local generation and shedding non-priority loads, high levels of reliability can be guaranteed for priority loads. However, this is a complex task. First, the fault condition in the distribution grid must be detected. Then the microgrid must transfer to the islanded mode in a controlled manner with the cooperation of a variety of DERs. Much work is required to develop reliable flexible and affordable microgrid controllers to realize the goals.

Microgrids are expensive to install and set up for research purposes. It is necessary to rely on software tools that can model and predict the performance of microgrids accurately.

In a microgrid, current can flow in several directions depending on the state of the microgrid at the particular time. This makes planning and operation of protection systems a challenging task. Furthermore, faults may not result in high currents, which makes it difficult to detect faults in a microgrid.

Business Models

The commercial benefits of installing a microgrid need to be evaluated in formulating a business model. In a simplistic analysis, the role of a microgrid may be assumed to provide uninterruptible power following an electrical outage. The financial benefit from this service can be compared to the costs of providing insurance against a grid failure. Microgrids can also participate in demand side response and thus provide a return for the services rendered. However, estimating the actual returns for providing this service is quite difficult as demand side markets can fluctuate widely based on jurisdiction, location, and the incentive programs in place. The varying fuel prices and the risks inherent in a new technology complicate the estimation of the business return from microgrids.

The costs of DER compared with generation in a central generating station tend to be higher. The energy costs of DER could be high in addition to the costs related to maintenance and system operation costs. However, recent studies indicate that the cost of generating power in a microgrid is comparable with the costs involved with central generating stations. The fact that the installation of DERs reduces the transmission costs (involving the cost of the losses) and also could postpone the required expansion in the transmission network is a plus point.

The reduction in greenhouse gases and carbon footprint with the installation of renewable energy and reduction of transmission losses (as DERs operate close to the loads they supply) is a major benefit. However, the computation of the benefits and their comparison with installation and operation and maintenance costs can be a challenging exercise.

A key benefit of the microgrid is to improve the resiliency of the grid when subjected to major outages due to storms and cyclones. However, the estimation of the risks and their consequences is a difficult task. It is also possible that stakeholders may underestimate the benefits of microgrids.

Regulation and Standards

As the microgrid concept is relatively new, it is not surprising that the regulatory framework for integration of microgrids into the existing distribution grids is still in a nascent stage. Developing an appropriate framework is essential to avoid operational problems. For example, the IEEE Standard 1547 (2003) requires that a grid-connected converter will shut down after the detection of a grid fault. Consequently, commercial inverters had no incentive to offer transition to an islanded state and operate as UPS for essential loads. This standard is promoting research in a static switch which can disconnect and reconnect in subcycle times.

Research into standards is being conducted for the operation of microgrids and a new generation of EMS is being designed to meet the requirements of International Standard IEC61970.

Clustered Microgrids

With the proliferation of microgrids, a number of microgrids can be coordinated to function together, using a hierarchical structure. However, care has to be taken to avoid too many levels of hierarchy that could result in slowing down the response and add overheads.

There is also the possibility of clustered microgrids adding to management complexity in spite of the fact that they bring improved flexibility.

Resources Networks

In many applications, electricity is one of the key resources (in addition to water, gas, and other resources having distribution networks). There could be interactions between them and the microgrid. Recent research suggests that a microgrid should encompass all the key resources and functions. This may include the provision of reliable electricity and natural gas in addition to heating and cooling. Although research on microgrids of multiple resources is relatively in its infancy, it is an established fact that an electrical microgrid is dependent on other resources such as gas supply. Factoring in such dependencies is crucial to reliable system operation.

Study of Reliability Options and Restoration Strategies

A paper published in 2014 [27] proposed a possible approach to providing differentiated reliability options to customers in urban and suburban distribution networks. To offer different reliability options, a utility can operate DG, sectionalizing switches (normally closed), and tie switches (normally open). Reliability is one of the parameters that can determine pricing.

System Restoration

Optimal distribution system restoration strategies are essential to improve grid resiliency. In a recent paper [28] a bi-level optimization strategy was suggested where maximizing the total apparent power at a set of isolated buses is the primary objective while minimizing the number of switching pair operations is the secondary objective. Since maximum load recovery is the major objective of any restoration plan, the primary objective can be viewed as a constraint of the secondary objective.

Grid of the Future [29]

In [29], the authors propose a framework of the utility and grid of the future. This involves four layers and their components:

1) Foundation layers
   a) Foundational infrastructure and resources
   b) Organization and processes
   c) Standards and modules
   d) Business and regulatory
2) Enabling layers
   a) Infrastucture
   b) Incremental intelligence

3) Application layers
   a) Grid and customer analysis
   b) Real time awareness and control
   c) Customer interaction
4) Innovation layers
   a) Pilot and demonstration projects
   b) Research and development
   c) Partnerships
   d) Training

12.A A Three-Phase Model of an Induction Machine [30]

The model presented here is applicable to a squirrel cage induction machine or even a wound rotor induction machine with shorted rotor windings. The assumptions made are (i) uniform air gap, (ii) no magnetic saturation, (iii) negligible hysteresis and eddy current losses, and (iv) sinusoidal distribution of the air gap flux. Furthermore, it is assumed that the rotor has two short-circuited windings α and β in quadrature rotating at a speed of ω_r (see Figure 12.A.1).

The stator and rotor equations can be expressed as:

$$\frac{d\psi_s}{dt} + [R_s]i_s = V_s \tag{12.A.1}$$

$$\frac{d\psi_r}{dt} + [R_r]i_r = V_r = 0 \tag{12.A.2}$$

where

$$V_s^t = \begin{bmatrix} V_a & V_b & V_c \end{bmatrix}, \quad i_s^t = \begin{bmatrix} i_a & i_b & i_c \end{bmatrix}, \quad i_r^t = \begin{bmatrix} i_{r\alpha} & i_{r\beta} \end{bmatrix},$$

$$[R_s] = R_s[U_3], \quad [R_r] = R_r[U_2]$$

$[U_3]$ and $[U_2]$ are unit matrices of order 3 and 2, respectively. The stator and rotor flux linkages are given by

$$\psi_s = [L_s]i_s + [M]i_r \tag{12.A.3}$$

$$\psi_r = [M]^t i_s + [L_r]i_r \tag{12.A.4}$$

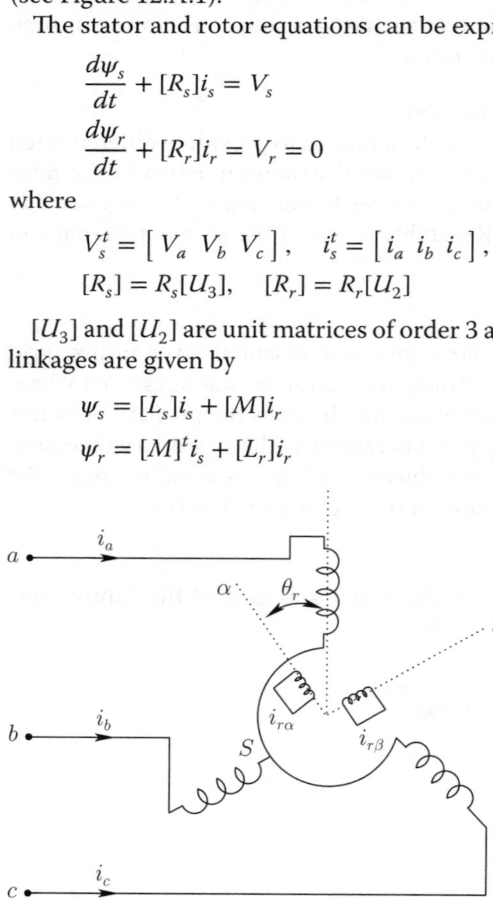

Figure 12.A.1 Three-phase induction machine model.

where

$$[L_s] = \begin{bmatrix} L_{aa} & L_{ab} & L_{ab} \\ L_{ab} & L_{aa} & L_{ab} \\ L_{ab} & L_{ab} & L_{aa} \end{bmatrix}, [L_r] = L_r[U_2]$$

$$[M] = L_{12} \begin{bmatrix} \cos\theta_r & \sin\theta_r \\ \cos\left(\theta_r - \frac{2\pi}{3}\right) & \sin\left(\theta_r - \frac{2\pi}{3}\right) \\ \cos\left(\theta_r + \frac{2\pi}{3}\right) & \sin\left(\theta_r + \frac{2\pi}{3}\right) \end{bmatrix}$$

From (12.A.4), we have

$$i_r = L_r^{-1}(\psi_r - M^t i_s) \tag{12.A.5}$$

Substituting the above equation in (12.A.3), we get

$$\psi_s = [L_s - ML_r^{-1}M^t]i_s + [M]L_r^{-1}\psi_r \tag{12.A.6}$$

Defining $[L_s']$ and I_s as

$$[L_s'] = [L_s] - [M]L_r^{-1}[M^t], \quad I_s = [L_s']^{-1}[M]L_r^{-1}\psi_r$$

We can express (12.A.6) as

$$\psi_s = [L_s'](i_s + I_s) \tag{12.A.7}$$

$[L_s']$ is a 3×3 cyclic symmetric matrix with diagonal elements L'_{aa} and off-diagonal elements L'_{ab} defined by

$$L'_{aa} = L_{aa} - \frac{L_{ab}^2}{L_r}, L'_{ab} = L_{aa} + \frac{L_{ab}^2}{2L_r}$$

Equation (12.A.7) defines an equivalent circuit for the three-phase stator windings of the induction machine (see Figure 12.A.2). This shows a symmetrical three-phase current source in parallel with three symmetrical coupled inductances, which are in series

Figure 12.A.2 Equivalent circuit for the three-phase stator windings of an induction machine.

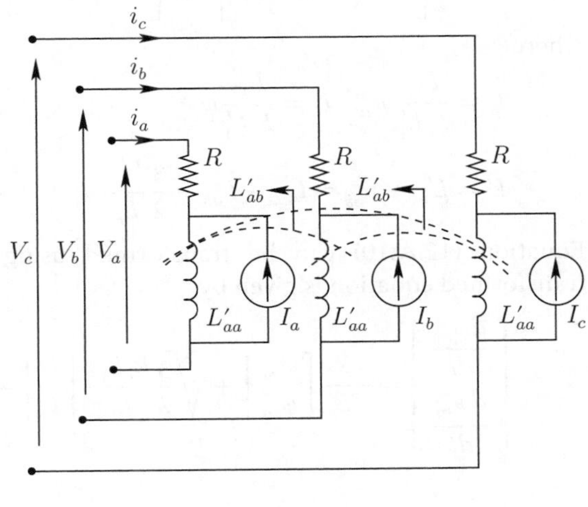

with the stator resistances. The current source I_s can be expressed as

$$I_s = I_\alpha \begin{bmatrix} \cos\theta_r \\ \cos\left(\theta_r - \frac{2\pi}{3}\right) \\ \cos\left(\theta_r + \frac{2\pi}{3}\right) \end{bmatrix} + I_\beta \begin{bmatrix} \sin\theta_r \\ \sin\left(\theta_r - \frac{2\pi}{3}\right) \\ \sin\left(\theta_r + \frac{2\pi}{3}\right) \end{bmatrix} \tag{12.A.8}$$

where

$$I_\alpha = \frac{L_{12}}{L_r(L'_{aa} - L'_{ab})}\psi_{r\alpha}, I_\beta = \frac{L_{12}}{L_r(L'_{aa} - L'_{ab})}\psi_{r\beta}$$

The stator voltage equation (12.A.1) can be rewritten as

$$V_s = [R_s]i_s + [L'_s]p(i_s + I_s) \tag{12.A.9}$$

where $p = \frac{d}{dt}$.

Substituting (12.A.5) in (12.A.2), we obtain

$$\frac{d\psi_r}{dt} = -R_r L_r^{-1}(\psi_r - M^t i_s) \tag{12.A.10}$$

Since $[M]$ is a time-varying matrix, it is advantageous to transform the rotor equations using d-q variables (which implies a synchronously rotating reference frame) so that the transformed variables become constants in steady-state operation with sinusoidal excitation of the stator windings. The transformation of the rotor flux linkages is defined by

$$\begin{bmatrix} \psi_{r\alpha} \\ \psi_{r\beta} \end{bmatrix} = \begin{bmatrix} \cos(\theta_o - \theta_r) & \sin(\theta_o - \theta_r) \\ -\sin(\theta_o - \theta_r) & \cos(\theta_o - \theta_r) \end{bmatrix} \begin{bmatrix} \psi_{rd} \\ \psi_{rq} \end{bmatrix} \tag{12.A.11}$$

In terms of d-q variables, I_s is defined by

$$I_s = I_1 \begin{bmatrix} \cos\theta_o \\ \cos\left(\theta_o - \frac{2\pi}{3}\right) \\ \cos\left(\theta_o + \frac{2\pi}{3}\right) \end{bmatrix} + I_2 \begin{bmatrix} \sin\theta_o \\ \sin\left(\theta_o - \frac{2\pi}{3}\right) \\ \sin\left(\theta_o + \frac{2\pi}{3}\right) \end{bmatrix} \tag{12.A.12}$$

where

$$I_1 = \frac{L_{12}}{L_r L'_1}\psi_{2d}, \ I_2 = \frac{L_{12}}{L_r L'_1}\psi_{2q}$$

$$L' = L'_{aa} - L'_{ab} = L_{aa} - L_{ab} - \frac{3}{2}\frac{L_{12}^2}{L_r}$$

Equation (12.A.10) can be transformed using d-q variables and (12.A.11). The transformed equation is given by

$$\begin{bmatrix} \frac{d\psi_{rd}}{dt} \\ \frac{d\psi_{rq}}{dt} \end{bmatrix} = -\frac{R_r}{L_r}\begin{bmatrix} \psi_{rd} \\ \psi_{rq} \end{bmatrix} + \sqrt{\frac{3}{2}}\frac{R_r L_{12}}{L_r}\begin{bmatrix} i_d \\ i_q \end{bmatrix} - S\omega_o \begin{bmatrix} 0 & 1 \\ -1 & 0 \end{bmatrix}\begin{bmatrix} \psi_{rd} \\ \psi_{rq} \end{bmatrix} \tag{12.A.13}$$

where

$$i_d = \sqrt{\frac{2}{3}}\left(i_a \cos\theta_o + i_b \cos\left(\theta_o - \frac{2\pi}{3}\right) + i_c \cos\left(\theta_o + \frac{2\pi}{3}\right)\right)$$

$$i_q = \sqrt{\frac{2}{3}}\left(i_a \sin\theta_o + i_b \sin\left(\theta_o - \frac{2\pi}{3}\right) + i_c \sin\left(\theta_o + \frac{2\pi}{3}\right)\right)$$

and $S = \frac{\omega_o - \omega_r}{\omega_o}$ is the slip of the induction machine.

Torque and Speed

The speed of an induction machine is expressed in terms of the slip and is governed by the equation

$$\frac{dS}{dt} = \frac{T_m - T_e}{J\omega_o} \tag{12.A.14}$$

where T_m is the mechanical torque and T_e is the electrical torque in the direction of rotation. J is the moment of inertia. The electrical torque T_e is given by

$$T_e = -\frac{1}{2} i^t \left[\frac{dL}{d\theta_r}\right] i \tag{12.A.15}$$

where $i^t = [i_s^t \quad i_r^t]$ and $[L]$ is the combined inductance matrix, including stator and rotor coils. It can be shown that

$$\frac{dL}{d\theta_r} = L_{12}\begin{bmatrix} 0 & N \\ N^t & 0 \end{bmatrix}$$

$$[N] = \begin{bmatrix} -\sin\theta_r & \cos\theta_r \\ -\sin\left(\theta_r - \frac{2\pi}{3}\right) & \cos\left(\theta_r - \frac{2\pi}{3}\right) \\ -\sin\left(\theta_r + \frac{2\pi}{3}\right) & \cos\left(\theta_r + \frac{2\pi}{3}\right) \end{bmatrix}$$

It can be shown that

$$T_e = \sqrt{\frac{3}{2}}\frac{L_{12}}{L_r}(\psi_{rq} i_d - \psi_{rd} i_q) \tag{12.A.16}$$

Remarks

1) It should be noted that the circuit model shown in Figure 12.A.2 is most general and defines a fifth-order model including (12.A.14). The model can be directly used for digital simulation of machine transients when it is connected to an external circuit (instead of a voltage source). The equations for the external circuit have to be included for the system simulation.
2) In the absence of the zero-sequence current, the circuit model is simplified when $[L_s']$ can be replaced by a diagonal matrix $\left(L_{aa}' \quad [U_3]\right)$. This implies that there are no mutual inductances in the circuit shown in Figure 12.A.2. It should be noted that when the stator windings are wye connected with isolated neutral, the zero-sequence current flowing in the stator is zero.

3) The model can be simplified by neglecting stator transients, in which case there are three state variables (i_{rd}, i_{rq}, and S).
4) The model has been used for digital simulation of the transients and also for simplified analysis of a voltage-controlled induction motor [30, 31].
5) It is also possible to device a circuit model where the equivalent circuit is obtained for the rotor and the stator is connected to a voltage source. This model is expected to be useful in the analysis of doubly-fed induction generator.

References

1 Lasseter, R.H. and Paigi, P. (2004) Microgrid: A conceptual solution, in *IEEE Power Electronics Specialists Conference*, Aachen.

2 Hatziargyriou, N., Asano, H., Iravani, R., and Murray, C. (2007) Microgrids. *IEEE Power and Energy Magazine*, **5** (4), 78–94.

3 Kroposki, B., Lasseter, R., Ise, T. *et al.* (2008) Making microgrids work. *IEEE Power and Energy Magazine*, **6** (3), 40–53.

4 Lopes, J.A.P., Moreira, C.L., and Madureira, A.G. (2006) Defining control strategies for microgrids islanded operation. *IEEE Transactions on Power Systems*, **21** (2), 916–924.

5 Katiraei, F., Iravani, R., Hatziargyriou, N., and Dimeas, A. (2008) Microgrid management control and operation aspects of microgrids. *IEEE Power & Energy Magazine*, **6** (3), 54–65.

6 US Department of Energy (2010) Value and technology assessment to enhance the business case for the CERTS microgrid. Final Project Report prepared by Lasseter R. and Eto J.

7 Lasseter, R.H. (2011) Smart distribution: Coupled microgrids. *Proceedings of the IEEE*, **99** (6), 1074–1082.

8 US Department of Energy (2014) The advanced microgrid: Integration and interoperability. Report prepared by Sandia National Laboratories.

9 Adamiak, M., Bose, S., Liu, Y. *et al.* (2014) Tieline controls in microgrid applications, in *Proceedings of International Workshop on Trustworthiness of Smart Grids(ToSG)*, Atlanta.

10 International Electrotechnical Commission (2014) White paper on microgrids for disaster preparedness and recovery. Geneva.

11 Kaur P, Jain S, and Jhunjhunwala A (2015) Solar-DC deployment experience in off-grid and near off-grid homes: Economics, technology and policy analysis, in *IEEE International Conference on DC Microgrids*, Atlanta.

12 Gonen, T. (1986) *Electric Power Distribution System Engineering*, McGraw-Hill, New York.

13 Chen, A.C.M. (1982) Automated power distribution. *IEEE Spectrum*, **19** (4), 55–60.

14 Dimeas, A.L. and Hatziargyriou, N.D. (2005) Operation of a multiagent system for microgrid control. *IEEE Transactions on Power Systems*, **20** (3), 1447–1455.

15 Hatziargyriou, N., Kleftakis, V., Papaspiliotopoulos, V., and Korres, G. (2016) Adaptive protection for microgrids, in *IEEE PES General Meeting*, Boston.

16 Villalva, M.G., Gazoli, J.R., and Filho, E.R. (2009) Comprehensive approach to modeling and simulation of photovoltaic arrays. *IEEE Transactions on Power Electronics*, **24** (5), 1198–1208.
17 Kanellos, F.D., Tsouchnikas, A.I., and Hatziargyriou, N.D. (2005) Micro-grid simulation during grid-connected and islanded modes of operation, in *International Conference on Power Systems Transients*, Montreal.
18 Ellis, M.W., Von Spakovsky, M.R., and Nelson, D.J. (2001) Fuel cell systems: efficient, flexible energy conversion for the 21st century. *Proceedings of the IEEE*, **89** (12), 1808–1817.
19 Zhu, Y. and Tomsovic, K. (2002) Development of models for analyzing the load-following performance of microturbines and fuel cells. *Electric Power Systems Research*, **62** (1), 1–11.
20 Renjit, A.A., Illindala, M.S., and Klapp, D.A. (2014) Graphical and analytical methods for stalling analysis of engine generator sets. *IEEE Transactions on Industry Applications*, **50** (5), 2967–2975.
21 Slootweg, J.G., Polinder, H., and Kling, W.L. (2003) Representing wind turbine electrical generating systems in fundamental frequency simulations. *IEEE Transactions on Energy Conversion*, **18** (4), 516–524.
22 Padiyar, K.R. (2002) *Power System Dynamics: Stability and Control*, BS Publications, Hyderabad, 2nd edn.
23 Perez-Arriaga, I.J. (2016) The transmission of the future: The impact of distributed energy resources on the network. *IEEE Power & Energy Magazine*, **14** (4), 41–53.
24 Torres, M. and Lopes, L.A.C. (2013) A virtual synchronous machine to support dynamic frequency control in a mini-grid that operates in frequency droop mode. *Energy and Power Engineering*, **5** (3), 259–265.
25 Pulcherio, M., Renjit, A.A., Illindala, M.S. *et al.* (2016) Evaluation of control methods to prevent collapse of a mixed source microgrid, in *IEEE/IAS Industrial and Commercial Power Systems Technical Conference*, Detroit.
26 Illindala, M.S., Khasawneh, H.J., and Renjit, A.A. (2015) Flexible distribution of energy and storage resources: Integrating these resources into a microgrid. *IEEE Industry Applications Magazine*, **21** (5), 32–42.
27 Junlakarn, S. and Ilic, M. (2014) Distribution system reliability options and utility liability. *IEEE Transactions on Smart Grid*, **5** (5), 2227–2234.
28 Yuan, C., Illindala, M.S., and Khalsa, A.S. (2017) Modified Viterbi algorithm-based distribution system restoration strategy for grid resiliency. *IEEE Transactions on Power Delivery*, **32** (1), 310–319.
29 Aguero, J.R., Khodaei, A., and Masiello, R. (2016) The utility and grid of the future: Challenges, needs, and trends. *IEEE Power and Energy Magazine*, **14** (5), 29–37.
30 Padiyar, K.R. and Prabhakaran, N. (1982) Digital simulation of three-phase induction motors using hybrid models. *Journal of the Institution of Engineers(I)*, **63** (pt EL,1), 23–31.
31 Padiyar, K.R. and Prabhakaran, N. (1988) Simplified analysis of steady-state performance of a voltage controlled induction motor drive. *Electric Machines and Power Systems*, **15** (3), 149–162.

A

Equal Area Criterion

This is a direct method to check the stability of a synchronous generator (connected to an infinite bus) that is subjected to disturbance(s). The method does not require the solution of the nonlinear equation. The assumptions in applying the equal area criterion (EAC) are:

1) classical machine model described by the swing equation
2) rotor damping is neglected.

The application of the EAC assumes that if the system is stable in the first swing, it continues to remain stable although the rotor may oscillate about its final steady state.

Consider the swing equation given by

$$M\frac{d^2\delta}{dt^2} = T_m - T_e \approx P_m - P_e = P_a \tag{A.1}$$

where $M = \frac{2H}{\omega_B}$ is the inertia, δ is the rotor angle with respect to the synchronously rotating reference, and P_m and P_e are the mechanical and electrical (output) power, respectively. It is assumed that the operating speed is assumed to be at rated value (1 pu).

Multiplying both sides of (A.1) by $\frac{d\delta}{dt}$ and integrating with respect to time (t), we get

$$M\int_{t_o}^{t} \frac{d\delta}{dt} \cdot \frac{d^2\delta}{dt^2}\, dt = \int_{t_o}^{t} P_a \frac{d\delta}{dt}\, dt \tag{A.2}$$

This is simplified to

$$\frac{1}{2}M\left(\frac{d\delta}{dt}\right)^2 = \int_{t_o}^{t} P_a\, d\delta \tag{A.3}$$

It is assumed that at $t = t_o$ the system is at rest (in equilibrium state) and $\frac{d\delta}{dt} = 0$. The right-hand side of the above equation can be interpreted as the area between the curves, P_m versus δ and P_e versus δ. If the system is to be stable, then

$$\left.\frac{d\delta}{dt}\right|_{t=t_p} = 0, t_p > t_o \tag{A.4}$$

It can be observed that $\delta = \delta_{max}$ at $t = t_p$ and δ starts reducing for $t > t_p$. Assuming that there is some damping (both electrical and mechanical), then the oscillations will finally die out and the rotor angle will reach an equilibrium which may be identical to δ_o or a new equilibrium which depends on the post-disturbance system.

Dynamics and Control of Electric Transmission and Microgrids, First Edition.
K. R. Padiyar and Anil M. Kulkarni.

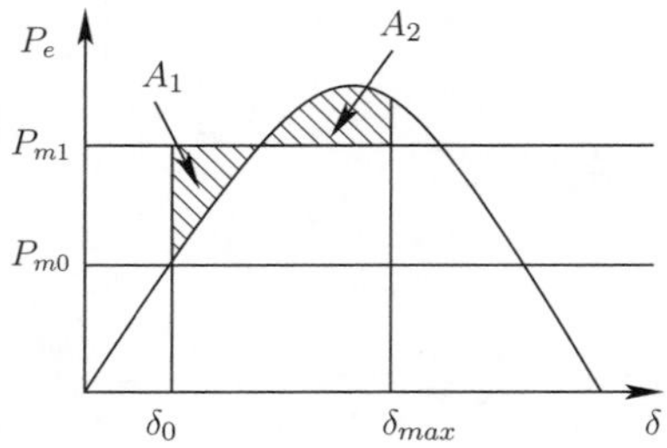

Figure A.1 Illustration of EAC.

From (A.4), we note that the area defined by

$$A = \int_{\delta_0}^{\delta_{\max}} (P_m - P_e)\, d\delta \tag{A.5}$$

must have a positive portion A_1 for which $P_m > P_e$, and a negative A_2 for which $P_m < P_e$. For the system to be transiently stable the two areas A_1 and A_2 must be identical in magnitude and $A = A_1 - A_2 = 0$. Hence, the nomenclature of EAC. This is illustrated in Figure A.1 assuming P_m is raised from P_{mo} to P_{m1} (at $t = t_o$ when $\delta = \delta_o$).

Remarks

1) The EAC is also applicable for a two-machine system (without an infinite bus) as it can be converted into a single-machine equivalent (SIME).
2) Mathematically, the problem of determination of transient stability can be viewed as checking whether the initial system state (δ_o) lies in the region of stability surrounding the post-disturbance stable equilibrium. It should be noted that every stable equilibrium point (SEP) has a region of stability in which a trajectory approaches SEP as $t \to \infty$. A trajectory starting outside the region of stability will not approach SEP and may even be unbounded.
3) EAC is a special case of the direct method for stability evaluation using energy functions [1].

An Interesting Network Analogy [2]

Consider a circuit consisting of a linear capacitor connected to a nonlinear inductor and excited by a DC current source I (see Figure A.2).

The initial current in the inductor is I and the initial voltage across the capacitor is v_0. The nonlinear inductor is defined by

$$i = I_m \sin \lambda \tag{A.6}$$

where λ is the flux linkage of the inductor. The equations for this circuit are given by

$$C\frac{dv}{dt} = I - i, \quad \frac{d\lambda}{dt} = v \tag{A.7}$$

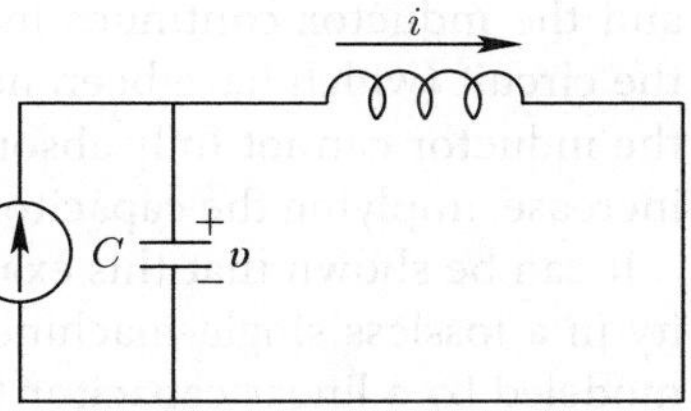

Figure A.2 A linear capacitor connected to a nonlinear inductor.

The energy associated with the capacitor is $\frac{1}{2}Cv^2$ while the change in energy of the inductor is given by

$$\Delta W_L = \int_{\lambda_0}^{\lambda} i\, d\lambda \tag{A.8}$$

Substituting from (A.6), we get

$$\Delta W_L = I_m[\cos\lambda_0 - \cos\lambda] \tag{A.9}$$

The variation of the current with λ is sinusoidal, as shown in Figure A.3. The initial value of $\lambda = \lambda_0$ is given by

$$\lambda_0 = \sin^{-1}\left(\frac{I}{I_m}\right) \tag{A.10}$$

The maximum increase in energy of the inductor is given by the area ABCDE in Figure A.3 (which shows current in the inductor as a function of flux linkages). This is given by (A.9) when we substitute $\lambda = \lambda_u = \pi - \lambda_0$.

However, the constant DC current source I is contributing energy ΔW_S given by

$$\Delta W_S = I(\lambda_u - \lambda_0) \tag{A.11}$$

Note that the maximum energy transfer from capacitor to inductor is only $(\Delta W_L - \Delta W_S)$. If this energy is greater than the initial energy stored in the capacitor $\left(\frac{1}{2}Cv_0^2\right)$, the capacitor discharges completely to the nonlinear inductor and the maximum flux linkage (λ_m) is less than or equal to λ_u. The oscillation of the energy between the capacitor

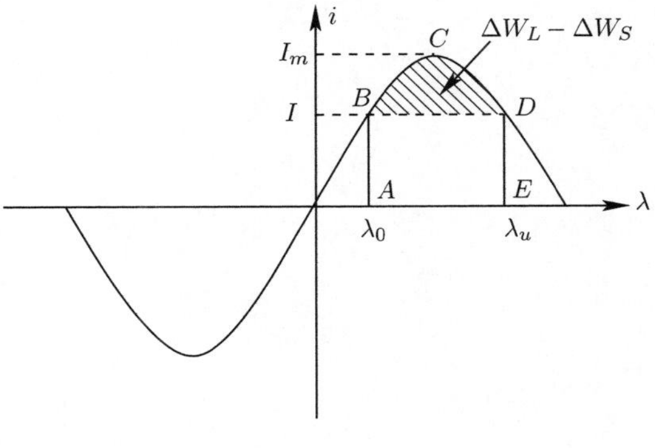

Figure A.3 Current vs flux linkage.

and the inductor continues indefinitely unless damped by dissipation in resistors in the circuit (which have been neglected in this analysis). If $\frac{1}{2}Cv_0^2 > (\Delta W_L - \Delta W_S)$ then the inductor cannot fully absorb the initial energy in the capacitor and λ continues to increase, implying the capacitor voltage continues to increase.

It can be shown that this example is similar to the determination of transient stability in a lossless single-machine infinite bus (SMIB) system. The SMIB system can be modeled by a linear capacitor (analogous to the machine inertia) connected to a nonlinear inductor (analogous to the net series reactance of the system connected between the generator internal bus and the infinite bus). The loss of synchronous stability in the SMIB system is analogous to the failure of the capacitor to completely transfer the stored energy to the nonlinear inductor. The current in the inductor is analogous to the power flow in the line and the voltage across the capacitor is analogous to the deviation in the rotor speed of the generator following a disturbance. Thus, the stored energy in the linear capacitor is analogous to the kinetic energy of the generator rotor moving with respect to a synchronously rotating reference frame.

The network analogy helps in deriving the total energy for detailed, structure-preserving system models. Apart from direct evaluation of system stability, the energy functions can also be used for online detection of loss of synchronism and the mode of instability accurately. In addition, it becomes feasible to determine accurately the optimum locations and the control laws for the network (flexible AC transmission system) controllers for enhancing system stability [2].

References

1 Pai, M.A. (1989) *Energy Function Analysis for Power System Stability*, Kluwer, Boston.
2 Padiyar, K.R. (2013) *Structure Preserving Energy Functions in Power Systems: Theory and Applications*, CRC Press, Boca Raton.

B

Grid Synchronization and Current Regulation

In many cases DER is connected to the microgrid through an inverter. The DER control consists of two cascaded loops. There is a fast current loop which regulates the current injected and an external voltage loop which controls the DC link voltage. The current loop is responsible for power quality and enforcing current limits. The current harmonics and dynamics are affected by the current regulator. The DC (link) voltage controller is designed for balancing the power flow. It is also feasible to have a DC link voltage loop in cascade with the inner power loop. Here, the injected AC current is indirectly controlled.

Choice of Reference Frames

There are three types of control [1] based on:

a) synchronous reference frame (*d*-*q* control)
b) stationary reference frame
c) natural frame

a) Synchronous Reference Frame (SRF) Control

This is also called *d*-*q* control and uses a reference frame transformation from *abc* to *d*-*q*. The structure of synchronous reference frame (SRF) controller is shown in Figure B.1. Here the DC link voltage is controlled to obtain the desired power output. The output of the DC link voltage controller is the reference quantity for the active current controller. Similarly the reactive power reference determines the reference for the reactive current output.

The *d*-*q* control structure is normally associated with proportional-integral (PI) controllers since they have satisfactory response in regulating DC variables. The matrix transfer function of the controller in *d*-*q* variables can be expressed as

$$G_{dq}(s) = \begin{bmatrix} K_p + \dfrac{K_I}{s} & 0 \\ 0 & K_p + \dfrac{K_I}{s} \end{bmatrix} \tag{B.1}$$

Dynamics and Control of Electric Transmission and Microgrids, First Edition.
K. R. Padiyar and Anil M. Kulkarni.

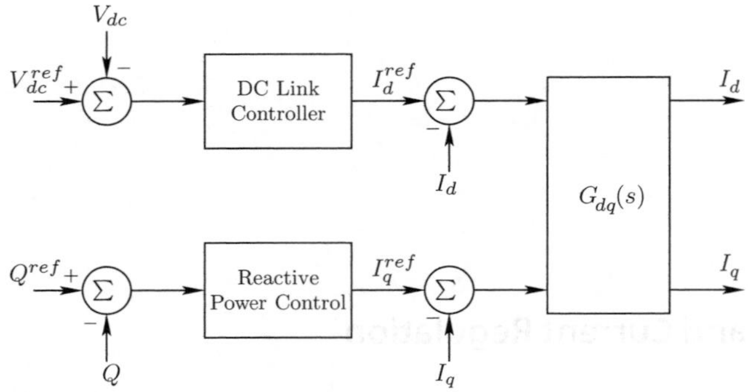

Figure B.1 Synchronous reference frame control

where K_p is the proportional gain and K_I is the integral gain. Since i_d has to be in phase with the grid voltage (and corresponding i_q is in quadrature), the phase angle θ used in the *abc* to *d-q* transformation requires a phase-locked loop (PLL). The SRF controller also requires ωL decoupling branches on both *d-* and *q*-axis current controllers. The major advantage of the SRF controller is that it can achieve zero steady-state error by acting on DC signals in the rotating reference frame. However, the *d-q* control is more complex, requiring satisfactory PLL operation, which can be difficult in the presence of low-order harmonics. In such situations, it is advantageous to use stationary reference frame control.

b) Stationary Reference Frame Control

Here, the phase currents are transformed from *abc* to α-β variables. Since the control variable are sinusoidal, the PI controllers are not able to eliminate the steady-state errors. Thus proportional resonant (PR) controllers are used [2]. This has several advantages in certain applications [3]:

1) Elimination of PLL removes errors that would be introduced by the synchronization problems.
2) In microgrids, the elimination of PLL implies that the converter can never lose synchronization with the grid. Thus stationary reference frame controllers will never trip due to synchronization errors caused by extreme frequency or phase variations.
3) The control hardware becomes simple with reduced cost. This is of importance in connecting PV panels to the grid.
4) Improved response to unbalanced faults and significant reduction in DC ripple under grid imbalances (that leads to reduced DC capacitor size).
5) Convenient to use in single-phase systems.

A typical SRF controller structure is shown in Figure B.2. Here, the PR controller is used to regulate the injected currents i_α and i_β. Although a PLL is used, it is applied only for the determination of reference currents i_d and i_q in contrast to Figure B.1 where the PLL determines the feedback currents, i_d and i_q.

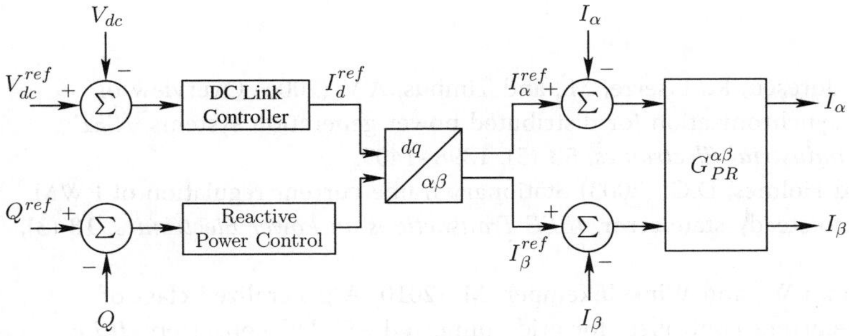

Figure B.2 Synchronous reference frame control.

The controller matrix for the PR control is given by

$$G_{PR}(s) = \begin{bmatrix} K_p + \dfrac{sK_I}{s^2 + \omega^2} & 0 \\ 0 & K_p + \dfrac{sK_I}{s^2 + \omega^2} \end{bmatrix} \tag{B.2}$$

In reference [3], a stationary frame controller is mathematically derived that is an exact equivalent of the commonly used synchronous frame controller with ωL decoupling. This is called a PRX2 controller and has the same frequency response as the decoupled synchronous frame PI controller. A PRX2 controller consists of a proportional (P), a resonant (R), and two cross-coupled ($X_{control}$ and $X_{feedback}$) components. The $X_{control}$ branch arises due to the generalized integrator which is centred at $s = j\omega_n$ and therefore contains a complex component in the denominator. Actually, the equivalent stationary reference controller, $G_{PRX}(s)$, is defined as

$$G_{PRX}(s) = K_p + \frac{K_I}{s - j\omega_n} = K_p + \frac{sK_I}{s^2 + \omega_n^2} + \frac{j\omega_n K_I}{s^2 + \omega_n^2} \tag{B.3}$$

It should be noted that K_p and K_I are identical to the proportional and integral gains used in the SRF controller.

c) Natural Frame Controller

Here, the current in each phase (a, b, and c) is controlled individually. In isolated neutral systems, the phases interact with one another and hence two controllers are adequate. The abc controller structure usually employs nonlinear controllers like hysteresis or dead beat due to their dynamic characteristics. The implementation of PR controllers in each phase is straightforward as the controller is in the stationary frame (a, b, and c) instead of α and β. A survey of current control techniques for three-phase voltage source converters is given in reference [4].

References

1 Blaabjerg, F., Teodorescu, R., Liserre, M., and Timbus, A.V. (2006) Overview of control and grid synchronization for distributed power generation systems. *IEEE Transactions on Industrial Electronics*, **53** (5), 1398–1409.
2 Zmood, D.N. and Holmes, D.G. (2003) Stationary frame current regulation of PWM inverters with zero steady-state error. *IEEE Transactions on Power Electronics*, **18** (3), 814–822.
3 Hwang, J.G., Lehn, P.W., and Winkelnkemper, M. (2010) A generalized class of stationary frame-current controllers for grid-connected AC–DC converters. *IEEE Transactions on Power Delivery*, **25** (4), 2742–2751.
4 Kazmierkowski, M.P. and Malesani, L. (1998) Current control techniques for three-phase voltage-source PWM converters: A survey. *IEEE Transactions on Industrial Electronics*, **45** (5), 691–703.

C

Fryze–Buchbolz–Depenbrock Method for Load Compensation

C.1 Introduction

We consider here a multi-terminal load fed by nonsinusoidal and unbalanced voltages at the terminals. It is assumed that the source voltages are sinusoidal, but the load bus voltages are distorted due to the imbalance in the lines feeding the load and the nonlinear load characteristics. The compensator is connected in parallel with the loads and is designed to inject currents that will compensate the non-active currents drawn by the load and this results in sinusoidal voltages at the load bus, if the source voltages are sinusoidal. The basic approach is to compute the desired currents in the line that will result in supplying the required (active) power to the load, which will lead to the algorithm for generating the compensator currents.

The method of compensation is based on the Fryze–Buchbolz–Depenbrock (FBD) approach [1] using time-domain power theory [2].

C.2 Description of FBD Theory

In a single-phase circuit (see Figure C.1) supplying a linear load (say an R-L load made up of resistance and inductance), the power and reactive power supplied by a sinusoidal voltage source are well defined and are given below:

$$P = VI\cos\phi \qquad Q = VI\sin\phi \tag{C.1}$$

Here V and I are the rms quantities of the sinusoidal voltage and current, respectively. ϕ is the angle by which the current lags the voltage.

Even for this simple circuit the analysis is complicated if the voltage is nonsinusoidal. The load current is also nonsinusoidal and can be expressed by Fourier series. It can be shown that, in steady state, the average power (P) supplied to the load is given by

$$P = \sum_{k=1}^{n} P_k = \sum_{k=1}^{n} V_k I_k \cos\phi_k \tag{C.2}$$

where V_k and I_k are the magnitudes (rms values) of the kth harmonic of the voltage and current, respectively. ϕ_k is the phase difference between the voltage and current for the kth harmonic component.

Dynamics and Control of Electric Transmission and Microgrids, First Edition.
K. R. Padiyar and Anil M. Kulkarni.

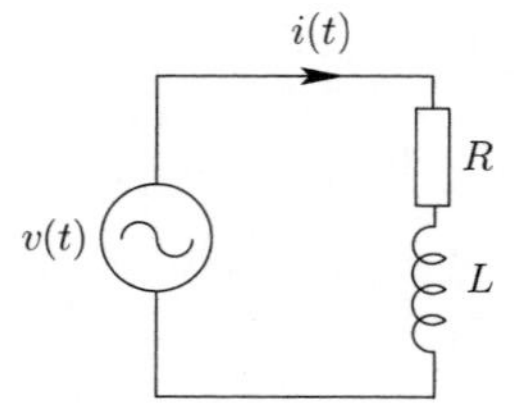

Figure C.1 A single-phase circuit.

It would be tempting to define the reactive power in a similar fashion, namely

$$Q_B = \sum_{k=1}^{n} Q_k = \sum_{k=1}^{n} V_k I_k \sin \phi_k \tag{C.3}$$

As a matter of fact, C. Budeanu, a Romanian engineer defined the reactive power (Q_B) in this way. However, it can be shown that, generally,

$$S^2 \neq P^2 + Q_B^2 \tag{C.4}$$

where the apparent power (S) is defined by

$$S = VI \tag{C.5}$$

and

$$V^2 = \frac{1}{T}\int_0^T v^2\, dt, I^2 = \frac{1}{T}\int_0^T i^2\, dt \tag{C.6}$$

Note that V and I are the rms values of the voltage and current, and are defined in the time domain for any periodic voltage and current waveform. T is the period of the waveform.

It can also be shown that

$$V^2 = \sum_{k=1}^{n} V_k^2, I^2 = \sum_{k=1}^{n} I_k^2 \tag{C.7}$$

where n is the highest harmonic in voltage and current.

To balance the equation (C.4), a distortion power (D_B) was added such that

$$S^2 = P^2 + Q_B^2 + D_B^2 \tag{C.8}$$

Using Fourier components, it can be shown that

$$D_B^2 = \frac{1}{2}\sum_{k=1}^{n}\sum_{m=1}^{n}[V_k^2 I_m^2 + V_m^2 I_k^2 - 2V_k V_m I_m I_k \cos(\phi_k - \phi_m)]$$
$$= \frac{1}{2}\sum_{k=1}^{n}\sum_{m=1}^{n}[(V_k I_m - V_m I_k)^2 + 2V_k V_m I_k I_m[1 - \cos(\phi_k - \phi_m)]]$$

The distortion power is zero only if

$$\frac{V_k}{I_k} = \frac{V_m}{I_m} \quad \text{and} \quad \phi_k = \phi_m$$

Thus, a pure resistor does not have any distortion power. The definition of reactive power (Q_B) given by Budeanu has been adopted by ANSI/IEEE Standard 100-1977. Czarnecki

[3] has shown the drawbacks of using Q_B. First and foremost of these is that even when $Q_B = 0$, individual components (Q_k) are nonzero, signifying energy exchange (between the source and the load) taking place at different harmonic frequencies. The reduction of Q_B is accompanied by increase of D_B. It should be noted that D_B is not a measure of current distortion. It is possible that when both voltage and current waveforms are identical, D_B is not zero. On the other hand, $D_B = 0$ does not guarantee that the current waveform is not distorted.

Both Q_B and D_B are difficult to instrument. However, the major problem is the uselessness of Q_B as an index of power factor improvement.

Remarks

1. It should be noted that generally the load can change with time. Thus it is convenient to use instantaneous load current even for a single-phase circuit as there is no need to introduce harmonic components, which are defined in steady state.
2. Consider a circuit consisting of an ideal switch across a voltage source in series with an impedance. If the ideal switch is considered as load, then it can be seen that the instantaneous power (vi) is zero at all times as either $i = 0$ when the switch is open and $v = 0$ when the switch is closed. However, the harmonic components of the load voltage and current are nonzero. Fryze represented a single-phase load as a parallel combination of an equivalent conductance (G) in parallel with a current source that represents the non-active component of the current, as shown in Figure C.2. Thus

$$\begin{aligned} i(t) &= i_p(t) + i_q(t) \\ &= G(t)v(t) + i_q(t) \end{aligned} \tag{C.9}$$

The instantaneous active current (i_p) determines the instantaneous power (p) given by

$$p = vi_p = G(t)v^2 \tag{C.10}$$

From the above equation, we obtain

$$G(t) = \frac{p}{v^2}$$

It should be noted that the active current i_p is directly proportional to the voltage across the load. The difference between the load current and its active component (i_p) is called the non-active or powerless current i_q. Under sinusoidal conditions (sinusoidal voltage and current), i_q is called as the reactive current. Here, the compensation can be provided by a variable capacitor or inductor which varies with the load. It should be noted that when the voltage and current are periodic functions

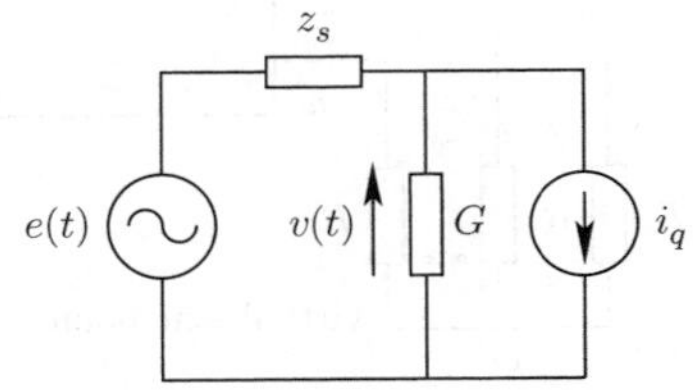

Figure C.2 Representation of a single-phase load.

(not necessarily sinusoidal), the average power (P) supplied to the load in a cycle is given by

$$P = \frac{1}{T}\int_0^T p\,dt = \frac{1}{T}\int_0^T vi\,dt = \frac{1}{T}\int_0^T vi_p\,dt = \frac{1}{T}\int_0^T Gv^2\,dt \tag{C.11}$$

It can be shown that

$$\int_0^T vi_q\,dt = \int_0^T v(i - Gv)\,dt = 0 \tag{C.12}$$

Defining I_p an I_q as rms values of i_p and i_q, the following equations apply

$$I_p^2 = \frac{1}{T}\int_0^T i_p^2\,dt, \quad I_q^2 = \frac{1}{T}\int_0^T i_q^2\,dt \tag{C.13}$$

It can be shown that

$$P = VI_p, \quad S = VI, \quad \text{and} \quad Q_F = VI_q$$

where S is the apparent power and Q_F is the reactive power according to Fryze. Q_F can be also expressed as

$$Q_F^2 = S^2 - P^2 \tag{C.14}$$

C.3 Power Theory in Multiconductor Circuits

The work of Fryze was extended to multiconductor circuits by Buchholz, who considered only periodical quantities, and subsequently Depenbrock, who considered instantaneous quantities.

Virtual Star Point

Consider an n conductor circuit as shown in Figure C.3. We can define voltages across any two conductors. If we define v_{ij} as the voltage drop from node (bus) i to node j,

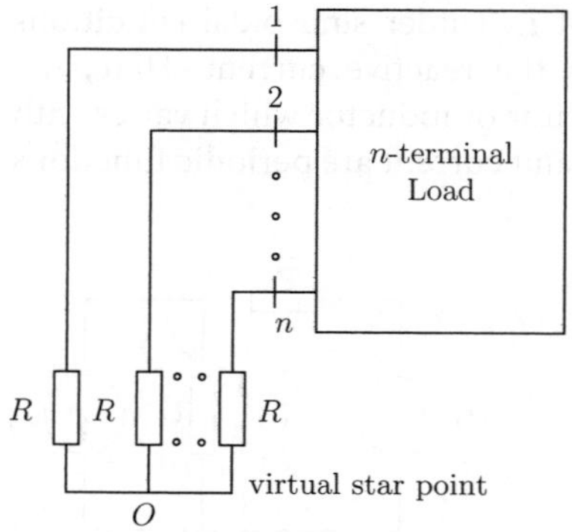

Figure C.3 A multiconductor circuit.

then choosing a reference node r we can measure the voltages of all the conductors with reference to the node r.

From Kirchhoff's current law (KCL), we obtain

$$\sum_{k=1}^{n} i_k = 0 \tag{C.15}$$

It would be desirable to have a virtual star point labelled as 0 and measure all the conductor voltages w.r.t the virtual star point (similar to the neutral point). If we consider n resistors of equal value (say R) connected to a common point 0 at one end (a star point) and the other ends connected to the n conductors, the star point becomes the virtual star point. We can show from KCL that

$$\sum_{k=1}^{n} i_{k0} = 0, \sum_{k=1}^{n} v_{k0} = 0 \quad \text{as} \quad v_{k0} = Ri_{k0} \tag{C.16}$$

Since $v_{k0} = v_{kr} + v_{r0}, k = 1, 2 \ldots .n$, we obtain

$$\begin{aligned} 0 &= \sum_{k=1}^{n} v_{k0} = \sum_{k=1}^{n}(v_{kr} + v_{r0}) = nv_{r0} + \sum_{k=1}^{n} v_{kr} \\ v_{r0} &= -\frac{1}{n}\sum_{k=1}^{n} v_{kr} \end{aligned} \tag{C.17}$$

Collective Quantities

We define the following two collective quantities

$$i_\Sigma = \sqrt{\sum_{k=1}^{n} i_k^2}, \quad v_\Sigma = \sqrt{\sum_{k=1}^{n} v_{k0}^2} \tag{C.18}$$

These are defined as the norm (length) of the instantaneous current and voltage vectors. The rms value of the collective instantaneous values is called the collective rms value defined by

$$V_\Sigma = \sqrt{\frac{1}{T}\int_{t-T}^{t} v_\Sigma^2 dt} = \sqrt{\sum_{k=1}^{n} V_{k0}^2} \tag{C.19}$$

$$I_\Sigma = \sqrt{\frac{1}{T}\int_{t-T}^{t} i_\Sigma^2 dt} = \sqrt{\sum_{k=1}^{n} I_k^2} \tag{C.20}$$

Defining the voltage vector v and current vector i of dimension n, the instantaneous power supplied to the load is the inner product of the current and voltage vectors. Thus,

$$p_\Sigma = v^t i, i = i_p + i_z \tag{C.21}$$

where i_p is defined as

$$i_p = G_p(t)v \tag{C.22}$$

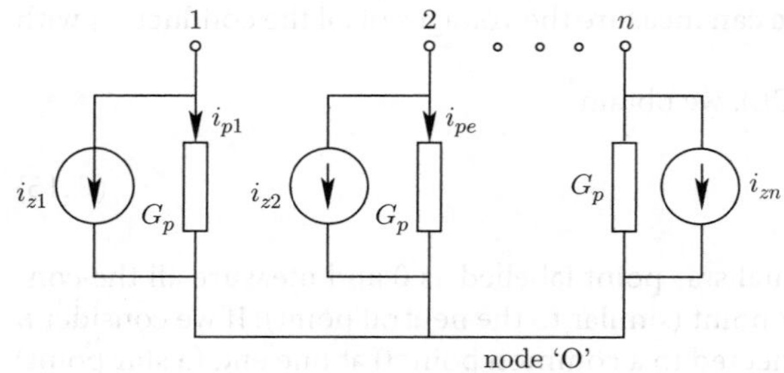

Figure C.4 Equivalent circuit of n terminal load.

and i_z does not result in instantaneous power, which implies

$$i_z = i - i_p, \quad v^t i_z = 0 \tag{C.23}$$

The n terminal load circuit can be represented by the equivalent circuit shown in Figure C.4. We can define two more collective quantities, collective instantaneous apparent power ($s_\Sigma(t)$) and collective instantaneous non-active (powerless) power (p_z):

$$\begin{aligned} s_\Sigma^2(t) &= v_\Sigma^2 i_\Sigma^2 = v_\Sigma^2 i_{p\Sigma}^2 + v_\Sigma^2 i_{z\Sigma}^2 \\ &= p_\Sigma^2 + p_{z\Sigma}^2 \end{aligned} \tag{C.24}$$

Remarks

1) $s^2(t) > p_\Sigma^2$ and $s^2(t) = p_\Sigma^2$ only if $i(t) = G_p(t)v(t)$.
2) For a specified set of voltages $v(t)$, the power is maximum when $i = i_p = G_p(t)v$. This implies $i_z = 0$.
3) Note that $i_z = 0$ implies minimum losses in the conductors feeding a multi-terminal load.
4) The compensation of powerless currents (i_z) in the load is easily implemented by a shunt-connected compensator which generates i_z from the knowledge of instantaneous values of the voltages across the load and currents flowing in the lines feeding the load (see Figure C.5).
   The compensator injects the current (vector) i_C computed from

$$i_C = i_z = i_L - i_p = i_L - G_P(t)v \tag{C.25}$$

$$G_P(t) = \frac{v^t i_L}{(v^t \cdot v)} = \frac{p_\Sigma}{v_\Sigma^2} \tag{C.26}$$

5) It should be noted that the input signal required by the compensator can be determined from the measurements of the voltages and currents at the terminals of the load. No additional information (from the line side or load side system parameters) is

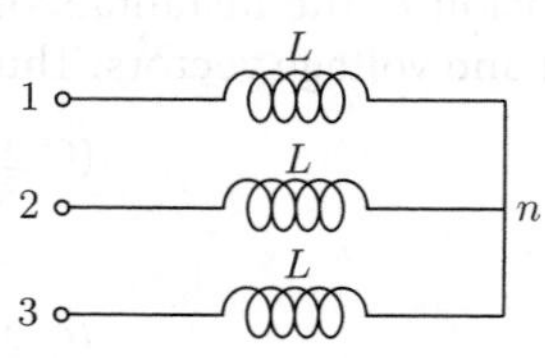

Figure C.5 A balanced reactive load.

required. If the source voltages are sinusoidal, FBD compensation leads to sinusoidal load voltages in steady state even though the load currents may be distorted and line impedances may be unbalanced. When the equivalent source voltages are distorted (due to other nonlinear, uncompensated loads), the line currents supplied with the compensator have the same waveforms as the source voltages. The line currents are always proportional to the terminal voltages with the common factor of proportion ($G(t)$) remaining constant in steady state.

C.4 Examples

Example C.1 Consider a three-phase load consisting of three identical lossless reactors having inductance L in each phase (see Figure C.5). The reactors are not coupled. If the voltages are sinusoidal and balanced, it can be shown that the current drawn by the load is purely powerless (instantaneous power is zero). If applied voltages are

$$v_1(t) = \sqrt{\frac{2}{3}}V\cos\omega t$$

$$v_2(t) = \sqrt{\frac{2}{3}}V\cos\left(\omega t - \frac{2\pi}{3}\right)$$

$$v_3(t) = \sqrt{\frac{2}{3}}V\cos\left(\omega t + \frac{2\pi}{3}\right)$$

the currents flowing to the load are

$$i_1(t) = \sqrt{\frac{2}{3}}I\sin\omega t$$

$$i_2(t) = \sqrt{\frac{2}{3}}I\sin\left(\omega t - \frac{2\pi}{3}\right)$$

$$i_3(t) = \sqrt{\frac{2}{3}}I\sin\left(\omega t + \frac{2\pi}{3}\right)$$

It can be shown that $p = v_1i_1 + v_2i_2 + v_3i_3 = 0$. The vector of instantaneous currents (i) can be expressed as $i = i_p + i_z$. Here, $i_p = 0$ and $i = i_z$.

Comment: If the inductors are not equal in magnitude $i_p \neq 0$ although the average power drawn by the load in a period is zero.

Example C.2 It should be noted that i_z can also exist in purely resistive networks. For example, consider a three-phase load with resistors in each phase as shown in Figure C.6. The load is not balanced. While $i_p \neq 0$, it can be shown that $i_z \neq 0$ depending on the applied voltages.

The voltage at the neutral bus is obtained from the relation

$$i_1 + i_2 + i_3 = 0$$

Since

$$i_1 = G(v_1 - v_n), i_2 = \frac{G}{2}(v_2 - v_n), i_3 = \frac{G}{2}(v_3 - v_n)$$

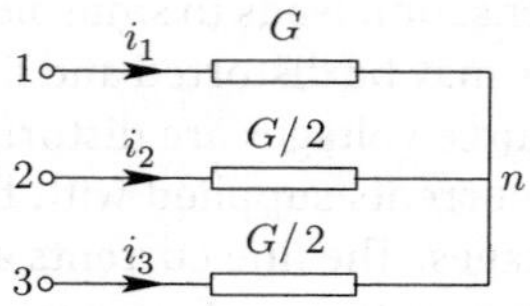

Figure C.6 An unbalanced resistive load.

we can obtain

$$v_n = \frac{v_1}{2} + \frac{1}{4}(v_2 + v_3)$$

If we select $v_1 = V$ and $v_2 = v_3 = -V$, then $v_n = 0$ and

$$i = GV \begin{bmatrix} 1 \\ -\frac{1}{2} \\ -\frac{1}{2} \end{bmatrix}, i_p = \begin{bmatrix} 1 \\ -1 \\ -1 \end{bmatrix} G_{eq} V, i_z = GV \begin{bmatrix} \frac{1}{3} \\ \frac{1}{6} \\ \frac{1}{6} \end{bmatrix}$$

where

$$G_{eq} = \frac{p_\Sigma}{V_\Sigma^2} = \frac{2}{3} G$$

This example clearly shows the existence of current components that do not contribute to load power. Hence a shunt compensator to eliminate i_z from the lines is required to minimize the power losses.

C.5 Load Characterization over a Period

Equation (C.24) defines the total instantaneous power supplied to the load through n conductors. The mean value over a period, defined at time instant t, is given by

$$P(t) = \frac{1}{T}\int_{t-T}^{t} p_\Sigma \, dt = \frac{1}{T}\int_{t-T}^{t} (v^t \cdot i) \, dt = \frac{1}{T}\int_{t-T}^{t} \sum_{k=1}^{n} (v_{k0} i_k) \, dt \tag{C.27}$$

If voltages and currents are periodic functions with period T, the active power P is constant. Otherwise, a sliding integration interval, as shown in (C.27), is used which results in $P(t)$ varying with time. $P(t)$ can also be expressed as

$$P(t) = \frac{1}{T}\int_{t-T}^{t} (v^t \cdot i_p) \, dt = G \cdot \frac{1}{T}\int_{t-T}^{t} v_\Sigma^2 \, dt = GV_\Sigma^2 = V_\Sigma I_{p\Sigma} \tag{C.28}$$

where $V_\Sigma^2 = \frac{1}{T}\int_{t-T}^{t} v_\Sigma^2\, dt$ and $G(t)$ is obtained from (C.28) as

$$G(t) = \frac{P(t)}{V_\Sigma^2} \tag{C.29}$$

It should be noted that $G(t)$ is not the mean value of $G_p(t)$ unless V_Σ is a constant. As a matter of fact, i_p can be split into two components, i_a and i_v (active and variation currents), such that

$$P(t) = \frac{1}{T}\int_{t-T}^{t} v^t \cdot i_p\, dt = \frac{1}{T}\int_{t-T}^{t} v^t \cdot i_a\, dt \tag{C.30}$$

$$\frac{1}{T}\int_{t-T}^{T} v^t \cdot i_v dt = 0 \tag{C.31}$$

The variation currents contribute to p_Σ, but their contribution to average power (P) over a period is zero. It should be noted that the rms values of i, i_a, i_v, and i_z satisfy

$$I^2 = I_a^2 + I_v^2 + I_z^2 = I_p^2 + I_z^2$$

C.6 Compensation of Non-Active Currents

The compensator current (i_C) is given by

$$i_C = i_L - i_a = i_L - G(t)v = i_z + i_v \tag{C.32}$$

It should be noted that the compensator provides the sum of two currents (i_z and i_v). In steady state, compensation of i_z does not require energy storage. However, compensation of i_v involves energy storage. The energy storage capability is based on the energy variation in a period of alternating currents. Under dynamic conditions, larger energy storage may be required.

Remarks

1) The decomposition of load currents (into active and variation currents) is based on the Schwarz inequality given by

$$\int_a^b fg\, dt \leqslant \left[\int_a^b |f|^2 dt\right]^{\frac{1}{2}} \left[\int_a^b |g|^2 dt\right]^{\frac{1}{2}} \tag{C.33}$$

where f and g are functions of time.

2) For sinusoidal currents and voltages in a n conductor system, the following inequality applies

$$\left|\sum_{k=1}^{n} S_k\right|^2 \leqslant \left[\sum_{k=1}^{n} |V_k|^2 \cdot \sum_{k=1}^{n} |I_k|^2\right] \tag{C.34}$$

when $S_k = V_k I_k^* = P_k + jQ_k$.

The left-hand side of the above inequality can be expressed as

$$P_{\Sigma}^2 + Q_{\Sigma}^2,\quad P_{\Sigma} = \sum_{k=1}^{n} P_k,\quad Q_{\Sigma} = \sum_{k=1}^{n} Q_k$$

Discussion

1) Akagi *et al.* [4, 5] proposed the instantaneous reactive power (IRP) theory for the three-phase three-wire and four-wire systems. For a three-wire system, it was shown (based on α, β components) that an instantaneous reactive power compensator comprising switching devices without energy storage components is feasible [4].
2) Depenbrock *et al.* [6] demonstrated the deficiencies of the IRP theory as applied to four-wire systems. Since the FBD theory is applicable to any number of wires (conductors) feeding a n terminal load, it always gives correct results.

References

1 Depenbrock, M. and Staudt, V. (1998) The FBD-method as tool for compensating total nonactive currents, in *8th International Conference on Harmonics and Quality of Power*, Athens, pp. 320–324.

2 Staudt, V. (2008) Fryze-Buchholz-Depenbrock: A time-domain power theory, in *International School on Nonsinusoidal Currents and Compensation*, IEEE, pp. 1–12.

3 Czarnecki, L.S. (1987) What is wrong with the Budeanu concept of reactive and distortion power and why it should be abandoned. *IEEE Transactions on Instrumentation and Measurement*, **IM-36** (3), 834–837.

4 Akagi, H., Kanazawa, Y., and Nabae, A. (1984) Instantaneous reactive power compensators comprising switching devices without energy storage components. *IEEE Transactions on Industry Applications*, **IA-20** (3), 625–630.

5 Akagi, H., Ogasawara, S., and Kim, H. (1999) The theory of instantaneous power in three-phase four-wire systems: A comprehensive approach, in *IEEE Industry Applications Conference and 34th IAS Annual Meeting*, Phoenix, pp. 431–439.

6 Depenbrock, M., Staudt, V., and Wrede, H. (2003) A theoretical investigation of original and modified instantaneous power theory applied to four-wire systems. *IEEE Transactions on Industry Applications*, **39** (4), 1160–1168.

D

Symmetrical Components and Per-Unit Representation

D.1 Symmetrical Component Representation of Three-Phase Systems

The generation, transmission, and distribution of electrical power is predominantly done using balanced three-phase sinusoidal AC systems, except at the downstream end of distribution systems that cater to small, low-voltage single-phase loads. Three-phase systems offer several advantages over single-phase AC systems. The additional complexity in the analysis, design, and engineering of three-phase systems is modest and manageable. Polyphase systems of order higher than three can also be built, but they generally involve greater complexity and therefore have not found widespread application. The advantages of balanced three-phase AC systems over single-phase AC systems are:

1) The total power in a three-phase system is constant while it pulsates at twice the AC frequency in a single-phase system. This also manifests as pulsating electrical torque in single-phase machines, while the torque is constant in three-phase machines.
2) Space and material can usually be better utilized in three-phase equipment. A separate return conductor for each phase in the transmission line is not necessary because the sum of currents in a three-phase balanced system is zero.
3) Some lower-order harmonics in power-electronic converters can be eliminated due to the three-phase connections.
4) Three-phase motors have a rotating magnetic field which is advantageous for starting. These motors generally have better efficiency, power factor, and a smaller size (for the same rating).

These advantages are available only when the systems are balanced, that is, the three-phase quantities (a, b, c) in the sinusoidal steady state are of the form:

$$f_a = \sqrt{2}F_{rms}\sin(\omega_o t + \phi), \quad f_b = \sqrt{2}F_{rms}\sin(\omega_o t + \phi - 120^\circ)$$
$$f_c = \sqrt{2}F_{rms}\sin(\omega_o t + \phi - 240^\circ) \tag{D.1}$$

where F_{rms} is the root mean square value of the sinusoidal waveforms. f represents three-phase quantities like current, voltage, flux, and charge. The usual convention is that the phase c waveform lags the phase b waveform, and phase b lags phase a. These waveforms may also be represented as complex quantities (phasors), as given below:

$$\hat{F}_a = F_{rms}\angle\phi \quad \hat{F}_b = F_{rms}\angle(\phi - 120^\circ) \quad \hat{F}_c = F_{rms}\angle(\phi - 240^\circ) \tag{D.2}$$

Dynamics and Control of Electric Transmission and Microgrids, First Edition.
K. R. Padiyar and Anil M. Kulkarni.

Since currents and voltages should be balanced under normal conditions (to derive the advantages of three-phase systems), three-phase equipment is designed to be symmetrical about the three phases. For example, the three windings corresponding to the three phases in a two-pole synchronous machine are identical except that the axes of the windings are geometrically displaced by 120° from each other. In overhead transmission lines, overall symmetry can be attained by transposing the conductors at regular intervals.

Three-phase balanced systems may be analyzed as single-phase systems since the waveforms in the b and c phases are merely phase-shifted versions of the a phase. In this context, it is useful to consider the following transformation of variables (also called the sequence transformation):

$$\begin{bmatrix} \hat{F}_p \\ \hat{F}_n \\ \hat{F}_o \end{bmatrix} = k_s \begin{bmatrix} 1 & e^{j120^\circ} & e^{-j120^\circ} \\ 1 & e^{-j120^\circ} & e^{j120^\circ} \\ 1 & 1 & 1 \end{bmatrix} \begin{bmatrix} \hat{F}_a \\ \hat{F}_b \\ \hat{F}_c \end{bmatrix} = T_s \begin{bmatrix} \hat{F}_a \\ \hat{F}_b \\ \hat{F}_c \end{bmatrix} \tag{D.3}$$

The inverse transformation is given by:

$$\begin{bmatrix} \hat{F}_a \\ \hat{F}_b \\ \hat{F}_c \end{bmatrix} = \frac{1}{3k_s} \begin{bmatrix} 1 & 1 & 1 \\ e^{-j120^\circ} & e^{j120^\circ} & 1 \\ e^{j120^\circ} & e^{-j120^\circ} & 1 \end{bmatrix} \begin{bmatrix} \hat{F}_p \\ \hat{F}_n \\ \hat{F}_o \end{bmatrix} = T_s^{-1} \begin{bmatrix} \hat{F}_p \\ \hat{F}_n \\ \hat{F}_o \end{bmatrix} \tag{D.4}$$

$\hat{F}_p$, $\hat{F}_n$, and $\hat{F}_o$ are called the positive-sequence, negative-sequence, and zero-sequence variables (or the symmetrical components), respectively. Note that $T_s^{-1} = \frac{1}{3k_s^2}(T_s^*)^t$, where t denotes the transpose. As a result, the apparent power expression is given by

$$S_{abc} = \hat{V}_a\hat{I}_a^* + \hat{V}_b\hat{I}_b^* + \hat{V}_c\hat{I}_c^* = \frac{1}{3k_s^2}(\hat{V}_p\hat{I}_p^* + \hat{V}_n\hat{I}_n^* + \hat{V}_o\hat{I}_o^*) = \frac{1}{3k_s^2}S_{pno}$$

where $*$ denotes the complex conjugate. The benefits of the transformation (which will be discussed presently) are not dependent on the choice of the constant k_s, as long as it is non-zero. While $k_s = \frac{1}{3}$ is commonly used in power system literature, the choice $k_s = \frac{1}{\sqrt{3}}$ results in T_s^{-1} being equal to $(T_s^*)^t$ and $S_{abc} = S_{pno}$ (power invariance).

The key benefits of the sequence transformation are:

1) If there is a symmetrical coupling of parameters between the phases, then the system is decoupled in the transformed variables. For example,

$$\begin{bmatrix} \hat{V}_a \\ \hat{V}_b \\ \hat{V}_c \end{bmatrix} = \begin{bmatrix} Z_s & Z_{m1} & Z_{m2} \\ Z_{m2} & Z_s & Z_{m1} \\ Z_{m1} & Z_{m2} & Z_s \end{bmatrix} \begin{bmatrix} \hat{I}_a \\ \hat{I}_b \\ \hat{I}_c \end{bmatrix} \Rightarrow \begin{bmatrix} \hat{V}_p \\ \hat{V}_n \\ \hat{V}_o \end{bmatrix} = \begin{bmatrix} Z_p & 0 & 0 \\ 0 & Z_n & 0 \\ 0 & 0 & Z_o \end{bmatrix} \begin{bmatrix} \hat{I}_p \\ \hat{I}_n \\ \hat{I}_o \end{bmatrix}$$

   where
   $Z_p = Z_s + Z_{m1}\, e^{-j120^\circ} + Z_{m2}\, e^{j120^\circ}$, $Z_n = Z_s + Z_{m1}\, e^{j120^\circ} + Z_{m2}\, e^{-j120^\circ}$, $Z_o = Z_s + Z_{m1} + Z_{m2}$.
   Although the voltages and currents in the sequence domain are dependent on the value of k_s, the impedance matrix in the sequence domain is *not* dependent on it.
2) Three-phase transformers can also be represented in a decoupled fashion in the sequence domain. For example, consider the three-phase star (grounded)-delta

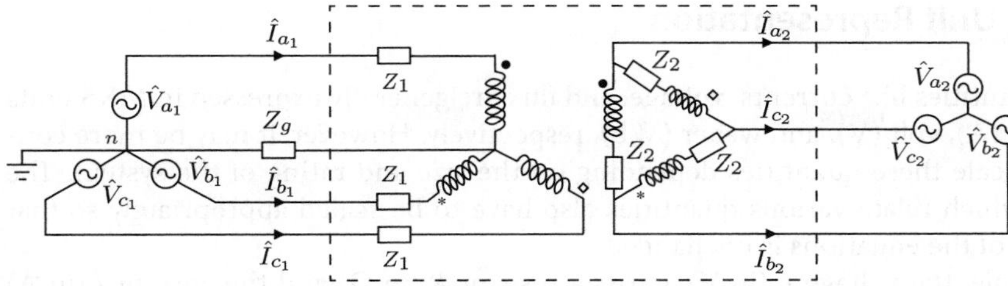

Figure D.1 Star-delta transformer.

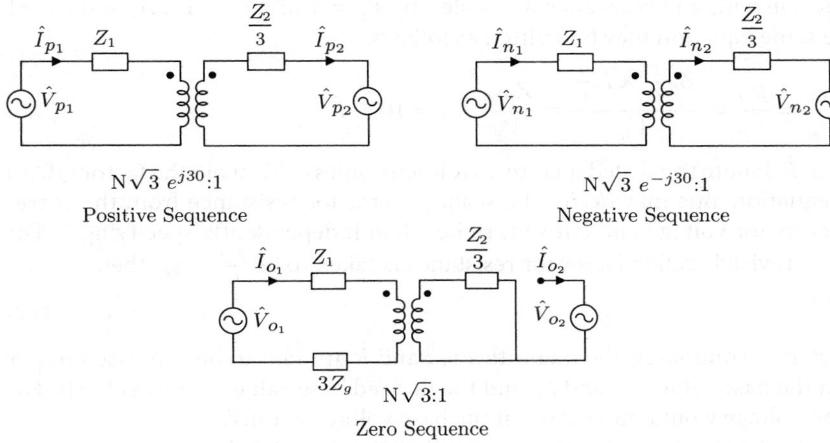

Figure D.2 Star-delta transformer in the sequence domain.

transformer shown in Figure D.1. The primary and secondary windings of each phase have N_1 and N_2 turns, respectively, and $N = \frac{N_1}{N_2}$. This transformer can be represented in the sequence domain, as shown in Figure D.2.

3) If a balanced system is excited by unbalanced voltages and/or currents, then the analysis can be done in a decoupled fashion by transforming the voltages and currents to the respective sequence components. The network solution can be computed in the sequence domain and can be converted back into the phase domain by the inverse transformation.
4) Under normal conditions, a power system has balanced excitation (currents or voltages) and symmetrical parameters. In these conditions, only the positive sequence variables are non-zero and only these need to be considered. For example, if $\hat{V}_a = 400\angle 5°$ kV, $\hat{V}_b = 400\angle{-115°}$ kV, and $\hat{V}_c = 400\angle 125°$ kV, then $\hat{V}_p = 400\sqrt{3}\angle 5°$ kV, $\hat{V}_n = 0$ kV, and $\hat{V}_o = 0$ kV.
5) Unbalanced parameters may exist in a small part of the system, say, due to a single-line-to-ground fault at a bus. This situation can be analyzed quite conveniently by representing the rest of the system by Thevenin equivalents for each sequence component at the faulted bus [1].

D.2 Per-Unit Representation

Physical quantities like currents, voltage, and flux are generally expressed in MKS units like ampere (A), volt (V), and weber (Wb), respectively. However, it may be more convenient to scale these quantities depending on the size and rating of the system. The equations which relate various quantities also have to be scaled appropriately so that the essence of the equations is unchanged.

For example, the voltage v (in V) across a resistor R (in Ω) and the current i (in A) through it are related by the equation,

$$v = R\,i \tag{D.5}$$

If the voltage, current, and resistance are scaled by $V_B = 1$ mV, $I_B = 1$ mA, and $Z_B = 1$ kΩ, then the same equation may be written as follows:

$$\tilde{v} = \frac{v}{V_B} = \frac{R\,i}{V_B} = \frac{Z_B\frac{R}{Z_B} \times I_B\frac{i}{I_B}}{V_B} = \frac{Z_B I_B}{V_B}\,\hat{R}\,\tilde{i} = 1000\,\hat{R}\,\tilde{i} \tag{D.6}$$

where $\tilde{v}$, $\tilde{i}$, and $\hat{R}$ denote the scaled quantities (dimensionless). To avoid the factor (1000) in the final equation, one may *derive* the scaling factor for resistance from the corresponding factors for voltage and current, rather than independently specifying it. For example, if the revised scaling factor for resistance is taken to be $\frac{V_B}{I_B} = Z_B$, then

$$\tilde{v} = \tilde{R}\,\tilde{i} \tag{D.7}$$

In power system terminology, the quantities $\tilde{v}$, $\tilde{i}$ and $\tilde{R}$ are said to be expressed in per units (pu) on the base values V_B and I_B, and the derived base value Z_B, respectively. For example, 2 pu voltage would mean 2 mV if the base voltage is 1 mV.

As discussed below, a per-unit representation (with appropriate base quantities) simplifies computations, avoids a large disparity in the numerical values of voltages and currents in the system, and gives a better insight, including the ability to quickly spot errors.

1) If the base values are chosen to be rated values, then the per-unit apparent power and voltage normally lie in a range from 0.0 to near 1.0 pu. The per-unit values can thus be correlated to the capability of the equipment. For example, a 1.2 pu voltage would indicate an significant over-voltage situation irrespective of the size of the equipment.
2) The per-unit parameters of equipment (e.g., reactances, resistances) lie in a narrow range for a class of equipment of similar design if the base values are chosen to be the rated values.
3) By choosing the ratio of voltage bases on the primary and secondary side of the transformers to be equal to the nominal open-circuit line–line voltage ratio, the transformer ratio effectively becomes unity in the per-unit system. If a transformer is operating with an off-nominal tap then its effect can be separately accounted for by adding a π network in cascade with the transformer, without disturbing the base values on either side [1].

In a power system, the specified base quantities are generally the volt-ampere (VA) base S_B, the voltage base V_B (on one side of a transformer in the system), and the frequency base ω_B, while the other base values are derived from these. The VA base which is used for scaling active/reactive/apparent power is taken to be the same throughout the system.

Network Calculations using the Sequence and Per-Unit Representations

The use of the sequence transformation and the per-unit representation for network calculations is illustrated in Figure D.3. The VA and voltage bases are taken to be the nominal three-phase MVA and the nominal line–line voltages, respectively. The derived base quantities are $I_B = \frac{S_B}{V_B}$ (the current base) and $Z_B = \frac{V_B}{I_B} = \frac{V_B^2}{S_B}$ (the impedance base). Note that the sequence transformation uses $k_s = \frac{1}{\sqrt{3}}$.

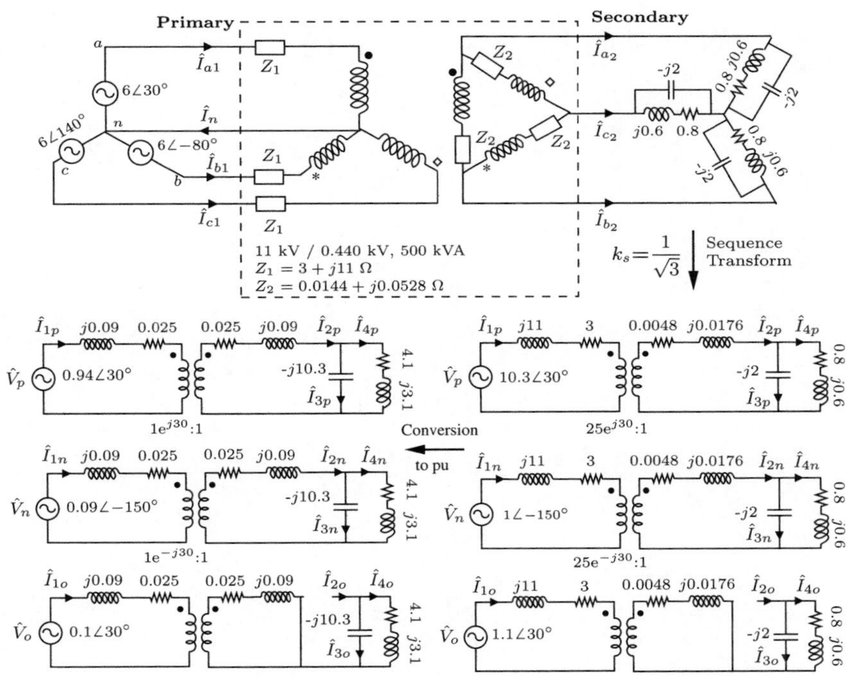

	Sequence Currents in Per-Unit		
	p	n	o
$\hat{I}_1$	0.145∠21°	0.014∠−159°	0.53∠−44°
$\hat{I}_2$	0.145∠−9°	0.014∠−129°	0
$\hat{I}_3$	0.09∠88°	0.008∠−32°	0
$\hat{I}_4$	0.18∠−39°	0.017∠−158°	0

Conversion to A

	Sequence Currents in A		
	p	n	o
$\hat{I}_1$	13.2∠21°	1.3∠−159°	48.2∠−44°
$\hat{I}_2$	330∠−9°	32∠−129°	0
$\hat{I}_3$	205∠88°	18∠−32°	0
$\hat{I}_4$	409∠−39°	39∠−158°	0

Transform to phase domain

	Base Values	
	Primary	Secondary
V_B	11 kV	0.44 kV
S_B	1 MVA	1 MVA
I_B	90.9 A	2273 A
Z_B	121 Ω	0.19 Ω

	Phase Currents in A		
	a	b	c
$\hat{I}_1$	31.4∠−33°	33.5∠−55°	19.8∠−44°
$\hat{I}_2$	182∠−14°	182∠−124°	209∠111°
$\hat{I}_3$	114∠84°	114∠−27°	129∠−152°
$\hat{I}_4$	226∠−44°	225∠−154°	259∠81°

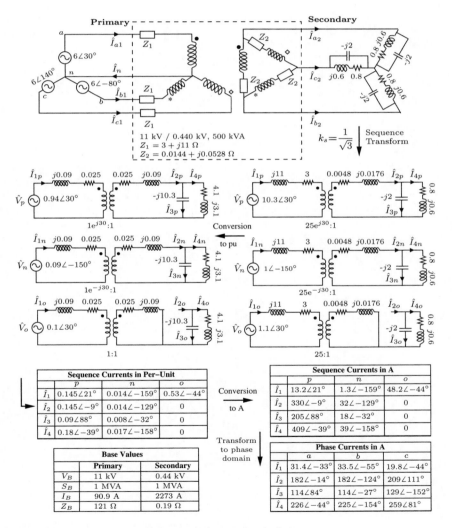

Figure D.3 Sequence transformation and per-unit calculations.

Alternative choice of k_s and base values: A reader can verify that if $k_s = \frac{1}{3}$ and the base values are taken as S_B = the per-phase nominal MVA ($\frac{1}{3}\times$ the three-phase MVA) and V_B = the respective nominal phase voltages = ($\frac{1}{\sqrt{3}}\times$ the respective nominal line-to-line voltages), then this gives us the same per-unit network (in the sequence domain) and the same solution in the original units.

References

1 Elgerd, O.I. (1973) *Electric Energy Systems Theory: An Introduction*, Tata McGraw-Hill, New Delhi.

Index

Dynamics and Control of Electric Transmission and Microgrids, First Edition.
K. R. Padiyar and Anil M. Kulkarni.